NOURAGUES

MONOGRAPHIAE BIOLOGICAE

VOLUME 80

Series Editors

H. J. Dumont and M. J. A. Werger

The titles published in this series are listed at the end of this volume.

Nouragues

Dynamics and Plant-Animal Interactions in a Neotropical Rainforest

edited by

FRANS BONGERS

PIERRE CHARLES-DOMINIQUE

PIERRE-MICHEL FORGET

MARC THÉRY

KLUWER ACADEMIC PUBLISHERS

DORDRECHT / BOSTON / LONDON

A C.I.P. Catalogue record for this book is available from the Library of Congress.

ISBN 1-4020-0123-1

Published by Kluwer Academic Publishers,
P.O. Box 17, 3300 AA Dordrecht, The Netherlands.

Sold and distributed in North, Central and South America
by Kluwer Academic Publishers,
101 Philip Drive, Norwell, MA 02061, U.S.A.

In all other countries, sold and distributed
by Kluwer Academic Publishers,
P.O. Box 322, 3300 AH Dordrecht, The Netherlands.

Printed on acid-free paper

Printed in the Netherlands.

CONTRIBUTORS

Martine Atramentowicz. Laboratoire d'Ecologie Générale, UMR 8571 CNRS-MNHN, 4 Avenue du Petit Château, 91800 Brunoy, France.

Françoise Bayart. Laboratoire d'Ecologie Générale, UMR 8571 CNRS-MNHN, 4 Avenue du Petit Château, 91800 Brunoy, France.

Pierre Belbenoit. Laboratoire d'Ecologie Générale, UMR 8571 CNRS-MNHN, 4 Avenue du Petit Château, 91800 Brunoy, France.

Patrick Blanc. Laboratoire d'Ecologie Générale, UMR 8571 CNRS-MNHN, 4 Avenue du Petit Château, 91800 Brunoy, France.
Current address: Université Pierre et Marie Curie, 12 Rue Cuvier, 75005 Paris, France.

Frans Bongers. Silviculture and Forest Ecology Group, Department of Environmental Sciences, Wageningen University, P.O. Box 342, 6700 AH Wageningen, The Netherlands.

Marga Born. Silviculture and Forest Ecology Group, Department of Environmental Sciences, Wageningen University, P.O. Box 342, 6700 AH Wageningen, The Netherlands.
Current address: Hof van Versailles 51, 1064 NX Amsterdam, The Netherlands.

Thierry Boujard. Laboratoire de Nutrition, Centre de Recherche Hydrobiologie, INRA, BP. 3, Saint-Pée sur Nivelle, 64310 Ascain, France.

André Brosset. Laboratoire d'Ecologie Générale, UMR 8571 CNRS-MNHN, 4 Avenue du Petit Château, 91800 Brunoy, France.

Pierre Charles-Dominique. Laboratoire d'Ecologie Générale, UMR 8571 CNRS-MNHN, 4 Avenue du Petit Château, 91800 Brunoy, France.

Stéphanie Chauvet. Laboratoire d'Ecologie Générale, UMR 8571 CNRS-MNHN, 4 Avenue du Petit Château, 91800 Brunoy, France.

Anya Cockle. Laboratoire d'Ecologie Générale, UMR 8571 CNRS-MNHN, 4 Avenue du Petit Château, 91800 Brunoy, France.

Jean-Jacques de Granville. Dept of Botany, Centre IRD de Cayenne, BP. 165, 97323 Cayenne cédex, Guyane Française.

John A. Endler. Department of EEM Biology, University of California, Santa Barbara 93016, USA.

Christian Erard. Laboratoire d'Ecologie Générale, UMR 8571 CNRS-MNHN, 4 Avenue du Petit Château, 91800 Brunoy, France.

François Feer. Laboratoire d'Ecologie Générale, UMR 8571 CNRS-MNHN, 4 Avenue du Petit Château, 91800 Brunoy, France.

Pierre-Michel Forget. Laboratoire d'Ecologie Générale, UMR 8571 CNRS-MNHN, 4 Avenue du Petit Château, 91800 Brunoy, France.

Jean-Pierre Gasc. Laboratoire d'Anatomie Comparée, UMR 8570 CNRS-MNHN, 55 Rue Buffon, 75005 Paris, France.

Bongers *et al.* (eds.), Nouragues. Dynamics and plant-animal interactions in a neotropical rainforest.
2001 Kluwer Academic Publishers. Printed in the Netherlands.

Philippe Gaucher. Mission Parc de la Guyane, 81 Rue Christophe Colomb, BP 275, 97300 Cayenne, Guyane Française.

Marc Gayot. Laboratoire de Conservation des Espèces Animales, MNHN, 57 Rue Cuvier 75005 Paris, France.

Michel Grimaldi. Unité des Sciences du Sol, ENSA – INRA, 65 route de St Brieuc, 35042 Rennes, France.

Marie-Laure Guillemin. Laboratoire d'Ecologie Générale, UMR 8571 CNRS-MNHN, 4 Avenue du Petit Château, 91800 Brunoy, France.

Isabelle Hardy. Laboratoire d'Ecologie Générale, UMR 8571 CNRS-MNHN, 4 Avenue du Petit Château, 91800 Brunoy, France.

Olivier Henry. Laboratoire de Conservation des Espèces Animales, MNHN, 57 Rue Cuvier 75005 Paris, France.

Patrick A. Jansen. Silviculture and Forest Ecology Group, Department of Environmental Sciences, Wageningen University, P.O. Box 342, 6700 AH Wageningen, The Netherlands.

Didier Julien-Laferrière. Laboratoire d'Ecologie Générale, UMR 8571 CNRS-MNHN, 4 Avenue du Petit Château, 91800 Brunoy, France.
Current address: MNHN Galerie de l'Evolution 36 Rue Geoffroy St Hilaire 75005 Paris, France.

Mathilde Jullien. ENS, Laboratoire d'Ecologie UMR 7625 CNRS, 46 Rue d'Ulm 75231 Paris cédex 05, France.

Catherine Julliot. Laboratoire d'Ecologie Générale, UMR 8571 CNRS-MNHN, 4 Avenue du Petit Château, 91800 Brunoy, France.
Current address: Centre IRD, Silvolab, BP. 165 97323 Cayenne cédex, Guyane Française.

Denis Larpin. Laboratoire d'Ecologie Générale, UMR 8571 CNRS-MNHN, 4 Avenue du Petit Château, 91800 Brunoy, France.
Current address: Service des Cultures, MNHN, 43 Rue de Buffon , 75005 Paris, France.

François J. Meunier. Laboratoire d'Ichtyologie, MNHN, 43 Rue Cuvier 75005 Paris, France.

Elisabeth Pagès-Feuillade. Laboratoire d'Ecologie Générale, UMR 8571 CNRS-MNHN, 4 Avenue du Petit Château, 91800 Brunoy, France.

Odile Poncy. Laboratoire d'Ecologie Générale, UMR 8571 CNRS-MNHN, 4 Avenue du Petit Château, 91800 Brunoy, France
Current address: Laboratoire de Phanérogamie, MNHN, 16 Rue Buffon 75005 Paris, France

Marie-Françoise Prévost. Centre IRD de Cayenne, BP. 165, 97323 Cayenne cédex, Guyane Française.

Riéra Bernard. Laboratoire d'Ecologie Générale, UMR 8571 CNRS-MNHN, 4 Avenue du Petit Château, 91800 Brunoy, France.

Toon Rijkers. Silviculture and Forest Ecology Group, Department of Environmental Sciences, Wageningen University, P.O. Box 342, 6700 AH Wageningen, The Netherlands.

Daniel Sabatier. Institut de Botanique, 163 Rue Auguste Broussonnet, 34000 Montpellier, France.

Corinne Sarthou. Laboratoire d'Ecologie Générale, UMR 8571 CNRS-MNHN, 4 Avenue du Petit Château, 91800 Brunoy, France.

Bruno Simmen. Laboratoire d'Ecologie Générale, FRE 2323 CNRS-MNHN, 4 Avenue du Petit Château, 91800 Brunoy, France.

Franck J. Sterck. Silviculture and Forest Ecology Group, Department of Environmental Sciences, Wageningen University, P.O. Box 342, 6700 AH Wageningen, The Netherlands. Current address: Department of Plant Ecology, Utrecht University, P.O. Box 80084, 3508 TB Utrecht, The Netherlands.

Marc Théry. Laboratoire d'Ecologie Générale, UMR 8571 CNRS-MNHN, 4 Avenue du Petit Château, 91800 Brunoy, France.

Jean-Marc Thiollay. Laboratoire d'Ecologie UMR 7625 CNRS, ENS, 46 Rue d'Ulm 75231 Paris cédex 05.

Peter J. Van der Meer. Silviculture and Forest Ecology Group, Department of Environmental Sciences, Wageningen University, P.O. Box 342, 6700 AH Wageningen, The Netherlands. Current address: Alterra, P.O. Box 47, 6700 AH Wageningen, The Netherlands.

Dionisos Youlatos. Laboratoire d'Anatomie Comparée, MNHN, 55 Rue Buffon, 75005 Paris, France.

Shuyi Zhang. Laboratoire d'Ecologie Générale, UMR 8571 CNRS-MNHN, 4 Avenue du Petit Château, 91800 Brunoy, France.
Current address: Institute of Zoology, Chinese Academy of Sciences 19 Zhong-Guan-Cun Road Haidian, Beijingg 100080, China.

Some canopy trees have extremely large and extended buttresses. *Photo Gerrie Offerman.*

CONTENTS

Bongers *et al.* (eds.), Nouragues. Dynamics and plant-animal interactions in a neotropical rainforest.
2001 Kluwer Academic Publishers. Printed in the Netherlands.

APPENDICES

The arboreal tree frog *Osteocephalus taurinus* is a common species in the Nouragues forest. *Photo Gerrie Offerman.*

EDITORS PREFACE

The idea of writing this book arose out of discussions between the four editors, who agreed that, twelve years after the creation of the Nouragues field station, it was time to present a synthesis of the scientific output of the research undertaken there.

The Nouragues site was originally selected through the initiative of a small group of researchers who needed an area of pristine forest for their investigations. It had to be an area untouched by human activities such as hunting, fishing, slash-and-burn agriculture or the gathering of natural products. Back in 1986, scientists in French Guiana had access only to areas already within the reach of hunters, and this was a serious hindrance to the study of the larger fruit-eating vertebrates, which are hunted wherever the forest is accessible. The study site had thus to be located far from settlements and access routes such as roads and waterways. It also involved the setting up of a permanent camp where researchers would find accommodation and field work facilities.

The selection of the Nouragues region took into account these requirements as well as the need to be able to undertake long-term projects. Many studies in tropical ecology have been initiated by individuals or small groups of people whose choice of study site was determined by what was currently available. In many cases, these sites unfortunately suffered severe degradation after the researchers left, which made further monitoring impossible. As the time scale of ecological mechanisms sometimes considerably exceeds the life-span of a researcher, one way to reach a better understanding of the mechanisms at work over time, although this cannot be done by one and the same person all along, is the constitution of diachronic biological archives from a particular site. In the course of the last two decades, ecological researchers have become increasingly organised and most of the fieldwork now takes place in field research stations with a guaranteed continuity. This is the case with Nouragues, where conservation measures have been taken to protect the site, resulting in the creation, in 1995, of a 1000 km^2 wilderness reserve that includes the research area.

Soon after its creation, the Nouragues field station attracted other teams anxious to find a tract of pristine forest for their research work. The resulting complementarities and cooperation of researchers, at both national and international level, provided an opportunity to widen the fields of research. Some of these developments are too recent, however, to permit their inclusion in this volume.

The topics covered by this book do not provide an exhaustive overview of the forest ecosystem, but reflect the scientific research programmes that have been implemented at the Nouragues field station since 1986. Whereas there is no chapter dealing with insects, 16 chapters concern vertebrates, among which nine focus on the role of mammals and birds in seed dispersal processes. This is because the station was originally set up to meet the needs of a research team studying the role of vertebrates in the processes of forest regeneration.

The book contains three chapters describing the physical environment, four chapters on plant communities, nine on vertebrate communities and evolutionary ecology, five on fruit-

Bongers *et al.* (eds.), Nouragues. Dynamics and plant-animal interactions in a neotropical rainforest.
2001 Kluwer Academic Publishers. Printed in the Netherlands.

vertebrates relationships and their function in forest regeneration, and a further six on forest dynamics and recruitment. Plant and vertebrate species lists, along with notes on life-forms, are included in the appendices. An identification key to palms is provided, with accompanying plates. Other data not easily publishable but nonetheless useful for comparative studies may be found on the Nouragues web site 'www.cnrs.fr/nouragues'.

The questions addressed in this book are complex and far from having been entirely resolved. They concern neotropical forests as well as forests in Africa, Asia, Madagascar or Australia, where the mechanisms of forest regeneration are also widely determined by interactions between plants and animals. Within each of these ecosystems, long evolutionary processes lead plant and animal communities of different phyletic lineages to establish complex interactions that govern forest dynamics and the distribution of biodiversity. Each biogeographical region harbours several thousand plant species, close to 1,500 vertebrate species and a number of invertebrate species as yet impossible to quantify with precision, all interacting with one another. At present, researchers in ecology are tending to concentrate on the study of a few functional models that involve easily identifiable species. But one of the long-term goals of scientists engaged in tropical research is to count and identify, region by region, the different plant and animal species involved in each of the functional levels of the ecosystem and to uncover the mechanisms that link these living organisms to one another. The task is a long and difficult one, maybe even unrealistic, but each step forward brings new insights into the way these ecosystems operate, and thereby helps to better orientate further research. The magnitude of the undertaking is such that we must not progress at random, and constantly renew our hypotheses so as to incorporate newly acquired knowledge. The use of such up-to-date hypotheses in conjunction with modelling should make it possible to obtain coherent data more rapidly. Similarly, species inventories need to be coupled with ongoing investigations in functional ecology.

This scientific approach is based on a network of field research stations, each one representative of its biogeographical region. A full understanding of the processes at work in tropical ecosystems can only be approached through the sampling of particular sites, but this will require a relative harmonisation of the methodologies used in the different field stations. Comparing data from different sites and analysing exploratory 'test' surveys from their periphery are fundamental procedures that can help determine, *a posteriori*, on which bases and at which geographical scale the results obtained from a single site can be extrapolated. The present book provides an overview of the main scientific results of one such site.

This book was possible due to the help of a lot of people. First of all we thank all the authors for their contribution.

All chapters have been sent for independent review to specialists around the world. We are very thankful for the constructive comments of (in order of alphabet):
R.O. Bierregaard, Nick Brokaw, André Brosset, Guy Caballé, Jean-François Cosson, Marc Debussche, Julie Denslow, Dick van Dorp, Joost van Duivenvoorden, Christian Erard, Joe Fragoso, Anne Gautier-Hion, Russ Greenberg, David Hammond, Henry Hooghiemstra, Marinus Hoogmoed, Yvan Ineich, Pedro Jordano, Elizabeth Kalko, Francis Kahn, Kaoru Kitajima, Bill Laurance, Pieter Ketner, Robert D. Martin, Bernard de Merona, Kathy Milton, Marilyn A. Norconk, Jan den Ouden, Thijs Pons, Thierry Rosique, Dominique

Schwartz, Liesbeth Sterck, Françoise Viénot, Jean-Francois Villiers, Sip van Wieren, Sander Wijdeven.

We are indebted to Desmo and Wemo Bétian, who are for a large part of their time working at the Nouragues biological station. CNRS (UMR 8571 and UPS 656) is acknowledged for their financial support, and the Laboratory of General Ecology (MNHN) in Brunoy. Last, but certainly not least, colleagues at the Brunoy laboratory are thanked for their general support: Anya Cockle, who translated several chapters from French to English, and Adeline Caubère, Roland Flaccus, Isabelle Hardy and Sylvie Jouard for technical help.

A special, and warm, thanks to Mireille Charles-Dominique, for the incredible amount of energy and work she has put into this book. Without her, this book would never have been finished.

F. Bongers
P. Charles-Dominique
P.-M. Forget
M. Théry

A bending river in the forest zone of French Guiana. *Photo Frans Bongers.*

PART 1

INTRODUCTION

CHAPTER 1

THE FIELD STATION

Pierre Charles-Dominique

The Nouragues tropical forest research station[1] was established in 1986 to implement research on the natural mechanisms of forest regeneration. We chose to select a place relatively removed from rivers (navigable water courses) but yet not too far from Cayenne (for obvious financial reasons) and close to a pre-existent light research camp established by the National Museum of Natural History on the Pararé rapids of the Arataye river. Both the fauna and flora of this area had been under study since 1978, and this provided a valuable scientific data base on which to build further research.

The selection of the actual site, about 6 Km from the Pararé rapids as the crow flies, was determined using the IGN (National Geographical Institute) maps and the indications of a knowledgeable aeroplane pilot. The area is covered by extensive primary forest uninhabited since over two centuries, and is particularly notable for its inselberg (rocky outcrop towering above the forest and reaching 430 m above sea level). In March 1986, a group of six researchers and two technical assistants undertook an 8-day preliminary exploration of the area, prospecting by foot from the Arataye river. The area appeared rich and diversified, and the behaviour of the larger animals indicated negligible hunting pressure. A freshwater spring was also located which could be tapped. The local topography was found compatible with the opening of an helicopter landing platform, while an expanse of ' rock savanna ' at a 20 min walk from the site allowed access from the air in the mean time. The helicopter landing platform was cleared in September that same year, with the help of the Army (the 32nd RIMA), and the first elements of the camp were soon erected by French Guianan technicians.

The name ' Nouragues ' was chosen as a reference to the Nouragues (or Norak) Native Amerindians who were present in the area when the Westerners arrived and who disappeared during the eighteenth century, after their last members joined the Emerillons ethnic group near Camopi (Grenand 1979). Traces of fireplaces, sherds, groving stones and funerary urns discovered under a rocky ledge are tokens of this - or of a more ancient -

[1] The Nouragues station is a CNRS (Centre National de la Recherche Scientifique) structure (UPS 656) and a constitutive element of the ' Silvolab ' GIS (Group of Scientific Interest), a joint venture grouping different institutions in French Guiana. The Silvolab facilities open to European and international scientific teams include the Cayenne Herbarium (IRD), the Paracou experimental station (CIRAD, ENGREF), 30 km from Kourou and accessible by dirt road, where mostly forestry research programmes are carried out, and the Nouragues station is more specifically designed for the study of the undisturbed tropical forest ecosystem. This volume presents a fair part of the results obtained from research undertaken on the Nouragues station.

Bongers *et al.* (eds.), Nouragues. Dynamics and plant-animal interactions in a neotropical rainforest, 1-7.
2001 Kluwer Academic Publishers. Printed in the Netherlands.

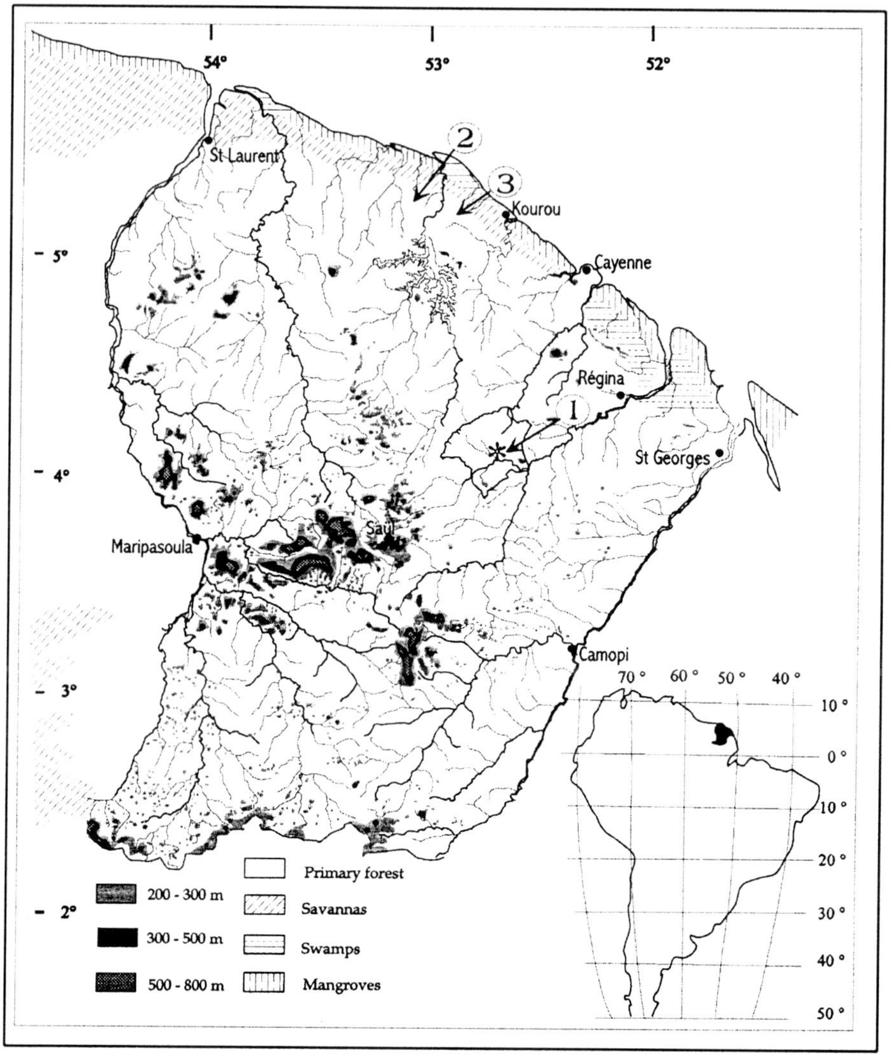

Figure 1. French Guiana, relief, main vegetation formations, and location of the following study areas : Nouragues (1), Piste St Elie (2), Paracou (3).

occupation. Remains of an ancient habitat at the exact location of the present research camp were carbon-dated around 1000 years BP. The area was thereafter totally deserted until the 1855 gold rush, followed by the exploitation of *Manilkara bidentata* rubber and *Aniba rosaeodora* spirit, until the 1930's. This later human impact was essentially concentrated along the navigable water courses, and no traces of it were found on the station grounds and around the inselberg. However, this does not rule out occasional visits at that time.

The camp sits on a minor ridge 400 m from the inselberg, which steeply dominates an intricate pattern of narrow plateaus and valleys between 40 and 100 m a.s.l. The clearing of the vegetation on the campsite was kept to a minimum (about 80 m x 40 m) and, right from the start, it was agreed to leave plants, animals and fish unmolested in the entire region. This rule has now been officialized by the creation decree of the Natural Reserve in December 1995 (Fig. 1, 2).

Photo 1. View of the Nouragues area from the top of the inselberg. *Photo Pierre Charles-Dominique.*

In ten years, the station has been equipped progressively[2]. Local characteristics of the ecosystem and housing facilities provided good working conditions for both teams and individual researchers, and scientific investigations quickly diversified. Complementary studies and collaborations between different research teams were encouraged right from the start. Over the last five years, the station has welcomed an average of 45 researchers and students each year, including one third from foreign countries (both from European Community countries and from outside countries). Projects can be accepted as long as they do not involve the collection of important quantities of material and/or heavy disturbance. The collection of herbarium samples and animal specimens (for diagnosis or identification, when not possible otherwise) are accepted, while massive collection methods (tree felling, ultra violet light traps, fogging, etc.) are not authorized. However, massive collections can be undertaken if need be on the borders of the Reserve as long as they remain compatible with the general legislation of French Guiana. Research undertaken on the Nouragues station today concerns the following topics: - the mechanisms of forest regeneration, in particular plant growth and mortality, - the modalities of seed dispersal (which involve

---

[2] see <www.cnrs.fr/nouragues>

many fruit-eating animals), - gap formation, - the biology of various groups of plants and animals, - the social organization of several species of mammals and birds, - the quantitative and qualitative variations of the light environment, - the study of biodiversity and its distribution, - the genetics of several populations, - the history of the environment during the last ten thousand years (fossilized pollen and seed collected in core samples, charcoal fragments embedded in the soil, distribution of indicator plant species, etc.). The data obtained (publications, intermediary reports, species lists, meteorological data, etc.) are centralized at the Brunoy laboratory, Laboratoire d'Ecologie Générale, Muséum National d'Histoire Naturelle.

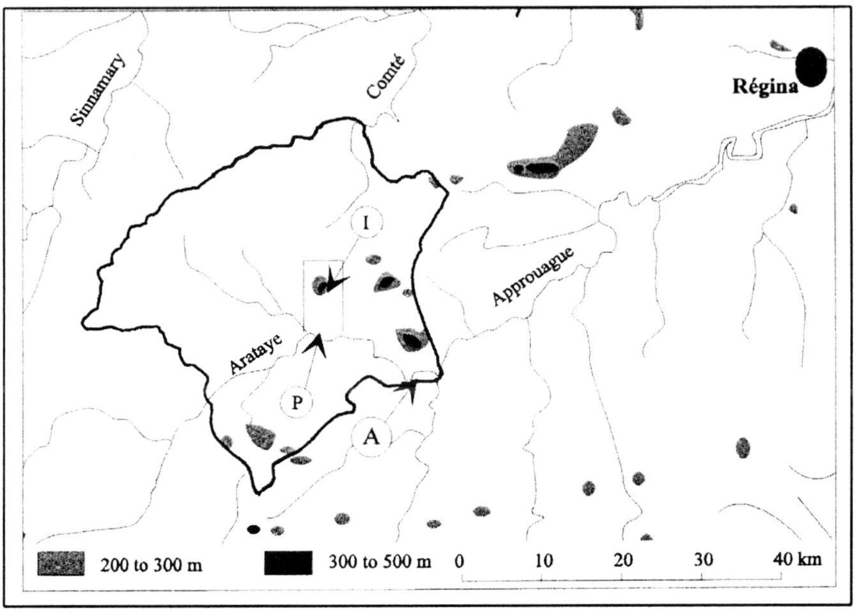

Figure 2. Location of the Nouragues Natural Reserve, with the area devoted to research (rectangular area). I = Inselberg site (present station), P = Pararé site (COPAS) device, A = Arataye site (eco-tourism camp).

Part of the Natural Reserve has recently been opened to ecotouristic activities on the initiative of the DIREN (Regional Environment Board) and the local councils ofRoura and Regina, on the land of which the Reserve is established. With this in perspective, avisitors camp (the Arataye camp) was constructed at the entrance of the Reserve, on the Arataye river close to its confluence with the Approuague river, and thus easily accessible by pirogue. This new structure is independent from the Nouragues scientific station and can welcome small groups of visitors who wish to view a tract of pristine forest where fauna and flora are still intact. An initiation course in tropical biology fieldwork will soon be offered based on results obtained in the scientific station. The trails open to touristic activities remain strictly separate from the perimeter specifically affected to research activities, but researchers who wish to be based in the visitors' camp are welcomed to do so.

Photo 2. The Nouragues camp, a view from a helicopter. We can notice the dropping zone for the helicopter and the solar panels. *Photo Pierre Charles-Dominique*

Photo 3. A canopy walkway at the Nouragues station. *Photo Pierre Charles-Dominique*

The creation decree of the Les Nouragues Natural Reserve provides for a perimeter of 6 km by 8 km specifically dedicated to scientific research and which will include two base campsites - the ' inselberg site ' which corresponds to the present scientific station, and the ' Pararé site ' where the COPAS device will shortly be erected. This new project is centred

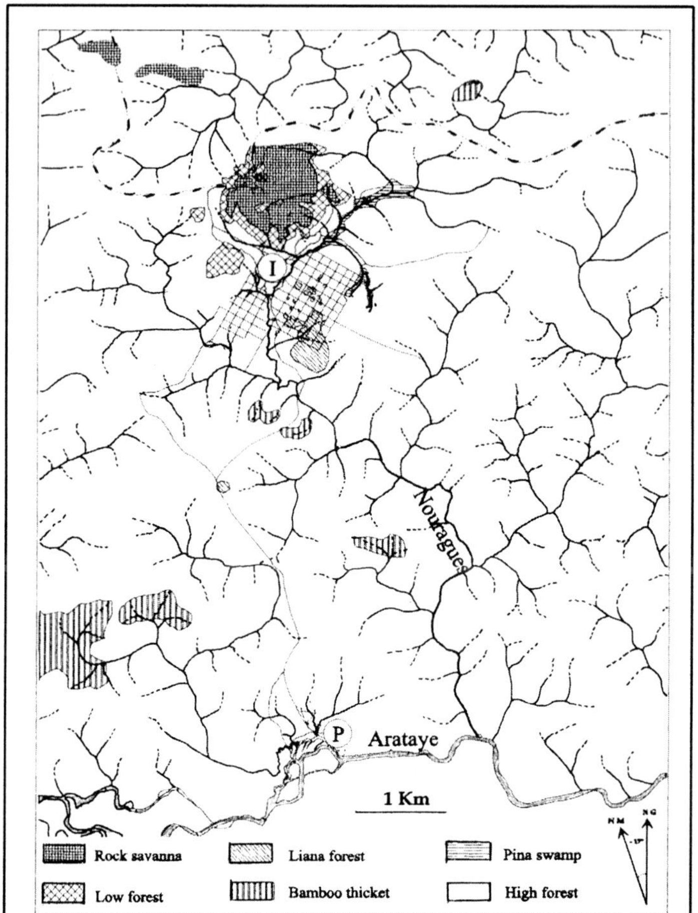

Figure 3. Research area, and principal plant formations, location of Inselberg site (I) and Pararé site (P).

on the construction of a permanent canopy access device (Canopy Operation Permanent Access System) and will give an even broader scope to the Nouragues station. The development of a research station comes up against two opposing needs: 1- the need to intensify research activities in order to gain an overall vision of the ecosystem, to make an economically efficient use of the facilities and to ensure the presence of a minimum number of researchers, 2- the need to limit the use of the study areas in order to prevent their degradation through over-use. One answer to the problem of these conflicting requirements is the creation of complementary research sites, each with its own ecological specificities and within a few kilometers of each other. The ' Inselberg site ' (the present Nouragues station) will be particularly suited to the study of forest gradients (rock savannas, 'transition' forests, liana forests, bamboo thickets or ' cambrouzes ', plateau forests) and to the study of population dynamics and large vertebrates. The ' Pararé site ' will be more suited to the study of riverine forests and particularly to the study of canopy epiphytes and microfauna. As the two sites are only 7 km apart (2h1/2 hiking) and located within the radio relay's range (communication using walkie-talkies), research activities will be able to develop inside the Nouragues research perimeter without imposing undue pressure on the environment. Moreover, it will be possible for research projects to use both sites for comparative purposes (Fig. 3, 4).

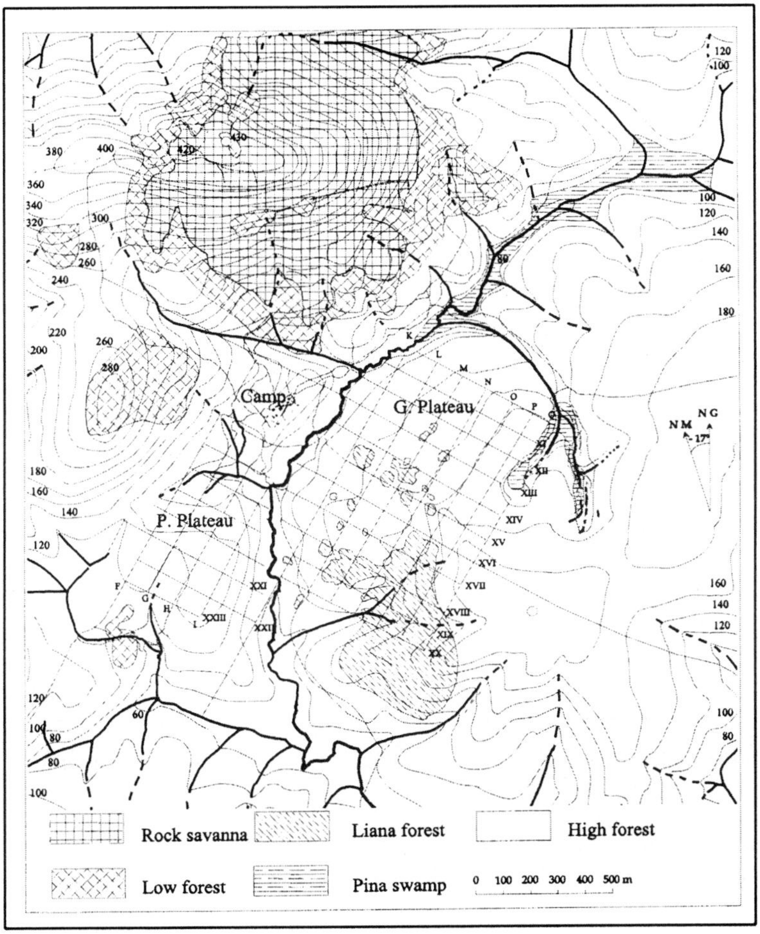

Figure 4. Detail of the 'Inselberg site' study area with position of the camp, the 'grand plateau' and 'petit plateau'. Topographic curves are equidistant of 20 m and trails are named by letters or latin numbers.

A nest of night wasps during their daytime rest. *Photo Pierre Charles-Dominique.*

CHAPTER 2

GEOGRAPHY AND CLIMATE

Michel Grimaldi and Bernard Riéra

1. Introduction

The Nouragues station is located within the 'Montagnes Balenfois' massif to the Southwest of the commune of Régina. The landscape of hills covered by dense tropical rainforest is dominated by a partly denuded granitic inselberg which reaches an elevation of 430 m above sea level. The limit between the River Comté drainage basin in the north and the River Approuague basin passes across the summit of this inselberg. The Nouragues creek (a tributary of the River Approuague) flows past the foot of the inselberg, forming the boundary between two main types of geological substratum: the 'Caribbean' type granite and the metavolcanic rocks of the Paramaca series. The lithology of these terrains has an influence on the relief and on the soils. Thus, the geology, the geomorphology and the soil types are presented here in succession. Finally, an analysis is given of the climatic data that has been collected since the setting up of the station.

2. Geology

French Guiana is situated in the north-eastern part of the Guiana Shield, which is composed mainly of Paleoproterozoic terrains (2.2-2.0 Ga). The Guiana Shield is generally younger than other cratons, the only Archaean ages being found in the Imataca granulitic complex (> 3.4 Ga) in north-eastern Venezuela (Fig. 1). Volcanic and sedimentary rocks are particularly abundant in northern Guiana, where they make up the greenstone belts of the Transamazonian orogeny. The syn- and post-orogenic periods are characterised by various phases of deformation and metamorphism that accompany the progressive intrusion of the different types of pluton. Over a very long time span, erosion has exposed vast expanses of granitic terrain while reducing the importance of the intruded country rocks (volcanic and sedimentary origin).

The lithostratigraphic succession proposed by Milési *et al.* (1995) for the greenstone belts of northern Guiana is based on the recognition of two major units lying structurally above a substratum composed of highly metamorphosed volcanic rocks, the Île de Cayenne Formation:

- a lower volcanic and sedimentary unit, variably metamorphosed, comprising the volcanic rocks of the Paramaca Formation (older than 2.13 Ga and up to 2.1 Ga) overlain by flyschoid facies of the Armina Formation;
- a sand-conglomerate unit, known as the 'Upper Detrital Formation', which was laid down after 2.02 Ga in the 'North Guiana Trough'.

Bongers *et al.* (eds.), Nouragues. Dynamics and plant-animal interactions in a neotropical rainforest, 9–18.

The Paleoproterozoic also contains granitic and metamorphic terrains of medium and high grade, including gneisses and amphibolites, as well as granitoid intrusions of 'Guianan' and 'Caribbean' type dated at between 2.14 and 2.09 Ga in the Régina area. The 'Guianan' granitoids are generally transformed into gneisses, which are associated with migmatites. Their compositions vary between quartz-dioritic and a granitic end-members. Among the 'Caribbean' granitoids, it is possible to distinguish disparate types of migmatitic gneiss, coarse granitoids of porphyritic affinity and two-mica granites associated with abundant pegmatites.

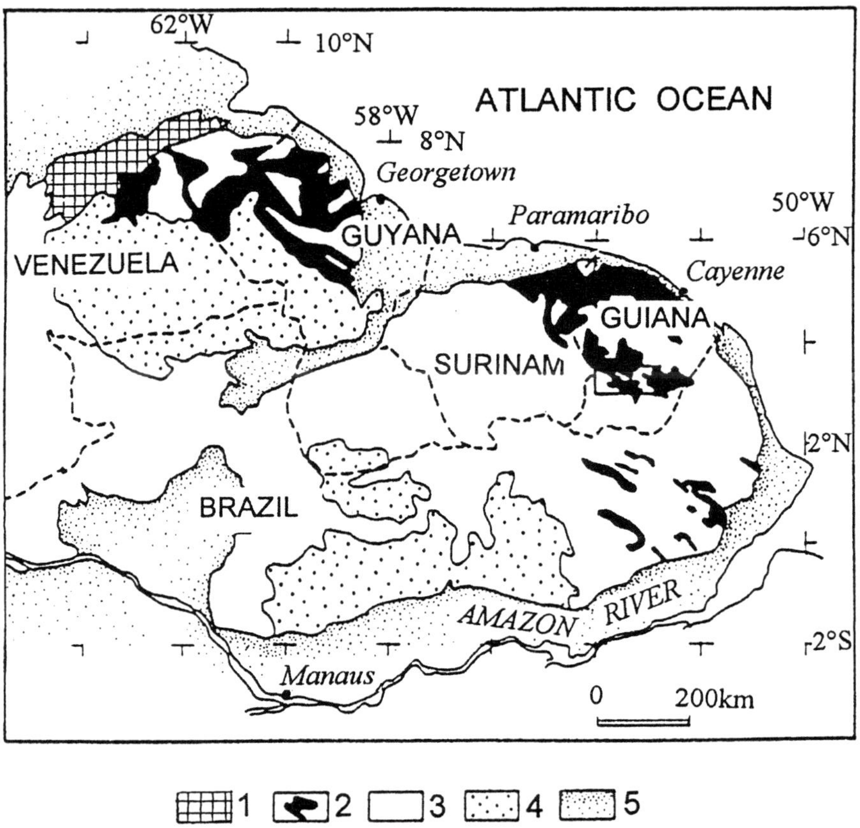

Figure 1. Geological sketch map of the Guiana shield (after Gibbs & Baron 1983, and Gruau *et al.* 1985). Legend: 1: Imataca granulitic complex; 2: Greenstone belts; 3: Paleoproterozoic granitoid and metamorphic terrains; 4: Mesoproterozoic volcanic and detrital formations; 5: Younger sediments.

The 'Montagnes Balenfois' massif is made up of metavolcanic terrains of the Paramaca Formation which are cross cut by 'Guianan' and 'Caribbean' type granites (Fig. 2). The Nouragues inselberg is composed of a tabular outcrop of 'Caribbean' granite. The table-topped hills on the other side of the Nouragues creek, situated to the south and east of the inselberg, are formed of rocks belonging to the Paramaca Formation. These volcanic rocks (lavas and pyroclastic deposits) cover a wide range of composition from basaltic to rhyolitic, and are generally metamorphosed in the greenschist facies. Sericitic schists are intercalated within the metavolcanics. However, the relative abundance and distribution of the different volcanic members remains poorly known. The observed metamorphic recrystallization is very irregularly developed and is interpreted as a contact metamorphism related to the emplacement of the granites. Some dolerite veins, accompanied by small massifs of gabbro and basalt, are seen to cut across the Paleoproterozoic basement (Fig. 2). These bodies are unaffected by any metamorphic or tectonic transformation, and their emplacement is associated with Permo-Triassic distension linked to the opening of the Atlantic Ocean.

The Nouragues inselberg is made up of a pinkish monzonitic-type granite, containing 27%K-feldspar (orthoclase) and 37% plagioclase, along with 33% quartz as coarse-grained crystals and 2% accessory minerals (pyroxene, corundum, apatite). The whole-rock chemical analysis (Sarthou 1992) shows that the granite is highly siliceous (76.4% SiO_2) and rich in alkalis (4.6% K_2O, 4.2% Na_2O). The granite has a non-negligible surface porosity as measured by water absorption, attaining 3.1% of the total rock volume in a slightly weathered sample and 7.8% in a more weathered sample (Sarthou 1992).

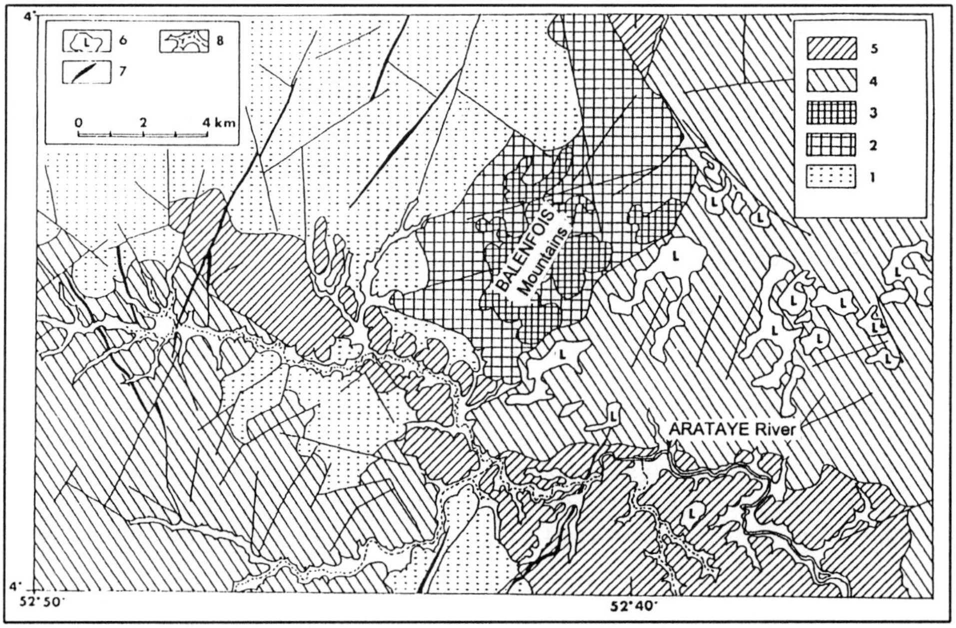

Figure 2. Geological map of the Balenfois Mountains region. Legend: 1: 'Guianan' granites; 2: 'Caribbean' migmatites; 3: 'Caribbean' granites; 4: 'Paramaca' Formation; 5: 'Ile de Cayenne' Formation; 6: Lateritic domains; 7: dolerite veins; 8: alluvial deposits.

3. Geomorphology

Although the Guiana Shield has been a stable block since 1.7 Ga, it has nevertheless been affected by slow and non-deformational uplift (epeirogenic movements). According to Bigarella and Ferreira (1985), these movements occurred in several different phases: around 100 Ma (end-Cretaceous), at 23 Ma (Oligocene/Miocene boundary) and at 2 Ma (end- Pliocene). The shield also underwent repercussions from the Andean orogeny, especially between 25 and 20 Ma, and again between 8 and 2 Ma (Bourgeois & Jeanjou 1981, Haffer 1987), thus producing the reactivation of faults, as well as the horizontal and vertical movements of certain blocks. These epeirogenic phases produced renewed episodes of erosion by lowering of the marine base level (tectono-eustatic effects). According to Bigarella and Ferreira (1985), the present-day inselbergs of the Guiana Shield began to form following the epeirogenic phase that occurred 23 Ma ago.

The Quaternary glaciations are the cause of the glacio-eustatic variations in marine base level. These climatic variations also led to changes in vegetation cover produced by alternations of moist conditions associated with expansion of the forest duringinterglacials, and dry conditions associated with expansion of the savanna during glacials. This alternation of climatic regimes may explain the lightly weathered accumulations of sand and gravel at the foot of the continental rise (Damuth & Fairbridge 1970, Damuth 1977, Pujos & Odin 1986), as well as the formation of inselbergs by a process of etchplanation (Twidale 1980, Sarthou 1992). This process comprises two stages:

- intense weathering under prolonged tropical humid conditions in a forest environment, forming a cover of loose weathered deposits (regolith) of variable thickness according to variations in the tectonic and textural characteristics of the parent rock.
- The weathered and decayed rock is incised and the more resistant core-massifs are dismantled by erosion during one or more phases of oscillation in base level. The relief is then enhanced by runoff which dries out the denuded rock massif and leads to a concentration of humidity around its rim. The weathering is also much less intense by runoff on a bare rock surface than beneath forest cover in gently sloping terrains.

Apart from having different elevations, the hills around the Nouragues station show quite distinct landforms according to whether the substratum is granitic or metavolcanic (Fig. 3). The average slopes for these different terrains are 32 and 11%, respectively. The hills on the south-eastern side of the Nouragues creek, which are underlain bymetavolcanic rocks, and are termed plateaus due to their wide flat summits. They display convex-upward profiles on their flanks with gentle and regular slopes. By contrast, the inselberg and the surrounding granitic hills make up a terrain with manybreaks of slope. A series of benches ring the inselberg to the south and east, continuing up to an elevation of 100 m above the Nouragues creek.

The inselberg is dome-shaped and elongate about an E-W axis. It lacks any forest cover on the eastern flank, where a particular type of denuded or rock-savanna vegetation is developed (Sarthou 1992, Larpin 1993). On one of the first transects studied by Sarthou (Fig. 3), the base of the inselberg is made up of two gently sloping areas that are termed terraces: the lower one, at an elevation of 80-100 m, has relics of a lateritic crust and is covered by low-canopy forest, while the higher one (at 120-200 m) ischaracterized by rock savanna vegetation. An abrupt break of slope separates this latter terrace from the denuded flank of the inselberg, which forms a veritable cliff rising from 200 to 320 m where rainwater runs of in deeply scoured grooves. The gradient then flattens off from 320m to the summit of the inselberg. Here, grooves are less abundant than on the cliff face, channelling the sheet wash into initial branches of the drainage network. Thenorth-eastern flank is covered with forest and is undergoing soil formation processes, while the southern flank is subject to runoff. Erosion on the rock gives rise to a chaotic pattern of domes and troughs that are several metres in size. Exfoliation and flaking of the granite have produced slabstones that appear to rest on the rock surface as well as smaller slabs with a tendency to slide.

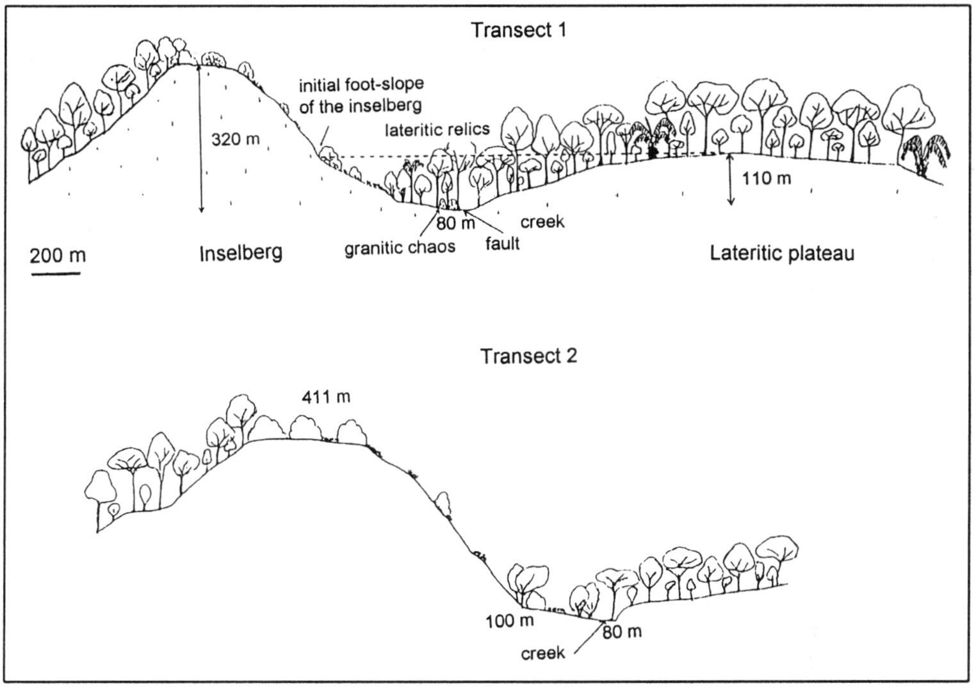

Figure 3. Topographical transects (after Sarthou 1992).

The lateritic relics developed on granite at the foot of the inselberg may represent part of an ancient weathering mantle whose surface could have attained the same level as the lateritic plateaus developed on the metavolcanics. In fact, the highest point of this plateau is at the same elevation as the second break of slope on the inselberg at the foot of the cliff (Fig. 3), thus possibly corresponding to the extent of the initial inselberg. In this way, the terrace areas would have been cleared by headward erosion. On the other hand, the upward-convex slope of the cliff would correspond to an area that had been denuded by prior erosion.

4. Pedology

The soils developed on the rocks of the Paramaca Formation are termed ferralsols; they are yellowish brown in colour under a thin litter and grade progressively into yellowish red varieties. These soils display a clayey texture (Table 1) and are very rich in ferruginous nodules a few mm to several cm in size occurring at all depths in the profile (abundances greater than 35% by weight), except at the tops of hills. Blocks of iron pan are common on the valley sides. The microaggregate structure of biological origin, which is a characteristic feature of ferralsols, is generally well developed to a depth of several metres. This structure ensures a deep vertical drainage of the soil as well as a deep penetration of tree roots. Thus,

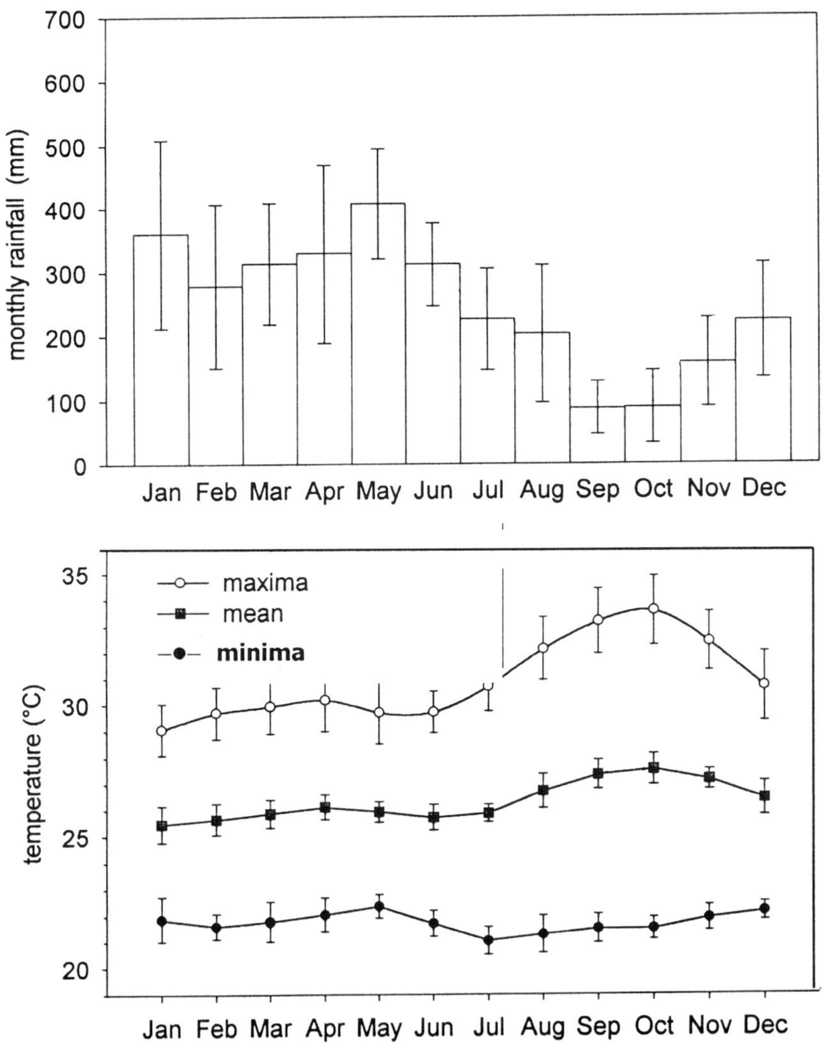

Figure 4. Mean monthly rainfall (a) and temperature (b) between 1987 and 1997 at the Nouragues station.

the trees are not subject to excess water during the wet season and can find sufficient water reserves stored in the void space between microaggregates (representing more than half the total porosity). These soils are acidic, having an average water pH of 4.5, while exhibiting a strong depletion of bases with a low cation exchange capacity varying from 15 down to 1 meq/100 g between the humic horizon and a depth of 50 cm (Table 1).

TABLE 1. Chemical and granulometrical properties of two pedological profiles at the Nouragues station.

	Depth	Water pH	CEC (a)	Gravels >2mm	Organic Mater	Clay <2μm	Fine Silt 2 - 20μm	Coarse Silt 20 – 50μm	Fine Sand 0.05 - 0.2mm	Coarse Sand 0.2 – 2mm
	cm		meq/100g	g/100g		g/100g (100 g of fine earth: < 2mm)				
Ferralsol	0 - 5	3.9	16.3	51.7	12.4	44.4	16.9	1.2	2.7	9.3
on	5 - 10	4.3	8.6	59.8	4.9	69.5	8.2	1.2	2.2	7.2
Paramaca	10 - 20	4.5	5.6	62.8	3.1	70.6	11.9	1.6	2.6	5.0
formation	20 - 30	4.8	3.7	33.7	2.2	52.3	9.2	3.5	17.1	14.6
	30 - 40	4.8	1.9	42.0	1.5	42.8	8.4	4.8	20.9	20.5
	40 - 50	5.1	1.5	38.5	0.9	37.5	11.2	7.7	27.4	14.7
	90 - 100	5.3	1.5	72.5	0.4	43.1	20.7	13.4	16.7	4.7
	140 - 150	5.3	1.5	78.9		40.5	20.0	13.8	16.7	9.0
	190 - 200	5.1	1.5	42.6	0.7	72.4	15.2	3.5	2.8	5.0
	290 - 300	5.5	1.4	43.9	0.3	46.9	14.9	6.0	8.8	22.9
Ferralsol	0 - 5	4.3	8.4	33.3	7.9	26.4	3.7	1.9	5.7	51.1
on	5 - 10	4.3	6.6	25.7	5.5	34.8	4.3	2.1	6.2	45.5
granite	10 - 20	4.1	4.7	16.5	4.1	38.5	6.0	2.9	7.1	40.7
	20 - 30	4.4	4.2	18.3	2.8	47.9	5.9	3.4	7.4	32.0
	30 - 40	4..5	3.4	17.9	2.4	49.7	6.7	4.0	8.0	28.7
	40 - 50	4.8	3.1	21.4	1.9	38.9	9.3	5.2	11.3	32.9
	90 - 100	4.6	2.4	28.1	0.6	53.4	6.7	3.2	6.7	29.2
	140 - 150	4.6	2.6	25.9	0.6	51.1	7.9	3.7	6.9	29.6

(a) Cation exchangeable capacity determined at pH 7

The soils of the granitic terrain have a highly variable thickness that shows no relation with slope or topographic position except at the summit of the inselberg. Here, a relatively gradual transition is observed between the bare rock and the forest-covered zone, with commonly organic–rich cambisols passing laterally into ferralsols. The spatial variability in soil thickness is the result of the spheroidal weathering of the granite as well as the influence of erosive episodes. The cambisols are black to dark brown in colour and are very sandy. With increasing thickness, the soil acquires a yellowish brown to intense brown coloration and a sandy clayey to clayey texture (Table 1). These soils are practically devoid of ferruginous nodules; constituents coarser than 2 mm make up 15-30% of the total soil, being composed of quartz grains and partially weathered fragments of granite. The intergranular porosity (or void space between clayey microaggregates) is high. During rain showers, there is no slowing down of percolation through the soil before the water reaches the parent rock. On steep slopes, the infiltrating water flows over the sharp boundary between soil and parent rock. In the case of gentle slopes, water stagnates within the profile and the soil becomes more or less waterlogged above the rock. At the summit of the inselberg, on gently sloping surfaces, a particular type of forest is developed on a greyish brown sandy soil of variable thickness which is saturated with water during the rainy season. To summarise, the hydric behaviour of soils on granite, particularly regarding the risk of hydric excess or deficit, is mainly dependent on the local topography and the depth to unweathered bedrock. The soils developed on granite are chemically very poor, being more acidic than soils on metavolcanics and comprising an even more strongly depleted cation exchange complex (Table 1).

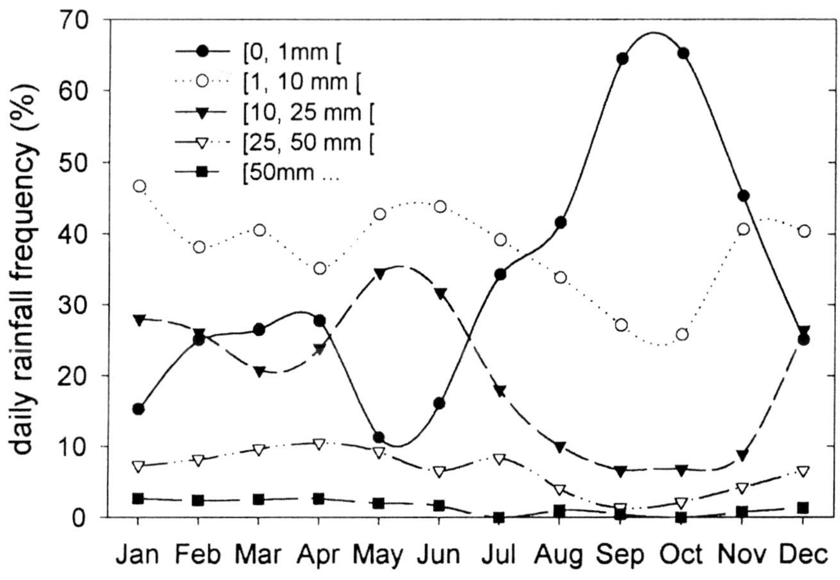

Figure 5. Monthly distribution of daily rainfall frequencies for five different rainfall classes.

5. Climate

The climate of French Guiana is of equatorial type (Boyé *et al.* 1979). Situated in the equatorial zone of the Northern Hemisphere between latitudes 2 and 5°N, this region is subject to trade winds from the Northeast and Southeast that are controlled by the Azores and Saint-Helena anticyclones, respectively. These two trade wind systems come together along a low-pressure zone known as the Inter-tropical Convergence Zone (ITCZ) which is characterized by intense cloudiness. The movements of the ITCZ depend on the relative activity of both anticyclonic systems and determine the seasons in the region. The rainy season starts in November when the ITCZ moves over French Guiana from the north, and ends in August when it moves away towards the Lesser Antilles. During the rainy season, French Guiana experiences periods of fine weather due to oscillations of the ITCZ across the region, especially when the ITCZ approaches to the Equator between February and

April; this is the little dry season, which is marked to a variable degree in different years. The main dry season is characterized by low precipitations, with local precipitation of stormy type, and sometimes there is a total absence of rain for more than a month. The annual rainfall varies widely throughout French Guiana, from 2 000 mm in the north-west (Saint Laurent) to 4 000 mm in the north-east (Kaw area). The variability between years is of the same order of magnitude, in such a way that the average values (whether monthly or annual) are only a partial reflection of the true rainfall regime.

The meteorological station at Nouragues enables the daily recording of precipitation, minimum and maximum temperatures, as well as evaporation (Piche evaporometer). The nebulosity, or amount of cloud expressed in oktats, is noted at 7h and 18h local time. Although the available series of data runs since November 1987 to December 1996, only 88 months out of a total of 110 in this period have a complete record. The least well recorded months are January and December.

On the annual rainfall map published in the Atlas of French Guiana (Boyé *et al.* 1979), the Nouragues station is located between the 3000mm and 3250mm isohyets. Over the period of records, the measured annual precipitation attains an average of 2990 mm, the total being spread over 310 days of rainfall. The nearest meteorological station, which is at Régina, has one of the highest rainfalls in French Guiana, yielding an annual average of 3940 mm between 1971 and 1990. The mean seasonal distribution of rainfall at Régina is comparable with that observed at Nouragues. The driest months at Nouragues are September and October, with a mean precipitation of 88 and 89 mm, respectively. By contrast, May is the wettest month here with a total rainfall of 407 mm.

In statistical terms, the transitions with the two main seasons take place during July and August on the one hand and November (as well as December to a lesser extent) on the other. However, it should be noted that the limits of the dry season have varied appreciably over the nine years of records, starting between mid July and late September and ending between early November and early December. The distribution of daily precipitation in 6 water-level classes over 9 years yields an average of 55 days without rain, 67 days of very low rainfall (<1 mm), 139 days of low rainfall ($1 \leq P < 10$ mm), 74 days of medium rainfall ($10 \leq P < 25$ mm), 24 days of heavy rainfall ($25 \leq P < 50$ mm) and 6 days of very heavy rainfall ($P > 50$ mm). This distribution pattern is a function of the season (Fig. 5): the frequency of days with very little or no rain (<1 mm), as well as the frequency of days with medium rainfall can be used as good discriminants of the most characteristic months in each season. These are May and June for the rainy season, and September and October for the dry season.

The mean annual temperature calculated for Nouragues is 26.3°C. The mean monthly temperature varies from 25.5°C in January to 27.5°C in October, the difference being due to an increase in the maximum temperature during the dry season (Fig. 4). On average, the daily temperature range varies from 7°C in May to 12°C in September. The absolute minimum and maximum recorded temperatures are 17.8 and 37.0°C, respectively. It is noteworthy that extreme temperatures are observed in exposed environments, especially on the bare hillslopes of the inselberg facing south (Sarthou 1992) where maximum temperatures can attain 60°C. According to aspect, the daily temperature range on the inselberg (rock savanna) can vary from 18 to 33.1°C during the dry season and from 16.1 to 20.8°C during the rainy season. Under these same conditions, the relative humidity of the air is negatively correlated with temperature (Sarthou 1992), with a minimum of 22%

during the dry season and 65% during the rainy season. The daily range in humidity varies from 61-69% during the dry season to 27-35% during the rainy season.

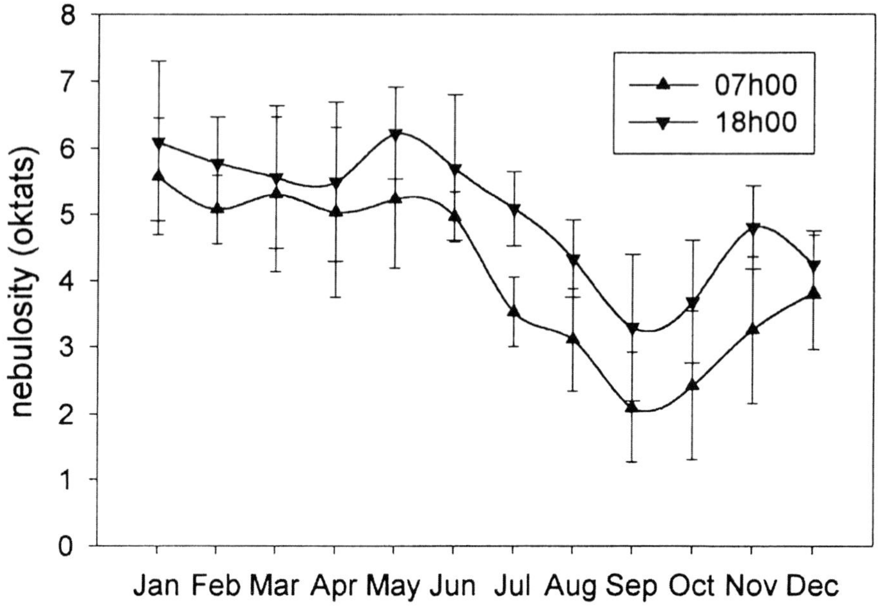

Figure 6. Mean monthly nebulosity at the Nouragues station.

Nebulosity is correlated with the amount of precipitation. Throughout the year, nebulosity is more marked at the end of the day (Fig. 6). During the rainy season, the average value is greater than 5 oktas, while the lowest values are observed in September (2.1 oktas at 7 h and 3.3 oktas at 18h).

Acknowledgements

We are grateful to M.S.N. Carpenter for translation of this chapter.

CHAPTER 3

SCALES OF AMBIENT LIGHT VARIATION

Frans Bongers, Peter J. van der Meer and Marc Théry

1. Introduction

In tropical rain forests light availability is extremely variable, at many scales of observation, in time as well as in space (Chazdon 1988, Yanhong & Naoki 1997, Clark *et al.* 1996, Endler 1993). Light is abundant in the upper part of the forest canopy and in (large) canopy gaps but it is scarce in the forest understorey. This is mainly due to the fact that the leaves intercept and filter light while the stem and branches reflect it. This leads to lower light intensities and drastic changes in light quality with shading.

Spatial variation in light is the result of spatial variation in forest structure and composition. This variation has large scale (forest types), medium scale (forest mosaic patches) and small scale (individual trees and within crown variation) components. Temporally, light varies at the scale of minutes to hours (sun flecks through small holes in the canopy), weeks to months (small canopy gaps that are quickly filled by lateral in-growth of neighbouring crowns), to years (large canopy gaps that make take many years to close).

Most forests consist of a mosaic of forest patches in different stages of development (Whitmore 1978, Oldeman 1983). Because these patches differ in species composition and structure light levels vary among the patches, leading to a spatial mosaic of light (both quantity and quality). These patches are not stable in time: plants may grow, leaves may drop, branches may fall, trees may topple over. The result of this spatial and temporal dynamics in forest structure is that the light environment changes over space and time.

Light environments in the forest can be grouped into four major light habitats, based on forest structure, weather, and time of day (Endler 1993). Using light colour spectra (i.e. the distribution of light intensity with wavelength) Endler distinguished (1) forest shade, characterised by a yellow-green light (2) woodland shade, characterised by blue-grey light (3) small gaps, characterised by reddish light, and (4) large gaps, characterised by white light. An additional light environment is associated with purple light at dawn and dusk. During cloudy weather the spectra of light habitats converge toward white (Endler 1993).

The variation of light environments affects germination (Vazquez-Yanes & Orozco-Segovia 1984), photosynthesis (Pearcy 1990), plant morphogenesis (Kendrick & Kronenberg 1994), communication among animals (review by Endler 1993; Craig *et al.* 1996, Endler & Théry 1996) and communication between animals and plants (Endler 1992). Each of the scales of light variation initiates its own responses, in plants (Chazdon

Bongers *et al.* (eds.), Nouragues. Dynamics and plant-animal interactions in a neotropical rainforest, 19-29.
2001 Kluwer Academic Publishers. Printed in the Netherlands.

1988, Gross 1986) as well as in animals (Endler 1992). Consequently, the irradiance level and the spectral distribution of ambient light potentially determines optimal habitats both for many animals and for plants. For many plants growing in the forest understorey, light is the main growth limiting resource.

In this chapter we describe the spatial distribution of light in the Nouragues forest. We focus on both light quantity and light quality. In the Nouragues forest, light quantity is mainly studied in relation to plant growth and performance (see Sterck 1997, Bongers & Sterck 1998, Rijkers *et al.* 2000a, b, Van der Meer *et al.* 1998, see also Sterck *et al.* Chapter 27). Light quality is mainly studied in relation to animal behaviour, notably birds (Théry 1990, 2000, see also Théry & Endler Chapter 14). Specifically we discuss light quantity (irradiance) at the scale of the forest mosaic, the forest patch (canopy gap), and the forest tree (individual tree crowns). Light quality is discussed at the scale of the community and of the forest mosaic.

2. Methods

2.1 LIGHT QUANTITY

Average light quantities at the forest floor were described using hemispherical photographs on 58 grid points in 12 ha of forest. The forest mosaic of the Nouragues forest consisted of four patch types: treefall patches (canopy gaps) with a vegetation lower than 2 m high, early building phase patches with a vegetation height between 2 and 10 m, late building phase patches with a height between 10 and 20 m, and (more or less) mature phases with vegetation over 20 m. We mapped these four patch types in the 12 ha plot. In the centres of a selection of the patches in total 166 hemispherical photographs were taken.

To analyse the relation between gap size and light availability we took hemispherical photographs at the centre of 100 canopy gaps.

All photographs were made at ca. 1m height. We used a Minolta X-700 body with a Sigma 8 mm Fisheye lens or a Canon Ti70 body with a Sigma 7.5 mm Fisheye lens. For each photograph the canopy openness ('percentage of sky not covered by vegetation', c.f Mitchell & Whitmore 1993), and the photosynthetically active photon flux density (PPFD, as a percentage of the PPFD above the canopy) was calculated using the program Hemiphot (Ter Steege 1993).

To show light variability across a canopy gap we measured light levels along a 70m long transect through a recently formed canopy gap. On the 5m points of a 70m long transect crossing the gap in east-west direction we measured light intensity in 10 minute intervals during 3 days. These data were compared to a data above the forest canopy. We used LI-COR SA-190 sensors (Lincoln, Nebraska, USA) with LI-COR LI-1000 dataloggers.

For a number of species light levels on individual crowns were estimated using the Dawkins index (Dawkins & Field 1978, Clark & Clark 1992, Sterck 1997). This is an estimate of the forest structure above the focal tree.

2.2 LIGHT QUALITY

In different forest types and forest patches of the Nouragues forest light spectra were measured in crisp shadow between 10:00 and 14:00 hrs, about 50 cm above forest floor, and in a circle 20 m in diameter to sample the local range of light variation. Pseudo-replication was avoided by keeping all non-overlapping spectra sampled at the same station, leading to unequal numbers of shade measurement at each site. Heterogeneity of light was sampled in six vegetation types at Nouragues in November 1994, March 1995, February 1996 and October 1996: (1) closed *terra firme* old-growth rain forest on a moderate slope (Reference, n=82), (2) old-growth rain forest along a creek (Creek, n=9), (3) closed *terra firme* rain forest with remains of old gaps at mid-level of the slope (Slope, n=15), (4) old-growth rain forest on the plateau (Plateau, n=10), (5) liana forest (Liana, n=17), and (6) ten years old gap on a moderate slope (Gap, n=7).

The light spectra was measured with a PS1000 portable spectroradiometer (Ocean Optics, Dunedin, Florida, USA) optimised between 300nm and 800nm, connected with an optic fibre to a 180° acceptance angle cosine-corrected irradiance sensor. Wavelength calibration was regularly checked with a Mercury-Argon light source. Irradiance was calibrated with the LI-COR (Lincoln, Nebraska, USA) LI-1800-02 optical radiance calibrator, and a program was written to convert the spectrometer voltage output to irradiance in micromoles per square meter per second.

Because of the difficulties of comparing spectra, we used a colour space developed by Endler (1990). For this colour space the spectra's hue and chroma are the most important characteristics. Hue depends on the wavelengths over which the transition between high and low intensity occurs, as well as the sign of its slope. For example, red light has a spectrum with high intensity long wavelengths and low intensity short wavelengths, with the transition around 650 nm (Endler 1993). Chroma (saturation or purity) depends upon the magnitude of the difference between the high and low parts of the spectrum, and also on the maximum slope in intensity. A white light has a flat spectrum (all wavelength with equal intensity) and thus a chroma of zero. A bright red light will have a high chroma, compared to a less bright red light (see Endler 1990 and 1993 for further details).

In colour space each spectrum is represented by a point with two polar co-ordinates: angle H, function of the spectrum's hue, and distance from the origin C, function of spectrum's chroma or saturation. A third dimension is given by total brightness Qt, actually total irradiance in $\mu mol/m^2/s$. Because wavelengths between 400 nm and 700nm are used, Qt equals photosynthetic photon flux density PPFD. We will also refer to red:far-red ratio R/FR as the ratio of quantum flux in 10nm wide wave-bands centred at 660nm and 730nm respectively (Holmes & Smith 1975).

We log-transformed variables to meet the normality requirements for parametric statistical tests, and used one-way analysis of variance to compare hue, chroma, PPFD, and R/FR among vegetation types.

3. Irradiance at forest and patch level

The forest consists of a mosaic of patches in different stages of development. Most sites in the forest receive very low levels of irradiance. Of the 58 grid points 51 were in closed forest. The average canopy openness was 3.9% and PPFD was 4.2% (Table 1). Only in canopy gaps the light levels were considerably higher. Forest development in gaps quickly results in a decrease of light intensity to those of the forest understorey (Van der Meer & Bongers 1996b, and Van der Meer *et al.* unpublished data). Stratification of the forest into distinct development patches (e.g. gap phase, early and late building phase and mature phase) was expected to lead to different light levels at the forest floor for the phases, where the levels in the building phase were expected to be between the gap phase and the mature forest phase. There was a general decline in canopy openness and in PPFD from gap phase via early and late building phase to mature phase. Gap phase and early building phase were significantly different from the others (Mann-Whitney *U*-tests for differences of medians, $p < 0.05$). Late building phase was not different from the mature phase, but both patch types were significantly different from gap and early building phase.

TABLE 1. Canopy openness (%) and total photon flux (% of above canopy) at the centre of different forest patches and on a grid in the whole forest. Letters indicate whether values are different (Mann-Whitney *U*-test, $p < 0.05$). (data from Jansen & Koning 1993)

Patch type	Vegetation height (m)	n	Canopy openness (%)		Total Photon Flux (%)	
			Mean	SD	Mean	SD
Gap phase	0 - 2	12	15.0^a	6.4	12.6^a	3.3
Early building phase	2 - 10	51	6.8^a	5.2	6.7^a	2.4
Late building phase	10 - 20	52	3.9^b	3.2	3.9^b	1.3
Mature patches	> 20	51	3.2^b	2.2	4.4^b	1.0
Total forest		58	3.9	4.3	4.2	1.9

This shows that, at the level of the forest floor, the development of the vegetation quickly results in low irradiance levels, with profound consequences for the opportunities for germination (Vasquez-Yanes & Orozco-Segovia 1984, 1993), growth and survival of light demanding pioneer species. Hence, at the forest floor the time that high irradiance levels are available is short. Van der Meer (1997) estimated that small gaps (smaller than 250 m^2, *sensu* Runkle 1981) in the Nouragues forest would have a closed canopy layer (between 25-30m) within five to six years.

The general light level we found in the Nouragues forest (3.9% canopy openness, 4.2% PPFD) is well comparable to other forests. Poorter & Arets (unpublished data) compared light levels in 6 forest sites where hemispherical photographs were used. The means vary between 2.8 and 4.3% canopy openness (with 0.2 and 15.3% as range values) with exception of one (extremely dark) site in Panama where mean openness was 0.3%. However, actual light levels seem to be more different, and mean values may not be the best way to compare light levels among sites. Nicotra and coworkers (1999), for instance, showed that in 8 forest types the mean light levels were not different, but frequency distributions of light availability were. Old growth forests showed more extreme levels, both at the low and at the high end of the light gradient. Also the overall variances were higher in old growth forest. This most probably is related to the sizes of individual tree crowns. Light availability was significantly spatially auto-correlated in all sites but the scale (patch size) was twice as large in old growth stands.

We were surprised not to find clear differences between forest patches of different stature, except for the early gap sites. This either means that light levels are reduced quickly after initial gap closure, or that the method is not fine enough to pick up the more subtle differences. Lack of clear differences in different aged secondary forests in Bolivia (Pena-Claros 2001), using the same method, may support both the first and the second possibility. Detailed and extensive measurements, using various methods, are needed to solve this problem.

4. Irradiance in gaps

In canopy gaps light levels are generally much higher than in forest understorey, but the increase is dependent on several factors among which the size and shape of the gap. In a study on canopy dynamics in the Nouragues forest Van der Meer & Bongers (1996b) analysed the relation between gap size and canopy openness in the centre of the gap (Figure 1). In that study a positive relation was found between the two, independent of the way gap size was measured, but the correlation (Pearson's *r*) values of between 0.48 and 0.69 indicate that other factors than mere gap size are also important determinants of light levels in the centre of the gap. These factors may include the shape of the gap, the orientation, the inclination and the height of the surrounding vegetation. The combination of these factors also lead to a high within-gap variation of light levels.

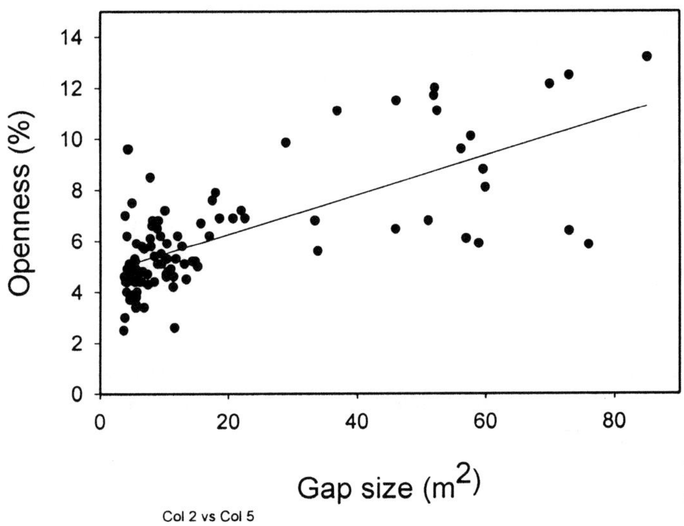

Figure 1. Canopy openness (%) measured at the centre of gaps of different size. Canopy openness = 4.70 + 0.07 * gaps size. R^2 = 0.47 n = 100. (after Van der Meer & Bongers 1996b).

Within the gaps, light changes gradually from the gap centre to the gap edge to the forest understorey (Van der Meer *et al.* 1998). Here we give the LI-COR based light intensity levels through one gap as an example (Figure 2). The size of that gap was 53 m^2 (*sensu* Brokaw 1982) or 284 m^2 (*sensu* Van der Meer & Bongers 1996a) or 965 m^2 (*sensu* Runkle 1981). Although this was a relatively large gap in the forest the canopy opennes in the gap centre was only 11.1% of the value above the forest. The eastern part of the gap was clearly receiving more light than the western part. This probably is due to the very heterogeneous forest structure around the gap: the western part was extremely closed, the eastern part more open. Irradiance levels above 100 µmols/10 min were nearly absent in the western locations. Even in the gap centre far the majority of 10-min intervals receive less than 100 μmols. For all transect points light levels were very low for the largest part of the day. The –35 m location received less than 1% of the above canopy irradiance during nearly all day. The higher levels in the centre were caused by only a short high light period, between 12 and 14 hrs. These results are indicative of the generally low light values in the generally small gaps in the natural forest.

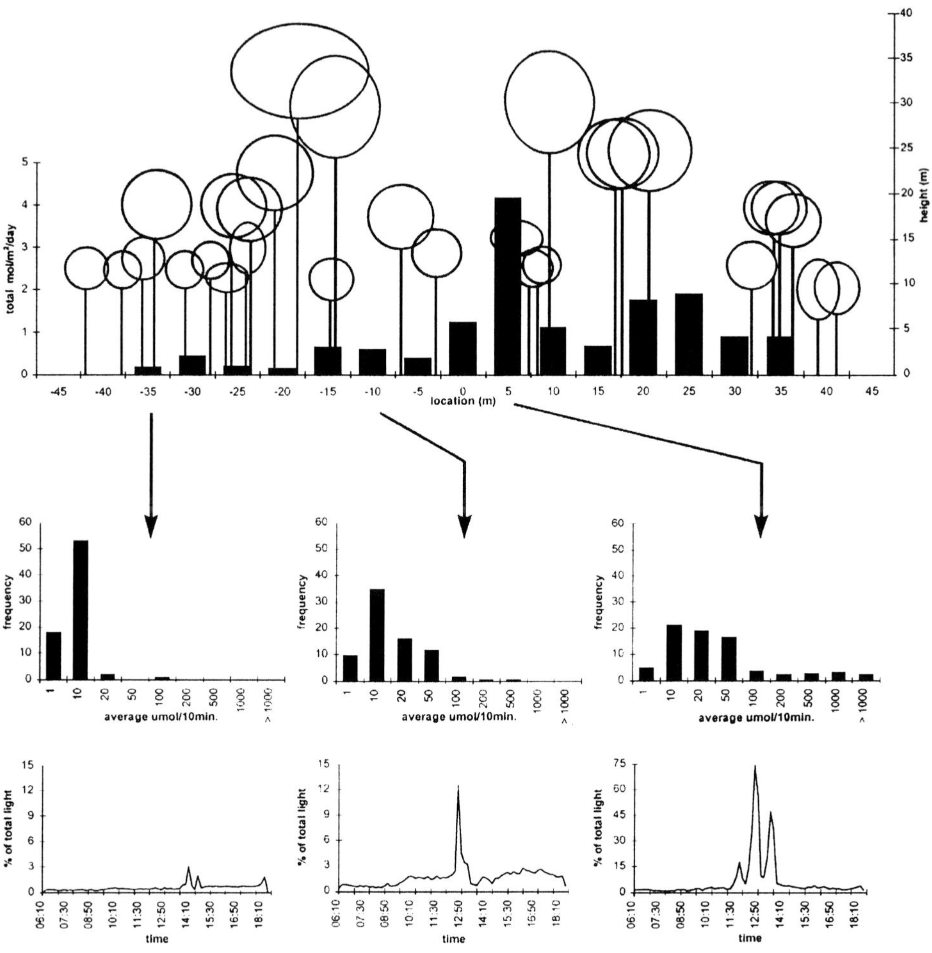

Figure 2. Irradiance levels along a transect through a gap in the Nouragues forest. Total daily PPFD values are given for every 5 m along a 70m long east (positive part) to west (negative part) transect across the gap (location 5 m is the gap centre). For three selected transect points at the western part of the transect details are shown: a frequency distribution (absolute values in number of 10-min periods) of irradiance classes, and the irradiance levels in 10-min interval (as a percentage of the above canopy level). (Bongers *et al.* unpublished data)

5. Irradiance on trees

An important question is whether trees are specialised with respect to the light levels they experience. Pioneers for instance are expected to live at high levels only, while shade tolerant species are expected to be found mostly at low light levels. For a number of species we have studied the trees' light levels (Bongers & Sterck 1998, Sterck 1997, Rijkers *et al.* 2000a, see also Sterck *et al.* Chapter 27) and how trees develop with respect to the light levels they receive. Although most individuals of the pioneer tree species *Cecropia* were found at higher light levels, also individuals were found at low light, contrary to our expectations. Individuals of the shade tolerant species *Dicorynia guianensis*, *Vouacapoua americana* and of the intermediate species *Pourouma bicolor* are found at all light levels as well. Rijkers *et al.* (2000a) showed, however, that of 4 species (*D. guianensis*, *V. americana*, *Duguetia surinamensis* and *Goupia glabra*) 95% of all individuals less than 20 m height had a canopy openness of less than 15%. In general the taller the individuals the higher the light levels, which is expected based on the general changes in light levels along vertical gradients in the forest (Yoda 1974, Koop & Sterck 1994).

Light levels on individual trees strongly influence the probability of survival, not only for small plants (Van der Meer *et al.* 1998, Svenning 2000), but also for larger ones (Clark & Clark 1992, Sterck *et al.* 1999). Growth and development is also strongly affected by light (Sterck 1999, Bongers & Sterck 1998, Sterck & Bongers 2001, Oberbauer *et al.* 1989, Poorter 1999). In Chapter 27 Sterck *et al.* go into detail on its consequences for trees.

Data on light levels of individual trees are still scarce because of the practical problems associated with measurements of tall trees. For larger trees the Dawkins index is being used (Dawkins & Field 1978, Clark & Clark 1992, Sterck 1997) to quantify the light environment. This in fact is an estimate of the forest structure above the focal tree. In some cases hemipherical photography is used for large trees as well (e.g. Sterck & Bongers 2001), but due to difficulties of aligning the camera to the north only canopy openness can be used and not the site factors. Consequently, for measurements of irradiance levels on trees over 6-7 m tall it still is very difficult to get detailed data.

A lower scale level is the light variability on individual tree crowns, which can be considerable as well. Oberbauer *et al.* (1989) compared crown light environments of (10 saplings of) two species, a slow growing shade tolerant *Minquartia guianensis* and the faster growing less shade tolerant emergent *Pithecellobium pedicellare*. The median total daily PPFD was less than 2% of full sun, and did not differ between species. Flux densities varied significantly from day to day and over the short distances within crowns, for both species.

6. Spatial variation of forest light spectra

Our ambient light spectra measurements can be grouped in 4 main light habitats (Figure 3). Forest shade is typically in the yellow-green region of the spectrum, because of absorbance of blue and red irradiance by leaves. This light habitat is quite variable in hue and saturation. On the contrary, the white light of open areas is nearly at the centre of the colour space. Direct sunlight mostly contributes to those gaps which correspond to an opening of > 1 m in diameter at the forest floor, or roughly 1° canopy openness. There is a whole gradation between large and small gaps. At the extreme of that continuum we measured 5 cm wide sun patches. Compared to large gaps, small gaps are progressively enriched in yellow and red (because of the filtering effect of the vegetation). Compared to forest shade, a higher proportion of sky visible from the ground leads to a shade enriched in blue. This woodland shade (*sensu* Endler 1993) is very rare at ground level at Nouragues. At dawn and dusk, ambient light is purple to blue-purple, but this could only be measured in the large opening of the camp because of electronic sensitivity. During cloudy weather, ambient light spectra of forest shade, woodland shade and small gaps converge towards the white spectra of large gaps.

At Nouragues, shade of the six vegetation types show significantly different hues, PPFD and R/FR, but chroma was not different (Table 2).

All forest shade hues are in the typical yellow-green region of the colour space, which was expected (Endler 1993). Gap and Creek have the highest hue angles (Table 2) with greener shade spectra (Figure 3) which may be caused by a higher proportion of sky visible from the ground, slightly enriching spectra in blue. The Reference forest shows medium hue angles, while the dense vegetation and closed canopy of Liana and Plateau have the lowest hue angles, with slightly yellower spectra. These low values may be related to their closed canopy, few green leaves in the understudy, or to red-violet leaves of extreme-shade plants (Lee & Graham 1986).

TABLE 2. Colour variables (mean ± S.E.) of forest shade spectra measured at the Nouragues natural reserve (Nouragues Reference and the forest types) and at Paracou (Paracou Reference) and Piste St. Elie (Piste St. Elie Reference), computed for light with wavelengths between 400nm and 700nm, according to the colour space of Endler (1990). Additionally R/FR is given. See text for definition of forest types. Number of spectra given between parentheses. F values and level of significance of one-way analyses of variance for differences among forest types and reference for Nouragues forests, and for differences among References, are given. *, $P < 0.05$, ***, $P < 0.001$, ^{ns}, $P > 0.05$.

Forest type (n)		Hue angle (°)	Chroma	PPFD (μmol/m²/s)	R/FR
Nouragues	Creek (9)	156.6 ± 4.4	0.13 ± 0.01	3.73 ± 0.32	0.14 ± 0.02
	Slope (15)	150.7 ± 3.9	0.20 ± 0.02	3.63 ± 0.17	0.09 ± 0.02
	Plateau (10)	145.9 ± 5.5	0.16 ± 0.01	2.97 ± 0.24	0.10 ± 0.02
	Liana (17)	141.1 ± 1.3	0.15 ± 0.01	4.09 ± 0.35	0.15 ± 0.01
	Gap (7)	166.5 ± 3.1	0.17 ± 0.02	3.67 ± 0.49	0.16 ± 0.04
	Ref (82 at 9 sites)	154.0 ± 1.7	0.15 ± 0.01	3.24 ± 0.08	0.16 ± 0.01
Paracou	Ref (39 at 4 sites)	160.3 ± 2.4	0.22 ± 0.01	4.37 ± 0.14	0.26 ± 0.01
St. Elie	Ref (34 at 4 sites)	155.0 ± 1.9	0.21 ± 0.01	3.82 ± 0.14	0.12 ± 0.01
F Nouragues forest types		3.46*	$1.09^{n.s.}$	3.23*	2.47*
F References		2.27^{ns}	27.11***	29.29***	17.16***

The total brightness (PPFD) also varies with forest types. The brightest shades of all forest types are Liana, Creek, and Gap, whereas the (higher) forests of Plateau and Reference show lowest PPFD.

Mean R/FR is low in every forest type, with values typical of forest shade (Lee 1987, Endler 1993, Turnbull & Yates 1993), but Slope and Plateau forest types show significantly the lowest values of R/FR. The minimal values of R/FR for each of the forest types are lower than 0.1, indicating that very dense forest shade is available in all forest types. Locally, in slope forest and the Nouragues reference forest, R/FR reaches extremely low values of 0.01 in darkest shade (see also Turnbull & Yates 1993).

Chroma of all vegetation types is relatively high, which may be related to a high proportion of ambient light filtered and/or reflected by leaves rather than reflected by the rest of the vegetation, to thin leaves, and/or to relatively little skylight coming through gaps in the canopy (Endler 1993). High values of chroma are found in gap shade, and therefore the last explanation is quite unlikely. Differences of chroma between forest types may also arise from different leaf thickness and transmission properties of sun and shade plants (Lee & Graham 1986, Lee *et al.* 1990, Poorter *et al.* 1995).

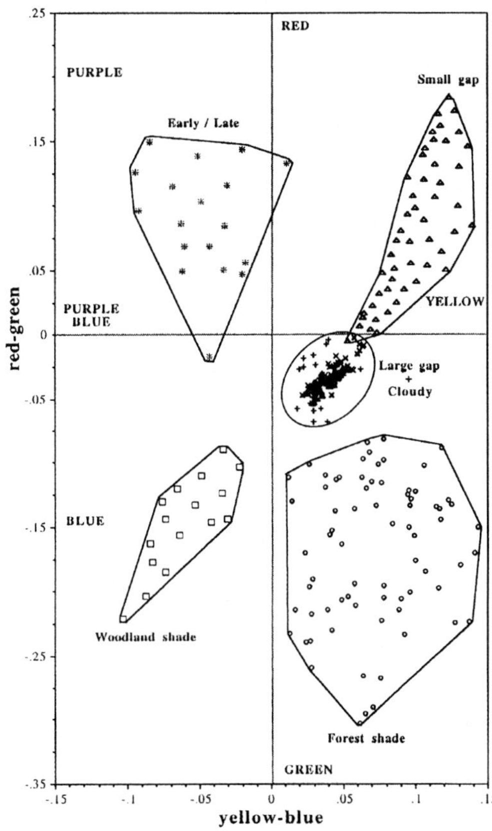

Figure 3. Colour variation of ambient light in the Nouragues primary tropical rain forest. Each point represents a single irradiance spectrum from which hue angle and chroma have been extracted using the segment classification method of Endler (1990) between 400 nm and 700 nm. The ordinate represents relative brightness of red and green spectral segments, and the absissa is the difference between yellow and blue segments. The distance from the origin is chroma (colour saturation) and the angle clockwise from the top of the ordinate is hue (aspect of colour identified by words like 'red', 'green', and 'blue'). The direction of vectors approximately corresponding to Munsell (1975), hues are indicated by colour names. Light was measured between 11:00 and 13:00 hrs about 50 cm above the forest floor. Light environments were selected before measurement, and outlines and names of light environments are indicated on the graph. No cloud was visible in clear sky conditions, and blue sky was totally absent in cloudy conditions. Small gaps measured 30-90 cm in diameter, the transition between small gaps and large gaps is shown by decreasing chroma of small gaps. The gradients between small gaps and forest shade, woodland shade and forest shade, early/late and large gaps, and clear sky and cloudy conditions are not shown here. (from Théry 2000)

These statistical differences in hue, PPFD, and R/FR separate forest types at Nouragues. We wondered whether reference forests at different sites would be comparable with respect to their ambient light spectra. Consequently, we compared the same undisturbed forest type between Nouragues, and two other sites in French Guiana: Paracou research station (5°15'N-52°55'W), located 20 km from the coast, and the Piste de Saint-Elie research station (5°14'N-53°03'W), 30 km from the coast in the same unbroken evergreen rain forest. These sites show similar climate, annual rainfall (Boyé *et al.* 1979), and dominant plant families (Sabatier & Prévost 1990, Bariteau 1993, Poncy *et al.* 1998). The undisturbed reference plots have different chroma, PPFD, and R/FR (Table 2). Hue angle of undisturbed *terra firme* forest is comparable among study sites (Table 2). This shows that light spectra vary considerable among forests and within forests.

7. Scales of ambient light variation

Light is highly variable in the forest, in quantity as well as in quality. Over the last decade irradiance levels have been studied in several forests. Currently equipment is ample available, although sensor equipment is still expensive and measurements at many locations simultaneously is very time consuming. Light quality data are far less available, but is increasingly receiving attention (see Endler 1993 and Théry 2000).

In the Nouragues forest several methods for measuring and estimating light levels have been used, at different spatial and temporal scales. Most extensively hemispherical photograph based measurements have been done, a method providing good results when light levels vary considerably (Whitmore *et al.* 1993, Valverde & Silvertown 1997). In a recent paper Nicotra *et al.* (1999) state clearly that both hemispherical photographs and direct irradiance measurements render different, complementary results. Choices have to be made depending on the specific questions asked.

Development patches within a forest differed in irradiance, gaps and early building phase had higher irradiance levels than late building and mature phases.

Gap size co-determines the irradiance levels in the centre of the gap, and thus that gap size may be used as a proxy of irradiance. Variability among gaps with similar sizes (due to differences in e.g. shape, orientation and slope exposition) is considerable, however, and for in-depth studies finer estimates or real irradiance measurements are needed. Irradiance variability within a gap is quite large, as we have shown here for one selected gap.

Light quality also varies considerably among forests and among the forest types within an area. Four major types of forest shade spectra have been distinguished (Endler 1993). We hoped to be able to characterise different forests and forest types based on these spectra and showed that indeed large differences existed, depending on the light parameters used. Within each of the forest types a large variety of shade spectra were available however, making it hard to generalise. Clearly, more research is needed on these aspects.

What are the consequences of this variability? For plants, small-scale variation plays an important role in maintaining high induction state in leaves (Chazdon *et al.* 1996). Sunflecks are the major light source for many understorey plants (Chazdon 1988, Rijkers *et al.* 2000b). At a larger scale, light variation in canopy gaps does affect seedling performance (e.g Van der Meer *et al.* 1998). Seedlings and saplings respond to the light fluctuations by changing growth, architecture and physiology. This will have large effects on survival of individuals.

In plants, R/FR is important in controlling growth and development (review by Casal & Smith 1989, Endler 1993). Because leaves absorb more red than far-red light, R/FR decreases with the intensity of shading. The R/FR serves for different signalling purposes, sometimes indicating the presence of neighbours (and then a plant may show adaptive responses), sometimes triggering germination (Vazquez-Yanes & Orozco-Segovia 1984, Smith 1995).

Because the colour of a light fleck depends both upon the reflectance spectrum of an object and the ambient light, animal species that signal only at certain times and places (e.g. when specific light conditions are available) should evolve a characteristic combination of colours, predictable from the encountered light habitat (Endler 1993). For example, fixed locations of display sites and colour patterns of three lekking forest-dwelling birds are non-randomly associated with specific light habitats (Endler & Théry 1996, see also Chapter 14).

Irradiance levels and light quality contribute to characterise and define optimal habitats both for plants and animals. As both spatial and temporal variability in these characteristics is high, this variability needs further attention in the Nouragues forest.

Acknowledgements

We are grateful to A. Hofmann, T. Rijkers, F. Sterck, J. Jansen, P. Jansen, K. Konings and Smeenge for fieldwork assistance. S. Wijdeven prepared Figure 2. We thank T. Rijkers, L. Poorter and an anonymous referee for their constructive comments on the manuscript, and J.A. Endler for calibration of the spectrometer. Spectrometric equipment was funded by the French Ministry of Environment.

Interior of the Nouragues camp site. *Photo Gerrie Offerman*.

PART 2

PLANT COMMUNITIES

CHAPTER 4

THE LOWLAND HIGH RAINFOREST: STRUCTURE AND TREE SPECIES DIVERSITY

Odile Poncy, Daniel Sabatier, Marie-Françoise Prévost and Isabelle Hardy

1. Introduction

Diversity studies provide data useful to the basic knowledge of biodiversity as well as a baseline for many other research fields. Forest dynamics, descriptive and functional ecology, conservation and sustainable development projects, need such data that are still few for neotropical forests, and especially so for French Guiana. Data from previous studies in this region come from 'Piste de Saint-Elie' (Sabatier & Prévost 1990b, Prévost & Sabatier 1996) and Saül (Oldeman 1974, Mori & Boom 1987). The research program at Nouragues aims to study forest dynamics and natural regeneration processes in the forest, including seed dispersal by vertebrates. The establishment of the station as a permanent research site has provided opportunity as well as appropriate conditions to undertake such studies, and to add one more piece to the jigsaw (Milliken 1998). The first surveys of floristic and tree diversity at Nouragues were based on preliminary results from 1989 (Sabatier & Prévost 1990a) and 1995 (Poncy *et al.* 1998).

In this chapter we present the tree diversity, as presently known, based on various inventories carried out on plots in lowland, old-growth *terra firme* dense rainforest. The particular type of low forest that covers the foothills, slopes or top of the inselberg is treated elsewhere in this volume (Larpin, Chapter 5).

Moreover, as floristic data were gathered (by two of us, D. Sabatier and M. Prévost) from several neighbouring one- hectare plots within the study area, some features of the variation of floristic diversity at the local scale will be presented. Such amounts of data from the same place are still rare, and they help to enhance our basic floristic knowledge. Further accurate analyses should help our understanding of the correlations between compositional patterns and local environmental factors. A first attempt was made by Prévost and Sabatier (1996), using the data collected at 'Piste de Saint-Elie', to suggest correlations between variations of species diversity and other parameters involved in forest dynamics, especially soil characteristics.

Bongers *et al.* (eds.), Nouragues. Dynamics and plant-animal interactions in a neotropical rainforest, 31-46.
2001 Kluwer Academic Publishers. Printed in the Netherlands.

2. Methods

2.1 THE 'SAUT PARARÉ' SITE

Ecological studies of trees in the area of 'Saut Pararé' were initiated in 1978 by the MNHN. The first was carried out on the border of the river Arataye, close to the rapids called 'Sauts Pararé' (Maury & Poncy 1986). A rectangle of 200 x 300 m, including 6 contiguous one-ha plots, was delineated in an area with uneven topography. It lies on hills with steep slopes and talwegs and includes little flat land, and no flooded ground. The forest was high, with a relatively open understorey, disturbed in some places by recent tree fall gaps. The tree inventory took place in several stages. The first one, carried out in 1978, consisted in numbering and mapping all trees with DBH ≥ 30cm. Later (1983-88) the inventory was continued in order to include all trees with DBH ≥ 10cm and to provide botanical identifications. By late 1985, the inventory itself (numbering, labelling, mapping) included 5 of the 6 one-ha plots. 2674 trees in total were inventoried. DBH was measured with a tape, or estimated above buttresses or stilt roots. As part of the first inventory in 1978, most trees with DBH ≥ 30cm were given a local name (of the *aluku* language). When available, these names help identification at least at the family level. Subsequently, botanical specimens were collected from most trees with DBH ≥ 10cm on the same five hectares, by climbing the trees. Two methods are used to climb the trees, depending on tree size and overall conditions: climbing spikes enable climbing the tree itself; harnesses for speleologists can be used to climb a rope previously installed between the ground and the biggest branches of a tall tree, and with this method, the climber is generally able to collect specimens from several trees around. Vouchers were deposited at the Cayenne herbarium, with duplicates in Paris and, if needed, sent to specialists. Identifications of the vouchers were conducted by the late eighties; sofar 1520 trees have been named, most for them at the species level (1149) and half of the total (1339 out of 2674) at the generic level. A reinventory was carried out in 1995, aiming to collect data about dynamics (growth and mortality).

2.2 THE INSELBERG SITE

The Nouragues area lies 7 km away from the 'Sauts Pararé' plot. A botanical and structural study of the forest was first carried out on five one-hectare transect-like plots (500 x 20 m) where all trees ≥ 10cm DBH were recorded in 1988/89 (Fig. 1, plots 1-5) (Sabatier & Prévost 1990a). Some of these plots were part of a 100-ha study area of the station, gridded in 1-ha quadrats. A comprehensive inventory of trees in this area was initiated in 1992, aiming at numbering, labelling, mapping all trees of DBH ≥ 30cm. The inventory also included all stems of DBH ≥ 10cm on 22 of these quadrats. Twelve of them formed a 12-ha area of the so-called 'Petit-Plateau' where forest dynamics have been studied (van der Meer *et al.* 1994, Van der Meer & Bongers 1996a, 1996b).

The botanical study of canopy trees (DBH ≥ 30cm) was carried out in a 9 x 1-ha strip of contiguous plots within this area (Poncy *et al.* 1998) and on the five plots mentioned above. Moreover, two additional plots were recorded by D. Sabatier and M.-F. Prévost in 1994

(Fig. 1, plots 6-7), also 1 hectare in area, but with a different shape (40 x 250 m). The botanical record on plot 7 was 80% completed[1].

Thus, the floristic knowledge of the trees at Nouragues presented in this chapter is based on data collected from 15.8 hectares of forest, with two distinct data sets recorded at

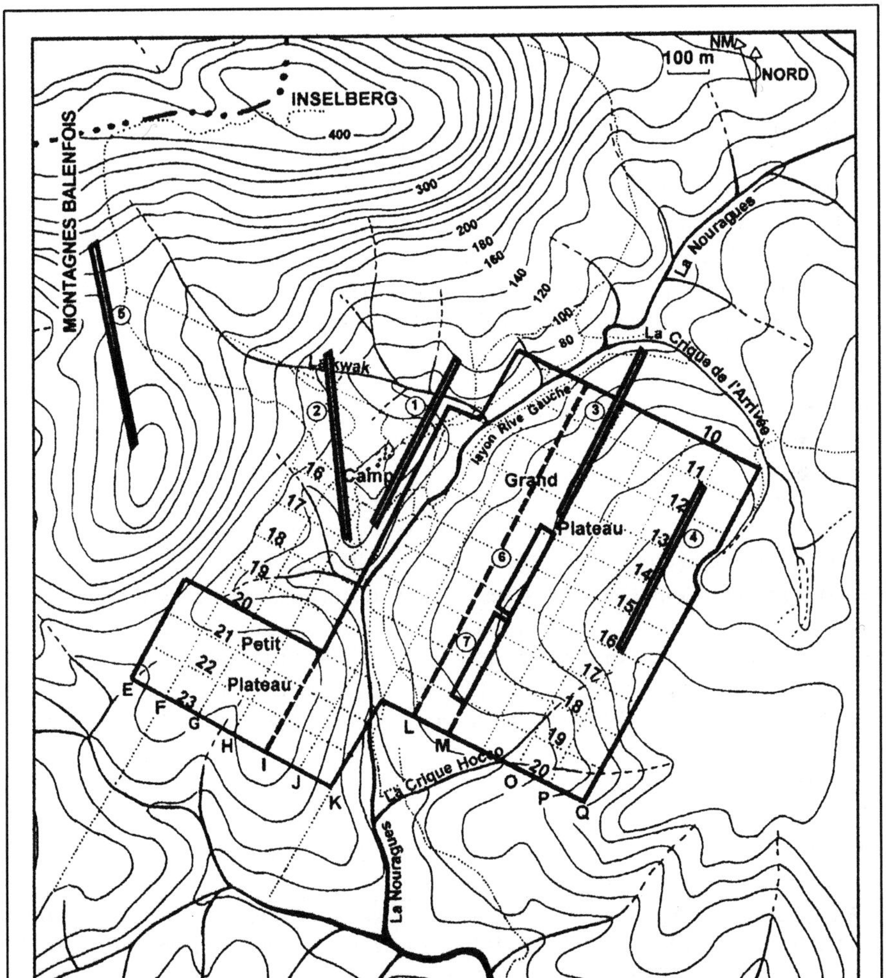

Figure 1. Map of the study area of the Inselberg site

two different scales: firstly, a survey of all trees with DHB $\geq$ 30 cm on a 10-hectares area; secondly, a study of trees with DBH $\geq$ 10 cm over 7 transect-like plots each of one-hectare (Fig. 1). Comparative data on specific diversity on these 7 plots, gathered by D. Sabatier

[1] This plot will not be deleted from this study because it includes a particular forest type, with many lianas. It will be taken into account only when pertinent, due to the lack of floristical data.

and M.-F. Prévost, will be presented in a separate paper, and only preliminary comments are given here.

Identifications were partly obtained in the field (based on habit, bark, leaves observed in the crown and collected from the ground), especially for the species that were common or recognised without any doubt. Local names provided by two Surinamese of the Saramaca ethnic group helped preliminary identifications, most often at a generic level. Collection of vouchers was done when needed for identification, and when climbing was possible. Vouchers were deposited in Cayenne (CAY, herbier de Guyane) and Brunoy (Laboratoire d'Ecologie, MNHN), and some duplicates were sent to specialists. Seven trees remain totally unidentified, and 10% of the total were given only a family or a genus name. Botanical names follow the Checklist of the vascular plants of the Guianas (Boggan *et al.* 1997)[2].

Floristic data are managed in Brunoy, using a multifile database (floristic checklist, tree inventory, reference collections); the database for herbarium specimens is related to the main database of the Cayenne Herbarium ('AUBLET': Hoff *et al.* 1989), now part of the integrated system of forest research databases in French Guiana, named 'ARLEQUIN'.

3. Results[3]

3.1 PARARE SITE

Structure and dynamics

The first inventory in 1978 recorded 777 trees of DBH ≥ 30 cm on 6ha (Maury & Poncy 1986) while the overall inventory (trees ≥ 10cm DBH) covered 2674 trees.

Densities (Tab. 1) range from 108 to 137 per hectare for the canopy trees (≥ 30 cm) and from 474 to 652 for all trees (≥ 10 cm). As a result of the reinventory conducted on 5 of the plots in 1995, mortality and recruitment of trees in the area can be estimated. The relative mortality for all trees with DBH ≥ 10cm after 10 years ranges from 9% to 14% in each plot, with a mean of 11.9 (Tab. 1). Similarly, the recruitment (in growth into the DBH ≥ 10 cm

TABLE 1. Density, mortality and recruitment of trees on 5 hectares at 'Pararé site', between 1983 and 1995 (percentage into bracket)

	DBH cm	1	2	3	4	5	total 5 ha
Density (1985)	≥ 30	115	116	125	136	127	619
	≥ 10	490	474	652	537	521	2674
Mortality (1985-95)	≥ 30	18	18	13	8	12	69
		(15.7)	*(15.5)*	*(10.4)*	*(5.9)*	*(9.4)*	*(9.2)*
	≥ 10	71	64	68	48	66	317
		(14.5)	*(13.5)*	*(10.4)*	*(9)*	*(12.7)*	*(11.9)*
Recruitment (1985-95)		62	82	56	34	41	275
		(12.7)	*(17.3)*	*(8.6)*	*(6.3)*	*(7.9)*	*(10.3)*

[2] Authorships of scientific names are not included in this chapter, they are available in the floristic list (see appendix).

[3] Further analyses of the data (especially completing voucher identifications) are currently in progress, in order to write a detailed study of this site.

size class) varies from 6.3% to 17.2%, with a mean of 10.3%. Based on these values, the mean mortality per year reached 1.19%. For the same lower DBH limit (10 cm), Spichiger*et al.* (1996) and Rankin-de Merona *et al.* (1990) found very similar percentages in Peru (Jenaro Herrera, 1.14%) and in Central Amazonia (1.13%) respectively, while Uhl (1982, cited by Rankin-de Merona, *l.c.*) gives 2.93% in Venezuelan Amazonia. In their study of mortality in forests in Venezuela, Carey *et al.* (1994) consider annual mortality rates (that range from 0.5% to 3.3%) to be random, 'quite variable from plot to plot and from year to year', and not related to the type of life zone or initial biomass.

Floristic composition and diversity of trees

Half of the trees with a DBH ≥ 10 cm inventoried on the 5 plots have yielded floristic information, since 1336 out of 2674 trees were given at least a family name. Because the data set was considered not complete enough to allow quantitative analysis, only a qualitative overview is given below.

Based on the inventory of the biggest trees (DHB ≥ 30cm) in the 6 ha sample (Maury & Poncy 1986), the most abundant family is Burseraceae (15% of the trees), then Lecythidaceae, Moraceae, Sapotaceae and Mimosaceae. When all trees are considered (DBH ≥ 10 cm), 46 families are represented. The highest number of trees is again in Burseraceae, but Meliaceae come out as the second most abundant family, then Sapotaceae, Lecythidaceae, Myristicaceae and Euphorbiaceae.

In total, 215 tree species were identified. This number clearly underestimates the real number of species present on the plots and the remaining unidentified trees are likely to belong to additional, unrecognised species.

The most abundant identified canopy species (DBH ≥ 30 cm) belonging to the leading families mentioned above, are *Protium sagotianum, Tetragastris altissima, T. panamensis* in Burseraceae and *Eschweilera coriacea, Lecythis persistens, L. poiteaui* in Lecythidaceae. The importance of Moraceae and Mimosaceae is due to their specific diversity (Moraceae, 16 species with *Brosimum rufescens* among the most abundant; Mimosaceae: 17 species, including 13 *Inga* species). Sapotaceae has the highest number of species (25), three of them are rather abundant: *Chrysophyllum sanguinolentum, Pouteria gonggrijpii, Pradosia ptychandra.* Other abundant species belong to additional families (*Dicorynia guianensis*, Caesalpiniaceae; *Carapa procera*, Meliaceae; *Virola michelii*, Myristicaceae; *Minquartia guianensis*, Olacaceae).

With respect to medium-size and understorey trees (DBH ≥ 10 cm), 20 species were represented by 10 trees or more. A few belong to families already mentioned (*Protium apiculatum*, Burseraceae; *Pseudolmedia laevis, Trymatococcus oligandrus*, Moraceae), while some additional important species in this patch of forest are *Unonopsis rufescens* (Annonaceae), *Thyrsodium spruceanum* (Anacardiaceae), *Quararibea turbinata* (possibly a synonym of *Q. duckei*, Bombacaceae), *Licania densiflora* (Chrysobalanaceae), *Tovomita cephalostigma, T. choisyana* (Clusiaceae), *Conceveiba guianensis, Drypetes variabilis, Mabea piriri* (Euphorbiaceae), *Ocotea rubra, Nectandra pisi* (Lauraceae), *Guarea silvatica, Trichilia septentrionalis* (Meliaceae), *Siparuna cuspidata* (Monimiaceae), *Eugenia coffeifolia* (Myrtaceae), *Iryanthera sagotiana* (Myristicaceae) and *Theobroma subincanum* (Sterculiaceae).

3.2 INSELBERG SITE

Structure and physiognomy

An analysis of structural data by Poncy *et al.* (1998) recognised the occurrence of physiognomically different forest zones in the Nouragues area. High mature forest with a fairly open understorey, can be divided into two subsets according to canopy height, the first with an average canopy height estimated at 30-35 m, and the second at 20-25 m. Low forest with many lianas was also subdivided into two zones: the first one with average canopy height of 12 m, few stems reaching 30 cm DBH (see below), many trees overloaded with lianas, and a dense understorey; the second one dominated by lianas, and imbricated liana stems at ground level making the understorey extremely difficult to penetrate. The last type consists of patches of swamp forest dominated by *Euterpe olacea* palms.

Densities for trees ≥ 30cm DBH ranged from 64 to 125 stems/ha for the 44 1-ha plots inventoried up to 1995 (Poncy *et al.* 1998). The variation reflects the forest types described above. Densities < 100 stems/ha indicate either a strong recent perturbation (large gap) or the presence of liana forest in the plot considered. For trees ≥10 cm DBH, densities range from 468 to 681 stems per ha on the 29 plots for which data are available (see also Table 4.4, for 7 plots). All plots but one have fewer than 600 trees. Such numbers accord well with other records from Amazonian *terra firme* forests (review by Spichiger *et al.* 1996, Matos & Amaral 1999), but are lower than those of Mori and Boom (1987) (619 trees/ha at Saül, ca 100 km), Gentry (1988) (ca/600 stems/ha), Rankin-de Merona *et al.* (1990) (600-700 stems/ha), Rankin-de Merona *et al.* (1992) (637 stems/ha). Very high densities (ca 2000 stems/ha or more) are mentioned in Northwestern Amazonian swamp forests (Duivenvoorden & Lips 1995).

The distribution of size classes based on stem diameters was given by Poncy *et al.* (1998) with respect to two forest types: 77% and 88% of trees below DBH 30 cm occur respectively in high forest and in forest with liana-forest patches, while the percentage of trees over 30 cm was lower in the liana forest. In the high forest, 12.2% of the trees are greater than 40 cm in DBH, and 2.2 % greater than 80 cm, *vs* 16.7% and 3.1%, respectively, on La Fumée Mountain near Saül, (Mori & Boom 1987).

Total basal areas, ranging from 30 to 45.3 m²/ha, are generally higher than those recorded in other Amazonian forests outside French Guiana (e.g. Valencia *et al.* 1994, Rankin-de Merona *et al.* 1990, Spichiger *et al.* 1996) but lower than the estimation of 53m²/ha, based on point-centred quadrat sampling by Mori and Boom (1987) in the vicinity of Saül, French Guiana.

Floristic composition

The data gathered to date, both from tree-inventories and sporadically records within the study area, have contributed to the floristic knowledge of the area. The different forest patches vary to some extent in their structure, age, and also in the families or the species that dominate each of them. The total number of tree species recorded over the inventoried area reaches 550, belonging to 63 families (see floristic list, appendix 1). *Eschweilera coriacea*, *Vouacapoua americana* and *Eperua falcata* are by far the most abundant species, but no species dominates the whole study area.

Families

The Family Importance Value (FIV) is an index now widely used to evaluate floristic composition at the family level, and it combines species richness, density and dominance[4] (Mori *et al.* 1983). For canopy trees (DBH ≥ 30cm), the highest values on the 9-ha studied initially (see figure 20.11 *in* Poncy *et al.* 1998) were for Caesalpiniaceae, Sapotaceae, Mimosaceae, Lecythidaceae, Burseraceae, Chrysobalanaceae and Lauraceae. Among the 33 families of Dicotyledons[5] recorded, 50% of the trees belonged to five families, with legumes (Caesalpiniaceae + Mimosaceae) totalling 26% ; Sapotaceae, Burseraceae and Lecythidaceae being the next most abundant. The high value for Caesalpiniaceae was mainly due to the great abundance of two species, *Eperua falcata* and *Vouacapoua americana.* The whole Leguminosae complex accounts for 31% of the identified species. For other families, high numbers of species were also responsible for high values.

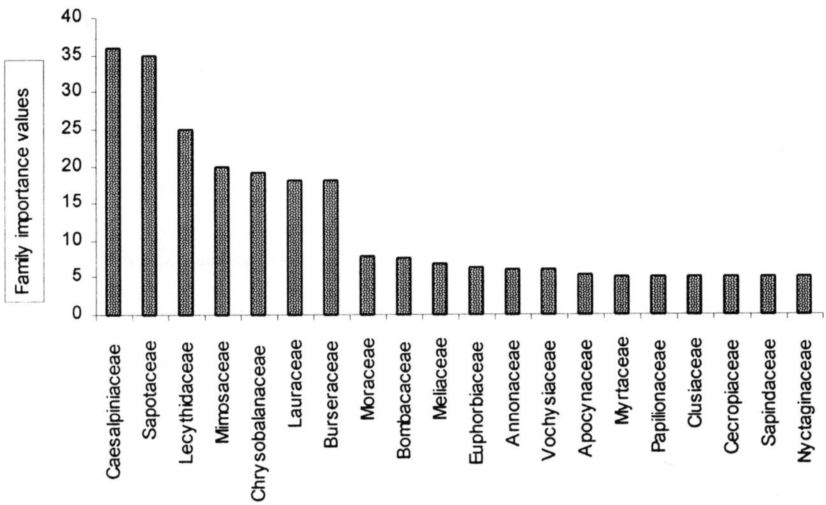

Figure 2. Family importance values for the most important families of trees on the Nouragues study site

Figure 2 shows the FIV pattern based on our 7 other plots, for all trees with DBH ≥ 10cm. The total 3714 trees recorded were assessed to 58 families (7 trees remain unidentified). This figure shows that the forest is markedly dominated by seven families with FIV > 15. They total ca 40% of the species and 50% of the trees. The four leading families are the same as in the canopy inventory sample (DBH ≥ 30 cm) (Poncy *et al.* 1998.), *i.e.* Caesalpiniaceae, Sapotaceae, Mimosaceae, Lecythidaceae. Again, the Leguminosae-complex, totalling 5% of the identified species and 18% of the trees, is important, although slightly less so than in the canopy inventory.

---

[4] FIV = (relative number of species + relative density + relative basal area)100

[5] In this paper, Leguminosae are considered as a suprafamilial taxon, including 3 separate families: Fabaceae, Caesalpiniaceae, Mimosaceae.

TABLE 2. The large trees of the Nouragues forest (DBH ≥ 70cm, height ≥ 40 m)

Family	Species
Apocynaceae	*Aspidosperma marcgravianum, A. oblongum, Geissospermum laevis*
Bignoniaceae	*Tabebuia serratifolia*
Bombacaceae	*Ceiba pentandra*
Boraginaceae	*Cordia goeldiana*
Burseraceae	*Protium apiculatum, P. araguense, P. demerarense, Tetragastris altissima*
Caesalpiniaceae	*Dicorynia guianensis, Dimorphandra pullei*
	Eperua falcata, E. grandiflora, Hymenaea courbaril, Peltogyne venosa Swartzia polyphylla, S. arborescens, Tachigali aff. albiflora, Vouacapou americana
Caryocaraceae	*Caryocar glabrum*
Cecropiaceae	*Pourouma bicolor, P. tomentosa*
Celastraceae	*Goupia glabra*
Chrysobalanaceae	*Licania alba, L. majuscula, Parinari excelsa*
Combretaceae	*Buchenavia grandis, Terminalia guyanensis*
Fabaceae	*Lonchocarpus aff. heptaphyllus, Dipteryx odorata, Dussia discolor*
Lauraceae	*Ocotea rubra*
Lecythidaceae	*Couratari guianensis, Eschweilera coriacea*
	Lecythis persistens, L. zabucajo
Meliaceae	*Carapa procera, Guarea sp.*
Mimosaceae	*Abarema jupunba, Balizia pedicellaris, Enterolobium oldemanii, Inga alba, Parkia nitida, P. pendula P. velutina, Pseudopiptadenia psilostachya, Zygia racemosa*
Moraceae	*Brosimum guianense, Bagassa guyanensis, Ficus nymphaeifolia*
Myristicaceae	*Virola michelii , V. kwatae, V. surinamensis*
Olacaceae	*Minquartia guianensis*
Rhamnaceae	*Ziziphus cinnamomum*
Rubiaceae	*Chimarrhis turbinata*
Sapotaceae	*Chrysophyllum lucentifolium, C. pomiferum, C. sanguinolentum, Micropholis melinoniana, M. guyanensis, Pouteria guianensis, P. gonggrijpii, P. glomerata, P. melanopoda, P. laevigata, Manilkara bidentata, M. huberi*
Sterculiaceae	*Sterculia pruriens*
Tiliaceae	*Apeiba petoumo*
Vochysiaceae	*Erisma uncinatum, Qualea rosea*

Among the 21 most important families listed in Fig. 2 (those with FIV > 4), 16 are also found in the canopy inventory (Poncy *et al.* 1998). The differences can be explained because the sample includes small to medium-sized trees, and the families that include many understorey or middle-size species and/or trees then come out among the highest FIV: Euphorbiaceae, Annonaceae, Sapindaceae, Chrysobalanaceae and Moraceae. Bombacaceae is the most striking example, because of the local abundance of a single species, *Quararibea turbinata.* The abundance of palms is a major physiognomical feature of the understorey in this forest, but since only the biggest individuals are captured in the inventory, the FIV for Arecaceae is only 3.2.

TABLE 3. Comparative importance (Family Importance Value) of the families in the 7 plots (plots 1-6 are 1 ha, plot 7 is 0.8 ha in size) for families with FIV > 10).

1	2	3	4	5	6	7
Caesalpiniace 73.4	Sapotaceae 48.8	Sapotaceae 50.4	Burseraceae 37.0	Sapotaceae 52.2	Mimosaceae 42.8	Mimosaceae 42.0
Lecythidacea 43.5	Caesalpiniaceae 45.2	Caesalpiniacea 40.9	Caesalpiniaceae 31.3	Chrysobalanaceae 36.1	Lecythidaceae 29.5	Bombacaceae 33.1
Sapotaceae 32.8	Lecythidaceae 31.1	Lecythidaceae 36.1	Chrysobalanacea 28.1	Caesalpiniaceae 27.4	Caesalpiniaceae 27.4	Sapotaceae 24.8
Burseraceae 17.9	Chrysobalanacea 24.8	Mimosaceae 18.4	Sapotaceae 27.8	Clusiaceae 20.1	Burseraceae 22.5	Lecythidaceae 24.0
Chrysobalanac ae 13.2	Burseraceae 12.4	Burseraceae 15.2	Lauraceae 18.3	Lauraceae 20.1	Sapotaceae 22.3	Meliaceae 18.7
Mimosaceae 12.6	Mimosaceae 11.7	Lauraceae 12.8	Annonaceae 16.1	Lecythidaceae 14.1	Lauraceae 17.4	Caesalpiniacea 14.0
	Lauraceae 10.7		Lecythidaceae 14.1	Fabaceae 13.4	Vochysiaceae 10..3	Nyctaginaceae 13.8
	Apocynaceae 10.7		Mimosaceae 10.4	Mimosaceae 10.4		Lauraceae 13.1
			Vochysiaceae 12.1	Myrtaceae 10.1		Moraceae 10.9
						Cecropiaceae 10..3
						Burseraceae 10.2

Differences in FIV are also due to local heterogeneity, since two different sets of plots are considered. With respect to the most important families, Table 3 shows the differences between plots within the study area (see also Sabatier & Prévost 1990a, for a preliminary analysis of 5 out of the 7 plots). Caesalpiniaceae and Sapotaceae are always present among the 5 major families per plot; other families with FIV > 10, that occur on all plots, include Lecythidaceae and Mimosaceae, while Lauraceae and Burseraceae occur on 6 plots, and Chrysobalanaceae on 4 plots. Vochysiaceae have a high FIV in 2 plots, while Annonaceae, Apocynaceae, Bombacaceae, Cecropiaceae, Fabaceae, Meliaceae, Moraceae and Nyctaginaceae do so only in one plot. Plot 5 is unusual with Clusiaceae and Myrtaceae coming out in the group of leading families. The second rank for Bombacaceae in plot 7 is due to the local abundance of *Quararibea turbinata*; the importance of Nyctaginaceae, Moraceae and Cecropiaceae also contribute to the distinctive pattern of this area. The occurrence on all plots of a number of families with at least middle Family Importance Value (3-10) contributes to homogeneity at this level: Anacardiaceae, Annonaceae, Apocynaceae, Icacinaceae, Meliaceae, Myristicaceae, Sapindaceae, Sterculiaceae, Tiliaceae. On the contrary, a number of families with lowFIV are found only on one or two plots. They include mainly the less species rich families, like Aquifoliaceae, Erythroxylaceae, Lacistemataceae, Linaceae, Proteaceae, Quiinaceae, Rhizophoraceae, Rutaceae, Styracaceae and Theaceae.

Species composition

The total number of woody species recorded on the 16 hectares includes 541 trees and a few large lianas.

As part of the botanical identification of mature canopy trees (DBH ≥ 30 cm) completed on 9 one-ha plots (100 x 900 m), 145 species were identified (Poncy *et al.* 1998). The species number per hectare ranged from 39 to 87. The 25 most abundant species (with 10 trees or more on the 9 hectares) represent ca. 20 % of the total. In addition to the three very abundant species mentioned above (*Eperua falcata, Vouacapoua americana, Eschweilera coriacea*), they include several members of the Leguminosae complex (*Swartzia polyphylla, Dicorynia guianensis, Sclerolobium melinonii, Sclerolobium* aff. *paraense, Tachigali* cf. *albiflora, Pseudopiptadenia psilostachya),* Sapotaceae (*Pouteria guianensis, P. gonggrijpii, Chrysophyllum lucentifolium, Manilkara bidentata, M. huberi, Micropholis guianensis),* Burseraceae *(Tetragastris altissima, Protium opacum, P.* aff. *demerarense*), Chrysobalanaceae *(Licania canescens, L. membranacea*), Lecythidaceae *(Lecythis persistens subsp. aurantiaca*), Myristicaceae (*Virola michelii, V. kwatae*), Rubiaceae (*Guettarda acreana*) and Nyctaginaceae (*Neea* aff. *floribunda*).

Not as abundant but rather frequent (present in more than half of the complete set of 16 studied plots) are *Poraqueiba guianensis, Dendrobangia boliviana* (Icacinaceae), *Qualea rosea* (Vochysiaceae), as well as several mimosoids *(Inga alba, I. huberi, I. rubiginosa, Abarema jupunba, Parkia nitida).*

In order to give another view of tree species composition, we listed the species of biggest trees, on the basis of DBH ≥ 70 cm and/or estimated height > 40 m, throughout the 15.8 hectares. They are often emergent trees dominating the canopy that averages ca 30mhigh. The species list comprises 75 species that fall into 25 families (Table 2).

With respect to medium-sized and understorey trees (trees with 10≤ DBH ≤ 30 cm on 7 non-contiguous plots, 6.8 ha), the most abundant species include:*Andira coriacea, Apeiba glabra, Bocoa prouacensis, Conceveiba guianensis, Coussarea racemosa, Duguetia surinamensis, Eugenia coffeifolia, Eschweilera grandiflora, E. micrantha, E. pedicellata, Gustavia hexapetala, Iryanthera sagotiana, Lecythis idatimon, L. poiteaui, Licania alba, L. canescens, Pourouma minor, Pouteria decorticans, Protium opacum ssp. rabelianum, P. sagotianum* and *Quararibea duckei.* Less abundant but frequently recorded are: *Mabea piriri, Swartzia polyphylla, Tovomita choisyana, Theobroma subincanum, Xylopia nitida.* Some other species are locally abundant but restricted to reduced areas, e.g. *Duroia genipoides, Eschweilera collina, Micropholis longipedicellata, Ocotea aciphylla, Oxandra asbeckii, Quiina oiapocensis, Siparuna cristata, Swartzia guianensis*, *Trichilia pallida, Unonopsis rufescens.* Similarly, *Guarea gomma* and *G. kunthiana* occur only in the forest with lianas (plot 7), as does *Inga acreana.* Rubiaceae, like *Capirona surinamensis* or *Guettarda acreana,* have striking smooth, peeling green or white bark that locally marks the physiognomy of the understorey.

Some taxa are important at the genus level but not at the species level, each species remaining poor in number of individuals, especially *Inga*, with at least 12 individuals per ha, and *Aspidosperma, Eschweilera, Lecythis, Pourouma, Pouteria, Licania, Ocotea, Talisia, Protium* and *Swartzia.* The commonest palm species scored is *Astrocaryum sciophilum* which is present on all but one plot, with a maximum density of 27 individuals per plot. Another abundant palm is *Jessenia bataua,* while *Oenocarpus bacaba* is rather frequent, and *Maximiliana maripa* is locally frequent in the western part of the study area.

Finally, 85 species were recorded only once, some of them being rare. While some are small-sized trees with only rare individuals reaching 10cm DBH (*Erythroxylum ligustrinum, Amphirrhox longifolia, Ouratea candollei, Posoqueria longifolia*...), others are known to be generally common species, but often with very low densities, and might be only locally rare (*Bagassa guianensis, Laetia procera, Macoubea guianensis, Swartzia panacoco*...). However, most are likely to be truly rare species in French Guiana (for instance, *Ephedranthus guianensis, Ilex* sp., *Kubitzkia mezii, Lacistema grandifolium, Licania rodriguesii, Ternstroemia dentata...).*

Patterns of specific diversity

Comparative data on specific diversity on these 7 plots are presented Table 4. In the 6 plots where botanical identification was completed, the number of species ranges from 171 to 217 per hectare. Values for the number of trees per plot are rather similar from plot to plot, as found by Duivenvoorden (1996) in NW Amazonia, based on the number of species in proportion to the number of trees. Here we use the reverse ratio (individual/species) as did Spichiger *et al.* (1996). In our plots, this ratio is rather similar for 5 of the six plots (2.6-2.9), and it is never as low as in the plot studied at Jenaro Herrera (Spichiger *et al.* 1996). This is consistent with the current assumption that the highest species richness is encountered in western Amazonia (Gentry 1988, Balslev *et al.* 1989, Valencia *et al.* 1994, Duivenvoorden 1994).

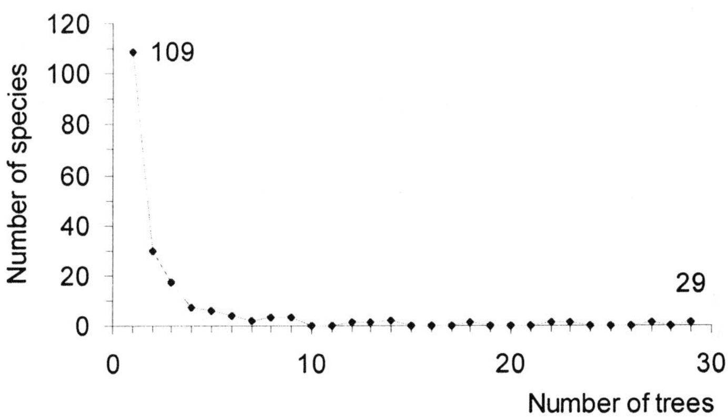

Figure 3. Species abundance distribution in plot 4

Species abundance in the plots was scored at two levels (Table 4): density ≥ 5 trees/ha and density ≥ 10 trees/ha, the latter considered as « abundant species ». More than half of the trees belong to a species having a minimum density of 5, on all plots but one. This figure (illustrated for plot 4 in Fig. 3) reflects the general pattern of species diversity in tropical rainforest, *i.e.* the relative abundance of a few species, and the scarcity of most species represented by less than 5 trees per hectare. This pattern is less marked in the plot at Jenaro Herrera, where there are lower percentages of abundant species (Table 4).

TABLE 4. Differential features of tree diversity measured on 7 one-ha plots, for all trees with DBH ≥ 10cm (JH = Jenaro Herrera: data of Spichiger *et al.* 1996 bring one point of comparison). * Identification was completed on 0.8 ha on plot 7

	1	2	3	4	5	6	7	JH
STRUCTURE								
Number of trees	468	575	471	539	681	504	489	504
Basal area (m²)	35.5	40.1	35.9	45.3	32.0	30.7	26.3	23.6
DIVERSITY								
Families with FIV ≥ 10	6	8	6	9	9	7	11*	12
Number of species	179	217	180	188	171	172	143*	227
Ratio nb trees / nb species	2.6	2.6	2.6	2.9	4.0	2.9	3.4	2.2
Abundant species (≥ 10 trees)								
Number of species	5	11	7	9	19	8	6*	5
% total number of species	*2.8*	*5.0*	*4.0*	*4.8*	*11.0*	*4.7*	*4.2*	*2.2*
Number of trees	142	175	125	172	313	145	174*	68
% total number trees	*30.5*	*30.4*	*26.5*	*32.0*	*46.0*	*28.8*	*35.6*	*13.4*
Species with ≥ 5 trees								
Number	20	31	21	26	42	24	25*	24
% total number of species	*11.2*	*14.3*	*12*	*14*	*24.5*	*14*	*17.5*	*10.6*
Number of trees	237	292	210	291	464	252	302*	178
% total number of trees	*50.6*	*50.8*	*44.6*	*54.0*	*68.1*	*50.1*	*62.0*	*35.3*
Species with a single tree								
Number	108	128	98	105	78	108	72*	125
% total number species	*60.0*	*59.0*	*54.5*	*56.0*	*46.0*	*63.0*	*50.3*	*55.0*

A general pattern similar to that of plot 4 would be obtained from our other plots, but the extreme values clearly make a difference. The highest number of trees per plot for the most abundant species varies from 25 (4.3%; plot 2) to 66 (14.1%; plot 1), and to 77 (for 0.8 ha) on plot 7 (15.7%), while the number of species with a single tree per hectare varies from 78 (plot 5) to 127 (plot 2). Indeed, the proportion of species with a single tree per hectare is over 50% in all plots but one, which is strikingly high.

Plot 5 clearly exhibits a different pattern compared to the other plots, since it is less species rich, despite a higher density (Table 4), as is also shown by the higher percentage of abundant species and lower percentage of 'single' species. This pattern is indicative of the particular forest type that occurs on the slope at the foothill of the inselberg (Fig. 1, plot 5), which is transitional between the high dense forest (plots 1-4, 6) and the low forests of the inselberg, described by Larpin (1993) and Poncy *et al.* (1998).

Contrasting with the overall distribution of species abundance (Fig. 3), qualitative differences affect the floristic composition from plot to plot. Evidence for this is given by an analysis of the occurrence of the species in the 7 plots (Fig. 4). Only four species occur in all 7 plots, while 257 occur in only one plot. The occurrence of about 100 species in more than 3 plots (less than 20 % of the total number of species recorded) suggests strong floristic heterogeneity at the local scale. The abundant species (*i.e.*, in the present study, those with more that 10 individuals on a plot, see Table 4) seem to contribute to this heterogeneity. Throughout the 7 plots, 31 species were recorded as abundant (≥ 10 trees) at least in one plot (Table 5). Only one is abundant in 4 plots (*Eschweilera coriacea*); another

Lecythidaceae (*Lecythis persistens* subsp. *persistens*) occurs with more that 10 trees in 3 plots; no other species is abundant in more than 2 plots. Thus, no species is abundant throughout the whole area.

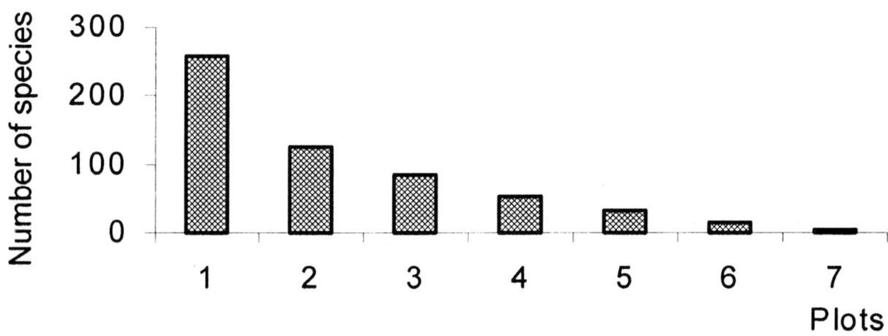

Figure 4. Species frequencies distribution

4. Discussion

4.1 THE DIVERSITY OF NEOTROPICAL FORESTS

Forest tree diversity has been worked on intensively over the last 20 years in the neotropics, especially in Central and Western Amazonia (Gentry 1988). The first results were summarised by Duivenvoorden (1995, 1996), Phillips *et al.* (1994), Spichiger *et al.* (1996), Richards (1996), Ek & van der Hout (1997), but few data were available from French Guiana. Our analyses show that the pattern of species richness and floristic composition at Nouragues is similar to that described from several inventories conducted in the Central Amazon region near Manaus (summarised by Rankin-de-Merona *et al.* 1992, see also Matos & Amaral 1999). As mentioned above, species richness is lower than in western Amazonia where the highest values have been recorded (Foster, 1990, Balée *et al.* 1989, Valencia *et al.* 1994, Spichiger *et al.* 1996, Duivenvoorden 1995, 1996).

Comparing the geographic distributions for a set of species, Mori (1991) recognised a distinct floristic ' province' as part of the large Amazonian biogeographic area. Recent studies in the nearby forests of Guyana (Comiskey *et al.* 1994, Johnston & Gillman 1995, Ek & Ter Steege 1997, Ek & Van den Hout 1997) have shown that these forests are species-poor and/or dominated by single species, such as Greenhart (*Chlorocardium rodiei*). In spite of some floristic affinity, for instance with respect to the dominant families (Leguminosae and Lecythidaceae in the large size-classes; Annonaceae and Bombacaceae in the smaller size-classes), major differences are apparent when species abundance are compared. At Kurupukari, Central Guyana (Johnston & Gillman 1995), 25% of species were represented by a single tree per hectare, while the percentage at Nouragues was 46% to 60%, and two to six species accounted for 50% of the trees at Kurupukari, while at Nouragues this was 20 to 30%. Ek (1997) surveyed the theories that may explain species richness in rainforests, including refuge theory and the stability–time hypothesis. She discussed the particular case of the Guianas and the assumption by Hammond and Brown (1995), that the Nouragues area shows a pattern of disturbance intermediate between that at Saül (most disturbed site), and Mabura Hill (least disturbed), based on the proportion of large-seeded families. However, Mori and Boom (1987) suggested that the forest at Saül has not undergone major disturbance, based on the high proportion of large trees compared with records from elsewhere. Using the same criterion would suggest that the forest at Nouragues was more disturbed (see below).

TABLE 5. The most abundant species (≥ 10 trees) in each of 7 plots at Nouragues. Number of trees are given in brackets.

Plot 1	Plot 2	Plot 3	Plot 4	Plot 5	Plot 6	Plot 7
Eperua falcata (66)	*Lecythis persistens* (16)	*Eschweilera coriacea* (26)	*Eperua falcata* (29)	*Brosimum guianense* (31)	*Quararibea turbinata* (30)	*Quararibea turbinata* (77)
Eschweilera coriacea (30)	*Licania membranacea* (11)	*Lecythis persistens* (22)	*Astrocaryum sciophilum* (27)	*Bocoa prouacensis* (28)	*Eschweilera coriacea* (24)	*Neea aff. floribunda* (34)
Lecythis idatimon (18)	*Eperua grandiflora* (11)	*Quararibea turbinata* (20)	*Licania membranacea* (23)	*Maytenus sp.* (25)	*Lecythis persistens* (19)	*Eschweilera coriacea* (31)
Lecythis persistens (15)	*Eschweilera micrantha* (13)	*Vouacapoua americana* (17)	*Licania kunthiana* (22)	*Ocotea aciphylla* (22)	*Sclerolobium paraense* (17)	*Apeiba glabra* (11)
Vouacapoua americana (13)	*Bocoa prouacensis* (12)	*Pouteria gonggrijpii* (15)	*Qualea rosea* (18)	*Licania laxiflora* (20)	*Pourouma minor* (16)	*Guarea kunthiana* (11)
		Eperua falcata (14)	*Unonopsis rufescens* (14)	*Manilkara bidentata* (19)	*Poraqueiba guianensis* (15)	*Trichilia surinamensis* (10)
			Licania canescens (14)	*Duroia genipoides* (15)	*Pouteria gonggrijpii* (13)	
			Trichilia pallida (13)	*Coussarea racemosa* (15)	*Inga huberi* (11)	
			Eschweilera grandiflora (12)	*Andira coriace* (15)		
			Micropholis longipedicellata (11)	*Astronium ulei* (11)		

4.2 FLORISTIC COMPOSITION AND ITS VARIATION AT THE LOCAL SCALE

Two interesting patterns are discussed here below. Firstly, patches of high lowland forest with similar structure and species richness may show important variation in their floristic composition, even at the family level. While at Inselberg site the forest is dominated by Caesalpiniaceae and Sapotaceae, Burseraceae is the most important family at Pararé site, only a few kilometres away. One species, *Tetragastris altissima,* is abundant here, as it is around Saül (Mori & Boom 1987), while *Vouacapoua americana* (Caesalpiniaceae) is absent, in contrast to Nouragues. These differences are consistent with the two main forest types recognised in central French Guiana by Sabatier and Prévost (1990b); each characterised by its dominant family, either Burseraceae or Caesalpiniaceae. Secondly, at a very local scale within the study site of Nouragues, the tree species composition is rather homogeneous at the family level but not at the species level. Some species are abundant but none dominates the whole area. Most species are neither abundant nor frequent, since only a few of them occur in all seven plots where trees with DBH $\geq$ 10 cm were scored, while most are found only in one or two plots. Variations of species richness also account for heterogeneity at the β-level. Two of the seven plots are less species rich than the remaining five, one of which (plot 5) is located at the base of the inselberg. Its particular structure and floristic composition reflect the transition between the high humid forests and the low forests found there (Larpin 1993, Poncy *et al.* 1998). Plot 7 is mainly situated in a forest dominated by lianas, and such liana-forest patches occur from place to place in French Guiana as well as in other areas of Amazonia. Their origin and dynamics remain controversial, although they are often considered to be of anthropogenic origin (Balée & Campbell 1990). This hypothesis was supported by Poncy *et al.* 1998, since evidence of previous human occupation, such as polishers, earthenware pieces and charcoal (Tardy 1998) has been found at several sites in the Inselberg site.

However, other factors might also be responsible for structural and floristic heterogeneity, including the occurrence of this liana forest. Sabatier and Prévost (1990b) favoured an edaphic origin, and according to preliminary pedological data reported by Poncy *et al.* 1998, a close association can be expected between liana forest and soil type. Moreover, as described by Grimaldi and Riéra (Chapter 2), the two parts of the study area, on either side of the river Nouragues, lie on different geological substrates and soils, and this might influence the distribution of some species. To the east of the river, the substrate is metamorphic rocks in the series 'Paramaca', called 'Roches vertes' by the local people and covered with clayey soil; the slope is gentle and ends on the 'Grand Plateau'. To the west, granite and crystalline rocks of the series 'Caraibe' are partly covered with sandy clay soil; its steeper slopes form the foothill of the inselberg. Most studies carried out in French Guiana (Lescure & Boulet 1985, Sabatier *et al.* 1997) and elsewhere (Duivenvoorden & Lips 1995) emphasise the importance of both topography and soil structure and composition (that determine hydrology and drainage) on species richness and on the distribution of tree species. Preliminary data observations from the forest plots at Nouragues on variation in floristic composition in relation to soil (Sabatier & Prévost 1990a,b) suggest that the relationship is complex. Climate changes in the recent past are also hypothesised to have influenced the distribution patterns of tree species. These changes are currently being studied using analyses of charcoals (Tardy 1998), of patterns of distribution of tree species that have low recolonisation rates and/or slow growth (Charles-Dominique, Chapter 7) and of the succession of pollen and seeds captured in the sediments (Ledru *et al.* 1997).

Other major topics of ecological research in the Nouragues area include the role of animals, especially vertebrates, as seed dispersers, and the dynamics of tree-fall gaps. As mentioned before by Sabatier and Prévost (1990b), both of these have effects at a small scale and may influence species composition and local variations within it at the Nouragues study site.

Further studies in tree diversity should include analyses for the whole site, based on the trees censused so far over the study area. Condit *et al.* (1996) consider 50-ha as a sufficiently large area for accurately characterising tropical tree communities. In addition, more detailed small-scale analyses, based on comparison of 1-ha or smaller plots would be valuable. Plot-shape could be taken into account, since plots of three different shapes were used in the present study. Recent work in Brazil (Laurance *et al.* 1998), however, showed that plot shape has only minor effects on estimates of tree-diversity. The general results presented in this chapter thus give a good overview on structure and diversity of the high rainforest at Nouragues.

Acknowledgements

Several colleagues participated in the field work at Pararé site. Completion of the inventory in 1983-85 is mainly due to S. Barrier, P.M. Forget and J.-F. Villiers; the floristic inventory would not have been undertaken without the strong determination of S. Barrier ; he and J.-F. Villiers, with others (C. Feuillet, O. Poncy, B. Riéra, D. Sabatier, A. Vieillescazes) climbed the trees and collected specimens; moreover, we owe to S. Barrier the most important part of the identification work, in the herbaria at Cayenne, Paris (MNHN) and St.Louis (Missouri Botanical Garden). The 1995 reinventory involved the Dep. Forestry of the Wageningen University (F. Bongers, P. Jansen, T. Rijkers, Jan van Ouden, and students) and two Museum staff at Brunoy (P.M. Forget, O. Poncy). The manuscript benefited greatly from the comments of P.-M. Forget, F. Bongers, an anonymous reviewer, and from corrections to the English by H. F. Hopkins. We wish to thank all of them.

CHAPTER 5

THE LOW FOREST (NOURAGUES INSELBERG)

Denis Larpin

1. Introduction

This study focuses on the low forest community found on the inselberg of the Nouragues area. Some authors have already indicated the occurrence of low forests in French Guiana, especially while exploring the vegetation of unknown or poorly known inselbergs, such as those of the south-western Tumuc Humac chain (Granville & Sastre 1974, Granville 1978) or more accessible inselbergs like 'Nouragues' and 'La Trinité' (Larpin 1993, 1994). Consequently, some floristic and ecological data on these plant communities can be found in the 'Aublet' data base (Hoff *et al.* 1989) from herbarium specimens.

However, since botanical and ecological surveys of the tropical inselbergs mainly focus on the vegetation of the rocky open habitats, only few studies have been done on these unique and poorly known forest communities; neither in the Neotropics nor in other tropical regions. Low forest communities are mostly just cited (Barthlott *et al.* 1993, Porembski *et al.* 1994, Porembski 1996) or more or less briefly described (Ibisch *et al.* 1995, Steyermark *et al.* 1995). Despite the insignificant amount of in-depth studies on low forests, some comparative botanical and ecological data on forests in South America, which have features in common with the inselberg low forest, does exist. Examples are foundin eastern Colombia (Duivenvoorden & Lips 1993, Duivenvoorden 1994, Duivenvoorden & Cleef 1994), in southern Venezuela (Bongers *et al.* 1985, Schnell 1987, Duivenvoorden & Cleef 1994, Steyermark *et al.* 1995) and in the Guayana Highlands and adjacent areas in eastern Venezuela (Steyermark *et al.* 1995).

The Nouragues inselberg is covered in its eastern part by a discontinuous, xeromorphic and low rock savanna vegetation (locally called 'savane-roche'), differing significantly from the neighbouring widespread rainforest (Sarthou 1992, Sarthou & Villiers 1998; Sarthou, Chapter 6). The western part of the inselberg is covered by a high forest faciès on a granitic substratum, which may be locally different in structure (lower canopy and higher densities) and floristic composition from the high forest of the lower lands surrounding the inselberg (Sabatier & Prévost 1990a, Poncy *et al.* 1998, Poncy *et al.* Chapter 4).

The low forest of the inselberg is much less widespread than the high forest and the rock savanna, and is characterised by a unique flora and distinctive features with regards to structure and physiognomy. It extends around the outer borders of the rock savanna as a continuous belt, of generally less than a hundred meters wide, between the rock savanna and the high dense forest (Fig. 1). A gradient in flora and structure is observed between

Bongers *et al.* (eds.), Nouragues. Dynamics and plant-animal interactions in a neotropical rainforest, 47–63.
2001 Kluwer Academic Publishers. Printed in the Netherlands.

these three plant communities. The low forest also occurs as isolated inclusions within the high forest or the rock savanna. Also, rocks and outcrops in the neighbouring high forest and the banks of the inselberg watercourses within the high forest may harbour a vegetation type similar to that of the inselberg low forest.

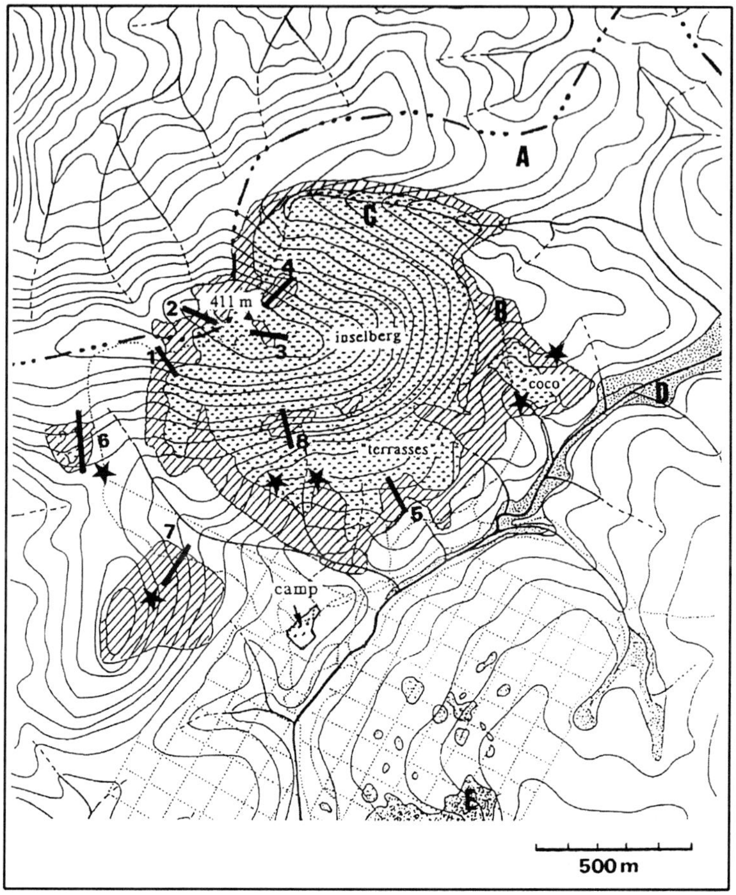

Figure 1. Study site location on the inselberg and types of vegetation in the Nouragues area (transects are noted 1 to 8; ★: main other sites prospected in this study; A. high forest; B. low forest; C. rock savanna; D. swamp forest; E. liana forest)

The aim of this study is to provide a description of the low forest with regards to edaphic, floristic, physiognomic and structural aspects. In the discussion, the floristic affinities of the low forest to other vegetation types encountered on inselbergs and different habitats in French Guiana are considered. Affinities of the low forest vegetation with vegetation of other regions of the Neotropics are also indicated. Several hypotheses, which not only concern these floristic affinities but also the adaptations of the low forest vegetation to ecological conditions and to the present and possible past vegetation dynamics of the low forest, are put forward.

2. Methods

Vegetation sampling using transects was considered as the most accurate method to describe plant composition and structure of the low forest community since a gradient is observed between the vegetation of the rock savanna, the low forest and the high dense forest. Sampling in transects was also conducted in isolated patches of low forest.

Eight representative sites of low forest were chosen for the study. In three of these sites – 1, 2 and 5 – transects follow a gradient from the rock savanna to the high forest, while in the others, they cut across patches of low forest enclosed either in rock savanna (3, 4 and 8) or in high forest (6 and 7). The transects were of four different sizes (called 'A', 'B', 'C' and 'P'), with a total area of 1.3 ha (Table 1). The transects have different lower size limits for individuals to be included (Table 1). A minimum length of 100 m was chosen for the transects. The length of the transect essentially depended on the size of the low forest patches and the width of the belt of low forest.

TABLE 1. Main features of the studied transects

Transect	A1, A2, A5	A3, A4	B7	B8	C1, C2, C5	C3, C4	C6	P1, P2, P5	P3, P4
Surface (ha)	0.02	0.02	0.12	0.1	0.12	0.12	0.40	0.0044	0.0044
Length x width (m)	100 x 2	100 x 2	120 x 10	100 x 10	100 x 12	100 x 12	200 x 20	(2 x 2)* x 11	(2 x 2)* x 11
Sampling limits	≥ 1m high	≥ 1 m high	≥ 5 cm dbh	≥ 5cm dbh	≥ 10cm dbh	≥ 10cm dbh	≥ 10cm dbh	< 1m high	< 1m high
Spatial form of the low forest **	belt	patch	patch	patch	belt	patch	patch	belt	patch
Surrounding vegetation ***	rs / hf	rs	hf	rs	rs /hf	rs	hf	rs / hf	rs / hf

* each $4m^2$ plot of the 'P' transects are located at a 10 meters interval
** belt: the low forest stands as a belt between the rock savanna and the high forest
patch: the low forest stands as a patch in the rock savanna or in the high forest
*** rs: rock savanna; hf : highforest

In each transect the full plant species composition was listed (Spermatophytes and Pteridophytes). In attempt to obtain an exhaustive species list for the vegetation of the low forest, additional sites were prospected and inventoried, though less systematically. In total 500 specimens were collected, identified and deposited at the Herbarium in French Guiana, Cayenne (CAY). Floristic and ecological data analysis was facilitated using the 'Aublet' database (Hoff *et al.* 1989), which allowed for comparison with the flora of other inselbergs and other habitats in French Guiana, in the Guianas and other regions of the Neotropics.

Borings were conducted for the measurements of soil depth and to obtain soil samples in transects A1, A2 and A3. A pit of 1.8 m depth was also dug near transect A2. Only soil samples from transect A2 and its nearby pit were analysed. Soil analysis was carried out at the chemistry laboratory of the ORSTOM Centre (IRD) at Cayenne. Analysis comprised of measurements in soil granulometry (clay, silt, sand, fragments larger than 2 mm and organic matter contents), pH (H20 and KCl), total C content, exchangeable bases concentrations (Ca, Mg, K, and Na), cation exchange capacity (CEC) and saturation rate. Hydrodynamic conditions were characterised by determining the soil pore size distribution using the mercury porosity analysis (Grimaldi & Boulet 1989, Chauvel *et al.* 1991) and completed with light-microscope observations.

3. Results

3.1 SOIL

Soil is generally shallow (0-180 cm), but occasionally can reach over 300 cm. Soil depth generally increases from the rock savanna, through the low forest, to the high forest. However, important discontinuities are noticed which depend on the subjacent geomorphological features of the granitic hill (grooves, domes, depressions). In all transects such features occur. At locations with deeper soils large trees can be found (Photo 1).

Photo 1. The forest patch on the top of the inselberg (site 3), with an emerging *Humiria balsamifera* (Humiriaceae). *Photo Denis Larpin.*

Soil colour and texture shows vertical differentiation, increasing from the rock savanna to the high forest. In all sites, the thin soil profiles of the low forest located near the rock savanna only show 1 or 2 organic horizons which are sandy and dark brown or black coloured. In the deeper soils of the low forest (such as in sites 1 and 2), soil colour changes successively from black or brown and organic, to yellowish brown, and then to ochre which becomes darker with increasing depth, and finally becomes increasingly lighter and granular. Very humid whitish, and at times waterlogged, horizons containing coarse quartz elements are found at the base of the profile on the underlying bedrock. In the low forest occurring on the top of the outcrop at site 3, soil profiles indicate coarse quartz elements in most horizons and a high degree of humidity in the deepest horizons. The dominance of the colour grey and brown of these horizons attests the hydromorphic characteristics of these soils.

Carbon concentrations are very high near the surface and in the thinnest soils due to the slow decay of the organic matter. Acidity also characterises the surface levels and the thinnest soils of the low forest, where the pH indicates a similar range (from 3 to 5) as that of soils under neighbouring thickets found in the rock savanna (Sarthou & Grimaldi 1992). Mg and Ca ions accumulate in the organic matter of the surface levels while K and Na ions are rather concentrated at the lower levels (Na is leached by the rain-water and K is abundant in the granite).

Hydrodynamic characteristics of the soil are indicated by a highly micro-aggregated structure and high porosity. Two pore size classes were revealed: micropores (of the size of about 10 nm) which constitutes a water reserve that is only slightly accessible to plants, and macropores (size ranging from 0,1 to 100 μm) of which those ranging from 1 to 10 μm retain water which is more readily available to plants. In the soil of the low forest, water penetrates easily through the largest pores between the clay aggregates whereas a reduction of water infiltration and a more compact soil is observed at the foothill of the inselberg which is encountered beyond this study's area (Grimaldi, personal communication).

Moreover, the overall water reserve is limited when the soil of the low forest is thinner or when the slope is steeper. In such areas during the dry periods, the vegetation suffers from a higher water-stress than in areas of the low forest with a deeper soil. Yet, during the rainy season excess water can also be locally encountered. This is observed in flat areas and depressions, as underneath the low forest patch on the top of the inselberg, where the soil can be waterlogged at important depths.

3.2 PHYSIOGNOMY AND STRUCTURE

Three main zones corresponding to vegetation changes in physiognomy, structure and floristics of the low forest are distinguished on a gradient (Fig. 2). Immediately after a continuous border of thickets in the rock savanna, where *Clusia minor* and *Clusia nemorosa* dominate (as they also do in isolated thickets of the rock savanna), a distinctive area of several meters wide is recognised in the low forest. This section identified as the 'rear border' zone in this study is characterised by a closed but clear canopy, relatively low stem densities and frequent areas of outcropping rock which alternate with areas of shallow soil. The canopy species are *Eriotheca surinamensis* (Bombacaceae) and old large individuals of *Clusia minor*. Additional species are locally found (e.g. *Micrandra brownsbergensis*, Euphorbiaceae at site 6, and *Syagrus stratincola,* Arecaceae at site 'Coco'). Essentially, this area is easily identifiable by the presence of the prevalent herb *Stelestylis surinamensis* (Cyclanthaceae) which forms dense and almost continuous populations on a band. Epiphytic species are also numerous and dense in this area (Photo 2).

Photo 2. Typical physiognomy of the 'rear border' zone of the low forest. A dense population of *Stelestylis surinamensis* (Cyclanthaceae) is seen in the foreground. The trees (mostly *Eriotheca surinamensis,* Bombacaceae) are abundantly covered with epiphytes, among which *Philodendron fragantissimum (*Araceae) is the most frequent species. *Stelestylis surinamensis* is also found as an epiphytic plant on the lower parts of the trees. *Photo D.Larpin.*

The 'low forest *(sensu stricto)*' zone follows the 'rear border' zone. In this area, main distinctive physiognomic features are the very high number of shrubs or small trees. The stems of this woody vegetation often have two or several leaning branches growing from the base, usually facing the direction of the rock savanna (Photo 3). This area also contains many other slender or twisted-stem species. Most Myrtaceae species have this multi-stemmed characteristic growth habit, as do other very common species such as in the Rubiaceae, Myrsinaceae and Nyctaginaceae families. Trees that have two or three large trunks emerging from the base are also observed in several species (e.g. *Eriotheca surinamensis* and *Micrandra brownsbergensis*). Other species including several myrts and other common taxa (e.g. *Cassipourea guianensis,* Rhizophoraceae, and *Ternstroemia dentata,* Theaceae) exhibit a single and large trunk that emerges above the canopy. The height of the low forest canopy is usually between 8 to 15 m. the low forest understorey is well lit with a high density of seedlings and saplings.

Herbaceous species proliferate on the ground level, and are also present as epiphytes on lower and upper branches of the woody plants. In particular, the lower branches and the lower parts of tree trunks of this forest are abundantly covered with mats of moss, liverworts and filmy ferns. These are especially well developed in the sites more exposed to humid winds and mist which are frequent on the northern side of the outcrop during the rainy season, and in the morning during the dry season.

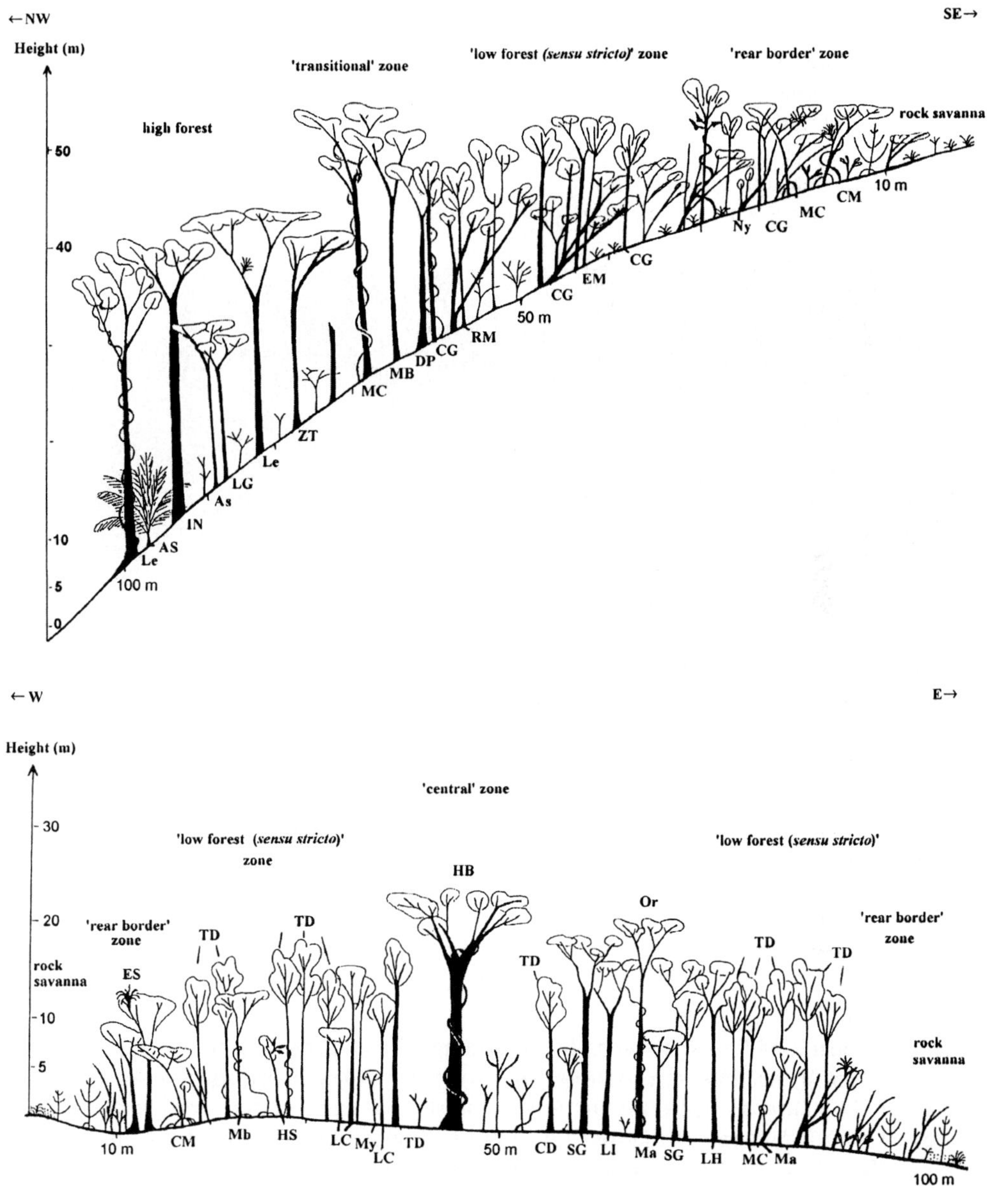

Figure 2. Profile diagrams of the vegetation of the low forest. A. forest on the uppernorthwestern slope of the inselberg (site 2), B. forest patch on the top of the inselberg (site 3). As. *Aspidosperma sp.*; AS. *Astrocaryum sciophilum*; CD. *Cupania diphylla;* CM. *Clusia minor*; CG. *Cassipourea guianensis*; ES. *Eriotheca surinamensis;* DP. *Dipteryx punctata*; EM. *Eugenia marowynensis*; HS. *Henriettea succosa*; IN. *Indet. sp.*; LC. *Licaria cannella*; Le. *Leguminosae indet. sp.*; LI. *Licania irwinii*; LG. *Licaria guianensis*; Ma. *Maytenus myrsinoides.*; Mb. *Manilkara bidentata*; MB. *Micrandra brownsbergensis*; MC. *Myrcia citrifolia*; My. *Myrciaria sp.*; Ny. sp *Guapira eggersiana*; Or. *Ormosia paraensis*; RM. *Roupala montana*; SG. *Sacoglottis guianensis*; TD. *Ternstroemia dentata;* ZT. *Zygia tetragona.*

The two preceding described zones constitute a confined environment, which retains humidity in the rainy season. However, after rainfall and in the dry season, the low forest dries up faster than the high forest; the low forest canopy being lower, the understorey being well illuminated and also because of its proximity to an open area i.e. the rock savanna. Woody and herbaceous species, typical of the low forest, may be found on blocks of rocks, rock outcrops and banks of watercourses which are located in the high forest on the inselberg, or in the neighbouring high forest areas.

Photo 3. Typical physiognomy of the 'low forest (*sensu stricto*)' zone, with a high number of multi-stemmed trees and shrubs and leaning branches. *Photo Denis Larpin.*

In the 'transitional' zone between the preceding area and the high forest trees are larger and taller than in the 'low forest (*sensu stricto*)' zone. Also, many species which show a leaning multi-stemmed habit in the lower forest now present tall and vertical stems. However, typical features of the low forest are still encountered such as the frequency of herb populations on the ground and the abundance of epiphytes at lower or medium height on the trees. This area is mainly characterised by several tree and treelet high forest species encountered in almost all the studied sites of the 'transitional' zone.

Structural data recorded in each of the six transects for stems ≥10 cm dbh were as follows (C1-C6): densities ranged from 867 to 1858 stems/ha and basal areas from 30.6 to 47.3 m^2/ha; average tree heights ranged from 12.3 to 15.5 m and average dbh from 14.3 to 21.8 cm (Table 2). Also, tree densities for size class above 1 m high (A transects) varied highly from 30150 stems/ha to 49200 stems/ha. Generally, the variation depended much on the local situation, position with respect to the summit, inclination, and presence of outcropping rock areas.

TABLE 2. Structural data of the 'A' and 'C' transects. Note that in transects A all individuals ≥ 1m high are included, and in transects C all individuals ≥ 10 cm diameter at breast height.

Transect	Surface (ha)	Number of stems	Density (number of stems. ha^{-1})	Maximal diameter (cm)	Average diameter (cm)	Maximal height (m)	Average height (m)	Basal area (m^2. ha^{-1})
A1	0.02	603	30150	35	2.61	25	3.26	54.20
A2	0.02	667	33350	35	2.05	30	3.32	44.25
A3	0.02	908	45500	60	2.24	25	3.05	73.90
A4	0.02	984	49200	27	2.26	18	3.50	57.50
A5	0.02	871	43550	24	1.84	28	2.76	56.93
C1	0.12	196	1633	68	15.68	45	13.22	39.56
C2	0.12	192	1600	77	17.45	38	15.55	47.28
C3	0.12	162	1350	60	15.76	25	13.05	30.65
C4	0.12	223	1858	33	14.30	23	12.27	33.39
C5	0.12	104	867	57	21.77	35	15.23	41.61
C6	0.40	479	1197	94	17.88	38	13.31	39.11

The low forest was extremely dense when compared to the nearby high forest that covers the lower lands at the base of the outcrop. Densities in this neighbouring high forest, for size class ≥ 10 cm dbh, ranged from 473 to 570 stems/ha (Sabatier & Prévost 1990a, Poncy *et al.* 1998). High stem densities encountered in the low forest are mainly due to the gregarism of many of its characteristic tree species. This phenomenon is also observed in areas of high forest located on the granitic substratum (Sabatier & Prévost 1990a). A density of 1197 stems /ha was recorded in a low forest patch enclosed in the high forest (C6). The density of the low forest is also high when compared to the densities of high forests in other areas in French Guiana (e.g. 800 stems/ha near Saül: Mori & Boom 1987) and western and central Amazonia (e.g. 734 stems/ha: Balslev & Renner 1989; between 600 and 700 stems/ha: Rankin-de-Merona *et al.* 1990; 637 stems/ha: Rankin-de-Merona *et al.* 1992; 600 stems/ha: Gentry 1988). Moreover, many small trees and shrubs which characterise the low forest (belonging to the most prevalent families such as the Myrtaceae, Rubiaceae, Nyctaginaceae, Erythroxylaceae) largely contribute to the very high densities recorded in the low forest since they often attain just above the sampling size of 10 cm dbh.

Basal areas for size class ≥ 10 cm dbh surprisingly reached relatively high values (even though shrubby habits and small trees dominate in the low forest). The basal areas recorded in the low forest are similar to those measured in the high dense forest. Basal areas from 30 to 45 m^2/ha were recorded in the high forest at the base of the outcrop (Sabatier & Prévost 1990, Poncy *et al.* 1998). Similar basal areas can be found in French Guiana (e.g. 38 m^2/ha at the Piste de St-Elie site: Puig 1979; 53 m^2/ha near Saül: Mori & Boom 1987) and Amazonia (from 27 to 34 m^2/ha near Manaus, Brazil: Rankin-de-Merona *et al.* 1990). The high values of the basal areas recorded in the low forest are mainly due to the fact that tree species, which attain large diameters (≥ 60 cm), are mostly located in the 'transitional' zone between the low and high forest and in the 'central' zone of the low forest patches. The very high proportion of stems that reach 10 cm dbh, or slightly over, characterises the 'low forest (*sensu stricto)*'zone. Despite basal area values similar to those of the high forest, the inselberg low forest remains a very distinctive type of vegetation.

3.3 FLORA

Five hundred and thirty-three species were recorded, belonging to 95 families and 274 genera. This comprises 386 woody species (including 10 palms) and 147 herbaceous species (including 24 Pteridophytes) (Tables 3 and 4). The most species rich flowering plant families are listed in Table 5. Only 10 genera are represented by more than four species: *Erythroxylum* (5 species); *Tillandsia* (6); *Inga, Licania, Maxillaria,* and *Myrcia* (each 7); *Eugenia, Philodendron,* and *Pouteria* (each 8); and *Psychotria* (10).

TABLE 3. Taxonomic distribution of the low forest taxa : * the total is not equal to 76 because two families - Rubiaceae and Rutaceae - comprise herbaceous and woody species.

	Woody	Herbaceous	Total
Dicotyledons			
Families	69	7	74*
Genera	171	13	184
Species	372	16	388
Monocotyledons			
Families	2	10	12
Genera	8	60	68
Species	14	107	121
Pteridophytes			
Families		9	9
Genera		16	16
Species		24	24
Total number of species	386	147	533

The inselberg low forest harbours nearly 200 species that are not found in the high forest. Among these are many of the commonest woody species, belonging to families not frequently or never present as big trees in the high forest.

The most common tree species of the low forest are myrts (*Myrciaria sp.*, *Myrcia citrifolia*, *Myrcia fallax*, *Calyptranthes lepida,* and *Eugenia marowynensis*), along with *Rudgea crassiloba* and *Alibertia myrciifolia* (Rubiaceae), *Cybianthus guianensis* (Myrsinaceae), *Erythroxylum ligustrinum* (Erythroxylaceae), *Guapira eggersiana* and *Neea ovalifolia* (Nyctaginaceae). The following species are also quite frequent: *Maytenus myrsinoides* (Celastraceae), three *Inga* species (*I. lateriflora, I. virgultosa,* and *I. cayennensis*), and *Roupala montana* (Proteaceae).

Some species such as *Ternstroemia dentata* (Theaceae), *Erythroxylum squamatum* (Erythroxylaceae), *Ouratea candollei* (Ochnaceae), *Micropholis mensalis (*Sapotaceae), *Syagrus stratincola* (Arecaceae), *Balizia pedicellaris* (Mimosaceae), *Esenbeckia cowanii* and *Esenbeckia grandiflora* (Rutaceae) are only encountered at few sites on the inselberg and illustrate a clumped distribution. In these sites they locally portray high densities of individuals.

Most common shrub or treelet species include *Psychotria ctenophora* (Rubiaceae), *Inga umbellifera* (Mimosaceae), *Picramnia guianensis* (Simaroubaceae), *Cordia sp.* (Boraginaceae), and an unidentified Thymeleaceae species.

TABLE 4. Lifeform distribution of the low forest taxa.

Lifeform	Number of species
Woody	386
Trees	263
Shrubs	57
(Palms included)	(10)
Lianas	53
Epiphytes	13
Herbaceous	147
Terrestrial	24
Epiphytes	85
Climbers	7
Saxicolous	31
(Pteridophytes included)	(24)

A few species occur in both low and high forest, but show distinct plant sizes or habits. Such species include *Eriotheca surinamensis* (Bombacaceae) and *Terminalia amazonia* (Combretaceae), both of which are medium height trees with large diameters in the low forest, and which grow on extremely thin soil or even on the granitic rock. On the contrary, in the high forest, both species although scarce, are encountered as very tall trees with trunks of substantial diameters. *Cassipourea guianensis* (Rhizophoraceae) is another of these species which is found as an emergent tree in the low forest whereas it is common throughout the high forest as an understorey species.

TABLE 5. The most speciose families of the low forest.

Family	Number of species	Number of genera
Woody		
Myrtaceae	33	5
Sapotaceae	21	6
Mimosaceae	19	6
Apocynaceae	13	8
Arecaceae	10	7
Euphorbiaceae	11	8
Annonaceae	11	6
Caesalpiniaceae	10	6
Clusiaceae	8	5
Flacourtiaceae	6	5
Herbaceaous		
Orchidaceae	50	29
Bromeliaceae	26	13
Araceae	14	6
Mixed		
Rubiaceae	36	15
Melastomataceae	15	7

In the 'transitional' zone between high and low forest, and in the 'central' zone of low forest patches isolated within the rock savanna, a set of frequently found high forest tree species is noteworthy: *Cupania diphylla* (Sapindaceae), *Duroia longiflora* (Rubiaceae), *Protium heptaphyllum* (Burseraceae), *Licania hypoleuca* and *Licania irwinii* (Chrysobalanaceae), *Inga lateriflora* and *Zygia tetragona* (Mimosaceae), *Lecythis idatimon* and *Lecythis poiteaui* (Lecythidaceae), *Sacoglottis guianensis* (Humiriaceae), *Brosimum guianense* (Moraceae), *Aspidosperma sp.* (Apocynaceae), *Manilkara bidentata* (Sapotaceae), and *Dipteryx punctata* (Fabaceae). Some species are clumped and are encountered at one or few studied sites including *Pogonophora schomburgkiana* (Euphorbiaceae), *Moronobea coccinea* (Clusiaceae), *Humiria balsamifera* and *Schistostemon dichotomum* (Humiriaceae).

Many shrub and small tree species are frequently found in the low forest and in the neighbouring thickets of the rock savanna including*Eugenia ramiflora, Myrcia guianensis,* and *Myrcia sylvatica* (Myrtaceae), *Miconia ciliata* (Melastomataceae), *Psychotria hoffmannseggiana* (Rubiaceae), *Ouratea leblondii* (Ochnaceae), *Croton tafelbergicus* (Euphorbiaceae), and *Piptocoma schomburgkii* (Asteraceae). Two species, *Clusia minor* and *Clusia nemorosa,* can reach important diameters in the 'rear border' zone of the low forest contiguous to the thickets forming the fringe of the rock savanna. In these thickets, these two species are dominant likewise in the isolated thickets of the rock savanna.*Myrcia saxatilis* (Myrtaceae) is always associated with *Clusia minor* and *Clusia nemorosa* in these thickets and is also found in the 'rear border' zone of the low forest. It was also observed that some species like *Rudgea crassiloba.*, *Alibertia myrciifolia, Guapira eggersiana*, and several myrts cited above, may be seen as trees or small trees in the low forest whereas these species are shrubs in the *Clusia* thickets of the rock savanna. It is also interesting to note that some low forest tree species such as *Roupala montana*, *Myrcia citrifolia, Myrcia fallax*, *Balizia pedicellaris,* and *Ouratea candollei* are present as juveniles in the shrubby thickets located at the fringe of the rock savanna in contact with the low forest. At times these same species also occur inside isolated thickets of the rock savanna (Sarthou, Chapter 6).

Myrtaceae dominate the low forest in terms of relative diversity (17% in transect 'A' inventories), where they represent the highest number of species of the woody plant families recorded (33 species which amount to 6% of the total number of species and 8.5% of the total number of woody species). Myrtaceae are also dominant in terms of abundance (30% to 50% relative densities depending on the site considered) where the highest number of stems occurred in all 'A' and 'C' transects. Other abundant families are Rubiaceae, Clusiaceae, Myrsinaceae, and Melastomataceae (in 'A' transects) and Clusiaceae, Rhizophoraceae, Nyctaginaceae, and Euphorbiaceae (in 'C' transects).

Liana species are few, but some woody species such as *Heteropterys sp.* (Malpighiaceae), *Securidaca uniflora* (Polygalaceae), and *Coccoloba sp.* (Polygonaceae) may be locally abundant. Other quite common woody liana species are *Salacia impressifolia* (Hippocrateaceae), *Malanea macrophylla* (Rubiaceae), *Moutabea guianensis* (Polygalaceae), *Souroubea guianensis* and *Norantea guianensis* (Marcgraviaceae), and the herbaceous *Paradrymonia campostyla* (Gesneriaceae) and *Vanilla ovata* (Orchidaceae). Epiphytes include some woody species, among which the ericaceous *Satyria cerander, Cavendishia callista,* and *Sphyrospermum cordifolium* are the most frequent.

Most of the herbaceous species belong to Monocotyledons. Orchidaceae, Bromeliaceae, and Araceae dominate the epiphyte and terrestrial herbs. Also common are Cyclanthaceae, Poaceae, and Marantaceae. Gesneriaceae and Rubiaceae are frequent among herbaceous Dicotyledons. Herbaceous epiphytes include 44 of the 50 orchid species recorded, among which *Aganisia pulchella, Batemannia colleyi, Elleanthus graminifolius, Elleanthus caravata, Koellensteinia graminea, Maxillaria crassifolia, M. uncata, M. violaceo-*

punctata, and *Stelis argentata* are well represented. Epiphytic bromeliads comprise *Araeococcus micranthus, Billbergia violacea, Streptocalyx longifolius,* and several species of *Aechmea, Guzmania, Tillandsia,* and *Vriesea.* Within the Araceae family *Philodendron fragantissimum* is prevalent in the low forest epiphytic stratum. Other frequent species in this family include *Philodendron billietiae, P. insigne, P. linnaei, P. pedatum,* and *P. squamiferum.* Pteridophytes also include many epiphytic species, such as *Asplenium juglandifolium* (Aspleniaceae), several *Elaphoglossum* (Lomariopsidaceae) and *Cochlidium* (Grammitidaceae), *Hymenophyllum polyanthos* and *Trichomanes crispum* (Hymenophyllaceae), and *Antrophyum guayanense* (Vittariaceae).

Saxicolous or terrestrial herbaceous species are represented by few orchids such as *Koellensteinia carraoënsis* and the rare and magnificent *Phragmipedium lindleyanum* with long spikes and large flowers, growing in rocky areas of the inselberg, along water courses or in other moist forest conditions. As for bromeliads, many species are clumped on the ground, or often on rocky places, in the low forest. Examples of frequently encountered species are among *Aechmea* and *Bromelia,* and the very imposing species *Disteganthus lateralis* with rigid and prickled leaves. Some other bromeliads occur very locally on the outcrop such as *Ananas ananassoides* and *Werauhia gladioliflora.* Many common bromeliad species are also found as lower epiphytes in the low forest whereas they are epiphytes in the high forest canopy. These include the impressive *Vriesea pleiosticha* and the splendid and renown *Vriesea splendens*, with its variegated leaves, red bracts and yellow flowers. *Stelestylis surinamensis* (Cyclanthaceae) is a prevalent herbaceous species in the low forest, encountered as very dense populations in the 'rear border' zone of the low forest, but isolated individuals are also found in the 'low forest (*sensu stricto*)' zone. Other herbaceous monocotyledon species include *Pariana campestris* and *Olyra obliquifolia* (Poaceae), *Calathea propinqua* and *Calathea squarrosa* (Marantaceae), which grow in the low forest understorey, and sometimes on the rock. Within the Dicotyledons, Gesneriaceae such as *Episcia sphalera*, *Paradrymonia densa,* and the monophyllous remarkable species *Lembocarpus amoenus*, generally grow under particularly moist conditions on rocky areas or in the low forest understorey.

Finally, habitats such as rocky outcrops, blocks of rocks and water outflows in the low forest are generally suitable for the instalment of herbaceous species such as *Anthurium jenmanii* (Araceae), *Pitcairnia sastrei* (Bromeliaceae), *Cheiradenia cuspidata* (Orchidaceae), *Selaginella producta* and *Selaginella radiata* (Selaginellaceae).

4. Discussion

4.1 FLORISTIC AFFINITIES

The vegetation of the inselberg low forest clearly shows a typical flora when compared to the nearby high dense forest, as only about 25% of its tree species are also found in the high dense forest. Several low forest species that do not occur in the high forest are typical of inselbergs and granitic outcrops. Examples include the bromeliads *Ananas ananassoides*, *Pitcairnia geyskesii,* and *Pitcairnia sastrei,* the palm *Syagrus stratincola* and *Sauvagesia aliciae* (Ochnaceae). Other low forest species are found on inselbergs and also in other habitats like clear or rocky areas, coastal forests on white sand soils, and forests peripheral to the savannas (e.g. in the coastal guianan savannas and in savannas of southern Guyana and Suriname). Examples include *Stelestylis surinamensis* and the orchid *Cyrtopodium andersonii* among herbs, and many woody species (*Tabebuia capitata*, Bignoniaceae, *Sapium montanum,* Euphorbiaceae, *Ernestia granvillei* and *Macrocentrum cristatum,* Melastomataceae, and the previously listed *Cochlospermum orinocense*, *Calyptranthes lepida, Inga virgultosa*, *Roupala montana*, *Esenbeckia grandiflora*, *Erythroxylum ligustrinum,* and *Ternstroemia dentata*). *Pitcairnia geyskesii* and *P. sastrei* are endemic of the southern inselbergs of French Guiana and Suriname whereas the other species are geographically restricted to the Guianas and adjacent areas (Granville 1978, Hoff *et al.* 1989, Larpin 1993, 1994, Boggan *et al.* 1997).

A few species, including *Inga lateriflora* and *Licania irwinii,* found in the low forest of the Nouragues inselberg were only previously known from forests on white sand soils in the Guianas. In such forests, as found along the coasts of French Guiana, *Humiria balsamifera* is often dominant and *Clusia nemorosa* is frequent (Granville 1988). Both species are also encountered in the 'savannah forests' in Suriname (Schnell 1987). *Humiria balsamifera* is likewise mentioned among shrub and low tree communities on white sand lowland areas and in the Guayana Highlands in Venezuela (Steyermark *et al.* 1995). *Inga cayennensis* and *Licania hypoleuca* have a wider geographic distribution, reaching southern and western Amazonia. A few other species are also found on rocky riparian habitats (river rapids), including *Calliandra aff. surinamensis* (Mimosaceae), or in very wet montane or cloud forests on ferrallitic soil or lateritic crust such as *Licania hypoleuca*, *Cybianthus guianensis*, *Inga virgultosa.* Some species such as *Miconia holosericea* (Melastomataceae) are found in a high diversity of habitats with dissimilar ecological conditions. This species is not only found in the low forest on inselbergs, but also on lateritic crusts and on open areas in the Guianas, and even in some types of semi-deciduous and xeromorphic 'caatingas' in north-eastern Brazil (Emperaire 1987).

Incidentally, the floristic inventory of the Nouragues low forest vegetation not only allowed to compare the floristic affinities of the low forest with other plant communities in other habitats, but it also provided the opportunity to collect a number of new, or extremely rare species in French Guiana. This includes some orchid species (*Koellensteinia carraoënsis, Bulbophyllum roraimense, Sigmatostalix huebneri,* and *Notylia yauaperiensis*), two mimosoids (*Zygia tetragona* and *Abarema barbouriana*), *Eriotheca surinamensis*, *Micrandra brownsbergensis*, *Esenbeckia cowanii*, *Cupania diphylla,* and the palm *Syagrus stratincola.*

4.2 ADAPTATIONS OF THE VEGETATION

Most of the previously mentioned habitats which have served as a comparison with the flora of the low forest, have a feature in common. They all portray constraining edaphic conditions and thereby specific vegetation adaptations. In the low forest of the Nouragues inselberg in particular, the characteristic composition and structure of the vegetation might be highly dependent on the periodic desiccation of the soil which occurs during the persistent or intermittent dry spells, especially when the soil is shallow and located on steep slopes. This hypothesis is supported by the occurrence of several tree species showing both a multi-stemmed habit (edaphic ecotype) in the low forest, and a tree-form near the high forest. Similar edaphic (soil-water) conditions that imply a periodic water deficit resulting to low forests formations have been reported in Amazonia. In eastern Colombia forests on white sand soil over hardrock (Duivenvoorden & Cleef 1994), and forests on deep podzolized white sands (Duivenvoorden & Lips 1993, Duivenvoorden 1994) show comparable features. Also, the low 'caatinga' forests and small scrub patches ('bana') on podzolised white sand sediments in southern Venezuela (Bongers *et al.* 1985, Steyermark*et al.* 1995) are governed by periodic water stress.

The dominance of Myrtaceae is a major aspect of the low forest on the Nouragues inselberg. The ability of members of this family to adapt to constraining edaphic and other ecological conditions in tropical climates has long been noted throughout the world: on other inselbergs in French Guiana (Granville 1978), in low forest of the Alto Parana region in Paraguay (Spichiger *et al.* 1992), in 'igapos' and other regularly flooded forest areas in Amazonia (P. Blanc, personal communication), in deciduous low forests in the 'caatingas' of the dry north-eastern Brazil (Schnell 1966, Emperaire 1987), in the shrubby sclerophyllous formations on ultrabasic substratum in New Caledonia (Morat *et al.* 1986, Jaffre & Veillon 1991), in the 'kerangas forests' on dry lowland podzols and in vegetation of the submontane areas on exposed sandstone rocks at high altitude, both in Sarawak and Brunei (Brunig 1974). In African Zambezian savannas shrubby myrt species seem to be successfully adapted to harsh edaphic conditions (seasonal hydromorphy and soil nutrient deficiency) as well as protected against annual savanna fires (by producing woody underground stocks and annual sprouts; White 1976). Examples of similar protective structures (production of shoots from lignotubers, rhizomes and roots; fire resistant main stems) have been described for many Myrtaceae species (especially Eucalypts) in areas of Australia where man-made fires are frequent. Also, the multi-stemmed habit of the Australian Myrt trees seems to be linked to unfavourable edaphic factors, whereas these species occur as forest trees on more favourable soil conditions (Beadle 1968, Chattaway 1968, Lacey & Whealan 1976, Lacey 1974, 1983).

The constraining soil conditions not only directly affect floristics and structure of the vegetation, but also lead to indirect consequences such as important epiphytic stratum as observed in the low forest of the Nouragues inselberg. Indeed, the low canopy, the high stem density and the abundance of leaning axis, along with the moist microclimatic and high illumination conditions, are favourable features for the establishment and growth of epiphytes in the low forest. In fact in the Nouragues area, many of the epiphytic species characteristic of the low forest (principally among bromeliads, orchids, aroids and ferns) are found in the lower part of the high forest canopy (see also Granville 1978), a zone described as 'lower-canopy' by Steege and Cornelissen (1989).

In our study area, the forest sites with the highest epiphyte densities are not those that grow on the richest and thickest soils, as suggested by Gentry and Dodson (1988). On the contrary, epiphytes proliferate primarily in very thin or no soil areas, where the low forest completely or partly covers outcropping granitic rock, and where carrier trunks and branches are abundant (Photo 2). A similar proliferation of epiphytic plants is observed in many other low forests under very humid climates such as the 'tepui' summits of the Guayana Highlands in Venezuela (Steyermark *et al.* 1995).

4.3 VEGETATION DYNAMICS

The characteristic growth habit of many woody species in the low forest of the Nouragues inselberg (leaning multi-stemmed trees and basal shoot production) favours the spatial colonisation and the survival of these species on periodically unfavourable soil conditions. Such a growth habit also denotes that vegetative propagation is very efficient in this community. The dynamics of the low forest is also highly determined by the sexual propagation of low forest plants and neighbouring plants of the high forest. The seeds of the predominantly zoochorous woody species of the low forest are usually dispersed by small birds, like Tangaras or Pipridae. Most of these species germinate locally or in the nearby rock savanna thickets. Seedlings and juveniles of several high forest species are also found, though sparsely, in the understorey of the low forest when sufficiently shaded. Larger seed-dispersing animals (mostly birds and monkeys) also favour the penetration of high forest species into the low forest community, as attested by the results of a study on the Cock-of-the-rock, showing the influence of this large bird species on forest dynamics (Théry & Larpin 1993). Such observations and the actual wet climate of the Nouragues area would be favourable conditions for the high forest to gain on the low forest, but this dynamic process would occur at a very slow pace. The low forest may also very slowly progress on the rock savanna by means of germination and growth of some of its species inside the *Clusia* thickets of the rock savanna. Favourable places for seed germination would be in thickets located at borders between the rock savanna and low forest formations, and also inside the most dense *Clusia* thickets isolated in the rock savanna. This tendency is however reversed in favour of the rock savanna when the ecotone is disturbed by the fall of the oldest *Clusia* trees, or more precisely in the small gaps which frequently occur in the rock savanna thickets in contact with the low forest. Additionally, periodic and often intense runoff water on exposed rock and steepest slopes prevent most non pioneer species from establishing in many areas. This contributes to maintaining the open rock surfaces which alternate with the rock savanna vegetation types on the outcrop. Only areas well-protected from these extreme physical and topographic conditions would therefore be suitable for a possible progressive vegetation change towards a low forest or higher forest vegetation type.

The preliminary results of the paleobotanical data obtained by Tardy (1998) indicate that fire has played an important role in this vegetation. The low forest exhibits floristic composition and structural characteristics which are liable to burn much easier than those of the high forest (Charles-Dominique *et al.* 1998, 2000, Larpin *et al.* 2000). Moreover, the typical growth habit of many woody plants of the low forest (principally the myrts) indicate strong abilities to recolonise the environment after fire events, in addition to their successful adaptations to unfavourable edaphic factors.

Most of the taxa identified from the charcoal samples (Tardy 1998) belong to Myrtaceae and Rubiaceae (currently the most species rich and abundant families in the low forest) and to Melastomataceae which is another prevalent family, particularly in the low forest patch of the summit of the inselberg where the charcoal samples were collected and analysed. The fact that these taxa were recorded at all depths indicates that the vegetation which existed at the time when most fires occurred is probably very similar to the present low forest vegetation. This hypothesis is supported by the presence of many other current low forest taxa among the identified charcoal samples. Some of these taxa are not present today at the site where the charcoal samples were collected but do occur on the inselberg or in the closely surrounding areas.

Many high forest taxa have also been identified in the charcoal samples of the low forest soil on the summit of the outcrop. Among these taxa a few do not occur at present in this forest patch. Several high forest taxa were also recorded in the charcoal samples of the soil of a neighbouring dense rock savanna *Clusia* thicket, although not currently present in this thicket. These results suggest that the high forest might have occurred on the summit of the inselberg at certain periods of the past, or the existence of a low forest containing high forest species different from those currently found in the low forest.

In the light of these preliminary results (Tardy 1998), the possible more widespread distribution of a low vegetation type similar to the current low forest, during the past driest periods, can be hypothesised; at least for areas on the summit of the inselberg. A larger extension of the high forest at this site during the past alternating vegetation phases, due to more humid climate conditions, can also be envisaged. In addition to these results and interpretations (see also Tardy 1998), further investigation and analysis would help towards a better comprehension of the past vegetation changes and composition of the Nouragues inselberg during the late Pleistocene and Holocene.

« Four-eyes opossum » *Philander opossum. Photo Pierre Charles-Dominique.*

CHAPTER 6

PLANT COMMUNITIES ON A GRANITIC OUTCROP

Corinne Sarthou

1. Introduction

The term 'inselberg', proposed by Bonhardt (1900), characterises isolated rock outcrops consisting mainly of Precambrian granites or gneisses. The origin, geology and geomorphology of inselbergs have been widely studied (Bremer & Jennings 1978, Twidale & Bourne 1978, Twidale 1980). In South America, classical whale back forms dominate and are scattered throughout the Guiana and Brazilian Shield up to East Bolivia. These hills rise abruptly from the surrounding landscape and constitute singular habitats in tropical rain forest. Due to the poorly developed soils and extreme microclimatic fluctuations, they are colonised by vegetation drastically different from the surroundings. Thus, they reflect a clear habitat fragmentation and constitute functional 'terrestrial' islands (Prance 1996).

The origin of the present vegetation growing on inselbergs has been questioned. The vegetation in Amazonia has not been stable through the scale of geological times and fluctuations in climate have resulted in fluctuations in the balance between the drier types of vegetation and the wetter tropical rain forest. The current situation was sometimes reversed, patches of rain forest being islands within savanna (Haffer 1969, Prance 1973, 1983, van der Hammen 1974, Hooghiemstra & van der Hammen 1998). Thus, inselbergs are interpreted as isolated refuges for a xeric flora, dominated by the tropical rain forest, in the actual wet climatic phase (Hurault 1974, Descamps *et al.* 1978, Granville 1982). This flora was much more widespread through South America during the glacial episode, especially the Würm (22000 to 13000 BP), and also other dry shorter periods of the Holocene. Recent palynological data (Hooghiemstra & van der Hammen 1998) from eastern Amazonia evidenced the existence of three dry savanna periods following a rhythm of ca 20,000 years since 65000 BP. The composition of the vegetation present today on the granitic outcrops could result from a combination of several dispersal waves and the occurrence of corridors from the coastal and/or the North Amazonian savannas.

Some inselberg vegetation studies have been carried out in Surinam (Theunissen & Wildschut 1970), in French Guiana (Granville 1978, Sarthou 1992, Sarthou & Villiers 1998), in Venezuela (Huber 1980, Barthlott *et al.* 1993), in Brazil (Barthlott *et al.* 1993, Porembski *et al.* 1998) and in Bolivia (Ibisch *et al.* 1995). These investigations showed that herbaceous outcrop vegetation is generally dominated by monocotyledonous mats consisting mainly of Bromeliaceae, Velloziaceae and Cyperaceae. But, little information is available on the shrub vegetation.

Bongers *et al.* (eds.), Nouragues. Dynamics and plant-animal interactions in a neotropical rainforest, 65–78.
2001 Kluwer Academic Publishers. Printed in the Netherlands.

In French Guiana, inselbergs are constituted by Precambrian granites only (Choubert 1974, Hurault 1963, 1967). More than a hundred sites were spotted (Descamps *et al.* 1978). The highest density is located in the southern zone of the territory and also between the upper Arataye and the middle Oyapock rivers. Data regarding the inselbergs' flora are available in the Aublet data base (Hoff *et al.* 1989) but are still fragmentary. The open areas of exposed rocks are colonised by a very discontinuous, xeric and low vegetation commonly called 'savane-roche' in French Guiana. The first vegetation studies were based on qualitative field observations on inselbergs of the SW of French Guiana (Granville & Sastre 1974, Granville 1978). A recent study describes six new herbaceous associations and three sub-associations on Guianan inselbergs (Sarthou & Villiers 1998).

The present chapter deals with the floristical, ecological and phytosociological study carried out at the Nouragues area. A classification and description of the main herbaceous and shrubby epilithic plant communities and their associated habitats on the rocky outcrops are presented.

Photo 1. Aerial view of the rock savanna vegetation. Mat communities and insular shrub units are visible on granitic outcrops. *Photo Corinne Sarthou.*

2. Study area

The fieldwork was carried out on Nouragues' inselberg (410 m above sea level), located 100 km from the coast in tropical rain forest. The Nouragues inselberg included in the Balenfois Mountains near the Arataye river basin is dome shaped (whale-back type). Physical and chemical erosion is important in determining the surface character of outcrops. Their geomorphological features include grooves, flat surfaces, steep slopes and depressions. The inselberg investigated belongs to the 'cyanobacteria type' i.e. the exposed rocks are covered by cyanobacteria, in contrast to the 'lichen type' which is characteristic for drier climatic regions (Porembski & Barthlott 1992, Büdel *et al.* 1997). The rock surfaces are colonised by cyanobacteria belonging to genera *Scytonema*, *Stigonema*, *Schizothrix*, *Glœocapsa* and *Glœothece* (Sarthou *et al.* 1995), also found colonising the bare granitic outcrops in Venezuela (Büdel *et al.* 1994). These organisms give the red-brownish or blackish colour of the inselberg except on areas with intense water flow.

The typical 'savane-roche' xerophytic vegetation covers nearly half of the total area of the inselberg. This is composed of small and numerous scattered patches of herbs and shrubs separated by open rock areas (Photo 1). Drastic local conditions are associated with that vegetation: very high fluctuations in temperature and relative humidity in amplitude and frequency occur during the day cycle. The daily temperature may range between 18-55°C and the daily humidity between 20-100% (Sarthou 1992). The temperature of the bare rock surface reaches 50°C (up to 75°C in dry season). Cyanobacteria produce organic matter with high Ca^{2+}, Mg^{2+} and K^{2+} contents which are periodically washed down by runoff water (Sarthou & Grimaldi 1992).

These local conditions constitute the primary factors in the establishment of the nutrient cycles on the rock surface. The soil formation and the resulting plant settlements depend also on the interactions within the set 'granite-water-cyanobacteria' enhanced by the extreme physical conditions.

The characteristic features of the soils are their low depth (1-20 cm for the herbaceous patches, 10-40 cm for the shrub patches), acidity (pH ranging from 4 to 5.5), coarse materials (quartz grained skeleton) and high sand content (Sarthou & Grimaldi 1992). Stagnant waters are acid (pH 4-5). Na^+, Cl^-, Si O^{2-}, NO^{3-} are the most important components. Other cations (K^+, Ca^{2+}, Mg^{2+}) have a very low concentration (< 10 μmol. l^{-1}).

3. Methods

The field work was undertaken from March to June when species richness and diversity were more apparent, especially in the herbaceous plant communities characterised by an ephemeral flora rich in species of Lentibulariaceae. Three plant units were identified according to their homogeneous physiognomy and structure: two herbaceous units and one shrub unit.

A total of 134 relevés were made on the Nouragues inselberg. Plot sizes are based on the minimum area, defined on the basis of variation in species numbers with the surface of communities occurring as relatively small patches. In order to calculate the minimum area, a species list was made for a small plot which are then enlarged and the additional species are noted. This process was continued until fewer and fewer taxa were encountered in successively larger areas. When the species area curve thus produced reaches a horizontal asymptote, the minimum area for the community has been obtained. Thus, the minimum area is fixed to 2 x 2m for one of the herbaceous communities (61 relevés, n°s 1-61), 2 x 6 m for the second one (13 relevés, n°s 62-74), and 2 x 8 m for the shrub unit (60 relevés, n°s 75-134).

For each relevé, cover-abundance data for all pteridophytes and spermatophytes were recorded according to the Braun-Blanquet approach (Braun-Blanquet 1964, Westhoff & van der Maarel 1973). Voucher specimens are deposited in the Herbarium of French Guiana (CAY).

For the numerical analysis, the cover-abundance values in the scale of Braun-Blanquet were transformed into the 1-9 ordinal scale of van der Maarel (1979). Only species present in at least three stands were included in the data analysis. Hierarchical Analysis (using Euclidian distance) and Correspondence Analysis (CA) were applied to the floristic data set using the SPADN (CISIA, Centre International de Statistique et d'Informatique Appliquées, Saint Mandé, France) program in order to identify the plant units and sub-units. Species characterised by their cover-abundance values were used as active variables (contributing to factorial axes). Passive variables, describing environmental characteristics (soil depth, shady or exposed places) were added to measure their degree of relationship with the ordination.

4. Results

4.1. FLORA

A list of taxa collected in the outcrop communities is given in the Appendix 'list of plants' . Lichens, very scarce on granitic surfaces which are mainly colonised by cyanobacteria, are included in the epiphytic communities, often on isolated shrubs. Some mosses, such as *Campylopus richardii* and *C. savannarum* only grow on the granite in the running water sites.

One hundred seventy-six species of vascular plants (18 Pteridophytes, 158 Angiosperms) representing 53 families and 112 genera were recorded in the study area. Many ferns belong to the Polypodiaceae. Among the flowering plants, Orchidaceae, Myrtaceae, Bromeliaceae, Lentibulariaceae and Cyperaceae are the families with the highest representation: 19%, 8.9%, 7.6%, 7% and 5.1% of the species, respectively. The orchids are generally found as epiphytes in the thickets except the epilithic species *Cleistes rosea, C. tenuis, Cyrtopodium andersonii, Epidendrum nocturnum, Habenaria rodriguesii* and *Phragmipedium lindleyanum*.

The bromeliad *Pitcairnia geyskesii* is the most typical plant of the granitic outcrops. Species of *Aechmea (Aechmea aquilega, A. bromeliifolia, A. melinonii*) and species of *Vriesea (Vriesea splendens, V. pleiostocha*) grow on the thin soil of thickets. *Catopsis berteroniana* and species of *Tillandsia* (*Tillandsia bulbosa, T. flexuosa*) are important elements of the epiphytic flora in the shrub unit. Among the other herbaceous and suffrutescent plants, some common species belong to other families: *Stylosanthes guianensis* and *Chamaecrista desvauxii* var. *saxatilis* (Fabaceae), *Sipanea pratensis* and *Perama hirsuta* (Rubiaceae), *Irlbachya purpurescens* and *I. alata* (Gentianaceae), *Stelestylis surinamensis* (Cyclanthaceae), *Anthurium jenmanii* (Araceae), and *Ernestia confertiflora* (Melastomataceae).

Shrubs belong to Clusiaceae (*Clusia minor, C. nemorosa*), Myrtaceae (*Myrcia saxatilis, M. guianensis, M. silvatica*), Melastomataceae (*Miconia ciliata, Ernestia granvillei*),

Rubiaceae (*Psychotria hoffmannseggiana, Rudgea crassiloba, Alibertia myrciifolia*), Erythroxylaceae (*Erythroxylum citrifolium*), and Ochnaceae (*Ouratea leblondii*).

Phanerophytes (mainly micro and nanophanerophytes), epiphytes and hemicryptophytes are the most dominant life-forms and constitute 27.2%, 24.8% and 23.7% of the total flora respectively . They are followed by chamaephytes (10.4%). Therophytes (5.8%), belonging especially to Lentibulariaceae, only occurred during short periods of the wet season, one week to 3 weeks from March to May. Lianas (8.1%) are mainly represented by Ericaceae (*Cavendishia callista, Satyria cerander, Sphyrospermum cordifolium*), Marcgraviaceae (*Souroubea guianensis, Norantea guianensis*) and Asclepiadaceae (*Matelea sastrei, Ditassa sp.*). Two species of Loranthaceae (*Phtirusa pyrifolia, Psittacanthus redactus*) are parasites on branches of *Clusia minor* in the thickets.

4.2 HERBACEOUS AND SHRUB VEGETATION

Classification and ordination

The 74 relevés belonging to the herbaceous vegetation were classified into four entities in the first step of the Hierarchical Analysis (Fig. 1a). The dendrogram assembled the four clusters in two main branches. The first group of the dendrogram included 3 clusters (A1, A2, A3) and the second included one cluster (B). From a floristical point of view the

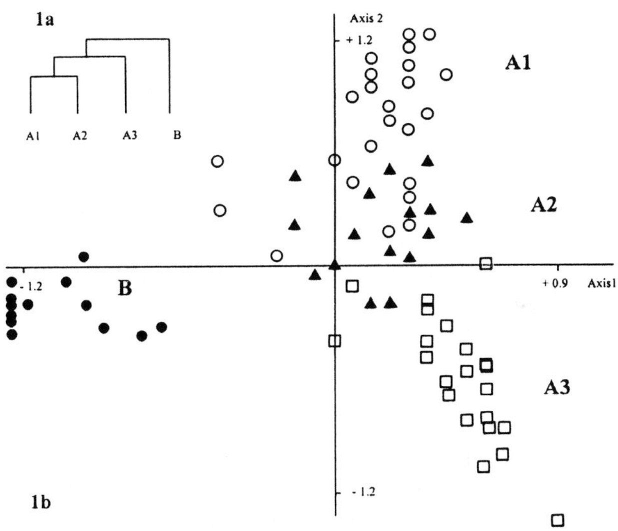

Figure 1. Herbaceous vegetation. **a.** Floristic classification dendrogram of 74 relevés and 28 taxa. **b.** Correspondence Analysis of the 74 relevés. - A. *Pitcairnia geyskesii* community and its sub-units. A1: *Utricularia subulata* and *U. amethystina* sub-unit. A2: *Stylosanthes guianensis* sub-unit. A3: *Sipanea pratensis* sub-unit. - B. *Axonopus ramosus* community.

segregation of these two groups is relatively easy to establish when considering the flowering components. Thus, the first groups are associated with the bromeliad*Pitcairnia geyskesii* and other suffrutescent species such as *Stylosanthes guianensis, Sipanea* *pratensis, Chamaecrista desvauxii* var. *saxatilis*, while the second is characterised by the grass species *Axonopus ramosus* and other herbaceous taxa such as *Cleistes rosea, Rhynchospora filiformis, R. tenuis, Panicum rivale* and *Polygala adenophora.*

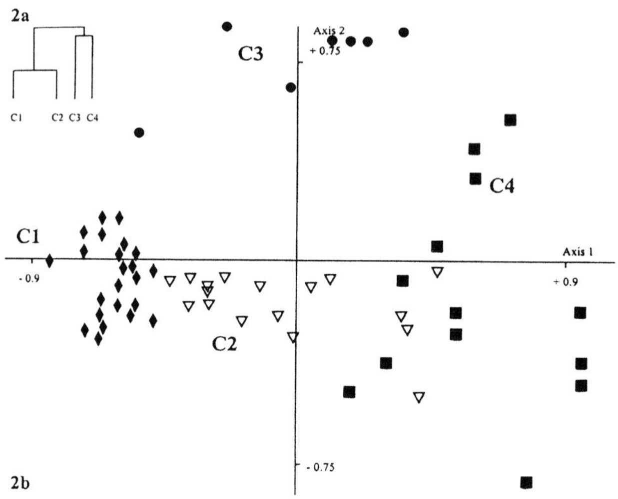

Figure 2. Shrub vegetation. **a.** Floristic classification dendrogram of 60 relevés and 56 taxa. **b.** Correspondence Analysis of the 60 relevés. C. *Clusia minor* community and its sub-units identified as successionnal stages: C1: stage 1. C2: stage 2. C3: stage 3. C4: stage 4.

The floristic relationships of these 4 clusters have been surveyed in a Correspondence Analysis where the two groups are clearly separated (Fig. 1b). The first axis can be interpreted as a soil depth gradient related with the rocky surfaces. Thus, group B relevés are restricted to the small depressions with deeper differentiated sandy soils (8-20 cm), while groups A1, A2, A3 relevés are associated with thin quartz soils (1-8 cm) and generally located on the slopes. The second axis reflects an insolation gradient and opposes group A3 relevés located near shrub thickets to those of groups A1 and A2 relevés occurred on exposed places. This gradient is coupled with a water availability gradient. Wetter conditions appeared in the positive part of axis 2 where group A1 relevés are characterised by free water during prolonged periods. These conditions enable the development of hygrophilous species of Lentibulariaceae associated to this group.

The hierarchical classification of the 60 shrub stands (Fig. 2a) resulted in four groups, each of which could be linked to a stage of a dynamical process in the shrub unit. Thus, in the ordination diagram (Fig. 2b), the first axis can be interpreted as a dynamic gradient coupled with an insolation gradient and a moisture gradient. Groups C1 and C2 relevés (40 relevés) in the negative part are characterised by a less dense open vegetation in drier conditions. Groups C3 and C4 relevés (20 relevés) are closed, dense, wetter, casting shade on the herb layer. The first groups relevés (C1-C2) mainly occurred in the centre of the outcrops and constituted isolated big thickets, while the others, and especially the group C4 relevés, come into contact with the bordering forest. We have not identified a particular ecological factor explaining the distribution of stands along the second axis.

Description of the plant units

Group A. *Pitcairnia geyskesii* community (Table 1, Photo 2). This plant community is the most widespread and important plant community on granitic outcrops of the inselberg and gives a typical aspect to the rock savanna landscape. It always forms dense patches, occurs on slopes as well as in gullies and shallow depressions. This unit, 30-50 cm tall, is physiognomically dominated by the bromeliad *Pitcairnia geyskesii*, an endemic species of the Guianan outcrops. This epilithic and xeric species builds homogeneous and dense mats on the rocky outcrops with high cover abundance. The plant community is characterised by thin quartzous soils kept in the well developed root system of the Bromeliaceae mats. The preferent species are *Utricularia subulata, U. amethystina, Rhynchospora barbata, Stylosanthes guianensis, Sipanea pratensis, Chamaecrista desvauxii* var. *saxatilis* and *Cuphea blackii.* Anemochory is the predominant way of dispersion of these species.

TABLE 1. Herbaceous vegetation: *Pitcairnia geyskesii* and *Axonopus ramosus* communities. Species abundance is listed according to a scale of + to 5 (+ = < 1% cover, 1 = 1-5 %, 2 = 6-25%, 3 = 26-50%, 4 = 51-75%, 5 = > 75%).

Relevé no	1	2	4	5	7	8	9	16	17	27	28	32	35	38	39	40	41	43	44	47	48	49	51	52	6	13	15	18	23	29	30	31	33	34	36	37	42	45	46	50	3	10	11	12	14	19	20	21	22	24	25	26	53	54	55	56	57	58	59	60	61	62	63	64	65	66	67	68	69	70	71	72	73	74
Area (m2)	4	4	4	4	4	4	4	4	4	4	4	4	4	4	4	4	4	4	4	4	4	4	4	4	4	4	4	4	4	4	4	4	4	4	4	4	4	4	4	4	4	4	4	4	4	4	4	4	4	4	4	4	4	4	4	4	4	4	4	4	4	8	8	8	8	8	8	8	8	8	8	8	8	8
Number of species	9	10	7	9	11	8	5	7	9	9	5	8	6	7	5	6	6	11	12	5	6	5	7	5	6	5	5	8	7	8	8	7	8	7	11	9	6	11	6	8	6	6	6	10	6	5	5	8	7	6	8	6	6	7	7	9	8	10	7	6	8	10	13	14	12	14	7	10	5	13	6	5	8	10
PEPINIA GEYSKESI COMMUNITY																																																																										
Pepinia geyskesii	5	5	5	3	4	4	4	3	4	4	4	4	4	4	4	4	3	3	3	3	3	4	4	4	3	4	5	4	3	4	4	4	3	3	3	3	4	2	3	3	4	4	4	4	3	4	4	3	4	4	3	3	3	3	4	4	3	4	3	3	3		+					1		+				1
Utricularia subulata and U. amethystina sub-unit																																																																										
Utricularia subulata	+	+	+		+	+	+	+	+	+	+	+		+	+	+	+	+	+	2	1	+	1	1				+										+	+	+																						+		+	+	+				+				
Utricularia amethystina	+	+			+	+		+		1	1		1			1	+	+	1	1	+	+	1	+							+							+																																				
Selaginella producta	+	+	1	+	1	1		1	+										2				+																	+																						+				+		+	+				+	
Perama hirsuta				1	1	1		2		1				+			+	+																	+																												+		+				+			+	+
Burmannia capitata			+		+		+														+		+	+																																																		
Stylosanthes guianensis sub-unit																																																																										
Stylosanthes guianensis																			2						3	2	2	2	3				2	2	2	3	2	3	2	2											1			1		2	3								+	+								
Chamaecrista desvauxii						2			+																1	+	+	+	1	1			+	+	+			1						+			+		+				1										+	+	+									+
Sipanea pratensis sub-unit																																																																										
Sipanea pratensis				2															2																						2	3	2	3	3	2	3	1	2	1	+	1		1	2	3	1	1																
Cuphea blacki		+							2										1																1	+		1			1							1				1	1	2	1	1	1	3	2	2	2	+	+			+								
Croton hostmanii																																										2	2	2								3	1	1	1	1	1	2		2	1													
Irlbachya purpurescens														+																					+			+																	2		1	1	1	2	1													
Sauvagesia aliciae																			+																															1	+	1			1	1																		
AXONOPUS RAMOSUS COMMUNITY																																																																										
Axonopus ramosus				2						2								2									1							1										1				2	1													3	3	4	4	4	5	5	5	4	5	5	3	4
Cleistes rosea	+	+							+			+						+	+																																							+				+	+	+	+		+	+	+	+	+	+		
Rhynchospora filiformis																																																															+	+	+		1	1		+	+	+		
Polygala adenophora																																																																+		+							+	
Other taxa																																																																										
Scleria cyperina	2	+	1	3	+				2	2	2	2	2	2		+		2	1	+	1	1	+		2		2	3	1	2	2	+		1	3		1	1	1	1		2	3	1	2	3	2	1	2	2	1		2	2	3	1	1	2	2	2	1	2	2	3	3	2	1	2	1	2	3	2	2	2
Utricularia hispida	+	+	+	+	1	+	1	1	+	1	1	+	1	+	2	2	2	+	1	2	1	1	2	2	1			+	+	1	1	1	+	1	1	+	+	+	2	2	+		+	+				+	+	+	+							1	1		+	+	+	+	+	+	+	+		+			+	+
Clusia minor	+			+	+					+		+	+																		+	+	+			+		+		+		+		+	+		+									+						+			+	+	+	+	+					+
Panicum rivale			+		+	1				+		1	+		1	+	2	1											1	+		1			+	+					1				+													1	1		+	+	+	+	+	+				+			+	+
Rhynchospora barbata				1	2		2	2	1			1		2	2				+						1	+		+	1		1				2	+		2	2	+	2	+	+	1	+	+		1	+														+	+										
Rhynchospora tenuis																																1	+		+	+	2																										+	+	+	+	+	1		+	+	+		+
Rhynchospora globosa																		+										2		2	1		3	1		3										2		2		2	2	3	2	2		1	3			2	4	3	3		3	+				+			2	
Ernestia confertiflora																														3	2	3					2																										+											+
Borreria capitata		1																																										1							1																							
Myrcia saxatilis																																																																		+		+		+				

Additional taxa: *Ipomoea angulata*: 13: 1. *Irlbachya alata*: 29: +; 33: +. *Matelea sastrei*: 64: +. *Paepalanthus lamarckii*: 64: +; 71: +. *Poaceae sp.*: 58: 1; 59: 3. *Actinostachys pennula*: 43: +; 70: +. *Utricularia choristotheca*: 1:+; 2: +.

Three sub-units could be defined according to specific ecological conditions.

Photo 2. Typical xeric plant community characterised by *Pitcairnia geyskesii*, an endemic bromeliad of the Guianan outcrops occurring on rock surfaces covered by Cyanobacteria. *Photo Corinne Sarthou.*

Group A1. Utricularia subulata and U. amethystina sub-unit (Table 1). This unit is dominated by the hygrophilous species, *Utricularia subulata* and *U. amethystina, Selaginella producta*, *Perama hirsuta* and *Burmannia capitata*. This plant community is found in seasonally wet microhabitats with a high insolation, on thin skeleton soils (1-2 cm), associated with algae species belonging to the Chlorophyceae (*Mougeotia capucina, M. cherockeana, Zygogonium sp.*) and to filamentous cyanobacteria (*Schizothrix lutea, Hapalosiphon fontinalis*, *Calothrix gypsophila*). The mucilaginous sheaths of these organisms are able to stock water for a long time after rainfall, maintaining therefore the moisture in the habitat. The hygrophilous dominant Lentibulariaceae and *Burmannia capitata* are therophytes. These species form a low layer giving a typical aspect to this sub-unit. Hemicryptophytes and therophytes are the predominant life forms.

Group A2. Stylosanthes guianensis sub-unit (Table 1). This sub-unit is characterised by the chamaephytes *Stylosanthes guianensis*, *Chamaecrista desvauxii* var. *saxatilis* and *Ernestia confertiflora. Rhynchospora tenuis* and *R. barbata* are also well represented. The community is found on drier slopes and shallow depressions in xeric conditions. The thin quartzous soils are often covered with humiferous crusts rich in cyanobacteria (species of *Stigonema*, *Scytonema* and *Schizotrix*) keeping a relatively high moisture. The insolation is always high. Chamaephytes and hemicryptophytes constitute the predominant life forms.

Group A3. *Sipanea pratensis* sub-unit (Table 1). This community occurs near thicket borders and is easily recognised by the suffrutescent species *Sipanea pratensis* associated with the low shrub *Croton hostmannii* (1.5-2 m high), *Irlbachya purpurescens* and *Sauvagesia aliciae*. Other species such as *Cuphea blackii* and *Rhynchospora globosa* found in other sub-units, have their better development in these ecological conditions. The plant community is more or less submitted to shade of woody thickets and grow on the deeper soils (5-8 cm) with a high sand content and coarse gravel on the surface. Chamaephytes, hemicryptophytes and isolated nanophanerophytes characterise the sub-unit physiognomy.

Group B. *Axonopus ramosus* community (Table 1, Photo 3). This herbaceous unit, 25-50 cm tall, is characterised by the tufted perennial grass species *Axonopus ramosus*. Some of the stands (relevés 67, 68, 69, 71, 72) are very homogenous and form an uniform

Photo 3. A herbaceous community dominated by the tufled perennial grass *Axonopus ramosus* restricted to shallow depressions at the summit of the inselberg. *Photo Corinne Sarthou.*

'prairial' community dominated by the *Poaceae*. This unit also includes the preferent species *Cleistes rosea*, *Rhynchospora tenuis* and *R. filiformis*. Other Cyperaceae (the dominant family with 22%) such as *Scleria cyperina*, *Rhynchospora globosa*, the Pocaeae (14%) and the Lentibulariaceae (14%) are also well represented. This plant community typically occurs in the shallow depressions at the top of the inselberg, on deeper and more differentiated sandy soils (8-20 cm) with quartzous gravels in the lower part. It is particularly well adapted to extreme environmental conditions with sudden and extreme changes in water availability: complete dessication during prolonged dry periods and temporarily flooding during the wet season. Mosses sometimes grow between the grass tufts forming a discontinuous layer. They keep moisture for a long time after rainfalls. Few hygrophilous species such as *Utricularia subulata*, *Selaginella producta*, *Perama hirsuta* and *Polygala adenophora* occur in the plant community during the rainy periods.

Hemicryptophytes (61%) are the predominant life form. Some chamaephytes (22%) including *Chamaecrista desvauxii* var. *saxatilis*, *Pitcairnia geyskesii*, *Cuphea blackii*, *Ernestia confertiflora* and *Stylosanthes guianensis* found in other plant communities, are often restricted on the peripheral rocky zones of depressions. Seedlings, plantlets or young shrubs of the phanerophytes *Clusia minor* and *Myrcia saxatilis*, already appear in this community. Anemochorous dispersion is dominant.

Group C. Clusia minor community (Table 2, Photo 4). This is another widespread plant community on the granitic outcrops of the inselberg. It represents the shrub vegetation unit of the rock savanna forming thickets, 2-8 m tall. This community occurred both in depressions and on slopes, on sandy soils (10-40 cm). They are dominated by micro and nanophanerophytes, the most common being *Clusia minor*, *Myrcia saxatilis*, *Miconia ciliata*, *Psychotria hoffmannseggianna*, *Erythroxylum citrifolium* and *Clusia nemorosa*. Some herbaceous species such as *Stelestylis surinamensis*, *Epidendrum nocturnum*, *Encyclia ionosma* and *Anthurium jenmanii* are characteristic for this plant community. The orchid *Encyclia ionosma* can also be found on rocky surfaces, isolated or in very small groups. Ornithochory is the predominant way of dispersion for the Myrtaceae species; the other taxa are mainly anemochorous species.

Photo 4. Shrub granitic vegetation: *Clusia minor* community, organised in isolated thickets on open exposed rocks of the inselberg, including predominantly evergreen and sclerophyllous shrubs. Young *Clusia minor* thicket is seen in front and an older one behind. *Photo Corinne Sarthou.*

TABLE 2. Shrub vegetation: *Clusia minor* community. Species abundance is listed according to a scale of + to 5 (+ = < 1% cover, 1 = 1-5 %, 2 = 6-25%, 3 = 26-50%, 4 = 51-75%, 5 = > 75%).

Relevé no	75	77	82	76	81	80	97	95	96	108	99	106	98	93	104	109	90	100	92	94	110	91	89	101	103	119	78	105	107	102
Area (m2)	16	16	16	16	16	16	16	16	16	16	16	16	16	16	16	16	16	16	16	16	16	16	16	16	16	16	16	16	16	16
Number of species	12	8	12	10	7	12	11	18	15	15	13	19	16	14	15	16	12	12	10	16	18	15	12	16	21	16	13	22	18	21
***CLUSIA MINOR* COMMUNITY**																														
Clusia minor	5	4	4	5	4	4	5	4	4	4	4	5	4	5	4	5	4	5	5	4	4	5	4	4	4	5	4	4	4	4
Myrcia saxatilis	2		3	3	2	3	2	3	2	3	3	3	2	2	2	2	3		2	2	2	2	2	2	3	2	3	2	2	3
Miconia ciliata	1			1			1	1	1	1	1		1	2	1	2		2	2	1	2	1	2	2	1	2	1	2	1	1
Psychotria hoffmannseggiana	+	+	3	2	+	2	1	2	1	1		1	1		1		1	1			1			2	2	+	1		2	2
Stelestylis surinamensis	1	1	1					1	1		1	1	1	1		1	1			1	1	1	1	1	1	3	+	1	1	2
Clusia nemorosa	1	1	2	1	1	2	2	2	2	3	2	2	2	1	2	2	1	1	1	2	3	1	2	2	2	2	1	2	2	2
Erythroxylum citrifolium	1	3	1	1	3	3	2	2	2	2	2	1	2		2	2	1	1		2	2	2	1	1	2	1	2	2	2	1
Epidendrum nocturnum								+	+	+	+	+	+	+	+	+		+		+	+	+	+	+	+	+	+	+	+	+
Anthurium jenmanii	3	+		1		1		1	+	+	+	+	+	1											+	+	+	+		+
Myrcia guianensis					1		1	1	1		1	1	1	1	1	1			1	1	1			1	1			1	1	1
Eriotheca surinamensis						2	1	2	1	+	2					+		1		1	+				1			+	2	
Encyclia ionosma								+	+			+	+	+	+	+		+		+	+	+			+	+		+	+	+
Stage 1																														
Pitcairnia geyskesii	2	2	2	3	2	1	2	2	2	1	2	1	2	2	+	1	2	2	2	2	1	2	2	+	1		+	1	1	1
Ernestia granvillei			+	1		1		2		+	2	2		1	2	2	2	2	1	2	2	2	2		1		+	2		1
Irlbachya purpurescens								+	+	+		+	+	+	+		+	+	+	+	+	+	+	+	+	+		+		
Aechmea aquilega																1	1			1	1					1				
Stage 2																														
Satyria cerander															+										+	+		+	+	+
Myrcia silvatica										1					+	+					+				1			1	+	1
Cybianthus guianensis			+									+														1		+		
Guapira eggersiana										+															+			+		+
Vriesea pleiosticha															+				1						1					1
Poaceae sp.																														
Ternstroemia dentata			+														+					+							+	
Stage 3																														
Croton tafelbergicus																														
Rudgea crassiloba																														
Albertia myrciifolia																														
Tabebuia capitata																														
Rhynchospora cephalotes																														
Macrocentrum cristatum																														
Sauvagesia aliciae																														
Voyria aphylla																														
Guzmannia lingulata																														
Burmannia sp.																														
Calyptranthes lepida																														
Clusia grandiflora																														
Stage 4																														
Pariana campestris																														
Sphyrospermum cordifolium																														
Vriesea splendens																														
Myrcia fallax																											1	1		
Eugenia ramiflora																														
Myrcia sp. 1																														
Ouratea leblondii																														
Eugenia florida																														
Calathea squarrosa																														
Myrcia sp.2																														
Roupala montana																														
Other taxa																														
Scleria cyperina	+		1	+		1	1	1	+	+	1	+	+	1		1	+	1	+	1	1	1	1	1		+	+	+	1	+
Axonopus ramosus												+	1										1	1				+	1	+
Psittacanthus redactus	+					1		+				1	+								+									
Apteria aphylla																														
Matelea sastrei																										+				+
Humiria basalmifera												+												+	1					
Actinostachys pennula												+		+								+						+	+	
Coussapoa latifolia	1	1	1																											
Phthirusa pyrifolia							+	+																						+
Cavendishia callista																								+						

Relevé no	132	121	126	111	123	128	120	122	125	114	79	88	85	83	86	87	84	127	117	116	113	134	115	118	129	130	133	131	112	124
Area (m2)	16	16	16	16	16	16	16	16	16	16	16	16	16	16	16	16	16	16	16	16	16	16	16	16	16	16	16	16	16	16
Number of species	16	15	22	20	18	24	21	23	22	20	12	13	18	17	18	13	20	17	16	23	18	17	21	17	27	34	25	19	19	21
***CLUSIA MINOR* COMMUNITY**																														
Clusia minor	3	4	3	4	4	5	3	4	4	4	5	5	5	3	4	3	4	2	3	4	4	4	4	4	4	3	3	2	3	4
Myrcia saxatilis	3	4	3	2	3	2	3		3	2	3	4	4	2	4	4	2	2	1	+	2	3	3	3	4	4	4	2	1	2
Miconia ciliata	1	1	1	+	2	2	2	2	2	2				+			+	+	+	1	+	+	1	2	2	1	1	1	+	2
Psychotria hoffmannseggiana	+	+	1	1	+	1	1	1	2	+	2	4	2	3	3	+	4		+	+	+		+	+	2	+	1		+	2
Stelestylis surinamensis	2	2	2	3	2	2	3	+	2						1	2		4	2	1	3	3	4	3	3	2	2	2	2	+
Clusia nemorosa	1		1	1	1		+	2		2	1	2	2						+	3		1	2	1	1					
Erythroxylum citrifolium	1	1	2		1	2			1		2				1				+	2		+	2		1	1	+	1		
Epidendrum nocturnum		+	+	1	+	+	+	+	+										+	+	+	+			+	+	+			+
Anthurium jenmanii	+		+		+	+		+	+	+	3	1	1	2	1	2	+				+				+	+				+
Myrcia guianensis	1			1	1	1		1			1	2	+	5	1	1	5									1	2			
Eriotheca surinamensis						1	1	2				1	1	2	+		+	1			1		2	2	1	1	1	1		
Encyclia ionosma		1	+	1	1	2	2	2	3	3							1	1	2	2	+	2	3	+	1	2				3
Stage 1																														
Pitcairnia geyskesii		+				+		1	+		1	1	2																	
Ernestia granvillei		2	2						+											+										
Irlbachya purpurescens	+								+	+																				+
Aechmea aquilega			1			2																								2
Stage 2																														
Satyria cerander	+		+	+	1	+	1	+	+	2								+	+	+	+		+	+		+		2	2	1
Myrcia silvatica			1	1		1	2		+	1				+					2	+		1	+		1	+		2		1
Cybianthus guianensis																														
Guapira eggersiana					2		1						1	1		+	2								1	1		2		
Vriesea pleiosticha				1	2		1			1								2	2		+					1				1
Poaceae sp.	+		+			+	+											+		+						+	+	+		
Ternstroemia dentata				1					+	+								+		+						+	+			+
Stage 3																														
Croton tafelbergicus			2	1				1		1		1	1	1	1	1		2		1	2	1	1	1	1	1	1			
Rudgea crassiloba						1		1						1	+	3	+	+				+			1	1	3	3		
Albertia myrciifolia							+			1			+	+	+										1			1	+	1
Tabebuia capitata																1	1									1	1		1	
Rhynchospora cephalotes													+	1									1		+	+	1			
Macrocentrum cristatum												+	+				2							+	+	+	+			
Sauvagesia aliciae				+				+					+				2					1		2	1	2	1			
Voyria aphylla		+	+	+	+		+		+			+	+	+	+	+				+	+					+		+		+
Guzmannia lingulata											+		+		+	2	+						+							
Burmannia sp.														+	+		2									+				
Calyptranthes lepida														+	+	1														
Clusia grandiflora												2			1		1													
Stage 4																														
Pariana campestris						+	+			+								+		3			2		1	+	+	1	1	+
Sphyrospermum cordifolium	+					+	+		+	+								+	+	+	+							+	1	+
Vriesea splendens								+										2			1	2		2				3	2	
Myrcia fallax					+													1		+	1						2			
Eugenia ramiflora						2																1	1				1	3	3	1
Myrcia sp. 1								1		1																+		+	1	3
Ouratea leblondii																				+			1		1		2			
Eugenia florida						1																			1	1				
Calathea squarrosa								+										+			+			+					+	
Myrcia sp.2								1																	1			3		
Roupala montana											1																		+	2
Other taxa																														
Scleria cyperina	2	2	2	+	+	+		1		2	+	1	2				+		+		+	1	+	2		2	2			
Axonopus ramosus		+	1	1	2		+	+											4	+		4	2	+		+				
Psittacanthus redactus				+			1			+	+						1													
Apteria aphylla		+	+																					+			+			
Matelea sastrei	+	+	+					+																	+	+	1			
Humiria basalmifera			1			1														1										
Actinostachys pennula	+			+																										
Coussapoa latifolia																														
Phthirusa pyrifolia																														
Cavendishia callista									+																				1	

Additional taxa: *Aechmea melinonii*: 94: 1; 115: +. *Balizia pedicellaris*: 83: +; 84: 2. *Croton hostmannii*: 80: 1. *Cuphea blackii*: 134: 1. *Myrciaricia floribunda*: 129: 1; 130: 1. *Eugenia marowynensis*: 86: 1. *Ichnanthus nemoralis*: 130: +. *Inga umbellifera*: 112: +. *Miconia holosericea*: 125: +. *Myrcia citrifolia*: 122: +; 128: 1.

Ouratea candollei: 115: 1; 133: 2. *Phragmipedium lindleyanum*: 101: +; 91: +. *Piptocoma schomburgkii*: 103: +; 109: 1. *Polygala spectabilis*: 125: +. *Scleria secans*: 112: +. *Smilax sp.*: 125: +. *Voyriella parviflora*: 129: +.

The four groups identified reflect different dynamic changes in the development of shrub thickets.

Group C1 (Table 2, Stage 1). This group is composed of well isolated thickets dominated by the shrubs *Ernestia granvillei* and *Erythoxylum citrifolium*. These shrubs have greater cover-abundance values than in the other relevé groups and especially in C3 and C4. Herbaceous cover by *Pitcairnia geyskesii* is also higher in these plots. *Irlbachya purpurescens* and *Actinostachys pennula* encountered in the herbaceous and suffrutescent plant communities find in this community apparently enough light to grow.

Group C2 (Table 2, Stage 2). In these isolated thickets, shrubs such as *Myrcia silvatica, Cybianthus guianensis subsp. pseudoicacoreus, Guapira eggersiana* and the ligneous liana *Satyria cerander* occur.

Group C3 (Table 2, Stage 3). This sub-unit is characterised by thickets located not far from the forest border. The differential shrub species are *Croton tafelbergicus, Rudgea crassiloba, Alibertia myrciifolia, Tabebuia capitata* and *Calyptrantes lepida.* The herbaceous plant species such as *Macrocentrum cristatum, Guzmannia lingulata* and *Rhynchospora cephalotes*, which need moisture and shade, could develop here.

Group C4 (Table 2, Stage 4). This group includes relevés characterised by the appearance of shrubs and trees considered as typical elements of the low forest rich in Myrtaceae species (Larpin 1993, Chapter 5): *Eugenia ramiflora, E. florida, Myrciaria floribunda, Myrcia fallax, M. sp.1, M. sp.2* but also *Ouratea candollei*, *Roupala montana*, *Balizia pedicellaris*, and *M. citrifolia*. These new components are generally represented by juveniles and their occurrence can be explained by the contact of these thickets with the low forest. *Croton tafelbergicus* and *Rudgea crassiloba* are still well represented. The herbaceous layer is mainly compound of *Pariana campestris* and *Vriesea splendens.* The ecological conditions - more closed vegetation, casting shade on the herb layer, more moist - induce the death of the bromeliad *Pitcairnia geyskesii* prooved by disintegrating rhizomes on the thicket's soil. *Ernestia granvillei*, *Irlbachya purpurescens* and *Anthurium jenmanii* also disappear. The cover abundance values of *Psychotria hoffmannseggiana* and *Erythroxylum citrifolium* decrease while those of *Myrcia silvatica*, *Stelestylis surinamensis, Encyclia ionosma,* and *Vriesea splendens* increase.

5. Discussion

The floristically poor diversity and physiognomical simplicity of these plant communities produce an uniform landscape typical of French Guianan inselbergs. Many species encountered on the Nouragues inselberg show a broad distribution area, a feature also observed on Bolivian outcrops (Ibisch *et al.* 1995). The species recorded on the studied outcrops have also been generally collected on other French Guianan inselbergs (data base 'Aublet', Hoff *et al.* 1989). These taxa also include plants occurring as saxicole, epiphyte or terrestrial in various other types of vegetation: coastal savannas, ruderal habitats, river rocks, forest canopy, white sandy forest, secondary forest.

The number of species restricted to inselbergs, such as *Pitcairnia geyskesii*, *Axonopus ramosus*, *Ernestia confertiflora*, *E. granvillei*, *Chamaecrista desvauxii* var. *saxatilis*, *Sauvagesia aliciae*, *Cyrtopodium andersonii*, *Habenaria rodriguesii* is low in contrast to Sastre's (1977) estimate of 55% of species being confined to this habitat in the Guianas.

However, a more detailed biogeographical analysis should be carried out to assess the distribution of each species and their links with other regions where inselbergs occur.

In contrast, Venezuelan and Brazilian inselbergs harbour a large number of endemics with restricted distributions (Barthlott *et al.* 1993, Porembski *et al.* 1998). The high species richness of vegetation on East Brazilian inselbergs is possibly a consequence of a large species-pool (Porembski *et al.* 1998). But the processes that regulate regional and local diversity are not fully clear. It is assumed that historical (i.e. long term stability) and biotic (i.e. evolutionary interchange of taxa between canopies and rock outcrops followed by differentiation of local populations) conditions have promoted high rates of speciation and their coexistence in isolated habitats (Porembski *et al.* 1998).

Studies on the flora of South American granitic outcrops (Brazil, Bolivia, French Guiana, Venezuela) have already underlined the predominance of families such as Bromeliaceae, Orchidaceae, Lentibulariaceae, Cyperaceae and Eriocaulaceae (Granville 1978, Barthlott *et al.* 1993, Ibisch *et al.* 1995). However, Velloziaceae, one of the most typical families of the inselbergs in Brazil, Bolivia, Africa and Madagascar (Barthlott *et al.* 1993, Ibisch *et al.* 1995) is lacking in French Guiana. Porembski *et al.* (1998) interpret the distribution pattern of this family as indicative of an already Gondwanian presence on rock outcrops. Likewise, Cactaceae, another frequent element of inselberg flora of Venezuela, Brazil and Bolivia is not found here. However, scarce species have been encountered such as *Hylocereus triangularis* and *Epiphyllum phyllanthus* on Roche Dachine inselberg (Sarthou & Villiers 1998) and *Cereus hexagonus* and *Rhipsalis baccifera* on Tumuc Humac and Roche Koutou inselbergs (data base 'Aublet', Hoff *et al.* 1989). Other families occurring on Venezuelan inselbergs such as Poaceae, Rubiaceae, Melastomataceae, Gentianaceae, Clusiaceae, Marcgraviaceae, Bignoniaceae, Bombacaceae (Barthlott *et al.* 1993) are well represented on the studied outcrops, especially in the shrub community.

Genera also recorded on Bolivian and Venezuelan inselbergs (Barthlott *et al.* 1993, Ibisch *et al.* 1995) are *Ananas, Pitcairnia, Tillandsia, Encyclia, Anthurium, Philodendron, Rhynchospora, Norantea, Croton, Irlbachya, Tabebuia, Vriesea, Pleurothallis* and *Clusia.*

Many species colonising rock outcrops have developed succulent characteristics and strategies of vegetative propagation yielding clonal populations as a response to the selective pressure involved by xeric conditions. But some other life's style such as homorhizy (Bolivia), fire resistance (Bolivia, Venezuela), tubers (Venezuela) or poikilohydry (desiccation tolerance typical of Velloziaceae) are not found here.

The vegetation of granitic outcrops is represented by three main plant communities: *Pitcairnia geyskesii, Axonopus ramosus* and *Clusia minor* communities. The two first units have been recently described as two associations characteristic of the Guianan inselbergs (Sarthou & Villiers 1998). These communities form a mosaic in which the herbaceous *Axonopus ramosus* unit occupies wet and sunny depressions while the suffrutescent *Pitcairnia geyskesii* unit is always present at the drier places such as slopes. *Clusia minor* thickets develop both in depressions and on slopes. The *Pitcairnia geyskesii* community builds dense mats. Other species of Bromeliaceae are already known to form mats like *Pitcairnia bulbosa, P. armata* and *P. pruinosa* in Venezuela (Barthlott *et al.* 1993), *Ananas ananassoides* and *Deuterocohnia longipetala* in Bolivia (Ibisch *et al.* 1995) and *Alcantarea spp.* and *Encholirium gracile* in Brazil (Porembski *et al.* 1998). However, the small mats of Selaginellaceae, Poaceae or Orchidaceae and the large mats of Velloziaceae characteristic of Bolivian and Brazilian outcrops are lacking on the Nouragues' inselberg.

The shrub vegetation dominated by *Clusia minor* shows affinities with the complex 'shrub mats' described for Bolivian outcrops. They consist of characteristic woody species mainly belonging to Clusiaceae, Myrtaceae, Erythroxylaceae, Melastomataceae and Rubiaceae. They build well isolated thickets in places preoccupied by *Pitcairnia geyskesii* or *Axonopus ramosus* communities.

It seems that physiognomically similar vegetation units are present on South American inselbergs. The floristic species composition, however, generally is different although several genus and family are often in common.

Similarities also exist with African inselbergs. The *Axonopus ramosus* community, for instance, shows affinities with the vegetation in shallow depressions on Ivorian granite inselbergs (Porembski *et al.* 1995). Both communities are characterised by their extreme environmental conditions (complete dessication during long periods and temporary flooding), acid and sandy soils with a very low cation content (Sarthou & Grimaldi 1992, Porembski *et al.* 1995). In both, Poaceae and Cyperaceae are the predominant families. However, therophytes (65.2%) followed by geophytes (15.2%) and hemicryptophytes (13.6%) are the main life-form in Ivory Coast, while hemicryptophytes (61%) and chamaephytes (22%) are predominant on the Nouragues inselberg. The ephemeral flush vegetation described by Dörrstock *et al.* (1996) from Ivory Coast as a seasonal low-grown herb community, and also found on inselbergs of Guinea (Porembski *et al.* 1994) and Malawi (Porembski, 1996), can be found only in part on the Nouragues outcrops. Indeed, Lentibulariaceae (*Utricularia subulata, U. amethystina, U. choristotheca*) and Burmanniaceae (*Burmannia capitata*), which are characteristic elements of one of the *Pitcairnia geyskesii* sub-units, constitute an ephemeral flora during the rainy season.

All these peculiar vegetation types on the Nouragues inselberg are clearly a very interesting and attractive element in an area dominated by high rain forest. As 'terrestrial islands', inselbergs could be also suitable models for studying questions of island ecology.

Acknowledgements

This research was supported by a grant from 'Ministère de l'Environnement et de la Recherche' for my doctorate thesis achieved in the Laboratoire d'Ecologie générale CNRS-MNHN URA 1183 (presently UMR 8571). I am grateful to J.F. Villiers and J.P. Gasc for their field assistance and their helpful suggestions. I also thank M.C. Boisselier which gave me access and help to use the computer program for data processing.

CHAPTER 7

PALAEOCLIMATES AND THEIR CONSEQUENCES ON FOREST COMPOSITION

Pierre Charles-Dominique, Patrick Blanc, Denis Larpin, Marie-Pierre Ledru, Bernard Riéra, Thierry Rosique, Corinne Sarthou, Michel Servant and Christophe Tardy

1. Introduction

The analysis of most plant populations in natural forests clearly retraces the mechanisms of sylvigenesis, based on the occurrence of small-scale incidents - in particular treefall gaps - more or less regularly distributed in time and space (van der Meer *et al.* chapter 24). However, a number of 'anomalies' detected in the population structure or in the distribution of some species cannot be adequately explained by the internal dynamic processes which occur on the century time scale. Plants react at different speeds to perturbations, and major events, even very ancient ones such as the perturbations recorded in the sediments, have probably also left a durable mark in the present organisation of the vegetation. We can therefore hypothesise that relatively ancient events, which would have occurred on a much larger scale than treefall gaps, would have caused large modifications of the forest ecosystems and left long-lasting tell-tale signs.

This article constitutes a brief synthesis of a multidisciplinary approach of the forest ecosystem viewed on a several thousand years scale.

2. The past recorded in the sediments

2.1 PALAEOFIRES

The charcoal layers found in the soil are the most obvious indicators of palaeofires (Soubies 1980, Saldarriaga 1994, Tardy 1998). In the hyperhumid forest zone of French Guiana (Boyé *et al.* 1979, Puig *et al.* 1981), annual rainfall exceeds 3000 mm distributed over 280 days. In the less humid areas of Amazonia, extensive forest fires can develop during exceptionally severe dry seasons - like the fires of the Roraïma in 1998, or those of 1988, when 80,000 km² were destroyed (J. Ronchail personal communication) - but in French Guiana no-one has such an incident in memory. The traditional agriculture, like elsewhere in tropical forests, is based on slash-and-burn techniques which proceed by the felling of trees on a plot and by the on-the-spot burning of the woody debris at the end of the following dry season. Tardy (1998) shows that charcoal layers are generally independent of the levels of human occupation. The presence of Man in America is attested since 30,000

Bongers *et al.* (eds.), Nouragues. Dynamics and plant-animal interactions in a neotropical rainforest, 79–87.
2001 Kluwer Academic Publishers. Printed in the Netherlands.

BP at least, and agriculture since 6,000 BP (Grand Atlas de l'Archéologie 1990, Lorenzo 1994,Vilhena Vialou *et al.* 1994, Behling 1996, Roosvelt *et al.* 1996), but evidence for tropical rain forest agriculture is only found since 2,500 BP in Meso-America (Linares 1984). Saldiarraga (1994), based on indirect proofs, estimates the agriculture in Colombia (areas of Amazon and Rio Negro) to 4,000 BP. However, in French Guiana, intensive archaeological searches undertaken in the Petit Saut area over 275 archaeological sites (Puaux 1993, Tardy 1998) have not yielded any remains anterior to 2,000 BP, while palaeofires appeared a long time before this, during the periods 10,000-8,000 BP, 6,000-4,000 BP and 2,000 BP to present time.

The fact that these fires, whether they are of natural or human origins, have been able to develop in hyperhumid forests, suggests that the dry seasons and the resulting water losses had then been extremely severe.

2.2 EROSION UNDER FOREST COVER

Today, the physical erosion of the forest floor is minimal under primary forest cover. The thick vegetation considerably reduces the impact of heavy rains. Moreover, the forest floor is entirely colonised by a dense mat of surface roots which 'filter' particles. During heavy rains, the flow of small streams and rivulets rises sharply but very small amounts of solid matter are actually torn from the forest floor, and the water flow merely reorganises ancient sediments. Large-size elements are never driven from the forest floor into the stream beds. However, recent investigations (Charles-Dominique *et al.* 1998, Charles-Dominique & Servant 2000) along the course of the Nouragues River have uncovered remains of alluvial terraces that are quite unlikely under the present-time vegetation cover and which could only have arisen during periods of intensive erosion (Fig. 1). The lower terrace (1 m

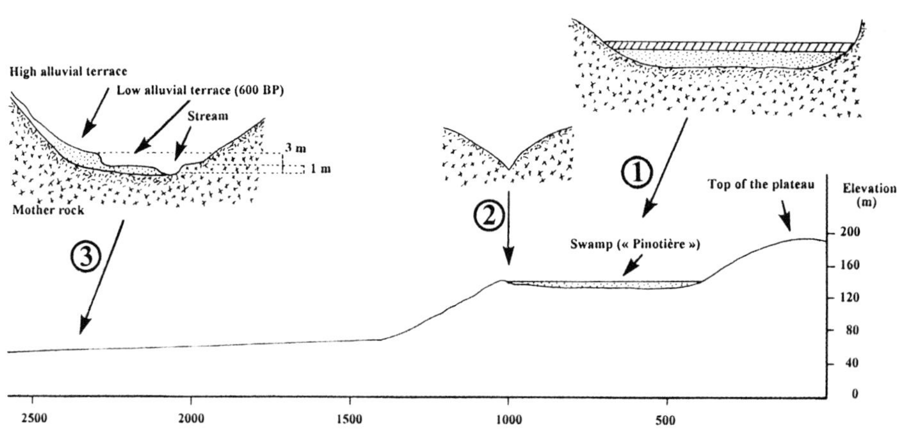

Figure 1. Longitudinal profile of the pina swamp, and transversal profile of the Nouragues stream at the levels 1, 2 and 3.

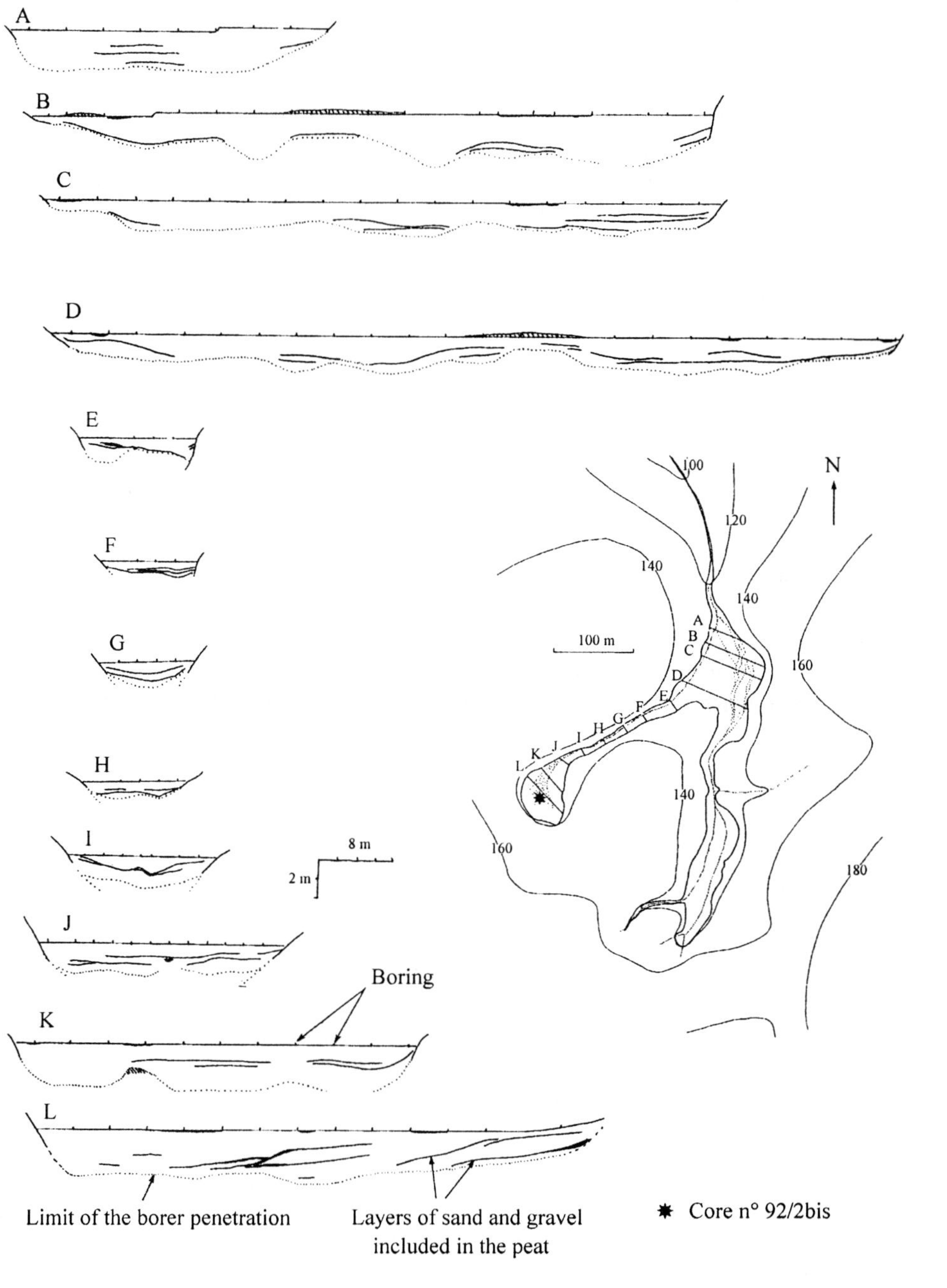

Figure 2: Transversal profiles of the pina swamp (*cf.* level 1 of Fig. 1) showing the layers of sand and gravels included in the peat.

high) sits on a bed of clay and sand containing charcoal fragments, seeds and dead leaves. The examination of the seeds reveals a great abundance of pioneer species: *Henriettea spp*, *Miconia spp*, Melastomatacea, *Cecropia obtusa*, *C. sciadophylla*, Cecropiaceae, *Goupia glabra*, Celastraceae, etc. These taxa are characteristic of important disturbances of the forest cover at the time.

Ancient tree stumps still rooted in place were discovered on the border of and under this terrace, during the excavation of the river banks on 80cm width. The carbon datings of these woody remains are difficult to interpret: 810 ± 60 years BP, 1250 ± 50 years BP, 1390 ± 40 years BP, 2380 ± 50 years BP and 4310 ± 50 years BP. Successive events may have preserved these stumps from decomposition by successively embedding them in silt and uncovering them several times. The latest sedimentation event would correspond to the formation of the lower terrace and would be contemporary with the dead leaves deposit, dated 530 ± 70 years BP, which is undoubtedly too fragile to have undergone several reorganisations. A higher (3 m) and older terrace has also been discovered but has not been dated so far.

Geomorphological and sedimentological studies recently undertaken on the lower terrace completed these data and made it possible to identify several major trends of the morphogenetic evolution of the Nouragues stream catchment area throughout the Holocene. The ^{14}C datings of several charcoal debris and organic matter found in the deposits will soon yield a precise chronology of these events. Recurring acute erosion phases were reconstructed, materialised on one hand by an increased input of detritic matter in the valleys (coarse and sandy alluvia) and, on the other hand, by a severe destabilisation of the slopes (mass landslides and colluvial deposits). Such morphogenetic events seem clearly correlated with disturbances of the forest ecosystem. The episode of superficial scouring for which evidence can be found on the right bank of the Nouragues stream necessarily implies a significant degradation of the vegetation cover (palaeofire?) at the time. The event began around 530 ± 70 years BP (^{14}C dating undertaken on the above-cited lower terrace of the Nouragues stream). In two of the Nouragues tributaries, the Cascade and Kwak streams, the presence of sandy hydromorphic clays and organic clays interbedded in the alluvial sands attest a radically different orientation of the morphogenesis. It signals one to several episodes of low sedimentary activity linked with the development of poorly drained habitats: wandering meanders depositing their load of suspended matter in floodplains probably covered by dense hygromorphic vegetation. No such processes can be found at work today in either catchment area: on the contrary, evidence supports that a decisive reversal of the dynamic trend has taken place since, reactivating the alluvial water channels. On the other hand, these once-ruling processes share some characteristics with the mechanisms governing sedimentation in present-time pina palms alluvial basins (Rosique*et al.* 2000).

The study of a pina palm swamp, a little further upstream and lodged in a plateau depression, also displays traces of several periods of intensive erosion (Fig. 2). In this case, layers of sand and gravels a few centimetres thick have been found included within clayey-organic sediments or peat. The reconstitution of the soil profiles reveals up to three such layers one above the other, with a slight slope towards the centre of the swamp, suggesting transport of material by runoff. One of these layers, with its upper limit at –85 cm, has been dated just under 1,570 BP. Another, located between -131 and –140 cm (Ledru *et al.* 1997), would be about 3,000 years old. In order to explain such alluvial deposits, knowing that the

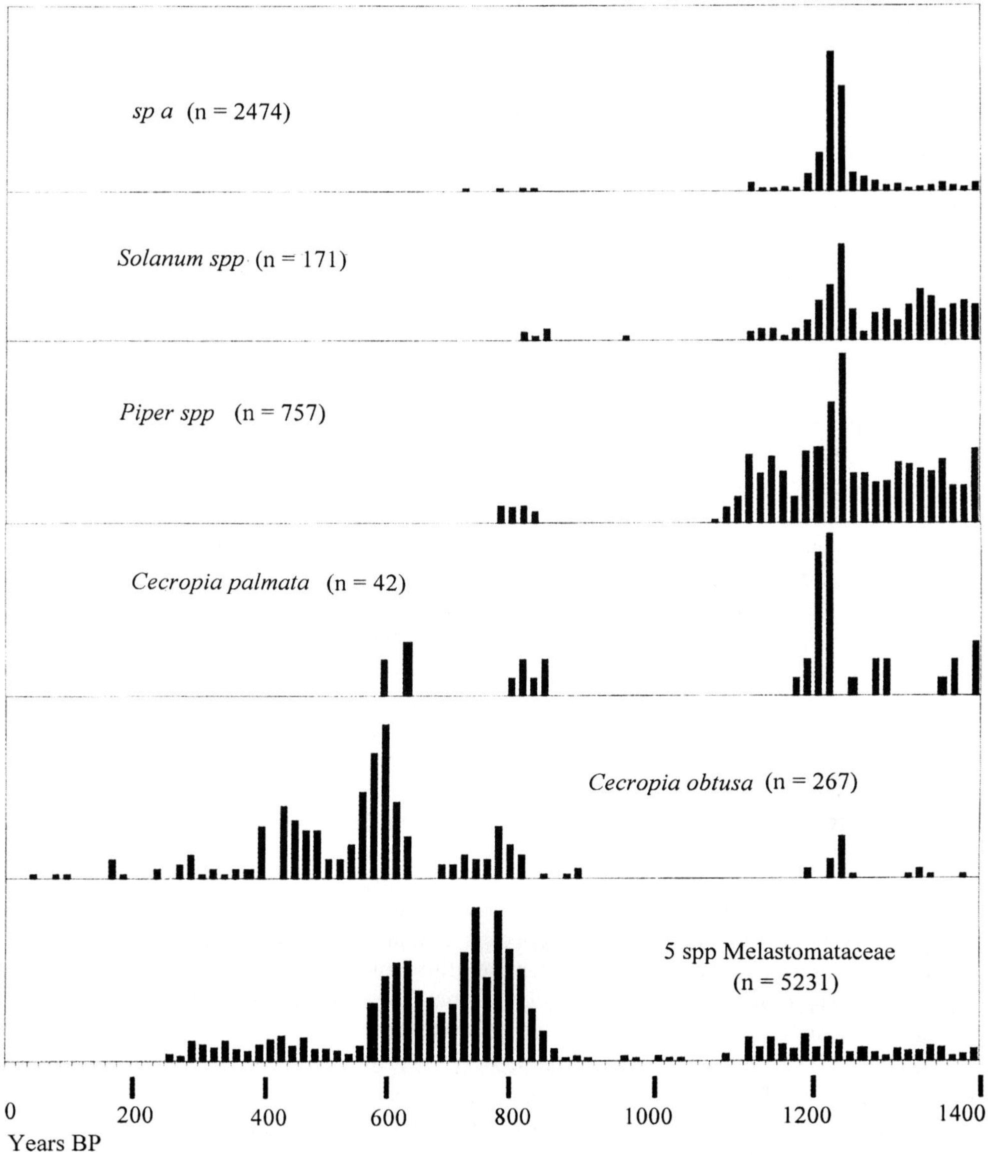

Figure 3a. Seed analysis of the peatbog core (in number of seeds).

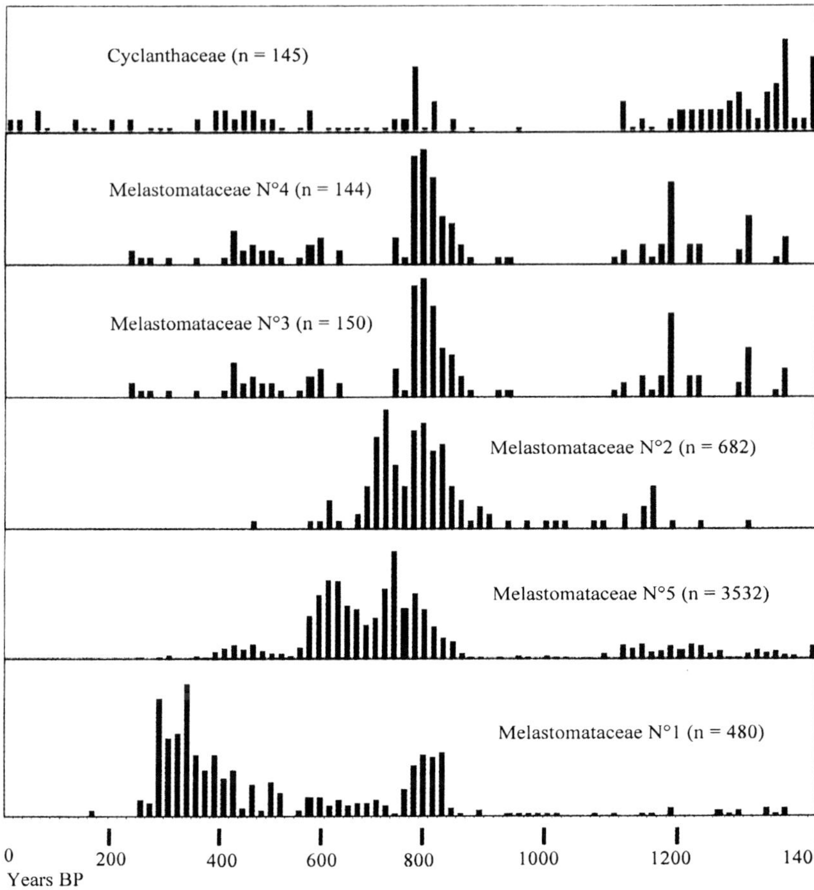

Figure 3b. Seed analysis of the peatbog core (in number of seeds).

swamp does not receive at present any coarse particles during heavy rains, one must consider massive destructions of the forest cover which would have induced intensive soil erosion over large areas. The same goes for the alluvions accumulated further down along the stream bed. Such erosion/alluvion cannot be explained outside large-scale deforestations, such as palaeofires may have generated.

2.3 THE PAST VEGETATION

Past vegetation was studied by analysing the pollen (Ledru *et al.* 1997) and seeds (Charles-Dominique & Servant 2000) contained in core samples collected from a pina swamp. The peatbog cores go back 3,000 years BP for pollen and 1,400 years BP for seeds (Fig. 3). Overall, between 3,000 and 2,000 BP, the forest appears to have been more humid than today, with a higher taxonomic diversity (Ledru *et al.* 1997). Between 1,800 and 1,200 BP, the analyses suggest a marked drying-up and the opening of large areas favourable to pioneer species, in particular *Cecropia, Solanum,* and *Piper*, this phase peaking around 1,600-1,300 BP. Between 1,200 and 900 BP, the forest appears similar to what it is now, but between 900 and 600 BP new disturbances are indicated by the development of pioneer plant assemblages (in particular the family Melastomataceae). These subsided from 600 to 300 BP, and today's pollen and seed assemblages appeared around 300 BP (Fig 3a,b).

The appearance of pioneer species during several consecutive centuries suggest that brief disturbances must have occurred every 10-30 years during each of these periods. Indeed, the observation of forest recovery in a 'modern' gap shows that the short-lived pioneer tree species are progressively replaced within a few decades by mature forest species. Only repeated incidents can therefore explain the maintenance of a widespread secondary vegetation for several centuries.

No traces of herbs such as Poaceae and Cyperaceae have been found in the pollen profile to suggest the occurrence of open vegetation types such as woodland-savannas (Ledru *et al.* 1997). Moreover, histological analyses of charcoal remains (Tardy 1998) have always yielded some forest species ever since 13,000 BP. However, tree species diversity is much lower during the two more ancient periods (8,000-7,000 BP and 6,000-4,000 BP) - its specific composition recalling a semi deciduous forest - than from 3,000 BP onwards (Tardy 1998).

3. Traces of the past in present-day ecosystems

3.1 THE VEGETATION TYPES INCLUDED IN THE MAIN FOREST BLOCK

The open vegetation found on rocky outcrops - locally called 'rock-savannas' - and the low forests which surround it only occur on limited surface areas included within the main forest block (c.a. the primary forest extending from the coast to the inland). Most of these patches are found around inselbergs, in the south of French Guiana, between 100 and 300 km from the coast. Two other vegetation types, the liana forests and the bamboo thickets - or 'cambrouzes' - constitute other types of inclusions, the origins and dynamics of which are difficult to explain in the context of present-time sylvigenetic mechanisms. Although an ancient anthropic origin has been proposed in the case of the cambrouzes (Gasc & Sastre 1978), our present state of knowledge on their dynamics is still insufficient. Information can mainly be gathered from the savannas and low forests which have been intensively studied at Nouragues (Larpin Chapter 5, Sarthou Chapter 6).

Overall, in the course of the present relatively humid period, the forest is slowly expanding and gaining over the rock-savannas on its borders. The recovery of the forest is nonetheless extremely slow and even subject to reversals on steep slopes where water runoff is very important.

It is difficult to estimate the age of these open vegetation types, but they are probably very old (as they harbour distinctive plant species) and have undoubtedly suffered variations in surface area over time. Glacial episodes must have determined their maximum extension by inducing severe erosion of the forest soils on much larger areas than today. After the post-glacial forest recovery, more recent phases of erosion probably coincided with dry periods and forest fires, and must have periodically given a new development to these savannas, which would otherwise have disappeared altogether since the end of the latest glacial episode, absorbed by the surrounding forest.

The low forests - sometimes termed 'transition forests'- constitute an original vegetation type generally found on the periphery of rock-savannas. They occupy soils with drier edaphic conditions than in the surrounding high forest (Larpin 1993). The study of their dynamics has highlighted the importance of bird dispersal of seeds in the present colonisation phase of the rock-savannas by this vegetation (Larpin 1993, Théry & Larpin 1993). This present tendency is punctually counterbalanced by perturbations occurring on the ecotone, which are more favourable to the development of the rock-savanna vegetation. Similarly, the surrounding high forest tends to slowly gain over the low forest on its outskirts. Low forests display an original floristic composition: almost 200 plant species are specific to this vegetation type at Nouragues. This clearly indicates that they constitute a distinctive ecosystem as opposed to a 'transition' gradient. These types of forest, well adapted to a number of water constraints, may have known a greater extent in the past and their present distribution would thus correspond to relict patches, like the savannas included within the border zones of the great forest blocks.

3.2 ZOOCHORY, AUTOCHORY AND PLANT DISTRIBUTION

Many plant species in French Guiana disperse their seeds by zoochory, i.e. by means of fruit-eating animals which carry their seeds over a certain distance. The present floristic composition suggests frequent small-scale forest perturbations, as most plant species are capable of quickly recolonising lost ground. A small number of slow-spreading trees exist nonetheless which are not dispersed by animals (autochorous species). Individuals of these species are frequently patchily distributed, a pattern which again cannot be explained solely by today's mechanisms of regeneration. A favoured hypothesis is that these tree species have been eliminated from large areas during the latest palaeofires and that they are now in the process of recolonising from 'microrefuge' patches which these ancient fires have not reached (Riéra & Joly 1996). Tree mapping at Nouragues (85 ha so far) has outlined possible small-scale refuge patches around swamps and springs. More detailed studies on *Eperua falcata* (Caesalpiniaceae) show that the older individuals are situated in the centre of the patches, while the younger are distributed mainly on the periphery, as if the population was currently recolonising the plateau from the swampy lowlands (Riéra & Joly 1996). Although the species is not found on the plateau presently, charcoal remains identified to species level prove that it grew there some time ago (Tardy 1998).

4. Discussion

The study of the distribution of species in tropical forest ecosystems has so far approached the problem using two different scales of time and space.

On the local short-term scale, the mechanisms of the sylvigenetic cycle and in particular the key role of treefall gaps have been used to explain the structure and dynamics of the forest mosaic. The mechanisms examined here occur on a scale inferior to the kilometre and on a time scale of 1 to 10^2 years (Oldeman 1974, 1978, Whitmore 1978, Riéra 1995, Riéra *et al.* 1990, Pélissier & Riéra 1993).

On the continental and regional long-term scale, the Holocene refuge theory provides the theoretical framework for the interpretation of global species distributions. Here, the scales used range from 10^4 to 10^8 km² in space, and from 10^4 to 10^6 years in time (de Granville 1992, Haffer 1969, Prance 1973, 1982, Van der Hammen 1974, Vanzolini 1973).

These two axes of investigation can only bring satisfying answers at their own respective scale, and remain unable to explain abnormal population and community organisations (referring to the general distribution pattern). Our approach focuses on an intermediate scale and uses both functional arguments and local evidence for past events.

Exceptional events occurring at several centuries intervals should be taken into account as well as all the common constraints organisms have to face. Their exceptional occurrence explains that they have frequently been ignored in functional ecology models, which are often based on climatic parameter averages. Connell (1978) did consider however that perturbations occurring at a moderate rate could be favourable to the maintenance of a high species diversity. Sporadic catastrophes could give a new impulse to the dynamics of forest ecosystems, maintaining biotopes and species which would otherwise tend to disappear.

The forest fires which have left these numerous tell-tale signs seem to have been the only type of catastrophic event to have affected French Guiana. The crystallineprecambrian shield of the Guianas does not exhibit volcanic nor seismic activity and the region is located close to the equator, between 2° and 6° N, and thereby away from the cyclone routes (Grand Atlas de l'Archéologie 1990). In tropical hyperhumid forests, fires can only develop beyond a certain water stress threshold and have probably been triggered by climatic fluctuations which intensified the effects of the dry season, in particular its duration.

Comparisons made between Amazonia and Western Equatorial Africa underline a fundamental difference between these two regions: in Amazonia, the dry season is cloudless and imposes important water stresses on the vegetation (Guehl 1984), while in Africa (in particular Southern Cameroon, Gabon and Congo), the 'dry' season is overcast and misty, although rainless, and plants are less exposed to water stress (Maley 1996, Schwartz *et al.* 1986). This difference is thought to be caused by the surface temperature of the ocean, in particular the upwelling of cold polar water in the Golf of Guinea. This important contrast between the two regions could explain the presence of abundant charcoal fragments in Amazonian and Guianan soils, whereas remains of forest fires are rare in West Africa and are generally associated with human settlements (Schwartz 1993).

The tropical rain forests of Amazonia and the Guianas thus appear to be relatively exposed to fire because the intensification of the dry season increases the water deficit. Beyond a certain threshold, any increase in the severity of the dry season could lead to forest fires. Today, this type of catastrophe is prevented by the high air humidity and the short duration of the dry season, but it must be considered in the current context of global warming and increasing human pressure.

Acknowledgements

This collective study was supported by the Programme ECOFIT (GDR 489, CNRS PEVS, and IRD). We would like to thank our colleague Anya Cockle for the translation of the manuscript from French to English, and Sylvie Jouard who analysed the cores for seed determination.

A hummingbird is visiting a passion flower in the forest understorey. *Photo Frans Bongers.*

PART 3

VERTEBRATE COMMUNITIES AND EVOLUTIONARY ECOLOGY

CHAPTER 8

DIET AND POPULATION DENSITIES OF THE PRIMATE COMMUNITY IN RELATION TO FRUIT SUPPLIES

Bruno Simmen, Catherine Julliot, Françoise Bayart and Elisabeth Pagès-Feuillade

1. Introduction

Field studies in recent decades have shown that idea of an overabundance of food for consumers in tropical evergreen rainforests is too simplistic (*e.g.* Milton 1980, Terborgh 1986). There is considerable evidence of difficulties encountered by animals when seeking foods. Among these, food availability is known to vary between locations, seasons and years, often in an unpredictable manner (*e.g.* Sabatier 1985). Plant defences against consumers can reduce access to nutrients in subtle ways. For example, leaf allelochemicals may vary qualitatively and quantitatively according to light exposure within a single tree (Waterman & Kool 1994) and in relation to herbivore pressure (Feeny 1970). Following such variability in food availability, fluctuations in food choices and nutrient intake throughout the year are the rule among primates (Hladik 1988). In this context, species exhibit a wide variety of feeding strategies, depending on specific metabolic needs, digestive efficiency, sensory perception of the environment and cognitive abilities, sometimes leading to the development of feeding traditions (Hladik & Simmen 1996). Although primate species are variously constrained in their niche breadth (Milton 1981), it has become clear, in view of these findings, that primate diets are best described as flexible (Gautier-Hion *et al.* 1993).

In this paper, we provide an overview of the diet of the primate species occurring at Nouragues, followed by an examination of the structure of the community and its relationship to fruit supplies. Generally, within a community of arboreal primates, folivorous species and species that supplement their mainly fruit diet with leaves frequently have relatively large body sizes and exhibit high biomasses - in some cases, associated with high population densities (but see Robinson & Redford 1986) - compared to species feeding predominantly on fruits and invertebrates (Clutton-Brock & Harvey 1977, Hladik 1988, Oates *et al.* 1990). To assess whether this trend occurs at our study site, we estimated population densities of all species using a transect method. Three species among the Primates communities from Nouragues have previously been studied intensively and their densities estimated through repeated contacts. Results of these long-term counts will be compared to those obtained by the transect method. Crude species biomasses were also calculated, allowing us to compare estimated food requirements of the primate community with estimates of food availability. In this respect, we will focus on fruit production and its

Bongers *et al.* (eds.), Nouragues. Dynamics and plant-animal interactions in a neotropical rainforest, 89–101.
2001 Kluwer Academic Publishers. Printed in the Netherlands.

variability as recorded by Zhang and Wang (1995) during a one-year sampling period, since quantitative data for other potential food categories are lacking.

2. Overview of the primate species

In this section, emphasis is placed on diet, which has been the main subject of long-term studies undertaken on single species or on several species in parallel. Data relevant to estimates of population densities, such as home range size and group size, are also mentioned when available. Social characteristics, ecology of the most cryptic species, and other aspects that could not be documented with data collected *in situ* (*e.g.* body size), are described with reference to the literature.

The Nouragues area has had no resident human population since about 200 years ago and is still an undisturbed area for primates. Six primate species coexist: the red howler monkey (*Alouatta seniculus macconnelli*), the black spider monkey (*Ateles paniscus*), tufted and wedge-capped capuchin monkeys (*Cebus apella apella* and *Cebus olivaceus castaneus*), the white-faced saki (*Pithecia pithecia pithecia*), and the golden-handed tamarin (*Saguinus midas midas;* for taxonomical status, see Rylands *et al.* 1995). Since the establishment of the research station in 1986, local workers have observed squirrel monkeys (Saimiri sciureus sciureus) only 3 times. These sightings occurred more than 4 km away from the main study site, towards the Arataye river (a tributary of the Approuague); but, even at the research station of 'Saut Pararé' and further down from this point on the Arataye river where an ecotourism centre (Camp Arataye) is located, sightings of this small-bodied primate have rarely been reported. Accordingly, *Saimiri* is not included in the presentation of primate species.

Within the study site, the trail system consists of a grid of 100 x 100-m squares covering about 150 ha, as well as a few long pathways for studying primate groups with extended home ranges. Detailed knowledge of the behavioural ecology of *A. seniculus* and *C. apella* has been gained through long-term studies (19 months) carried out between April 1988 and May 1990 (*Alouatta*, Julliot 1992, Julliot & Sabatier 1993) and between August 1992 and July 1993 (*Cebus*, Zhang 1994, 1995). The diet of *A. paniscus* was compared with that of *A. seniculus* in parallel over several months during the rainy season (January/March 1989; June/July 1992; April 1993; February/March 1995) as well as with that of *C. apella* (same months except in 1989, Simmen & Sabatier 1996, Julliot & Simmen 1998, unpublished data). Socio-ecological studies of *Saguinus midas* have been undertaken in October/November 1988 and between 1994 and 1996 during different seasons (Baer 1995, Kessler 1995*a*, Bayart & Pagès 1998). In these studies, diets were characterised using either a feeding frequency method (as described in Struhsaker 1975) or a feeding time sampling method (Altmann 1974). Results on dietary overlap between *Alouatta*, *Cebus* and *Ateles* are based on simultaneous records using the feeding frequency method, allowing for comparison of dietary trends under identical trophic conditions.

2.1 LARGE BODY-SIZED PRIMATES

Alouatta seniculus and *Ateles paniscus* both belong to the Atelidae and are among the largest neotropical primates (Table 1). Social groups of *Alouatta* at Nouragues comprise 6 members on average and show signs of territoriality (*sensu lato*). Thirty percent of the total area used by one group (45 ha) overlaps with home ranges of five neighbouring troops.

TABLE 1. Socioecological characteristics of Guianan primates.

Species and study site	Mean body Weight (kg)[a]	Social system	Foods consumed	Group size	Home range (ha)
Ateles paniscus					
Nouragues	9.5	Multi-male/-female, fusion/fission	**fp,** l, fl, s, r,f	>13 (n=1)	-
Voltzberg, Suriname				18	255
Alouatta seniculus					
Nouragues	7.7	Harem (age graded multimale)	**l, fp,** s, fl, b	2-13 (n=8)	30-45 (n=3)
Voltzberg, Suriname				4-5	-
Cebus apella					
Nouragues	3.0	Multi-male/-female	**fp, i,** l, s, fl, v	12-17 (n=1)	355
Voltzberg, Suriname				6->20	-
Cebus olivaceus					
Nouragues	3.2	Multi-male/-female	fp, ?	Ca. 10-20	>150
Voltzberg, Suriname			fp, s, l, fl, (i?)	6->20[b]	-
Saguinus midas					
Nouragues	0,6	Extended family group	**fp, i,** s, fl (l?)	3-8 (n=4)	35-62 (n=4)
Voltzberg, Suriname				6	-
Pithecia pithecia					
French Guiana[c]	1.7	Family group	**df, fp,** l, fl, i	2.3 (n=35)	148 (n=1)
Voltzberg, Suriname			**s, fp,** fl	2.7	4-10

Note: data for Suriname come from Mittermeier & Roosmalen (1981) and Buchanan*et al.* (1981) ; [a]: body weight of wild specimens is taken from Guillotin*et al.* (1994); [b]: a range of 10-20 is given by Husson 1957, quoted by Freese and Oppenheimer 1981: socioecological data obtained at another study site of the French Guiana forest block (Petit Saut) by Vié (1998). Food choices are recorded from translocated animals. - not available. df: dry fruit, fp: fruit pulp, l: leaf, fl: flower, s: seed, r: root, f: fungus, i: invertebrate, v: small vertebrate, b: bark. Bold characters indicate the main dietary tendencies.

The occurrence of inter-troop confrontations, always accompanied by roaring vocalisations (Vercauteren Drubbel & Gautier 1993), indicates that a well-delimited territory is effectively defended within the home range (Julliot 1992). *Alouatta* can be roughly described as folivorous/frugivorous, supplementing its diet with flowers, old termitaria and wood. To date, the number of plant species used as food resources at Nouragues is 149 for fruits/seeds, 104 for leaves and 36 for flowers. There are large seasonal and annual variations in the relative proportions of fruits and leaves eaten (Julliot & Sabatier 1993). For instance, ripe fruits accounted for 45% to 59% of the feeding records in February/March 1989 and 1995, but only for 25% in June/July 1992 (Fig. 1). This proportion was much lower (< 2%) in June/July 1989. During these months of fruit scarcity, individuals reduce their locomotor activities and feed predominantly on leaves. Compared with *Alouatta* spp. living in other habitats (see review in Julliot & Sabatier 1993), At Nouragues, *Alouatta* show two dietary peculiarities that must be emphasised. First, the diversity of plant species used as food resources is very wide: one group feeds on 136 plant species over one annual cycle, compared, for instance, to only 70 plant species recorded in one group of *A. palliata* (Milton 1980) over the same period of time. Second, mature leaves account for only a minor part of the diet, with young leaves being highly preferred (54% of the global diet, almost half belonging to Mimosaceae, Fabaceae and Caesalpiniaceae; for comparison, 38% in Milton's study, which used a time sampling method). These patterns have been interpreted as reflecting both the high plant species richness at Nouragues and the high nutritional quality of leaves that flush when fruits and flowers are scarce in July and the following months (Julliot & Sabatier 1993, Julliot 1994*a*, Simmen & Sabatier 1996).

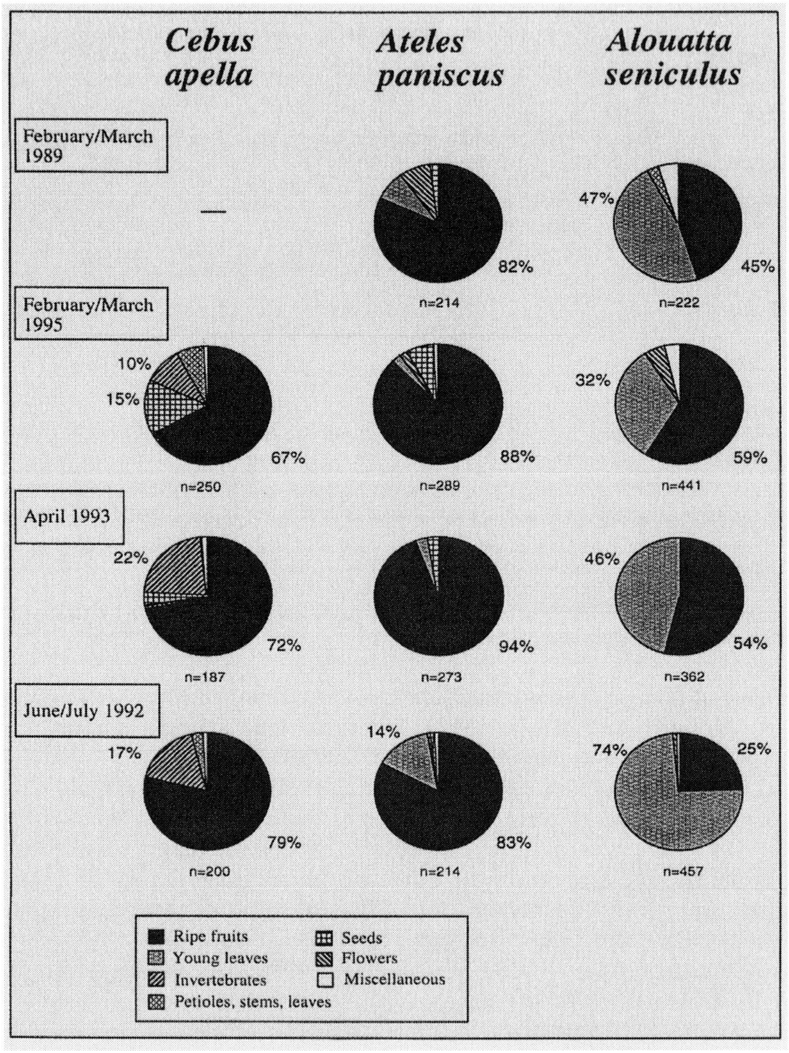

Figure 1. Dietary differences between howlers, spider monkeys and capuchins during the rainy season. For each species, food categories are expressed in terms of their relative proportions of the total feeding units (n) recorded during a given period (feeding frequency method).

Contrary to *Alouatta*, the size of *Ateles* foraging subgroups has been shown to fluctuate in relation to changes in fruit production (Roosmalen 1980, McFarland Symington 1988). Within their fission/fusion social system, subunits of 2-4 individuals are the most common associations encountered throughout the year, and the largest subgroups are observed during the rainy season (Roosmalen & Klein 1988). At Nouragues, meetings of up to 13 independently locomoting individuals have been seen around large food patches during the rainy season. Home ranges as large as 150-400 ha for a community of 15-40 members are reported in the literature (Roosmalen & Klein 1988). *Ateles* spp. are classically described as 'fruit-specialists', maintaining a predominantly frugivorous tendency throughout the year. At Nouragues, they also feed mainly on fruits during 6 months in the rainy season, supplementing this diet with young leaves and, to a lesser extent, flowers, seeds, stems, fungi and aerial roots of Araceae (Fig. 1). The number of identified plant species used as food resources so far is 103 for fruits/seeds, 8 for flowers and 5 for leaves. In a similar forest bordering inselbergs in Surinam, up to 171 fruit species are consumed throughout the year (Roosmalen & Klein 1988).

The overlap in fruit species selected by *Ateles* and *Alouatta* varies considerably during the rainy season (Simmen & Sabatier 1996, Julliot & Simmen 1998): taking into account the frequency of consumption of each fruit species (Pianka's index 1973), dietary overlap varies from a minimum of 38% (in February/March 1995) to a maximum of 95% (April 1993). It falls to 5% (June/July 1992) when *A. seniculus* shifts its mixed diet predominantly toward leaves (74%); at this time *Ateles paniscus* also increases leaf consumption, but to a much lower extent (14% of feeding records; Fig. 1). It is not clear whether feeding competition occurs between the two species: considerable overlap in the fruits consumed is observed when fruits are abundant (see below), especially those produced by mass-fruiting species (*e.g. Chrysophyllum lucentifolium*, Sapotaceae; *Bagassa guianensis*, Moraceae). During these periods, many feeding trees are visited simultaneously by groups of *Alouatta* and *Ateles*, and inter-species encounters may be peaceful (Simmen 1992). The large dietary differences observed between the two species in June/July may reflect their differential ability to cope with alternative resources when fruits are scarce, especially in relation to gut morphology and kinetics adapted to feeding on more or less fibrous plant items (Chivers & Hladik 1980, Milton 1981).

2.2 INTERMEDIATE- AND SMALL-BODY-SIZED PRIMATES

Cebus apella (Cebidae) has a smaller body size than the Atelidae and is one of the most widespread primate species throughout the neotropics. At Nouragues, one troop comprising up to 17 individuals has a remarkably large home range of 355 ha, largely overlapping (44%) with that of neighbouring groups (Zhang 1994). Recent unpublished observations indicate that this focal troop has split into two groups of approximately equal size, joined by a few migrant individuals. It is not known, however, whether this fission is temporary or permanent. The diet of this species is very eclectic, including mainly fruit pulps in addition to seeds, flowers, sap, green plant material, insects and small arboreal vertebrates (Fig. 1; Zhang 1994). Beside exploiting 10 leaf species and 5 flower species at Nouragues, the 180 fruit species recorded during the systematic, previous studies is certainly an underestimate as our recent observations reveal an ever-increasing dietary variety. According to Simpson's diversity index, food items are exploited much more evenly compared with *A. seniculus* and *A. paniscus* (Simmen & Sabatier 1996).

The fruit diet overlaps more or less extensively with that of spider and howler monkeys, being 15% (*Ateles*) and 42% (*Alouatta*) in February/March 1995 and around 75% in April 1993 (both species). Low overlap values are found in June/July 1992: 28% (*Ateles*) and 9% (*Alouatta*). Whereas both *Cebus* and *Ateles* maintain a predominantly frugivorous tendency throughout the rainy season (Fig. 1), they differ in the plant species composition of the diet, especially during the period of lowest fruit production. This can be explained by differences in habitat use, *Cebus* foraging in all forest types (Poncy *et al.*, Chapter 4, for description of spatial heterogeneity of the forest) and even at ground level. In contrast, *Ateles* is mostly found in the high forest. Interestingly, the ranging patterns of*Cebus apella* vary considerably in relation to fruit availability (Zhang 1995): when fruit production is very low and fruiting plants are patchily distributed, these monkeys spend much time foraging for invertebrates, travelling short daily distances. When fruits are widely available, they seek both fruits and invertebrates, also travelling short distances. By contrast, much longer distances are covered when fruit sources are evenly distributed and fruit biomass is at intermediate levels.

Unlike the case of the tufted capuchin monkey, little is known of the ecology of the wedge-capped capuchin, *Cebus olivaceus* (Cebidae) at Nouragues. One *Cebus olivaceus* group has been observed in the high forest as well as along streams, and its home range is suspected to be extremely wide, encompassing most of the 150-ha trail system. This group sometimes enters *Cebus apella* home range. When the two species meet, *C. apella* tends to avoid *C. olivaceus*. Troop size is estimated at 10 to 20 individuals (see also Robinson & Janson 1987). In Surinam, they feed on insects, fruits, seeds, flowers and flushes of young leaves (Mittermeier & Roosmalen 1981).

White-faced sakis (*Pithecia pithecia*, Atelidae) are also encountered infrequently. Sightings of 1 to 4 individuals occurred in the high forest as well as in the low forest close to the top of the inselberg. A recent study in primary forest in French Guiana (Petit-Saut; Vié, 1998) indicates that they live in small family groups averaging 2.3 animals and forage over a wide area (of about 150 ha) for their body weight (1.7 kg according to Guillotin*et al.* 1994). Curiously, much smaller home ranges have been reported in Surinam (below 10 ha; Buchanan *et al.* 1981). Data on their diet at Nouragues are lacking. According to the literature, *Pithecia* spp. are mainly seed predators, feeding on a large number of plant species, but also including fruit pulps, flowers, nectars, leaves and insects in their diet (Fooden 1964, Mittermeier & Roosmalen 1981, Peres 1993, Kinzey & Norconk 1993).

Golden-handed tamarins (*Saguinus midas*, Callitrichidae) use all types of forest habitats including the liana forest. They are frequently found in edge habitats and also around gap openings (Kessler 1995*b*). Their diet comprises insects, ripe pulps, nectars and occasionally seeds. Until now, *S. midas* has been seen feeding on at least 60 fruit and 10 flower species at Nouragues (F. Bayart unpublished results). The time spent foraging for each class of foods indicates that they are equally frugivorous and insectivorous during part of the rainy season (February/March 1996 and June 1995) as well as towards the end of the dry season (October/November 1988). In July 1994 and 1995, they spent more time foraging for insects (Fig. 2). A reverse seasonal pattern has recently been reported (Pack *et al.* 1999) based on stomach contents of *S. midas* sampled over different seasons and years in French Guiana. According to this study, the annual diet includes 47% fruit and 50% insects by weight, with large variations between months. Insects are predominant in the diet from February to May and in October/November. These results, however, are not strictly comparable with observations made at Nouragues, due to the distinct methods used to determine diet and because the data on food ingested were pooled from individuals sampled at various locations and over different years. Within a given forest site in French Guiana, fruit production varies enormously across years (Sabatier 1985), and the dietary tendencies

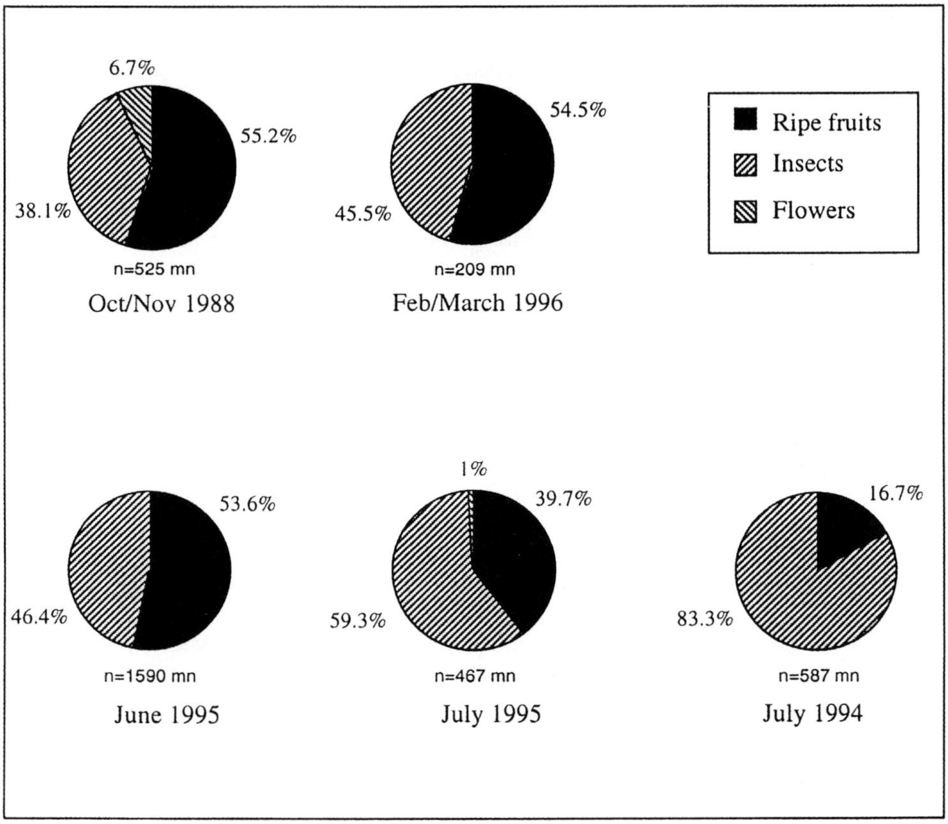

Figure 2. Feeding activity of *Saguinus midas* at different seasons. Food categories are expressed in terms of their relative proportions of time spent feeding.

of primate species can vary accordingly. Mean group size at Nouragues is 5.25 (n=4) for an average home range size of 51 ha (Bayart & Pagès 1998). Groups exhibit considerable home range overlap (up to 58% between two neighbouring groups) and vocal confrontations occur mostly in the overlapping area of the four groups studied (observed in October/November 1988, February/March 1996 and June 1995). During these three periods, figs (*Ficus* spp., Moraceae), pods of *Inga* spp. (Mimosaceae) and *Parkia nitida* (Mimosaceae) seed pod exudates accounted for most of the time spent feeding on plants (57.6%, 30.7% and 69%, respectively). Frequent intergroup encounters with intense vocal exchanges occur around these major food trees, probably a sign of resource defence against neighbouring groups as described in other *Saguinus* species (Garber 1993). It is also during such encounters that transfer of individuals between groups has been observed, probably a way of increasing gene flow within this small population. In contrast, confrontations were never observed in July 1994 and 1995, nor in August and September 1994 (P. Kessler and B. Baer, personal communication). At this time of fruit scarcity, the diet tended to be more insectivorous and groups barely encountered one another; any encounters that did occur tended to be along the border of the home range with parallel and silent locomotion. Thus, territoriality in *Saguinus midas* appears to be linked to the defence of specific food patches, circumscribed in time and space, thus reflecting seasonal variation in the diet as well as in fruit availability.

3. Population densities: long-term census versus transect method

Primate population densities are often difficult to assess in mature evergreen rain forests. The reliability of estimates may vary among species, depending for instance on ease of identifying individuals, visibility and shyness of the animals. In a previous study (Simmen *et al.* 1998), we compared the results obtained using one of the classical strip census methods with population densities already known from long-term studies of three out of the six species found at Nouragues. These data are presented here, including a discussion on environmental constraints that are likely to affect the structure of the primate community at this site, with emphasis on fruit supplies.

3.1 METHODOLOGY

The density of *Alouatta* is well known: home range, home range overlap and composition of three groups have been accurately estimated, and the location and composition of eight troops foraging within the trail system has been documented over several years by C. Julliot (from 1988 to 1996). It is important to note that population density remained stable within the 8-years duration of the survey, including the two successive years during which the transect method was applied (1995-1996). Data on group sizes and density of *Saguinus midas* are available, following surveys performed by E. Pagès and F. Bayart spanning several months at different seasons for three consecutive years (from 1994 to 1996). For *Cebus apella*, estimated density is based on home range size of one group (355 ha) of up to 17 members, corrected by substracting areas also used by neighbouring troops (Zhang 1994). *Ad libitum* observations of *Cebus apella* over several years indicate that neighbouring troops also contain 10 to 20 individuals and forage over wide home ranges. Figures for *Pithecia pithecia* and *Cebus olivaceus* are uncertain, because these monkeys are rarely seen and are not part of any focal study. For *Ateles paniscus*, which travel in temporary subgroups, *ad libitum* observations yield a confounding effect on density estimates if all members of one group are not individually recognised. Nevertheless, our experience at the study site permits us to comment with some confidence on the reliability of the transect results.

The method involves a visual census along a transect (Robinette *et al.* 1974). It was applied during the rainy season in March/April 1995 and repeated from February until the end of May 1996. The 2.5 km-long trail traversed the area over which the three species had been studied extensively. It was not possible to walk in a straight line, but the route was chosen so that the general direction could be as constant as possible. The transect passed through a variety of forest types and floristic compositions on hills, slopes and stream banks. It included high forests on lateritic soils and soils on a granitic layer (in relation to the inselberg proximity), liana forest and riparian forest, but excluded the low forest growing on the slopes of the inselberg.

The transect was walked bi-weekly starting from one end or the other, within an approximate 3 hour-period. Starting times were 7 a.m. or 3 p.m. (with approximately equal frequency) so that the probability of sightings being affected by differences in the activity rhythms between species was minimised. Transect sampling was performed 37 times, giving a cumulative distance walked by all four observers of 92.5 km. Usually, detection, counting and estimation of group composition of any primate species encountered necessitated a break of about 10 minutes. If monkeys fled before the observation was completed, a minimal estimate was noted and the census walk was continued to avoid encountering the same individuals again. The distance from the centre of the group to the observer and its perpendicular distance from the transect were recorded for later calculation of the number of individuals per unit area. Despite potential inaccuracy regarding the estimation of distances, Mendes Pontes (1994) showed that densities calculated on this basis were comparable to those derived from more sophisticated processing of the data, such as expanded Fourier series programs (Burnham *et al.* 1980) which take into account the differential probability of detecting individuals as the distance to the transect varies. This latter method requires a high contact frequency, which was not available in our case.

The calculation of densities based on sighting distance (King's method) and on perpendicular distance to the transect (Leopold's method) may yield different results (see Robinette *et al.* 1974). We found that Leopold's method gave more consistent results in relation to the known densities of *Alouatta*, *Cebus apella* and *Saguinus*. Thus, we provide only density estimates calculated according to the corresponding formula:

$$D = n / 2.L.R$$

where L is the total transect length (92.5 km), R the mean distance perpendicular to the transect from the centre of the group, and n the total number of individuals seen. The number of individuals sighted during contacts was often underestimated (see also Kessler 1998), since animals tend to hide in the upper canopy (*Alouatta*) or to flee in a very dispersed manner (*e.g.* large groups of *Cebus* spp.). Accordingly, n is corrected by multiplying the total number of contacts by mean foraging group size for each species. This latter variable is determined by combining accurate estimates of group size made during transect walks with complete counts performed outside strip censuses while investigating the feeding ecology of the different groups.

3.2 RESULTS AND DISCUSSION

Using the transect method, we find that *Alouatta seniculus* is the most abundant species, followed by the two *Cebus* species and then by *Saguinus midas* (Tab. 2). *Ateles paniscus* occurs at lower densities, while no sightings were recorded for *Pithecia pithecia*. For comparison, estimates of primate densities made by Guillotin *et al.* (1994) eight kilometers away from Nouragues along the Arataye river (Saut Pararé; riverine forests) are provided in Table 2.

Focusing on *Alouatta*, results obtained from the transect method and long-term censuses fully agree, despite the low number of contacts obtained along the transect line. For both *Alouatta* and *Ateles*, the results are consistent with data obtained in Surinam (respectively 17 ind/km^2 and 7.1 ind/km^2; Mittermeier 1977 quoted by Crockett & Eisenberg 1987,

Roosmalen & Klein 1988). Slight differences between results obtained from the two methods arise for *Saguinus* and *Cebus apella*. In a small species like *Saguinus*, which flees rapidly and silently, the location of groups sighted at a long distance from the transect might not be accurate. This can lead to variation in the surfaces calculated from transect observations, especially when combined with the low number of contacts (n=5). In *Cebus apella* (as well as in *Cebus olivaceus*), the constantly changing shape of the foraging group as well as the wide spacing between individuals in different feeding trees can easily yield a 10 m-error when estimating the centre of the group.

TABLE 2. Density and biomass estimates obtained from a transect method and long-term observations at Nouragues and Saut Pararé.

	Total number of: contacts	individuals	Mean group size[a]	Mean distance to the transect (m)	Density, transect/ effective (ind/km²)	Crude biomass [b] (kg/km²)	Density at Saut Pararé[c] (ind/km²)
Species							
Ateles paniscus	8	40	5	25.6	8.4/-	80	14.0
Alouatta seniculus	9	31	6	14.7	19.9/20.0	153	18.9
Cebus apella	5	33	12	25.2	12.9/8.6	39/26	29.1
Cebus olivaceus	4	26	15	21.2	15.3/-	49	2.0
Saguinus midas	5	18	5	12.0	11.3/14.0	7/8	37.1
Pithecia pithecia	0	0	-	-	-/-	(6)[c]	3.4

Notes. – not available. [a]: when calculating densities, mean group size is used rather than the number of individuals sighted during transect walks (see 'Methodology'). The figure indicated for *Cebus olivaceus* refers to data pooled from studies in Suriname (Mittermeier & Roosmalen 1981, and Husson 1957, quoted by Freese & Oppenheimer 1981) and *ad libitum* observations at Nouragues; [b]: figures for biomass refer to transect results and effective counts when available; [c]: derived from Guillotin *et al.* 1994.

Despite these minor discrepancies and given the consistency of results obtained for the largest species, the strip census method used is considered to be a reliable approach to estimating primate densities. Based on transect data alone or using available estimates, the total primate population density is 69-71 ind/km^2 (assigning a density of 3 ind/km^2 for *Pithecia*, as recorded at Saut Pararé by Guillotin *et al.* 1994). At Saut Pararé, the 6 species also occur (Table 2), but there the density of *Ateles* is more similar to that of *Alouatta* and the smaller primates are better represented, yielding an estimate of 100 ind/km^2 for the whole community (Guillotin *et al.* 1994). One cannot be sure at present whether the differences observed between the two Guianan communities should be considered as an artefact due to methodological differences or whether they reflect true habitat differences. Although geographically close, the two forests are not similar, Saut Pararé being located along the Arataye river and Nouragues bordering on the slopes of an inselberg (Poncy *et al.* Chapter 4). Whatever the factors involved, both sites are relatively poor in primates. In agreement with previous reports, *Alouatta seniculus*, like other congeneric species, occurs at relatively low density in evergreen forest compared with semideciduous habitats (Crockett & Eisenberg 1987). A similar pattern is found for *Saguinus* at Nouragues, where they have low population densities, as frequently observed in other primary forests (but see Guillotin *et al.* 1994) compared with those found in secondary and disturbed habitats (Lovejoy *et al.* 1985, Johns & Skorupa 1987). The low abundance of primates overall is notable when compared with undisturbed areas inhabited by the large species *A. seniculus* and *A. paniscus*, such as Manu National Park, Peru (288 ind/km^2 with 13 species; Terborgh 1983, Janson & Emmons 1990) and primate communities in terra firme forests of western Brazilian Amazonia subject to light hunting pressure: in the study of Peres (1997), 11-13 species were recorded and densities varied between 157 and 226 ind/km^2.

In terms of biomass however, the primate communities at Saut Pararé (397.5 kg/km^2) and at Nouragues (320-330 kg/km^2; Tab. 2) are comparable to those in Amazonian forests but are less productive than in Peru (655 kg/km^2; Terborgh 1986). In both FrenchGuianan communities, *Alouatta seniculus* contributes the most to total primate biomass, as was expected from its partly folivorous habits. Fruit specialists like *Ateles paniscus* and species with a very eclectic frugivorous diet like *Cebus spp.* contribute to a lower proportion of the biomass of the community. The lowest values are observed for the most insectivorous species, *Saguinus*, and for *Pithecia*, presumably, a seed predator. In this respect, the primate community structure appears to conform broadly to the classical notion of trophic levels, with different productivities and consumer biomass.

Low primate species richness may arise for historical or biogeographical reasons. Commenting on the low diversity of non-flying mammals on the Guiana Shield as compared with western Amazonia, Emmons (1984) argues that environmental factors rather than the lack of opportunity for speciation might be involved. The mosaic of habitats in terms of food species richness and food production may have different effects on the relative abundance of each primate species at Nouragues. Whereas coexistence of numerous species may be favoured by such a mosaic, a reduced area of high forest (*e.g.* locally replaced by liana or bamboo forests), including the border effect of the inselberg, may affect population densities of a species like *Ateles* which most frequently forages in the upper canopy and only rarely enters low forests.

The lack of data on species population dynamics and susceptibility to diseases, parasites and other mortality factors, currently precludes full understanding of what predominantly affects population sizes. Predation pressure, for instance, is likely to play a significant role: Julliot (1994 *b*) observed a successful attack by a crested eagle on a young *Ateles* despite the presence of adults. Attacks on *Saguinus* by birds of prey (Kessler 1995*b*) as well as by *Cebus olivaceus* (two attacks witnessed) have also been observed in the study area (F. Bayart, E. Pagès, P. Kessler, B. Baer, unpublished data). Since low primate densities might also be a simple reflection of poor nutritional quality of foods and food scarcity, these aspects are now being examined.

4. Food quality and food supplies as limiting factors for primate densities

At Nouragues, Simmen and Sabatier (1996) have investigated the nutritional quality of the diet in Alouatta and Ateles while seasonal variation in fruit production has been recorded by Zhang and Wang (1995). The analysis of the interaction between foods and the structure of the primate community will be based on this set of data, as estimates of the biomass of other food categories are lacking.

It has been shown that the proportions of soluble sugars (23.9%) and protein (10.4%) in the diet of *Alouatta seniculus* are similar to those determined for *A. palliata* at Barro Colorado Island (Hladik *et al.* 1971), a site supporting up to 4 times higher densities. Also, the screening of secondary compounds in leaves selected by *A. seniculus* reveals that respectively 19% (n = 16) and 21% (n = 14) of the leaf species are likely to contain significant amounts of alkaloids and phenolics. These proportions do not differ from those obtained from a random sample of plant species at Nouragues, which suggests that food choices of *A. seniculus* are frequently unrelated to the avoidance of these substances (Simmen & Sabatier 1996). In *Ateles*, the predominant frugivorous tendency maintained throughout the rainy season results in a diet rich in soluble sugars (34.7%) but low in protein (6.8%). In comparison to *Alouatta*, this species might have difficulties in optimising protein intake, especially during periods of fruit shortage: in June/July, leaves account for only 14% of the diet (26% on a dry weight basis versus 73% in *Alouatta* between June and September; Guillotin *et al.* 1994), and protein intake may not meet the requirements of an animal weighing 9.5 kg.

In these species as well as in the other primates in the community, which all include fruits in their diet, the hypothesis that fruit supplies limit population densities (with differences between species) can be tested. The biomass of fruits produced monthly between August 1992 and July 1993 was estimated by Zhang and Wang (1995) using three different methods, namely raked-ground survey, fruit-trap collection and platform observation. These methods yield different estimates and, as one cannot evaluate the accuracy of each estimate, results obtained with the platform observation method which yields the highest figure of fruit biomass) are considered here.

Throughout the 12-month study, the biomass of fresh fruits produced monthly at Nouragues averaged 156 kg/ha with important seasonal variations (Zhang & Wang 1995). This result does not adequately reflect the amount of fruits available to primates, which generally select fleshy parts, avoiding fibrous and woody fruits and anemochorous or barochorous species. For instance, primates underexploit fruits of Chrysobalanaceae and Lecythidaceae, representing two out of the five predominant plant families at Nouragues. According to Zhang's data (1995), less than 20% of the recorded production, on average, is suitable for *Cebus apella*. This probably applies to most other primate species as well, which not only share a common predilection for ripe fleshy fruits but also exhibit a lower diversity in the fruit species exploited. Accordingly, the mean availability of suitable fruits per hectare per month would be 6.2 kg dry weight (considering the 80% water content of most fleshy fruits; Simmen & Sabatier 1996). Even so, this figure is an overestimate as fruits are subject to significant rates of parasitism or abortion (i.e. *Chrysophyllum lucentifolium* on certain years, unpublished data) and the monkeys usually discard inedible parts. They only swallow pulpy materials and seeds, which frequently account for less than 50% of the overall fruit weight – *e.g.* in the plant families Sapotaceae, Leguminosae, Burseraceae, Hippocrateaceae and Myristicaceae, which provide most of the fruit species for the large primates. Following this figure, an annual mean of 3.1 kg dry weight per hectare per month of ingestible matter may be available to monkeys. In particular, between June and November, the biomass of ingestible dry matter falls down between 0.5 and 2 kg/ha/month.

A tentative calculation comparing fruit biomass with food intake of primate species can be made on the following grounds: the amount of food ingested by each species in the community has been extrapolated using the equation derived from the observation that food intake scales positively with body weight across a range of wild anthropoid primates exhibiting various dietary tendencies (Barton 1992). According to this equation (log daily intake, dry matter = 1.38 + 1.07 log body weight), the crude estimate of the food volume eaten by the community would fall between 2.5 and 2.7 kg dry weight per hectare per month.

Consequently, during half of the year, the biomass of consumable matter in fruit (below 2 kg/ha/month during 6 months) cannot ensure an all-fruit diet for the primate community (to the extent that Barton's equation holds across a larger number of species). The availability of suitable fruit may even be lower, given the requirements of potential competitors among the guild of pulp and seed-eaters (frugivorous birds and bats, kinkajou, opossum). For only a few months of the rainy season would the production of consumable matter largely satisfy predicted food intake – as in the 10 kg dry weight per hectare in April or May 1993 (calculation based on the figure of 100 kg/ha.month of fresh fruits suitable to capuchin monkeys during these months; Zhang 1995).

During periods of pronounced fruit scarcity, as little as 0.5 kg/ha/month dry weight may be available (between June and August). This would explain the dietary shifts observed at this time in all primate species studied. Interannual variations in fruit production certainly occur, making these estimates valid only for the period considered. Indeed, using a raked-trail survey method at another forest site in French Guiana (Piste de Saint Elie), Sabatier (1985) found that minimum fruit production (as well as maximum production) recorded over two successive years can vary by a factor 2.

Despite the crudeness of the estimates, the data seem consistent enough to support the notion that pulpy fruit production cannot sustain high densities of primate species for which such a resource is the staple food. Within the community, *Ateles paniscus* and presumably *Pithecia* are the species most concerned. But even *Alouatta* might be affected by low fruit supplies despite the ability to obtain energy from volatile fatty acids produced by leaf fermentation (Milton & Mc Bee 1983): howler monkeys do not reach the degree of specialisation of the most specialised Old World leaf-monkeys, and rely more upon fruits (46% of *A. seniculus* annual diet by stomach content analysis; Guillotin *et al.* 1994).

Low population density/low biomass appears to be a general trend in different groups of vertebrates in *terra firme* forests compared with more productive forests such as those periodically enriched by alluvial deposits (Peres 1997, see also Terborgh 1986, Janson & Emmons 1990). The striking similarity between densities of *Alouatta* and *Ateles* in undisturbed forests of Nouragues area and of the Voltzberg area in Surinam (with both sites located in proximity to an inselberg) might indicate the occurrence of similar trophic constraints within a wide region of the Guiana Shield. Estimates of primate densities have recently been made in Guyana (Sussman & Phillips-Conroy 1995), but the effects of habitat destruction and hunting pressure in the sites prospected preclude reliable comparisons.

In our study, the evidence that limited food supplies affect the primate community comes from an analysis of fruits suitable for primates. Other trophic parameters must be considered to account for different biomasses among species exhibiting dietary flexibility. Among these, estimates of leaf, nectar, and invertebrate productions should help clarify whether primate species that temporarily shift their diet toward a mainly folivorous tendency or species that feed largely on animal matter or flowers face additional constraining factors, including competition with other species.

Acknowledgments

We are grateful to the anonymous reviewers who thoroughly commented on the successive drafts of this paper and improved the quality of the English version.

Adult female of *Corallus hortulanus*, the Cook's boa. This species was frequently seen at the campsite. Nouragues, 1998. *Photo Marga Born.*

CHAPTER 9

COMPARATIVE POSITIONAL BEHAVIOUR OF FIVE PRIMATES

Dionisios Youlatos and Jean-Pierre Gasc

1. Introduction

Locomotion and posture allow animals to cope with some major survival problems, such as search and access for food sources, and escape from predators (Terborgh 1983, Garber 1991b, Cant 1992). Locomotion and posture seem to be the result of an interaction between the morphological features of an animal and the structure of the environment surrounding it. Most extant primates, and more specifically the New World monkeys or platyrrhines, are mainly or exclusively arboreal. Arboreal habitats are distinguished from terrestrial ones for the complex structure of tree crowns with the multiplanar setting of forests, the horizontal and vertical discontinuity, and the unsteady and randomly disposed supports. These features impose constraints on simple body displacement and arboreal animals are compelled to adopt various and more specialised locomotor and postural modes. In effect, New World monkeys show a wide range of locomotor and postural modes which are related to their body size, morphology, and dietary preferences (Rosenberger 1992). Small Callitrichids, with body weights ranging from 100 to 800 g, have claws on their hands and feet, and long hind limbs. They feed mainly on insects and tree exudates, and usually move in the discontinuous understorey, where vertical trunks dominate. They run quadrupedally on branches, are rapid and squirrel-like on vertical trunks where they use their claws, and can leap between trunks and terminal twigs (Garber 1992). Medium-sized monkeys, weighing between 1 and 5kg, tend to have higher intermembral indices. They feed less on insects and more on vegetal matter, mainly fruit, and are found in almost all forest layers. They use more quadrupedal walk, while clamber and leap between terminal branches may be equally shared. The largest monkeys, the Atelines with body weights from 5 to 12 kg, have even longer forelimbs, and prehensile tails with a ventral tactile surface. They are found in the upper forest layers and feed mainly on leaves and fruit. They clamber and suspend using their tails while moving and feeding (Rosenberger & Strier 1989).

In French Guiana, high primary rain forests often support six to eight sympatric platyrrhines: red howlers (*Alouatta seniculus*), black spider monkeys (*Ateles paniscus*), brown capuchins (*Cebus apella*), wedge-capped capuchins (*Cebus olivaceus*), bearded sakis (*Chiropotes satanas*), Guianan sakis (*Pithecia pithecia*), common squirrel monkeys (*Saimiri sciureus*) and golden-handed tamarins (*Saguinus midas*). Comparative studies of positional Behavior of micro-sympatric species can be valuable in many ways: the same observer is involved, habitat differences are partially eliminated, seasonal variation is

Bongers *et al.* (eds.), Nouragues. Dynamics and plant-animal interactions in a neotropical rainforest, 103-114.
2001 Kluwer Academic Publishers. Printed in the Netherlands.

avoided, the same mode definitions are applied, and the subjects are actually compared under the same conditions providing interesting results in habitat use and partitioning. In this context, we conducted a study of comparative locomotor and postural behaviour on five micro-sympatric platyrrhines in a primary *terra firme* rain forest at the Nouragues Reserve in central French Guiana.

2. Material and Methods

2.1 STUDY SITE AND SUBJECTS

The data presented in this paper were collected at the 'Station des Nouragues' (4°05'N, 52°40'W), a primary lowland wet rain forest site within the Nouragues Reserve, in French Guiana. This study site is mainly characterised by high forest, where the dominant plant families are Sapotaceae, Lecythidaceae, Chrysobalanaceae, Ceasalpiniaceae, and Burseraceae (Sabatier & Prévost 1990a). Annual rainfall varies from 3,000 to 3,250 mm, and the mean temperature is 26.1°C. The study site is described in detail by Julliot & Sabatier (1993), and Zhang (1995).

There are six sympatric platyrrhine species in the study site: red howlers (*Alouatta seniculus*), black spider monkeys (*Ateles paniscus*), brown capuchins (*Cebus apella*), wedge-capped capuchins (*Cebus olivaceus*), and golden-handed tamarins (*Saguinus midas*) are the five commonest species. Guianan sakis (*Pithecia pithecia*) are also present in the study site but are rare and shy. Red howlers and brown capuchins are entirely habituated, spider monkeys may sometimes react aggressively, tamarins are quite habituated but wedge-capped capuchins are usually aggressive and will readily flee. Therefore data on these latter will be interpreted with caution. During the observation period neither species was seen on the ground, therefore only arboreal locomotor and postural behaviour will be considered.

2.2 SAMPLING METHODS

We conducted focal animal instantaneous sampling on adult individuals (Altmann 1974). Focal individuals were followed for 15 min, and then we shifted to another individual, trying to collect the same amount of instants for both adult males and females. If the focal individual was lost from view before the 15-min period we shifted to another individual. The data presented were collected from April to July 1992 (rainy season, rainfall: 974 mm) and from June to September 1993 (transitional to dry season; rainfall: 356 mm).

In this paper we will only consider locomotor behaviour during *Travel* (movement from and to sleeping sites and between feeding sites) and postural behaviour during *Feed/Forage* (postures adopted during search, access, manipulation, and consumption of food items in or away from feeding/foraging sites). Locomotor behaviour was recorded at 20-sec intervals. Postural behaviour was recorded by time bouts measuring time duration when a bout ended when a variable changed (Cant 1987). The variables recorded in both locomotor and postural behaviour were: (1) positional (locomotor, postural) mode including tail use (only for howling, spider, and capuchin monkeys), (2) behavioural context, (3) support type (only in locomotion), size and inclination, and (4) forest layer.

The positional modes recorded were: *Quadrupedal Walk/Run* (horizontal quadrupedal progression above and along single horizontal and subhorizontal supports); *Bipedal Walk* (bipedal progression above and along and/or across one or multiple supports); *Vertical Ascent/Descent* (upward or downward vertical quadrupedal progression along single supports); *Clamber* (upward, horizontal, or downward quadrupedal progression above and across multiple, diversely oriented supports); *Bridge* (cautious gap-crossing mode through active or passive deformation of the initial and terminal supports always involving at least two limbs at a time); *Air* (gap-crossing modes involving an airborne phase such as leap, vertical leap, and drop), *Suspensory* (below support activities involving all possible combination of limbs, such as brachiation, arm- or tail-swing, hoist, and inverted quadrupedal walk); *Scansorial* (quadrupedal progression with the use of claws on vertical or steep single supports); *Sit* (crouched posture above one or multiple supports); *Stand* (standing on three or four limbs above one or multiple supports); *Bipedal Stand* (standing on two hind limbs above one or multiple supports); *Hang* (tail-assisted hanging below one or multiple supports using all possible combination of limbs); *Claw Cling* (vertical posture using claws).

The diameter (d) of the largest support used during an instantaneous observation or a postural bout was visually estimated and recorded to these size classes: *Small* ($d \leq 2$ cm), *Medium* (2 cm $< d \leq$ 10 cm), *Large* (10cm$<d\leq$20cm), and *Very Large* ($d >$ 20 cm). Support inclination classes were (relative to the true horizontal): *Horizontal* (0°-10°), *Moderate* (10°-45°), *Steep* (45°-80°), and *Vertical* (80°-90°). Support types recorded were: *Tree Branch, Liana, Palm Frond,* and *Vertical Trunk.* The forest was divided in three layers: *Emergents*, trees above 30 m height; *Main Canopy*, most of the trees ranging between 15 m and 30 m; *Understorey*, trees and shrubs from 50 cm to 15 m.

In travel locomotion sample sizes are expressed in instants of observations: *Alouatta*: 1898; *Ateles*: 1089; *C. apella*: 1218; *C. olivaceus*: 412; *Saguinus*: 464. In feeding and foraging postures samples sizes are expressed in minutes: *Alouatta*: 1187; *Ateles*: 258; *C. apella*: 243; *C. olivaceus*: 15; *Saguinus:* 42. Since such data form unordered contingency tables we used G tests for statistical analysis. *P* values of less than 0.05 are considered as significant. Table 7 summarises all statistical comparisons between the five Guianan monkeys studied.

3. Results

3.1 RED HOWLING MONKEYS (*ALOUATTA SENICULUS*)

Red howlers used very frequently the main canopy (Fig. 1). Travel occurred mainly on tree branches (87.5%, n=1898) and lianas were also used quite often (11.3%, n=1898). Clambering (Fig. 2C) to all directions was the most frequent locomotor mode and quadrupedal walk/run (Fig. 2A) ranked second (Table 1). Bridging was the main gap-crossing mode. Suspensory locomotion and leaping were used rarely (Table 1). Sitting (Fig. 2B) was the dominant feeding posture, while tail-assisted hang was used quite frequently (Table 2). The dominant hanging posture was tail-hindlimbs hang (Fig. 2D; 84.9% of changing postural subsample, n=279 min).

In travel locomotion, small and medium-sized supports predominated and were used almost equally (Table 3). During feeding postures, small supports were used for more than half of all bouts (Table 4). Moderately inclined supports dominated during travel locomotion (Table 5). During feeding horizontal supports were used frequently (Table 6).

Although the tail was used moderately in travel locomotion there was a significant increase during feeding postural behaviour (Fig. 4).

3.2 BLACK SPIDER MONKEYS (*ATELES PANISCUS*)

Black spider monkeys were found almost exclusively in the main canopy (Fig. 1). Tree branches were the principal travelling supports (92.4%, n = 1089) and lianas were used occasionally (5.7%, n = 1089). Suspensory locomotion and clambering (Fig. 2F) were the dominant travelling modes (Table 1). Suspensory locomotion was largely dominated by

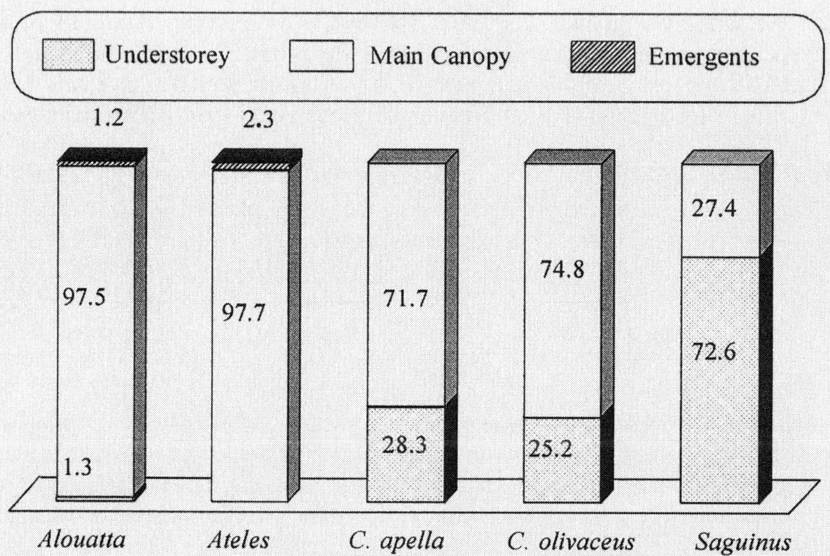

Figure 1. Percentages of use of forest layers in travel locomotion.

TABLE 1. Percentages of locomotor modes during travelling locomotion. Entries of columns are percentages for each species.

	Alouatta %	*Ateles* %	*C. apella* %	*C. olivaceus* %	*Saguinus* %
Quadrupedal Walk/Run	37.3	20.1	31.8	32.5	32.5
Bipedal Walk	0.0	1.6	2.1	0.0	0.0
Vertical Ascent/Descent	5.5	1.6	3.4	3.1	0.4
Clamber	38.4	28.3	25.8	27.7	25.0
Bridge	13.5	8.8	13.3	10.0	3.5
Suspensory	2.5	36.8	0.0	0.0	0.0
Air	2.8	2.8	23.6	26.7	26.5
Scansorial	0.0	0.0	0.0	0.0	12.1
Samples	1898	1089	1218	412	464

tail-arm brachiation (Fig. 2E) and arm-swinging. Gaps were crossed principally with bridging (Table 1). Hanging feeding postures predominated with sitting (Fig. 2H) ranked far behind (Table 2). Tail-only hang (Fig. 2G) was the most common hanging posture (47.1% of hanging postural subsample, n = 174 min). Small and medium-sized supports were used almost equally as both travelling and feeding supports, although there was a slight preponderance of small supports in feeding postures (Table 3,4). Moderately inclined supports predominated in both travel locomotor and feeding postural behaviours (Table 5,6). The tail was used extensively in locomotion and almost always during feeding postures (Fig. 4).

Figure 2. Locomotor and postural patterns of red howlers (A-D): Quadrupedal Walk (A), Sit (B), Clamber (C), Tail-hindlimbs Hang (D), and black spider monkeys (E-H): Tail-arm Brachiation (E), Clamber (F), Tail-only Hang (G), Sit (H).

Figure 3. Locomotor and postural patterns of brown capuchins (A-D): Quadrupedal Walk (A), Clamber (B), Sit (C), Stand (D), and golden-handed tamarins (E-H): Quadrupedal Walk (E), Leap (F), Sit (G), Stand (H).

3.3 BROWN CAPUCHIN MONKEYS (*CEBUS APELLA*)

Although brown capuchins were found mainly in the main canopy, they also used considerably the understorey (Fig. 1). Tree branches were the dominant travelling supports

(86.7%, n = 1218). Moreover, they used lianas quite often (8.4%, n = 1218) and were the most frequent users of palm fronds in Nouragues (3.5%, n = 1218). Quadrupedal walk/run (Fig. 3A) and clambering (Fig. 3B) were the most frequent travelling modes (Table 1). Leaping/dropping (Air) were the dominant gap-crossing modes although bridging represented a considerable proportion (Table 1). Almost all feeding and foraging was done by a seated posture (Fig. 3C) while standing (Fig. 3D) and hanging were used only moderately (Table 2). Small and moderately inclined supports predominated during travel locomotion (Table 3,5).

TABLE 2. Percentages of postural modes during feeding and foraging. Entries of columns are percentages for each species.

	Alouatta %	*Ateles* %	*C. apella* %	*C. olivaceus* %	*Saguinus* %
Sit	57.5	31.8	72.8	35.6	42.6
Stand	14.7	0.3	13.1	34.0	36.8
Bipedal Stand	4.3	0.5	4.9	13.2	9.3
Hang	23.5	67.4	9.2	17.2	0.5
Claw Cling	0.0	0.0	0.0	0.0	10.8
Samples	1187 min	258 min	243 min	15 min	42 min

TABLE 3. Percentages of size of supports used in travelling locomotion. Entries of columns are percentages for each species.

	Alouatta %	*Ateles* %	*C. apella* %	*C. olivaceus* %	*Saguinus* %
Small	40.1	41.9	59.2	56.6	55.0
Medium	39.7	40.3	34.6	33.3	24.1
Large	15.1	16.0	4.6	9.7	5.6
Very Large	5.1	1.8	1.6	0.4	15.3
Samples	1898	1089	1218	412	464

TABLE 4. Percentages of size of supports used in feeding and foraging postural behaviour. Entries of columns are percentages for each species.

	Alouatta %	*Ateles* %	*C. apella* %	*C. olivaceus* %	*Saguinus* %
Small	53.1	48.5	45.7	54.7	74.9
Medium	37.6	42.0	53.9	39.0	11.9
Large	5.9	9.5	0.4	6.3	2.4
Very Large	0.5	0.0	0.0	0.0	10.8
Samples	1187 min	258 min	243 min	15 min	42 min

On the other hand, medium-sized and horizontal supports were used quite frequently during feeding and foraging postural behaviour (Table 4,6). The tail was used moderately in locomotion and its use increased significantly during feeding and foraging (Fig. 4).

3.4 WEDGE-CAPPED CAPUCHIN MONKEYS (*CEBUS OLIVACEUS*)

Wedge-capped capuchins made extensive use of the main canopy and considerable use of the understorey (Fig. 1). Tree branches dominated as travelling supports (87.6%, n = 412) and lianas were also used quite often (9.7%, n = 412). Quadrupedal walk/run and clambering predominated in travel locomotion (Table 1). Leaping/dropping (Air) were the dominant gap-crossing modes (Table 1). Bridging was also used quite often (Table 1). Sitting and standing were the dominant feeding and foraging postures and were used almost equally (Table 2). Hanging postures also represented a considerable proportion, although these results are tentative due to the small sample size (15 min). Small and moderately inclined supports were used very frequently in both travel locomotor and feed/forage postural behaviours (Table 3,4,5,6). In locomotion, the tail was used moderately but there was a significant increase during feeding activities (Fig. 4).

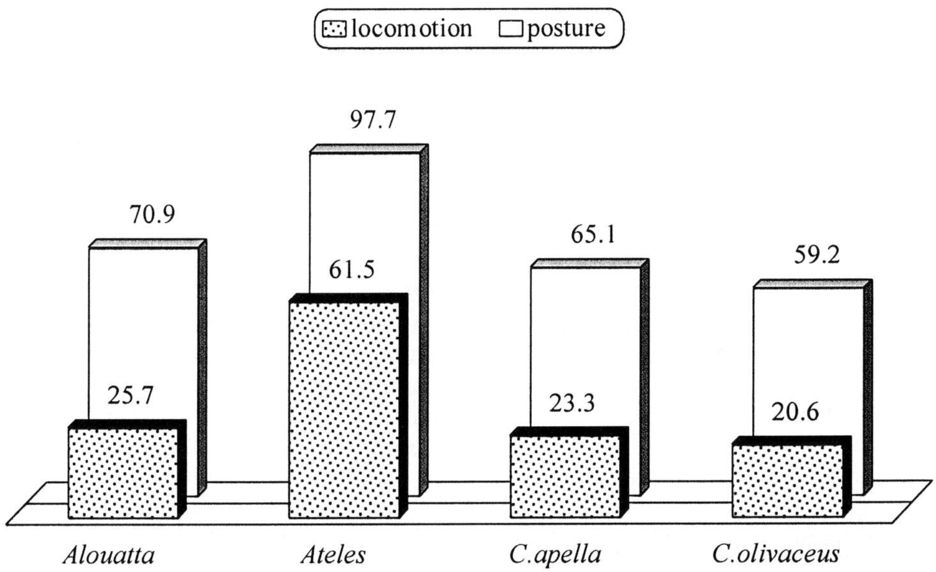

Figure 4. Percentages of tail use in travel locomotion, and feed/forage postural behaviour.

3.5 GOLDEN-HANDED TAMARIN MONKEYS (*SAGUINUS MIDAS*)

Golden-handed tamarins were found principally in the understorey (Fig. 1). Although tree branches were the main travelling supports (76.5%, n=464), they were the most frequent users of both lianas and vertical trunks (13.8% and 7.3% respectively, n=464). Quadrupedal walk/run (Fig. 3E) predominated while leaping/dropping (Air) (Fig. 3F) and clambering were almost equally used (Table 1). Scansorial locomotion was used moderately (Table 1).

TABLE 5. Percentages of inclination of supports used in travelling locomotion (a) and feeding and foraging postures (b). Entries of columns are percentages for each species.

	Alouatta %	*Ateles* %	*C. apella* %	*C. olivaceus* %	*Saguinus* %
Horizontal	30.5	26.5	28.4	24.8	26.3
Moderate	37.8	44.8	41.9	42.7	40.1
Steep	25.6	26.4	24.0	27.2	22.2
Vertical	6.1	2.3	5.7	5.3	11.4
Samples	1898	1089	1218	412	464

Sitting and standing (Figs. 3G,H) were the dominant feeding and foraging postures while claw-cling ranked behind (Table 2). Small and moderately inclined supports were used very frequently in both travel locomotor and feed/forage postural behaviours (Table 3,4,5,6). Moreover, golden-handed tamarins showed the highest percentages of use of very large and steeply inclined and vertical supports (Table 3,4,5,6). The tail was never used in a prehensile way.

4. Discussion

4.1 FOREST USE BY FRENCH GUIANAN PRIMATES

Mittermeier and Roosmalen (1981) and Terborgh (1983) had found for Surinamese and Peruvian primates respectively that larger primates tend to use higher forest layers while smaller ones move and feed in the lower layers of the canopy and in the understorey. The same pattern was observed in French Guiana: Large-bodied spider and howling monkeys were found almost exclusively in the main canopy and emergents. On the other hand, the proportion of understorey use increased for medium-sized capuchins while the small-bodied tamarins were found mainly in the understorey. Terborgh (1983) associated this stratification with respective strategies of predator escape and food preferences. In the neotropics, the large-bodied atelines feed mainly on fruit and leaves (Strier 1992) which are found in the main canopy, while their relatively large body size would discourage any potential aerial predators (although see Julliot 1995). Capuchins being more omnivorous, exploit all forest layers but given that potential predators are more numerous for their size (Janson & Boinski 1992) tend to use more often the lower parts of the forest. The small size of tamarins compels them to make extensive use of the understorey. Besides, they showed a high percentage of liana and vertical support use. Vertical supports are the main support in the understorey and liana entanglements do offer an abundant source of edible insects as well as cover from potential aerial or terrestrial predators (Terborgh 1983).

TABLE 6. Percentages of inclination of supports used in feeding and foraging postural behaviour . Entries of columns are percentages for each species.

	Alouatta %	*Ateles* %	*C. apella* %	*C. olivaceus* %	*Saguinus* %
Horizontal	42.0	30.5	38.5	10.2	22.0
Moderate	35.9	41.4	33.0	67.5	36.8
Steep	20.3	26.5	27.7	16.0	30.4
Vertical	1.8	1.6	0.8	6.3	10.8
Samples	1187 min	258 min	243 min	15 min	42 min

Fleagle and Mittermeier (1980) found that small primates use in general smaller supports than larger primates do. This pattern was also supported by our results (Table 3). There are usually two exceptions to the rule in the neotropics: the large-bodied suspensory atelines that use relatively high proportions of small supports as they travel suspended or by clambering, and the small-bodied callitrichids that use their claws (tegulae) for scansorial locomotion on large supports (Fleagle & Mittermeier 1980). In Nouragues, Guianan spiders used high proportions of medium-sized supports in both locomotor and postural behaviour in contrast to previous observations. On the other hand, golden-handed tamarins used relatively high percentages of very large supports in both locomotor and postural behaviour, and thus our results accord with those of Fleagle and Mittermeier's (1980).

4.2 LOCOMOTOR AND POSTURAL BEHAVIOUR

Spider monkeys used high percentages of suspensory locomotion and postures. All previous studies on spider monkey positional behaviour show that suspensory habits represent a considerable part in spider monkeys' positional behaviour (Mittermeier 1978, Cant 1986, Fontaine 1990). Spider monkeys usually move in the main canopy, and feed almost exclusively on fruit found in the terminal branches of tree crowns. Clambering and suspensory locomotion, and tail-limb-hang facilitate dealing with the abundant unsteady and flexible small supports.

The high percentage of tail use also shows its important role in supporting body weight so that spiders can gain access to the terminal fruit-bearing twigs. Furthermore, the moderate use of lianas, which usually interconnect tree crowns (Grand 1984), associated with the relatively low percentage of bridging may suggest that spiders use suspensory modes not only for within the crown travel but also for crossing gaps (see Bergeson 1996).

Clambering and quadrupedal walk were the most frequent locomotor modes of Guianan howlers, supporting previous results on the positional behaviour of other species of howlers (Mendel 1976, Schön Ybarra & Schön 1987, Gebo 1992). In effect, it appears that the forelimb anatomy of howlers restricts them to above branch quadrupedal activities (Schön Ybarra & Schön 1987) using the central parts of tree crowns where larger supports dominate. Furthermore, when they have to deal with small flexible supports, they would rather clamber above the supports distributing body weight on several supports while using their tails as a brake and for security. Gap crossing is usually performed by bridging as reflected by its high percentage, and the role of the tail is extremely important (Youlatos 1993). Furthermore, the relatively high percentage of liana use may suggest that they are used as bridges between tree crowns, and are especially useful for mammals that leap infrequently, such as howling or spider monkeys (Grand 1984). Access to young leaves and fruit located on the terminal twigs is achieved through hanging postures, where the tail is always used along with the hindlimbs. However, sitting was the dominant feeding posture with the tail anchored in order to secure the posture on the small terminal supports.

Quadrupedal walk and clambering were the main travelling modes for both capuchin species, with leaping/dropping (Air) ranked not far behind. Previous studies on capuchin positional behaviour have shown similar results (Fleagle & Mittermeier 1980, Gebo 1992). Capuchins moved quadrupedally above small supports and used their tails only during clambering and bridging. However, the relatively high percentage of air suggests that most gap crossings occurred by leaping or dropping. Such a behaviour may also be related to the more discontinuous nature of the understorey that capuchins use more frequently. Moreover, capuchins being of smaller size that the atelines and having longer hindlimbs (Erikson 1963, Fleagle 1988) and more powerful extensors (Stern 1971, Youlatos 1994) may prefer to jump instead of using active deformation of branches for bridging. Liana and palm frond use was also considerable, since capuchins forage for insects and feed on hard palm nuts (Zhang 1995). Feeding and foraging were mostly done in a seated posture.

In effect, these capuchins usually collect food items with a certain posture but manipulate and consume it seated on a medium-sized horizontal support in a different part of the same or other tree. This may explain further the frequent use of medium-sized and horizontal supports in feeding and foraging. Postural behaviour of wedge-capped capuchins is based on a limited sample size and therefore will be considered with caution. Robinson (1986) reports that this species appears to be more devoted to foraging than brown capuchins. As foraging requires more active search of food items, this may be related to the relative high percentage of stand and hang in *C. olivaceus*. This may also be supported by the fact that even brown capuchins used sitting less frequently during foraging (Youlatos 1998).

Quadrupedal walk, clambering, and leaping/dropping (Air) were the principal travelling modes for Guianan tamarins (see also Fleagle & Mittermeier 1980). Similar results have been reported for other tamarin species (*S. fuscicollis, S. mystax, S. geoffroyi*) although these species showed higher proportions of leaping (Garber 1980, 1991a). For small monkeys like tamarins, leaping is the major gap-crossing mode when moving in the discontinuous understorey. Tamarins were the only Guianan primates bearing claw-like tegulae. Tegulae enable tamarins to move on vertical and large supports that abound in the understorey. Moreover, tree-exudates, a supplementary seasonal food source can be acquired by claw-cling on vertical supports (see also Garber1992). However, the acquisition of other food items such as fruit or invertebrates was done merely by sitting and standing. Other tamarin species are also reported to use equally standing and sitting (Garber 1991) a fact that can be associated to a more active food search pattern (Garber 1992).

TABLE 7. Statistical comparisons between the Nouragues primates studied (*: p<0.05; ***: p<0.001).

	Alouatta	*Ateles*	*C. apella*	*C. olivaceus*	*Saguinus*
Locomotor modes					
Alouatta		***	***	***	***
Ateles			***	***	***
C. apella				*	***
C. olivaceus					***
Saguinu					
Postural modes					
Alouatta		***	***	n.s.	***
Ateles			***	***	***
C. apella				n.s.	***
C. olivaceus					n.s.
Saguinus					
Support size in locomotion					
Alouatta		***	***	***	***
Atele			***	***	***
C. apella				*	***
C. olivaceus					***
Saguinus					
Support size in postures					
Alouatta		***	***	n.s.	***
Ateles			***	n.s.	***
C. apella				n.s.	***
C. olivaceus					n.s.
Saguinus					
Support inclination in locomotion					
Alouatta		***	n.s.	n.s.	***
Ateles			***	*	***
C. apella				n.s.	***
C. olivaceus					***
Saguinus					
Support inclination in postures					
Alouatta		***	n.s.	n.s.	***
Ateles			n.s.	n.s.	***
C. apella				n.s.	***
C. olivaceus					n.s.
Saguinus					

5. Conclusions

In general, synecological studies of sympatric primates give an idea of how primates that live in the same forest share and compete for food sources. However, they do not report on how these sympatric primates manage to exploit different microhabitats or gain access to preferred food sources. Thus, the study of the positional behaviour of sympatric primates completes a primate profile, since locomotor and postural behaviour are the means for gaining access to food sources and are intimately related with body size and anatomy.

The last few years a certain number of studies of comparative positional behaviour have shown how primates tend to exploit different forest layers, preferring or avoiding a certain type of support, and preferring modes that are appropriate to the combination of both postcranial anatomy and habitat structure (Cant 1992, Dagosto 1995, Gebo & Chapman 1995, McGraw 1996). However, discerning preferences for certain modes or supports requires a preliminary study of habitat structure to measure its availability. The influence of habitat structure on positional behaviour has proven to be a very important factor as shown by studies of the same species in different habitats (Hunt 1992, Gebo & Chapman 1995, Garber & Preutz 1995).

Although our study did not include a quantification of habitat structural parameters, we managed to show certain differences in the strategies of the sympatric primates in Nouragues. The small-bodied tamarins which base their diet on invertebrates and small fruit were found in the lower parts of the forest. Quadrupedal walk and run and leaping provided quick travel within the discontinuous part of the forest. Moreover their tegulae enabled them to exploit the large vertical support niche providing them with fewer detours via the ground or upper forest layers as well as supplementary food sources (Garber 1992). On the other hand, there were no particular differences in the positional behaviours of both capuchin species. This may reflect the strong influence of the conservative nature of the postcranial morphology of the genus (Ford 1988). However, if the tentative results of wedge-capped capuchin postural behaviour reflect their stronger commitment to a more active foraging and feeding than their congeners then this lends support to the ecological distinction between them based on different food sources and suggested by Janson and Boinski (1992). Both sympatric Guianan atelines share the same body weight and exploit the same forest layers. However, the strong commitment of spider monkeys to ephemeral and widely dispersed food sources such as mature fruit requires faster travel within and between tree crowns and access to the terminal fragile branches. These requirements are satisfied in part by the particular postcranial features of spider monkeys which enable them to use suspensory locomotor modes and suspensory postures (Erikson 1963, Rosenberger & Strier 1989). On the other hand, the diet of howling monkeys based principally on leaves demands slower rates. The more generalised postcranium of howlers permits safer quadrupedal above-branch walk and clambering and secure crossings between tree-crowns by bridging (Schön Ybarra & Schön 1987). Furthermore, although sitting was the main food collecting and ingesting posture, suspensory postural behaviour enlarged the feeding sphere of howlers providing larger amounts of low quality food, such as leaves, per bout (Mendel 1976). It is clear therefore that only a complete study of the positional behaviour along with synecological data will help us comprehend the structure and function of any primate community.

Acknowledgements

We are greatly indebted to Dr. P. Charles-Dominique who facilitated access to the research station at Nouragues. Financial support was provided by the Muséum National d'Histoire Naturelle: 'Action Spécifique Guyane' and the CNRS-URA 1137 (presently UMR 8570), Laboratoire d'Anatomie Comparée. Many thanks go to Dr. B. de Thoisy, and Dr. B. Simmen for help in the field during data collection. The manuscript benefited from suggestions provided by Dr. H. Lantsbury, Dr. S. Bearder, and one anonymous reviewer.

CHAPTER 10

THE BAT COMMUNITY

André Brosset, Pierre Charles-Dominique and Anya Cockle

1. Introduction

There are few places in tropical America where bat communities have been systematically surveyed. Several sites in Suriname (Genoways & Williams 1979, Williams & Genoways 1980) and French Guiana (Brosset & Charles-Dominique 1990) have been documented, and Voss and Emmons (1996) also published lists of bat species collected at a few forested sites of Southern and Central America, including Arataye. The work of Simmons and Voss (1998) on the bats of Paracou greatly improved our knowledge on French Guianan bats. The longest and most complete study of a bat community was undertaken on Barro Colorado Island, Panama (Kalko *et al.* 1996). This chapter will present the bat community of Nouragues area based on data collected between 1988 and 1997.

2. Site and methods

The area prospected covers about 2 km² around the station and about 1 km² around the landing stage of the Pararé rapids on the Arataye river. This surface area is small in comparison to that of other studies; in Barro Colorado for example Kalko *et al.* (1996) surveyed 15 km². The Nouragues area is representative of a Guianan primary lowland forest, with no human interference within a radius of 50km. The station is surrounded by an exceptional range of habitats: the cliffs, patchily forested slopes and narrow gullies of an inselberg dominate a landscape of low forested plateaux extending to the banks of a large river, the Arataye. The primary forest covering the plateaux is a mosaic of plant community units varying in structure and composition. A number of caves and large boulder screes provide shelter for the species that spend the day in subterranean roosts. Such a range of habitats explains why all the main taxa, genera and families, are represented.

Bat research in Nouragues has mainly been related to the role of bats in seed dispersal (Charles-Dominique 1991, 1993). Nothing was known about the bat fauna of the area before this study and therefore capture for identification and individual marking was the first priority. The bat census was lead by P. Charles-Dominique, who regularly mist-netted in the different strata of the forest, from ground to canopy (at a height of 25-30 m), individually marked the bats and followed some of them using radio-tracking techniques. Harp-traps were also used with success. Mist-netting in different strata of the forest yielded a more complete picture of the bat community than mist-netting at ground level only (i.e. at a height of 0-5 m) in the understorey. Mist-netting was also carried out at the Arataye site

Bongers *et al.* (eds.), Nouragues. Dynamics and plant-animal interactions in a neotropical rainforest, 115–120.

by J.F. Cosson and L. Emmons. A. Brosset mainly focused on the location and examination of bat roosts, especially caves and boulder scree cavities. A. Cockle studied a foliage-roosting species over 20 months. Bat tents and young banana and *Heliconia* leaves around the station were systematically investigated (Charles-Dominique 1993). Some species were discovered by chance in the course of other studies: the cliff eyrie of a *Falco deiroleucus* falcon provided the skulls of several bat species, including that of 7 Molossids particularly difficult to catch since they roost in inaccessible crevices and forage in the open above the tree-tops. The census of the bat community at Nouragues was therefore carried out in different strata of the habitat using a wide array of methods, thereby reducing the bias of incomplete or inadequate sampling (Fleming *et al.* 1972, Wilson 1989).

3. Results

3.1 SPECIES RICHNESS AND COMPOSITION

In 1998, over 10,000 bats have been caught and identified by our team all over French Guiana. The total number of species recorded amounts to 101 (Brosset & Charles-Dominique 1990; Simmons & Voss 1998). At the Pararé site, 2119 bats were caught and 62 species identified, this being the highest figure ever reached on such a small surface area (about 3 km²). Moreover, since five species were only noted once at the site, we may assume that several additional species remain to be discovered and that the total number of bat species present at Nouragues is probably close to 70. The list of collected species is given at the end of the book. The following species were not found at Nouragues, but found elsewhere in French Guiana in primary forest: *Peropteryx leucopterus, Cyttarops alecto, Pteronotus gymnonotus, Pteronotus personatus, Phylloderma stenops, Choeroniscus intermedius, Desmodus yougi, Natalus tumidirostris, Thyroptera discifera, Lasiurus atratus, Lasiurus egregius.*

To date, the bats found at Nouragues belong to 8 families, 37 genera and 62 species. As in Barro Colorado (Kalko *et al.* 1996), the family Phyllostomidae contributed to about 70% of the community species richness. In decreasing rank of importance, we recorded: Phyllostomidae (24 genera, 44 species), Molossidae (3 genera, 6 species), Emballonuridae (5 genera, 6 species), Vespertilionidae (1 genera, 1 species), and a further 4 families each represented by a single species (Table 1).

In order to understand the evolutionary processes which may explain the composition of the bat community at Nouragues, it is useful to compare the local composition with that of other communities investigated in other parts of the country. As elsewhere in French Guiana, the bat species assemblage is primarily the result of the environmental factors that characterise the habitats. Because bats are able to fly over long distances, geographical factors play a minor role in their distribution, except on the very large continental scale. In French Guiana, only one pair of species, *Anoura caudifer* and *A. geoffroyi* seem separated by geographical limits, with *geoffroyi* in the North and *caudifer* in the South. After comparing the data from Brosset and Charles-Dominique (1990) and Simmons and Voss (1998), it appears that the bat communities of pristine *terra firme* forested areas are fundamentally the same everywhere on the Guianan plateau. The close examination of the differences found between four well studied sites in French Guiana (St Elie, St Eugène, Paracou and Nouragues) may give insights into how bat assemblages have been shaped.

Outside the forest, other physical structures and habitat components influence the species composition of bat communities. In French Guiana, the coastal swampy forest and its surrounding marshes harbour species such as *Noctilio leporinus*, *Tonatia carrikeri, T. brasiliense*, and *Platyrrhinus brachycephalus*, which have not been recorded elsewhere. Some species were only caught in modified habitats: *Sturnira lilium*, *Artibeus cinereus* and the species of the genus *Eptesicus.* A few specimens of *Glossophaga soricina,* a species very common in modified areas and generally absent from pristine habitats, were found on the inselbergs of Nouragues and La Trinité, thereby indicating that inselbergs may act as islands where isolated populations of non-forest bats may establish themselves.

Some bats form pairs of twin species in which one species is ecologically restricted to the pristine forest and the other to modified habitat. They may be externally almost identical, but are located in distinct habitats. *Artibeus cinereus* and *A. gnomus*, for example, can only be distinguished after examination of the molars. The former was collected in modified habitat only, the latter in primary forest, including in Nouragues where the species is common. In French Guiana, they only share the same habitat along forested corridors that connect scattered forest clumps in the coastal savanna to the main forest block of the interior (Brosset *et al.* 1996).

Comparisons dealing with the number of individuals recorded in the different biotopes, lead to the conclusion that, in a given biotope, the species richness is inversely proportional to the number of individuals for each species. The analysis of mist-netting results in different habitats shows that, technical conditions of captures being equal, the number of bats captured is over three times higher in the modified habitat of Cayenne than in the mature undisturbed rain forest of Nouragues (Brosset *et al.* 1996). The abundance of roosting sites in towns and man-made modified habitats may explain this difference (Brosset & Gaucher 1996). This suggests that, in pristine habitats, populations may be limited by roost availability rather than food. A good example is given by *Carollia perspicillata*, a species for which a demographic explosion was documented along a forest track, where the numerous storm drains provided vast roosting opportunities and were systematically colonised (Cosson 1994).

3.2 DIURNAL ROOSTS

Adequate daytime roosting opportunities constitute the prerequisite condition to the presence of a given species. Diurnal roosts include subterranean habitats and different types of shelters such as holes, buttresses, decayed bark, foliage, etc., depending on the species.

Caves

In each taxonomic group, some species need caves or crevices in cliffs or between boulders for their diurnal roost. These species are unable to establish populations in the absence of such roosts. Indeed, cave species such as the Phyllostomidae *Phyllostomus latifolius*, *Lonchorhina inusita*, *Lionycteris spurrelli* are either absent or rare at Piste de St Elie, where no subterranean habitat can be found, whereas they are well represented in Nouragues area, where they roost in caves. The results of the census of the populations observed in some of these caves are displayed in Table 1.

Crevices in the rock of the inselberg may host local populations of Molossids. A group of 17 *Nyctinomops laticaudatus* was observed under a slab, on the upper ledge of the inselberg cliff. Groups of 2-5 *Pteropteryx macrotis* and 2-10 *Lionycteris spurrelli* inhabit most of the great boulder screes around the foot of the inselberg.

TABLE 1. Species and the individuals in four caves close to the Nouragues Station

Grotte des parfums	Bassin du Tapir	Grotte du Jaguar	Grotte des urnes
Pteronotus parnelli c.a 1000	*Lonchorina inusita* c.a. 300	*Mimon benetti* c.a. 20	*Lionycteris spurelli* c.a. 120
Anoura geoffroyi several hundred	*Phyllostomus latifolius,* c.a. 50		*Carollia perspicillata* 10-12
Lionycteris spurelli > 1000	*Carollia perspicillata* c.a. 150		*Phyllostomus latifolius* 10-15
Peropteryx macrotis 5	*Peropteryx macrotis* 4		*Myotis nigricans* 20-30
			Peropteryx macrotis 5

Trees

The Emballonurid *Cormura brevirostris* roosts under the trunk of large fallen trees in *terra firme* forest. Along the river Arataye, another Emballonurid, *Rhynchonycteris naso* may be seen under sloping trunks and branches (or rocks) emerging from the water. Some Phyllostomids, such as *Micronycteris microtis, Carollia perspicillata,* were regularly observed inside the hollow trunks of fallen trees.

Bark, buttresses, and superficial crevices of standing trees give shelter to Emballonurids of the genus *Saccopteryx*. Small *Artibeus gnomus* have been observed roosting in the darkest recesses of the buttresses. Bunches of roots emerging from overhanging banks often host *Carollia perspicillata.* Large hollow trunks of standing trees are occupied by *Saccopteryx bilineata, Peropteryx macrotis* and several *Micronycteris* species. Molossids have been seen emerging at dusk from under the decayed bark of dead trees.

Most of the frugivorous Phyllostomids roost in foliage, where they are difficult to locate, because of the cryptic patterns of their heads, shoulders and backs. In spite of their abundance, the frugivorous species concealed in foliage were mostly observed by chance and the data collected remains scant. Tent building species, which modify large leaves to form shelters were more often observed and were more intensively studied (Charles-Dominique 1993). The species concerned were *Rhinophylla pumilio, Artibeus gnomus* and *Uroderma bilobatum*.

An individual of the poorly known Emballonurid *Diclidurus scutatus* was caught as it was roosting under the external edge of a terrestrial Cyclantaceae leaf, in the low forest at the foot of the inselberg. Harems of *Thyroptera tricolor* were found in the rolled terminal leaves of banana plants soon after a few of these plants were introduced in the camp.

3.3 DIET

In tropical America, bats exploit almost all types of food: pollen, nectar and fruits on the trees, insects flying in the open, or spiders on the vegetation, frogs calling from the banks of forest streams, fishes swimming near the surface of water, small birds, rodents and other bats taken from the trees or on the ground, and the blood of various vertebrates (Findley 1976, 1993, Fleming 1986). All these adaptations may be met in the Nouragues bat community in the super family Phyllostomatinae. This family is endemic to the neotropical region, and represented at Nouragues by the greatest number of species and individuals.

A considerable amount of data and material related to bat diets have been collected at Nouragues. The initial results of the analysis demonstrate that the frugivorous-pollinivorous species are specialised in the exploitation of particular structural elements of the habitat, where they select specific types of fruit or flower (Charles-Dominique 1986, 1991, 1993, Cosson 1994). The partition of the habitat into distinct ecological niches seems to be the rule amongst the frugivorous-pollinivorous species. On the other hand, considering insectivorous-carnivorous species, the study of widely separated populations (Brosset & Charles-Dominique 1990, Findley 1993, Gardner 1977), appears to indicate far more overlaps in the niches due to less specialised diets. Detailed information on the diet composition and feeding behaviour of the bats will be found in the relevant chapters of this book related to this subject.

3.4 BAT GUILDS

Guilds group local associations of taxa living under similar ecological conditions: habitat, foraging habits and diet. Habitat refers to the main foraging area. According to Kalko*et al.* (1986), the proximity of obstacles such as foliage, ground or water surface, determine the foraging area of each bat guild with the exception of*Pteronotus parnelli*. The constitutive species of a guild share wing shape and echolocation signal design, which indicate adaptation to specific ecological requirements. Species exploiting the open space high above the vegetation have long pointed wings for fast flight and use particular search phase echolocation signals that include narrow band, low frequency, shallowly-modulated pulses of long duration and long pulse interval. It is crucial to distinguish between frugivorous Phyllostomids and aerial insectivores since each have different strategies (picking up auditory or olfactory cues emitted by the prey versus actively detecting the prey by echolocation). The shape and size of the morphological traits related to echolocation may give a good indication of the nature of the ecological niche: for example the considerable size of the ears, tragus and nasal exfoliation of a foliage gleaner such as *Lonchorina inusita* (Fenton 1972, Fenton & Bell 1981, Findley & Wilson 1986, Brosset & Charles-Dominique 1990, Kalko *et al.* 1996).

Guilds were not studied *per se* in Nouragues. The data used here mostly came from radio tracking surveys, nocturnal observations and analysis of faecal contents. Indirect evidence was deduced from the morphology of the bats wings, ears and nose leaf, and from published data concerning the structure of ultrasonic emissions. This latter source of information is incomplete for most species, and quite conjectural for some of them.

In this paper, bat species are first classified according to foraging habitat and the feeding habits, following the classification criteria of Kalko *et al.* (1996) these being the most appropriate to the situation in Nouragues.

Classification of the bat guilds

Uncluttered space/ Aerial insectivores

Emballonuridae: *Diclidurus albus*

Molossidae: *Eumops hansae, Molossops greenhalli, Molossops planirostris, Molossus molossus, Molossus rufus, Nyctinomops laticaudatus.*

Background cluttered space/ Aerial insectivores

Emballonuridae: *Rhynchonycteris naso, Saccopteryx bilineata, Saccopteryx leptura, Cormura brevirostris.*
Noctilionidae: *Noctilio albiventris*; Vespertilionidae: *Myotis nigricans, Myotis riparius;* Furipteridae: *Furipteris horrens*; Thyropteridae: *Thyroptera tricolor.*
Highly cluttered space/ Aerial insectivores
Moormopidae: *Pteronotus parnelli*
Highly cluttered space/ Gleaning insectivores
Phyllostomidae: *Lonchorhina inusita, Macrophyllym macrophyllum, Micronycteris brosseti, Micronycteris hirsuta, Micronycteris nicefori, Micronycteris daviesi, Mimon benetti, Mimon crenulatum, Phylloderma stenops, Tonatia bidens, Tonatia schulzi, Tonatia sylvicola.*
Highly cluttered space/ Gleaning carnivores
Phyllostomidae: *Vampyrum spectrum, Chrotopterus auritus*
Highly cluttered space/ Gleaning sanguiniivores
Phyllostomidae: *Desmodus rotundus*
Highly cluttered space/ Gleaning frugivores
Phyllostomidae: *Carollia brevicauda, Carollia perspicillata, Rhinophylla pumilio, Ametrida centurio, Artibeus concolor, Artibeus obscurus, Artibeus gnomus, Artibeus jamaicensis, Artibeus lituratus, Chiroderma trinitatum, Chiroderma villosum, Mesophylla macconnelli, Platyrrhinus helleri, Sturnira tildae, S. lillium, Uroderma bilobatum, Vampyressa bidens, Vampyressa brocki, Vampyressa melissa, Vampyressa pusila, Vampyrodes caraccioli.*
Highly cluttered space/ Gleaning nectarivores
Phyllostomidae: *Anoura geoffroyi, Choeroniscus intermedius, Glossophaga soricina, Lionycteris spurelli, Lonchophylla thomasi, Phyllostomus discolor*
Highly cluttered space/ Gleaning omnivores
Phyllostomidae: *Phyllostomus hastatus, Micronycteris brachyotis*

We have no information concerning the possible place in a guild of the Phyllostomidae *Phyllostomus latifolius* and *Phyllostomus elongatus.* It is interesting to note that the classification of Kalko *et al* (1996) and our own differ very little as to the guild to which a same -or closely allied- species belongs.

4. Conclusion

Much remains to be done concerning the study of the bat populations and communities at Nouragues. Our present knowledge of individual species' population structures and dynamics is fragmentary and concerns but a small number of species; social behaviour and organisation -so varied and complex in bats- and important parameters such as mortality and fertility rates, longevity, sexual maturity, reproduction rhythms and determinisms would all constitute challenging new research topics. Composition and diversity are often the only quantified parameters of bat communities; guilds have rarely been studied*per se* and more in-depth research in the field is needed.

CHAPTER 11

THE MARSUPIAL COMMUNITY

Marie-Laure Guillemin, Martine Atramentowicz and Didier Julien-Laferrière

1. Introduction

Marsupials are an important component of the neotropical fauna, ranking third in species richness after rodents and bats (Wilson & Reeder 1993). The 69 species of marsupials (including 63 Didelphidae, order Didelphimorphia - Wilson & Reeder 1993) account for 6% of the neotropical mammal species richness. Didelphid marsupials are represented all over South America. In French Guiana, 8 genera and 10 species have been recorded, from the tiny mouse opossum, *Marmosops parvidens*, whose body mass does not exceed 20 g, to the large common opossum *Didelphis marsupialis* whose body mass may reach 2 kg.

Various ecological niches are occupied by marsupials in French Guiana, from the ground to the forest canopy, and from secondary regrowth to mature forest. Except for the semiaquatic species, the yapock (*Chironectes minimus*), which feeds on fish and crustaceans, all Didelphid species are partly frugivorous, and thus do play a role in seed dispersal. However, the smaller species are mainly insectivorous, while the medium size species are mainly frugivorous and large species are omnivorous (Charles-Dominique*et al.* 1981).

Previous studies at different sites in French Guiana showed that some species may be found in a large range of habitats, from secondary regrowth close to human activity to primary forest, whereas others are restricted to mature forest (Charles-Dominique *et al.* 1981, Julien-Laferrière 1991). Recent trapping sessions to identify marsupial species occurrence at the Nouragues station were performed. All these trapping results are summarized in this study. To date ten species of Didelphidae have been collected at the Nouragues study area, including a new species for French Guiana, *Marmosa lepida*. We hypothesized that a strong habitat partitioning exists between the ten species of marsupials observed at the Nouragues field station, with two major segregation patterns: horizontal between the different habitat types and vertical between the forest ground and the forest canopy.

2. Study area

Three habitat types were studied:

1- Surrounding of the field station: disturbed vegetation with a 4-10 m high dense undergrowth.

Bongers *et al.* (eds.), Nouragues. Dynamics and plant-animal interactions in a neotropical rainforest, 121–128.
2001 Kluwer Academic Publishers. Printed in the Netherlands.

2- Rock savanna: low, dense and dry stands dominated by Myrtaceae or Clusiaceae on thin soils and rocky outcrops at the base of the inselberg.
3- High mature forest (up to 40-50 m height): mostly uniform, with easily penetrable understorey and continuous canopy on plateau and gentle slope, but also opened by gaps, rocks, gullies or watercourses as well as patches of dense liana forests.
In this study, open area includes the habitat types 1 and 2. Trapping was performed on the ground in the three habitats, and in the canopy of the high mature forest.

3. Trapping methods

Between July 1991 and April 1997, Didelphids were sampled within all habitat types: open areas (~3600 trap-nights), forest ground (~7200 trap-nights) and forest canopy (~6200 trap-nights). One trap-night is equal to one trap being open during 24 h. Two types of live-traps were used in order to sample both small (<100 g) and large (100 g - 2 kg) marsupials: Sherman (23 x 8 x 9 cm) and Tomahawk (50 x 18 x 18 cm). On the ground, traps were placed at 20 to 50 m intervals along transect lines or in grids. In the canopy, traps were set and secured on tree branches, 20-25 m above the ground. Traps were baited each day with banana, apple or fish and checked the next morning. All marsupials were identified to species according to criteria and description of Emmons and Feer (1997). Opossums were lightly anaesthetized with ketamine to facilitate handling of large individuals and pouch young. Data recorded on each captured animal were species, ear cut number, sex, reproductive condition, age, body mass and head and body length. Age of the animals was determined according to dental stage based on rise and wear of the teeth (Atramentowicz 1986). Animals with a third premolar in its lacteal state (molar shaped) were considered as juveniles and those with a definitive third premolar (showing a single cuspid) as adults. This premolar switch occurs, on average, at the age of ten month in *Didelphis marsupialis*, *Philander opossum* and *Caluromys philander* (Atramentowicz 1986). All captured animals were released at the point of capture after recovering.

Nota: Tiny animals (body mass < 20 g) with a long tail, a sharp-pointed muzzle, large dark grey ears and grey fur except for the underparts which are pure white were considered as *Marmosops parvidens*. Nevertheless, this identification is tentative, as we are aware of the existence of other small marmosini which are difficult to separate from *M. parvidens* (Robert S. Voss, personal communication).

4. Results

4.1 SPECIES RICHNESS

In total, 257 captures of 128 individual marsupials were recorded, belonging to 10 didelphid marsupials species (Table 1). Among them, five species: *D. marsupialis*, *P. opossum*, *M. murina*, *M. nudicaudatus*, and *C. philander*, represented 92% of all captures-recaptures (Fig. 1). Mean body mass of marsupial species ranged from 20 to 1200g and the head and body length from 100 to 370 mm (Table 1). More than three third of captured animals were adults (Table 1). The proportion of females with nursing young for all the species was 39%.

Lactating females of five different species were captured, but 92% of these females belonged to only three species: *C. philander*, *D. marsupialis*, and *P. opossum*. Litter size in

TABLE 1. Population data in ten didelphid species trapped at the 'Nouragues' area in French Guiana, 1991-1997.

Species	Captured adults (n)	Captured juveniles[a] (n)	Males / females (n/n)	Body mass (g)[b]	Head and body length (mm)[b]	n Litters	Litter size
C. philander	15	1	7 / 8	287 ± 152	253 ± 27	3	3.9 ± 1.0
D. albiventris	2	1	0 / 3	560 & 690	290 & 270	-	-
D. marsupialis	27	13	24 / 14	1215 ± 420	373 ± 40	14	7.3 ± 2.6
M. brevicaudata	2	0	1 / 1	69 & 91	148 & 167	-	-
M. demerarae	3	0	2 / 1	71 ± 25	137 ± 11	-	-
M. murina	19	5	10 / 11	37 ± 15	110 ± 13	1	8
M. nudicaudatus	15	3	7 / 11	518 ± 149	246 ± 29	1	9
M. parvidens	2	1	3 / 0	17	100	-	-
P. opossum	14	4	7 / 11	371 ± 175	262 ± 24	5	3.8 ± 0.4
M. lepida	1	0	1 / 0	22	108		

[a] age based on dental state (Atramentowicz 1986).
[b] in adults only.

C. philander, D. marsupialis and *P. opossum* were respectively 3.9 ± 1.0, 7.3 ± 2.6 and 3.8 ± 0.4 (Table 1). General characteristics as diet, locomotory pattern, activity time and distribution in South America of these species are summarized in Table 2.

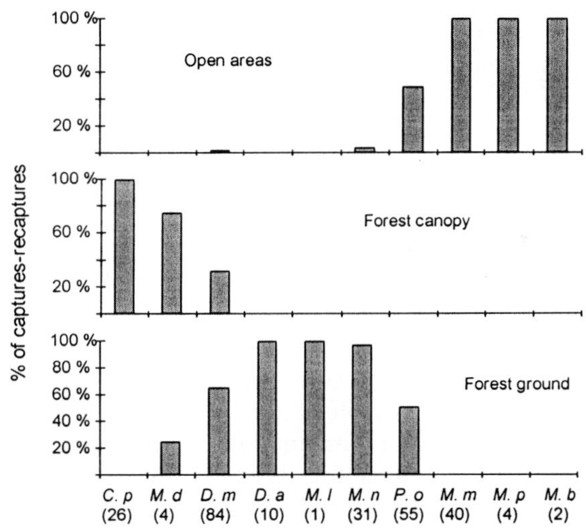

Figure 1. Percentage of captures of Didelphid species in different biotopes at the «Nouragues» field station: ground and canopy of mature tropical forest and open areas (surrounding of the camp and rock savanna). For all species, captures and recaptures are taken into account (total number is given into brackets). *C. p: Caluromys philander, M. d: Micoureus demerarae, D. m: Didelphis marsupialis, D. a: Didelphis albiventris, M. n: Metachirus nudicaudatus, P. o: Philander opossum, M. m: Marmosa murina, M. p: Marmosops parvidens, M. b: Monodelphis brevicaudata.*

4.2 HORIZONTAL AND VERTICAL SEGREGATION AND MICROHABITAT PARTITIONING

Important differences in community structure were obvious among the three habitats (Fig. 1). Three species were captured in the canopy: *C. philander*, *M. demerarae* and *D. marsupialis*. Six species were sampled on the ground of the mature forest: *M. demerarae*, *D. marsupialis*, *D. albiventris*, *M. nudicaudatus*, *P. opossum* and *M. lepida*. In open areas, we captured also six species: *D. marsupialis*, *M. nudicaudatus*, *P. opossum*, *M. murina*, *M. parvidens* and *M. brevicaudata*. *C. philander* was captured only in the canopy and *M. nudicaudatus* only on the ground. *D. marsupialis* is opportunistic and was present in the 3 different habitats. Three among the four smaller species, *M. murina, M. parvidens* and *M. brevicaudata*, were captured only in open areas. The fourth species, *M. lepida*, (one individual) was trapped in the mature forest ground.

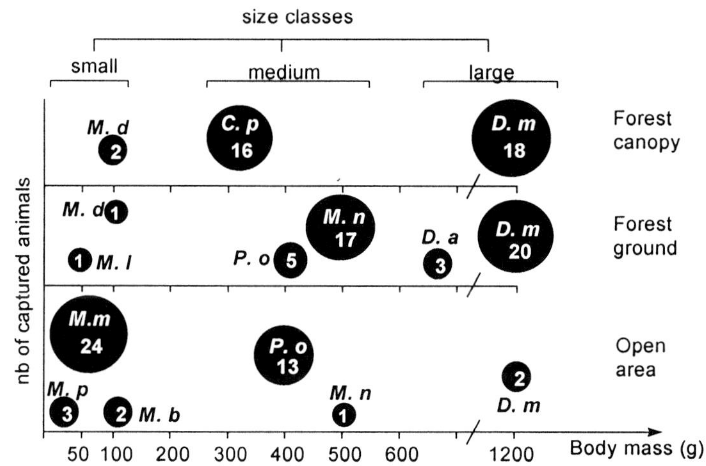

Figure 2. Number of captured animals for different size classes in the three different habitats sampled at the Nouragues field station. Abbreviations as in Figure 1.

In addition to vertical and horizontal segregation, a micro-habitat partitioning based on species body mass was noticeable in each stratum. The didelphid community was divided into 3 size classes according to body mass: small (20 – 100 g), medium (300 – 500 g) and large (1200 g). The canopy strata is the most representative of these micro-habitat partitioning phenomena, with each size class represented only by a single species: *M. demerarae* is small, *C. philander* medium and *D. marsupialis* large (Fig. 2). On the forest ground and in open area, two or three species were represented for almost all size classes (Fig. 2), but one species was relatively more abundant than the others in each micro-habitat. In open areas, *P. opossum* was abundant compared to *M. nudicaudatus*, as *M. murina* compared to *M. parvidens* and *M. brevicaudata*. On the mature forest ground, a similar pattern was observed: *M. nudicaudatus* and *D. marsupialis* were relatively more abundant than, respectively, *P. opossum* and *D. albiventris*.

5. Discussion

5.1 SPECIES RICHNESS

Many annotated checklists of South American mammal faunas are available in the literature, but most study sites have not been censused for long enough time to provide a complete list (Patton *et al.* 1997). We choose to compare species richness at the Nouragues station with three other South American study sites: La Selva (Costa Rica), Barro Colorado (Panama) and Cocha Cashu / Pakitza (Peru). These three sites have been studied for more than 20 years and probably few species are missing, their complete annotated checklists of mammal faunas are available in Voss & Emmons (1996). The presence of ten species of marsupials at the Nouragues study area represents a high but not unusual species richness for an Amazonian rain forest. By comparison, 12 species of marsupials were captured at Cocha Cashu, five species of marsupials were collected at Barro Colorado, and six at La Selva. The high species richness of marsupials observed at the Nouragues station may be due to the mosaic of various habitats present at the study site, which is located at the base of an inselberg. Gallery forests in central Brazil show, similarly a high species richness (Mares *et al.* 1986). These habitats are characterized by an interdigitation of grassland and scrub habitats, allowing rich broad ecological niche types (Mares *et al.* 1986).

Nine among the 10 marsupial species previously captured in French Guyana were trapped at the Nouragues field station. The only species lacking, *Chironectes minimus*, was seen near the river Arataye, nine kilometers apart from the station (Voss & Emmons 1996). *C. miminus* is probably present at the Nouragues station but our sampling procedure was not adapted to the capture of this species which requires partially submerged live traps baited with fresh pieces of fish or crabs. *Marmosa lepida* (Thomas 1888), collected at the Nouragues station, is a new species for French Guiana. The composition of the marsupial community at the Nouragues field station fits well the one proposed in Emmons and Feer (1997) for Didelphidae in South American lowland rain forests: one or two species of *Marmosa* occur together with one of *Micoureus*, *Marmosops*, *Didelphis*, *Caluromys*, *Metachirus*, *Philander* and sometimes *Monodelphis*. The difference in our study site was that two species of *Didelphis* were present but at different densities, with *D. marsupialis* far more abundant. Our data confirm the predictable composition of the Marsupial guild in South American rain-forest which could be explained by the population structure and resource partitioning.

5.2 HORIZONTAL AND VERTICAL SEGREGATION AND MICROHABITAT PARTITIONNING

The main mechanisms that result in the reduction of potential competition among species of the same guild are: distinct activity time, selection of a different portion of the habitat, and diet differentiation (Leite *et al.* 1994). The ecological segregation of neotropical marsupials did not seem to be determined by the species' diets, but better related to spatial differentiation (Miles *et al.* 1981). All marsupial species present are nocturnal (Table 2), except for *M. brevicaudata* which seems strictly diurnal (Charles-Dominique *et al.* 1981). Moreover, all these species feed essentially on invertebrates and fruits (Table 2). Previous work on Didelphid marsupials in French Guiana demonstrated an opportunistic diet, especially in fruit choice in the more abundant species: *Didelphis marsupialis*, *Philander opossum* and *Caluromys philander* (Charles-Dominique *et al.* 1981, Atramentowicz 1988).

Almost all available fleshy ripe fruits are consumed, without any selection based on size, colour or type. Seeds are swallowed if small enough, according to the size of the consumer: seeds of *Passiflora sp.* (< 1 cm) may be found in woolly opossums feces (body mass ~ 300 g), seeds of *Piper sp.* (1 mm) in mouse opossums feces (b. m.~50 g). As a result, Didelphid marsupials have an important impact on forest regeneration and seed dispersal.

An important horizontal segregation occurs in the marsupial community at the Nouragues station with very little overlap between forest species and those of open habitats, except for *P. opossum*. A similar segregation appears in rodent community in the Brazilian cerrado between forest and grassland habitats (Mares *et al.* 1986). In mature forest, *P. opossum* was preferentially found near streams and in old gaps. These zones have, like the rock savanna, a vegetation structure close to secondary forest with a low canopy level and a dense undergrowth. *M. parvidens* was captured only on the ground of the rock savanna, covered with small bush trees (*Clusia sp.*, Myrtaceae and *Pitcairnia sp.* Bromeliaceae). However, this species is also observed in mature forest, on sites where open areas like those occurring at the Nouragues are not available (Julien-Laferrière 1991). *M. murina* was also captured in rock savanna but the artificial opening created by the field station attracted many individuals around the camp. These results compare well with those from August (1984) who showed that *M. murina* preferred open field habitats with arborescent vegetation. The two individuals of *M. brevicaudata* were captured around the camp, two individuals had previously been captured at Cabassou (French Guiana) in a quarry edge and a garden (Atramentowicz & Julien-Laferrière unpublished data).

The vertical use of space has a considerable importance in spatial sharing of sympatric species (Leite *et al.* 1996), by amplifying the possibilities of resource partitioning in each habitat. At the Nouragues study site, seven marsupial species were found in mature forest. There is an important vertical heterogeneity dividing these marsupials species into arboreal and terrestrial communities, with *M. nudicaudatus* and *M. brevicaudata* exclusively terrestrial and *C. philander* exclusively arboreal. *M. nudicaudatus* could be the most terrestrial of any observed opossum, as suggested by body proportions, distribution of muscular masses and locomotory behaviour (Miles *et al.* 1981, Grand 1983). The results presented in this study confirm the arboreal habits of *C. philander* observed by several authors (Charles-Dominique *et al.* 1981, Leite *et al.* 1996). The boundary between terrestrial and arboreal communities was not stiff as some species are generalist (*D. marsupialis*) and more specialized species, like the canopy dwelling *M. demerarae*, occasionally venture into another stratum (Charles-Dominique *et al.* 1981). Different studies have shown that *D. marsupialis* forages in all vegetation stratum (Charles-Dominique *et al.* 1981, Fleck & Harder 1995, Leite *et al.* 1994). On the other hand, *D. albiventris* seems mainly terrestrial (Julien-Laferrière 1991) as observed in our study, where all the animals were trapped on the ground. *P. opossum* was only captured on the ground but this species is a good climber and uses scansorial and arboreal routes especially to escape (Charles-Dominique *et al.* 1981, Fleck & Harder 1995).

Body size differences between the studied opossums may be an important component in resource partitioning. For example, small species would move easily in a overgrown habitat and larger species only could tackle big or tough prey and fruit (Charles-Dominique 1981).

TABLE 2. General characteristics for the ten didelphid species captured at the Nouragues field station: diet, locomotory pattern, activity time and distribution in South America (Emmons & Feer 1997, Reid 1997). *C. minimus* was not captured during this study but is present at Nouragues (Voss & Emmons 1996).

Species	Diet	Locomotor adaptation	Activity	Distribution
Caluromys philander	FR/IN/ NEC	A	N	From Venezuela and Guianas to south central Brazil.
Micoureus demerarae	FR/IN/ NEC	A/S	N	From Colombia and to E Brazil and N Argentina.
Didelphis marsupialis	FR/IN/ CAR	T/S/A	N	From Mexico to Bolivia, NE Argentina and south central Brazil.
Didelphis albiventris	FR/IN/ CAR	T/S/A	N	Around the periphery of rainforest from Venezuela, Colombia and Guianas to Peru, Bolivia and NE Argentina.
Metachirus nudicaudatus	IN/FR	T	N	From Costa Rica to Paraguay and NE Argentina.
Philander opossum	IN/FR/ CAR/NEC	T/S/A	N	From Mexico to Peru, NE Argentina and south central Brazil.
Marmosa murina	IN/FR	S	N	From Venezuela and Guianas to Peru and south central Brazil.
Marmosa lepida	-	-	-	Colombia, Equator, Peru, Bolivia, Brazil and Surinam.
Marmosops parvidens	IN/FR	S	N	Venezuela and Guianas to Brazil and Peru.
Monodelphis brevicaudata	FR/IN	T	D	Colombia, Venezuela and Guianas to Brazil and N Argentina.
Chironectes minimus	P/IN	SA	N	From Mexico and Belize to SE Brazil and Argentina

FR = frugivorous, IN = insectivorous, CAR = carnivorous, NEC = nectarivorous, P= piscivorous.
A = arboreal, S = scansorial, T = terrestrial, SA = semiaquatic.
N = nocturnal, D = diurnal.

Body size variations results in differences in diet items proportion: smaller species are more insectivorous whereas larger ones are omnivorous. The marsupial communities in each stratum of each habitat are composed by three distinct size classes according to body mass: small (20 to 100 g), medium (300 to 500 g) and large (> 1000 g). In the mature forest canopy the three species present are of different size classes, and this size distribution could enhance a reduction of potential competition among these sympatric species. In the other two habitats, open areas and mature forest ground, species are organized by pairs in the majority of the different size classes, with one species relatively more abundant and one or two rarer species. Coexistence of rare and abundant species ecologically close is a general phenomenon in neotropical vertebrates, mammals and birds (Brosset*et al.* 1996). In French

Guiana the bat community is characterized by few common species and a string of rarer species, but fine scale ecological differences reduces the possible competition, as sympatric species are not syntopic (Brosset *et al.* 1996). In French Guianan marsupials, this type of coexistence with very different densities is well known for Didelphis,*D. marsupialis* being far more abundant than *D. albiventris* (Julien-Laferrière 1991, Catzeflis *et al.* 1997).

Acknowledgements

We are grateful to S. Colas and P. Charles-Dominique for allowing us to use their trapping data and to the two reviewers, S. Van Wieren and W.F. Laurance, for helpful comments.

CHAPTER 12

THE AVIAN COMMUNITY: AN OVERVIEW OF SPECIES COMPOSITION AND GUILD STRUCTURE

Jean-Marc Thiollay, Mathilde Jullien, Marc Théry and Christian Erard

1. Introduction

In spite of a growing interest in tropical forests, very few permanent field stations have been established, and still fewer for a time long enough to have produced complete lists of vertebrates with a significant knowledge of the status, distribution, and overall ecology, behaviour and biology of most species.

In the Neotropics (Gentry 1990), birds have been intensively surveyed in such areas in Central America: Barro Colorado Island in Panama (Karr *et al.* 1990) and La Selva in Costa Rica (Blake *et al.* 1990, Levey & Stiles 1994), and in South America: Manu National Park in Peru (Robinson *et al.* 1990, Robinson & Terborgh 1990, Terborgh *et al.* 1990), the Minimum Critical Size of Ecosystems Project area near Manaus in Brazil (Bierregaard 1990), also Tambopata in Peru and sites in Ecuador and Bolivia.

The Nouragues area emerges as another site where a fairly accurate synthesis can be made of the data obtained during a long-term census. The aim of the ornithological survey was a characterization of the bird community in order to understand more properly which species were important in the functioning of the tropical forest ecosystem present in this area, and also to evaluate the importance of birds in the local and regional biodiversity, particularly to outline implications for conservation.

A lot of results have already been published on the bird community of the Nouragues reserve (see particularly Erard 1986, 1989, Thiollay 1986, 1988, 1994, Tostain *et al.* 1992). The present paper is meant as a comment of the bird list given in Appendix 4. We give an overview of the avifauna of the Nouragues area with a special emphasis on the variations in composition and structure of the bird community according to the habitats present in the area and also according to the guilds that can be distinguished.

2. Study site and methods

In the present paper, we describe the bird community of an area of lowland rainforest extending over ca 100 km² from the rocky outcrops of the inselberg were the Nouragues station is located to the Arataye river. Several bird habitats constitute the forest mosaic. They include (1) 40-50 m high mature stands with open understorey, (2) lower stands with more irregular, moderately vine-loaded canopies, (3) heavily naturally disturbed areas, with

Bongers *et al.* (eds.), Nouragues. Dynamics and plant-animal interactions in a neotropical rainforest, 129–141.
2001 Kluwer Academic Publishers. Printed in the Netherlands.

thick vine tangles, a dense undergrowth and a low broken canopy, (4) swamp forest dominated by tall palms. These habitats cover respectively 55, 26, 10 and 4% of the 100-ha most intensively studied plot, to which must be added 5% of recent or older treefall gaps that are an important component of habitat dynamics and heterogeneity, and a source of bird species diversity. Two other localized bird habitats are found on the huge granitic inselbergs: a dense forest rich in Myrtaceae and Clusiaceae on the slopes, and a mosaic of grass, shrub and treelet patches on the rocky upper slopes. Some other openings occur (locally called 'cambrouzes') covered with a very dense bamboo-like vegetation. It is still unclear whether their presence is due to ancient human settlements or has an edaphic origin. Along the Arataye river, between Saut Pararé and Camp Arataye, one finds semi-open habitats, notably dense riparian vegetation, emerging rocks, fallen trees or branches overhanging water, seasonally flooded flats and low successional vegetation on disturbed riverbanks. This mosaic of different forest types, gaps and edges is a typical feature of the Guianan forest landscapes.

Bird species recorded in the whole area during almost 15 years are listed in the Appendix 4 with preliminary information on body mass, density, foraging location and behaviour, diet, social organization and habitat (for more details, particularly on census technique, density estimates and niche characteristics, see Thiollay 1994). However we are fully aware that these data are still scanty and that much further research is required before the biology of the various bird species can be better known. Abundances, in particular of territorial species, were more thoroughly evaluated in the forest mosaic of the 100 ha-area gridded with permanent 100 m x 100 m quadrats near the Nouragues field station. Our categories of social organization did not discriminate between species (e. g. vultures, some kites, parrots, nightjars, swifts, some trogons, toucans, swallows, martins and tanagers) that gather more or less regularly in single species flocks. These flocks have different origins and can vary greatly in size and composition. We do not know enough about them so we just consider here species that live solitary, in pairs or in family parties (MO), in permanent multi-adult groups (PO), promiscuous mating on leks (LE). We singled ant-following species (AN) and regular attendants of mixed flocks (FL, see chapter 13).

Though a number of ornithologists visited the area, few of them stayed there for long periods or were present in successive seasons; furthermore, those who spent long times in the area studied more particular groups or species than the bird community as a whole (see contributions by Erard, Jullien and Théry in this volume). Consequently the Nouragues area is still ornithologically incompletely surveyed and many species remain little known. Birds were particularly studied at the Nouragues station, at Saut Pararé and at Camp Arataye. In each site, though surveys extended over about 1,000 ha each, the most thorough and regular bird studies were conducted on less than 500-ha plots. However, though our list is probably still incomplete (some rare and occasional species will certainly be added in the future), it allows an overview of the bird community present in the Nouragues area.

3. Species richness

3.1 GLOBAL SPECIES RICHNESS

The present list of the birds of the Nouragues area includes 433 species which belong to 275 genera and 51 families. Three additional species recently observed are noted in Appendix 4 but are not considered in the analysis. These figures can be split into

Nonpasserines (180 species, 116 genera, 33 families) and Passerines (253 species, 159 genera, 18 families). Among Passerines, there are almost two times more Suboscines (106 genera, 164 species) than Oscines (53 genera, 84 species) though both are represented by 9 families. New World families (N = 25, i.e. only 49% of families present) account for only 58.2% of the genera and 56.1% of the species. These rates are different for Nonpasserines (39.4% of families, 37.9% of genera, 33.9% of species) and Passerines (66.7% of families, 73% of genera, 71.9% of species); the difference is due to the high proportion of Suboscines in the avifauna which all belong to New World endemic families.

3.2 BREEDING STATUS

Though actual breeding has not yet been proved for every species, we can confidently say that at least 362 species breed in the Nouragues area, and that a further 30 can be considered as potential breeders. The exact status of every species is still unclear. Among these 433 species, we can expect that at least 51 (25 Nonpasserines and 26 Passerines) are migrants or vagrants, wandering species not known to breed or to occur regularly in the area, though some may attempt to settle for instance on the inselbergs and along large rivers or clearings (for example *Tangara cayana* or *Icterus* species). In the Nouragues area true migrants are few: waders (e.g. *Tringa* and *Actitis*), some Tyrannidae (e.g. *Myiarchus* spp. and *Tyrannus dominicensis*), and some swallows and martins. Erratic species (i.e. birds coming from other Guianan areas, particularly from coast) include some vultures (e.g. *Coragyps* and *Cathartes aura*), *Phalacrocorax, Egretta, Bubulcus*, and several icterids and tanagers. Seasonal movements (i.e. erratic tracking of food resources or habitat shift) at least of various parrots, swifts, hummingbirds, toucans and cotingas do occur but are still poorly documented.

3.3 TAXONOMIC DIVERSITY

A few families or groups account for an important part of genus and species richnesses. For instance, Tyrannidae (New World flycatchers), Thamnophilidae (antbirds) and Emberizidae (honeycreepers, tanagers, seedeaters) represent 57.7% of Passerines and one third of the bird species recorded in the area. When diurnal raptors (Cathartidae, Accipitridae, Pandionidae and Falconidae), parrots (Psittacidae), nocturnal birds (owls, Strigidae; potoos, Nyctibiidae and nightjars, Caprimulgidae), hummingbirds (Trochilidae), and trunk specialists (woodpeckers, Picidae and woodcreepers, Dendrocolaptidae) are added, altogether, these 14 families amount to 167 genera (60.7% of all genera present) and 270 species (62.4% of species richness).

Interestingly 70.7% of Nonpasserine and 67.9% of Passerine genera (the percentage being the same for Suboscines and Oscines) are represented by a single species. Even though one may suspect some taxonomic oversplitting, this suggests a high degree of morphological diversification and fine-tuned adaptations to specialized niches.

3.4 SPECIES RICHNESS ACCORDING TO HABITAT TYPE

Table 1 shows the composition of the bird community according to the seven types of habitat defined in Appendix 4. Upland mature forest bird species represent only 52.7% of the avifauna, this percentage applying to both Nonpasserines and Passerines. Secondary forest and second growth species account for 22.9% with a slightly larger representation of

Passerines (23.7%) than Nonpasserines (21.7%). Aquatic or swampy habitats hold 16.4% of the bird species with a little fewer Passerines (14.2%) than Nonpasserines (19.4%). Treefall gap species account for 4.8% of the bird community (2.8% of Nonpasserines and 6.3% of Passerines) and inselberg species for 3.2% (same percent for Nonpasserines and Passerines).

TABLE 1. Bird species richness according to body size, diet and foraging site in each forest habitat type.

Global species richness	Primary forest	Gap	Disturbed forest	Stream, swamp	Large opening	River	Inselberg	Total
Total	228	21	41	29	58	42	14	433
Nonpasserines	95	5	17	12	22	23	6	180
Passerines	133	16	24	17	36	19	8	253
Body mass (g)								
< 8	15	2	2	3	5	1	-	28
8-16	47	4	11	8	9	9	2	90
16-32	40	7	11	7	15	7	6	93
32-64	39	3	8	1	16	9	1	77
64-128	29	3	1	2	3	3	1	42
128-256	20	2	2	1	1	6	1	33
256-512	17	-	5	2	3	1	3	31
512-1024	11	-	-	3	2	3	-	19
1024-2048	9	-	1	2	3	3	-	18
> 2048	1	-	-	-	1	-	-	2
Diet								
carnivore	20	1	1	3	8	1	2	36
frugivore	52	1	8	1	7	2	2	73
nectarivore	11	2	1	3	4	2	-	23
insectivore	108	13	29	16	23	23	10	222
omnivore	37	4	2	2	16	2	-	63
aquatic	-	-	-	4	-	12	-	16
Foraging site								
ground	25	4	3	5	4	-	3	44
low bark	2	-	-	-	-	-	-	2
low foliage	9	4	6	2	7	4	1	33
low perch	-	-	-	-	-	-	1	1
middle bark	17	1	1	2	1	-	-	22
middle foliage	32	4	10	7	5	6	3	67
middle perch	6	1	-	1	7	2	-	17
canopy bark	10	1	3	1	-	1	-	16
canopy foliage	112	5	15	6	17	5	-	160
canopy perch	8	1	3	2	7	3	2	26
water	-	-	-	3	-	15	-	18
aerial	7	-	-	-	10	6	4	27

All non resident species quoted above do not enter primary nor even old secondary forest but keep to open habitats and along large rivers. In French Guiana, true migrants are more common, diversified and regular in coastal habitats than in the forested interior.

4. Size structure

Though the distribution of body weight categories among the whole bird community (Fig. 1, Table 1) peaks between 8 and 64g and then regularly decreases to birds heavier than 2048 g, Nonpasserines and Passerines display different, almost opposite patterns. The former have representatives in all categories with an almost symmetrical distribution of weight categories on both sides of a peak at 128-256 g. On the other hand, Passerines weight distribution is strongly skewed between 8 and 32 g which account for 64.8% of the species (in fact 85.7% of Passerines are between 8 and 64 g while 52.8% of Nonpasserines are above 128 g).

The distribution of the species among the various categories of habitat according to their mass (Table 1) shows that there is no structural difference between habitats except that treefall-gap species are small (all species are under 256 g).

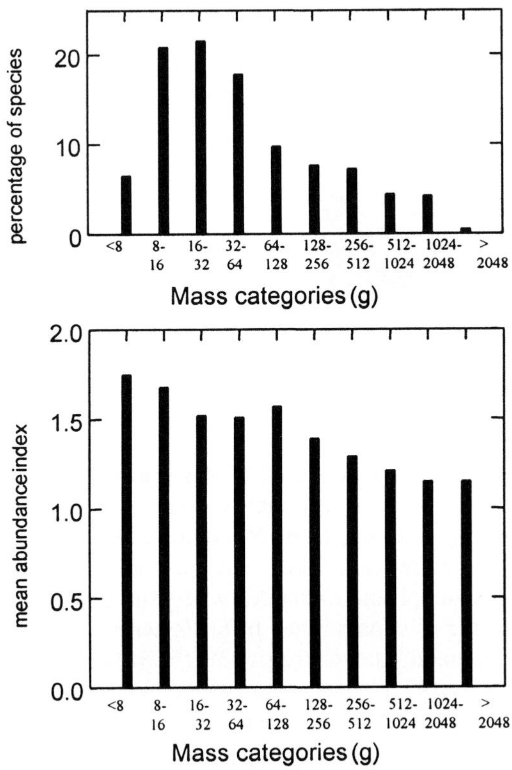

Figure 1. Upper graph: Distribution (in percent of species) of body mass categories in the bird community. Lower graph: Relationship between mean abundance index and body mass category.

5. Trophic structure

Insectivores are dominant, followed by frugivores and omnivores; specialists are represented by carnivores, nectarivores and species feeding on aquatic organisms (Table 1, Fig. 2). This pattern is quite different when Nonpasserines and Passerines are considered separately. The former are distributed among the six diet categories whereas the latter include by only three of these categories (there are no carnivores nor aquatic feeders among Passerines).

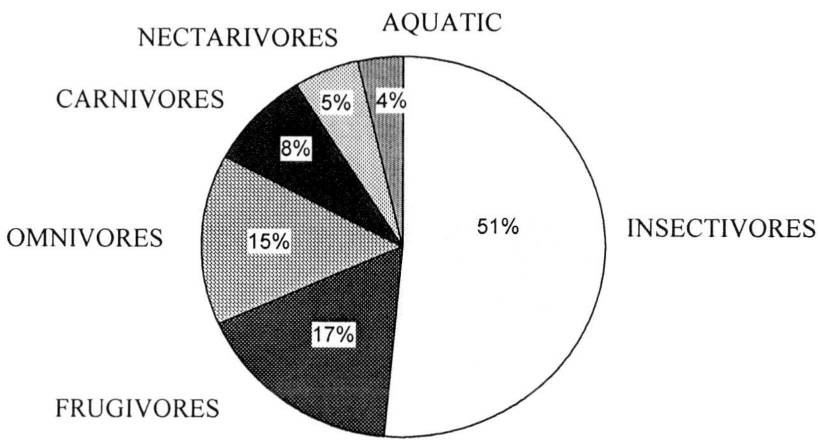

Figure 2. Trophic structure (in percent of diet categories) of the bird community.

When forest habitats (primary forest, disturbed forest and gaps; Table 1) are contrasted with swamps and riversides and with open habitats (large openings and inselberg), some significant differences appear ($\chi^2 = 89.98$, 10 df, $P < 0.001$): there are proportionately more frugivores in forest habitats, more carnivores and omnivores in open habitats, fewer frugivores and omnivores in swamps and riversides where the proportion of aquatic feeders is of course high. If we separate Nonpasserines from Passerines, the former include more frugivores (31.6%) in forest habitats, fewer frugivores (5.7%) and carnivores (11.4%) and of course more aquatic feeders (45.7%) in swamps and riversides, and more carnivores (35.7%) but fewer frugivores (10.7%) in open habitats ($P<0.001$, $\chi^2=47.94$, 8 df). Passerines in swamps and riversides are more insectivores (86.1% *vs.* 65.9%) but less frugivores (2.8% *vs.* 13.9%) and omnivores (11.1% *vs.* 20.2%) than in forest habitats while in open-habitats they are less insectivores (50%) but more omnivores (36.4%) ($P < 0.01, \chi^2 = 13.40$, 4 df).

Probably because of metabolic constraints, diet and body mass are related. Table 2 shows that carnivores (i.e. predators which catch vertebrate prey and need large home ranges) are medium or large-sized species (all weigh at least 62 g, 28% more than 1 kg). On the other hand nectarivores (which feed on localized sugar-rich resources highly variable in space and time) are very small species (all less than 16 g) whereas aquatic feeders (which feed on very diverse aquatic organisms which can be exploited using different foraging behaviours) are evenly distributed in a wide span of mass categories. While all are spread over several mass categories, 84.2% of insectivores and 77.8% of omnivores weigh less than 64 g. On the contrary, 63% of frugivores weigh more than 64 g, 37% more than 256 g (difference between frugivores and insectivores + omnivores, $P < 0.001, \chi^2=60.26$, 1 df). These differences are mainly due to those between Nonpasserines and Passerines. Whereas there is no important mass difference between diet categories in Passerines (though 51.6% of frugivores weigh less than 16 g) because they are more limited in weight range than Nonpasserines, the latter lack small frugivores but include small nectarivores, and insectivores are much smaller than frugivores (78.2% *vs.* 16.7% weigh less than 128g and 1.8% *vs.* 26.2% are above 512 g; $P < 0.001, \chi^2=61.68$, 4 df).

Another expression of trophic structure is how species are separated by their foraging behaviour, e.g. how they share or partition resources by means of differences in foraging location (Table 2). Aerial feeding is the concern of 6.2% of the avifauna but of only 3.6% of Passerines against 10% of Nonpasserines ($P < 0.02, \chi^2 = 6.41$, 1 df). Ten percent of Nonpasserines (i.e. 4.2% of avifauna) are associated with permanent water. Ground feeding is common and typical of 10.2% of the avifauna (7.8% of Nonpasserines, 11.9% of Passerines but difference not significant). Proportionately to the number of species present, ground feeding is most frequent in treefall gaps and on inselbergs (low vegetation) and in seasonally flooded forest, banks and swamps. Lower strata of the vegetation hold 8.3% of bird species: 3.3% of Nonpasserines but 11.9% of Passerines ($P<0.01, \chi^2=8.94$, 1df). On the other hand, 24.5 % of the species are restricted to middle strata (16.1% Nonpasserines *vs.* 30.4% Passerines; $P < 0.001, \chi^2 = 10.91$, 1) whereas 46.6% are typical of canopy (52.8% Nonpasserines *vs.* 42.3% Passerines; $P < 0.05, \chi^2=4.23$, 1 df). Foliage harbours 60% of the avifauna (50% Nonpasserines *vs.* 67.2% Passerines; $P < 0.001, \chi^2 = 12.25$, 1 df), barks and trunks 9.2% (8.9% Nonpasserines; 9.5% Passerines), while 12.8% of the bird community (13.3% Nonpasserines; 7.9% Passerines) are perch-hunters. Table 2 shows also that there are slightly more bark and trunk creepers in middle strata (5.2%) than in canopy (3.7%): this is due to fewer Passerines of this category in canopy (2%) than in mid-strata (6.7%) contrary to Nonpasserines (respectively 6.1% and 2.8%) (difference between Nonpasserines and Passerines, $P < 0.02, \chi^2 = 6.27$, 1 df). It is also clear that species which forage in foliage are more abundant in canopy than below, and fewer in lower than upper understorey. Discrepancies between levels are even sharper in mature primary forest where the number of foliage-gleaning species increases in successive strata by a factor of 5 for Nonpasserines against 3 for Passerines.

A further illustration of trophic structure is obtained when foraging location is considered in relation to diet (Table 2). Nectarivores are distributed in all vegetation levels, but they are better represented in the canopy than below it (56.5% *vs.* 17.4 + 26.1% in understorey; $P = 0.05, \chi^2 = 5.9$, 2 df). Globally, frugivores and insectivores include proportionately more ground feeders than omnivores (respectively 19.2%, 11.3% and 6.3%; $P = 0.05, \chi^2=5.9$, 2 df). However insectivores are more often (13.8% *vs.* 3.6%) and

TABLE 2. Bird species richness according to body mass, foraging site, social organization and abundance for each diet type.

	Carnivore	Frugivore	Nectarivore	Insectivore	Omnivore	Aquatic	Total
Body mass (g)							
< 8	-	-	19	9	-	-	28
8-16	-	16	4	59	10	1	90
16-32	-	5	-	68	19	1	93
32-64	-	6	-	51	20	-	77
64-128	3	8	-	22	7	2	42
128-256	6	11	-	9	3	3	32
256-512	8	16	-	3	3	2	32
512-1024	9	4	-	1	1	3	18
1024-2048	9	6	-	-	-	4	19
> 2048	1	1	-	-	-	-	2
Foraging site							
ground	-	14	-	25	4	1	44
low bark	-	-	-	2	-	-	2
low foliage	-	4	4	22	3	-	33
low perch	-	-	-	1	-	-	1
middle bark	-	-	-	22	-	-	22
middle foliage	7	6	6	39	9	-	67
middle perch	4	-	-	9	4	-	17
canopy bark	-	-	-	16	-	-	16
canopy foliage	14	49	13	43	41	-	160
canopy perch	5	-	-	19	2	-	26
water	-	-	-	3	-	15	18
aerial	6	-	-	21	-	-	27
Social organization							
monogamous	36	45	10	216	51	16	374
polygamous	-	10	-	2	10	-	22
lekking	-	18	10	-	2	-	30
ant-follower	1	-	-	7	-	-	8
mixed-flock	-	18	-	54	24	-	96
Abundance							
< 1 pair/km²	34	33	8	112	31	15	233
1-9 pairs/km²	2	37	12	98	27	1	177
> 10 pairs / km²	-	3	3	12	5	-	23

frugivores less often ground feeders (23.8% *vs.* 12.9%) in Passerines than in Nonpasserines. Though 9.5% of all insectivores are aerial, the difference between Nonpasserines and Passerines is significant: respectively 21.8% and 5.4% ($P < 0.01$, $\chi^2 = 11.2$, 1df). Nonpasserine insectivores are less evenly distributed among the different vegetation levels, being more concentrated in the canopy than Passerines (from lower understorey to canopy: 1.8%, 18.2% and 49.1% *vs.* 14.4%, 35.9% and 30.5%; $P < 0.001$, $\chi^2 = 14.75$, 2 df); they are also more bark creepers (29.1% *vs.* 14.4%; $P < 0.05$, $\chi^2 = 5.11$, 1 df), less foliage gleaners (14.5% *vs.* 57.5%; $P < 0.001$, $\chi^2 = 26.67$, 1 df) and more perch-hunters (25.4% *vs.* 9%; $P < 0.01$, $\chi^2 = 8.49$, 1df). Though represented by fewer species in canopy (38 *vs.* 51) and foliage (47 *vs.* 96) Passerine omnivores are proportionately more often found there (respectively 69.1% *vs.* 30.5% and 85.5% *vs.* 57.5%; in both cases P < 0.001, 1 df, χ^2 = 24.02 and 12.93 respectively) than Passerine insectivores.

6. Species abundance

There are 6 times fewer abundant bird species (> 10 pairs/km²) and 1.6 times more rare species (< 1 pair/km²) than expected from a random distribution (Table 3, Fig. 3). When Nonpasserines and Passerines are considered separately, it appears that about two thirds of the former are rare species against only half of the latter: in Nonpasserines 63.9% of the species have an estimated density lower or equal to 1 pair/km², 32.8% between 1 and 9 and 3.3% 10 or more whereas in Passerines the respective percentages are 46.6%, 46.6% and 6.8% (P < 0.01, χ^2 = 12.81, 2 df).

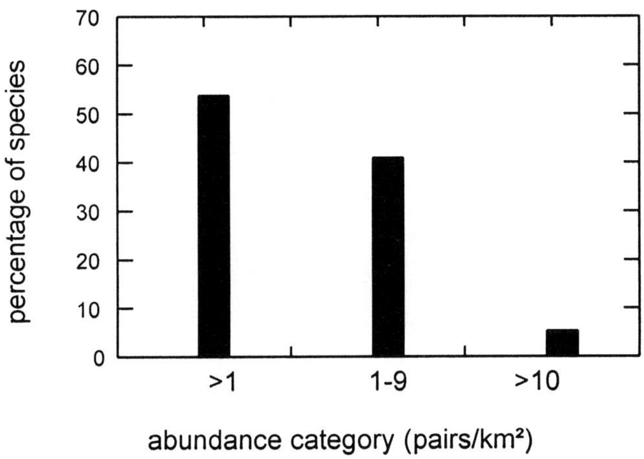

Figure 3. Distribution (in percent) of species among the abundance categories.

As expected, mean abundance index decreases with increasing weight (Fig. 1). The relationship should be stronger and the decrease sharper with real densities and mean species weights instead of coarse size and abundance indices (see Thiollay 1994).

When abundances are considered in relation to habitat (Table 3, Fig. 4), it appears that, except in mature forest, the proportion of rare species is much higher than other abundance categories in the bird community. In upland mature forest, if numbers of rare species do not differ from those derived from a random distribution, on the other hand those with 1-9 pairs/km² are almost two times higher while frequent species are just a quarter of those predicted by chance (P < 0.001, χ^2 = 90.06, 2 df). Riverine and particularly open habitats hold proportionately more rare species than forest habitats. Globally, in mature primary forest one finds proportionately more (68.4% *vs.* 41.9%) species with more than 1 pair/km² and fewer (31.6% *vs.* 58.1%) rare species (less than 1 pair per km²) than in treefall gaps and disturbed forest (P < 0.001, χ^2 = 13.52, 1 df).

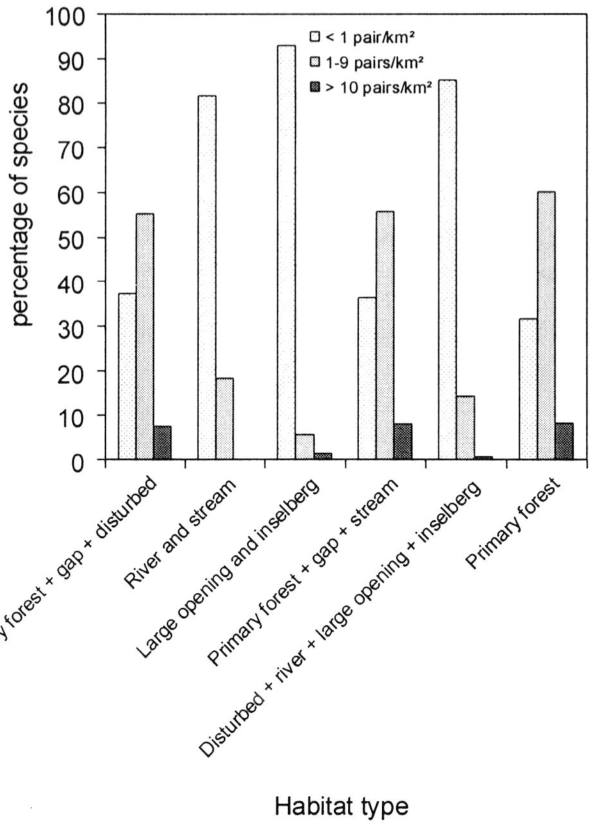

Figure 4. Distribution (in percent) of species among the abundance categories according to habitat type.

The distinction between Nonpasserines and Passerines shows that in mature primary forest Nonpasserines are equally distributed between abundance categories with less than 10 pairs/km² whereas Passerines include proportionately much fewer rare species (20.3% *vs.* 47.4%). Disturbed forests and treefall gaps hold proportionately more species with an abundance of more than 1 pair/km² than open habitats and inselberg or riverine and swampy forests; however this difference is significant for Passerines but not for Nonpasserines.

The 31 Passerines and 6 Nonpasserines with an abundance index of 3 or 2,3 in Appendix 4 are insectivores (23), frugivores (6), omnivores (5) and nectarivores (3); they are mainly forest birds (31) though some are associated with treefall gaps (3) or disturbed forest (2), or have their highest densities on inselbergs (1).

Some other differences appear also when abundance is considered in relation to diet (Table 3). Globally, predators and aquatic feeders have very low densities whereas frugivores and nectarivores include more common species: 94.4% of carnivores, 93.7% of species feeding on aquatic organisms, 50.5% of insectivores, 48.4% of omnivores, 45.2% of frugivores and 34.8% of nectarivores are rare species. Passerines differ from

Nonpasserines in that they include fewer rare insectivores (45.5 *vs.* 65.4%; $P < 0.02$, $\chi^2 = 5.81$, 1 df).

TABLE 3. Bird species richness according to social organization and abundance in each fores habitat.

	Primary forest	Gap	Disturbed forest	Stream, swamp	Large opening	River	Inselberg	Total
Socia organizatio								
monogamou	196	17	37	26	48	36	14	374
polygamou	10	2	3	-	5	2	-	22
lekkin	20	2	1	2	4	1	-	30
ant-followe	6	1	1	-	-	-	-	8
mixed-floc	77	2	9	2	3	3	-	96
Abundanc								
< 1 pair/km	72	10	26	19	56	39	11	233
1-9 pairs/km	137	8	15	10	2	3	2	177
> 10 pairs km	19	3	-	-	-	-	1	23

7. Social organization

Though the life history of all species is far from being well known, most of them can be placed in broad categories of social organization (see Appendix 4, and Tables 2 and 3). Seven Nonpasserine species were discarded from this analysis and we now deal with 173 Nonpasserines, 253 Passerines and a total of 426 species.

Overall, 87.8% of bird species of the Nouragues area are monogamous, 5.2% polygamous, 7% lekking. Among them 1.9% are ant-followers and 22.5% attend mixed flocks. Nonpasserines differ from Passerines in that they are less often lekking (5.8% *vs.* 7.9%), ant-followers (0.6% *vs.* 2.8%) and mixed flocks members (4.6% *vs.* 34.8%). The largest numbers of polygamous and lekking species are found in mature primary forest (Tab. 3); however, these categories are proportionately as well represented in treefall gaps as in secondary forest and large openings. The only significant difference between habitats is the higher number of mixed flocks species in mature primary forest than elsewhere ($\chi^2 = 47.58$, 1 df, $P < 0.001$).

When social organization is considered in relation to diet (Table 2) it appears that carnivores and aquatic feeders are monogamous whereas nectarivores include as many monogamous as lekking species. Fewer frugivores (61.6%) than omnivores (80.9%) and particularly insectivores (99.1%) are monogamous; frugivores and omnivores are more often polygamous than insectivores (respectively 13.7% and 15.9% *vs.* 0.9%), and frugivores include many more lekking species (24.7%) than omnivores (3.2%) and insectivores (none) ($P < 0.001$, $\chi^2 = 93.98$, 2 df). Mixed flocks include omnivores (38.1%), insectivores (24.8%) and frugivores (24.7%); it is particularly marked in Passerines.

8. Discussion

This list represents 61% of that for entire French Guiana by Tostain *et al.* (1992) who mention 708 species. Though important, the number of species recorded in the Nouragues area is still 122 (28%) less than that given for Cocha Cashu in Peru (Robinson *et al.* 1990). However, about half of the difference is due to the presence of early forest successions on ox-bow lakes in the Peruvian site, an habitat type which is lacking in that the Guianan site. The Nouragues list is richer than the Manaus one where 352 species were recorded from the forest study area and the surrounding clearings (Bierregaard 1990); it is close to that from La Selva (411 species *fide* Levey & Stiles 1994). The high number of breeding or potentially breeding species compare better with the breeding avifaunas of Peru, Ecuador and Colombia than with those of Central America (see Levey & Stiles 1994).

8.1 RARITY

On their study site at Cocha Cashu Terborgh *et al.* (1990) describe body size relationships close to those found in the Nouragues area. They emphasize the link between rarity and large size, particularly as large species that contribute most to biomass (e.g. raptors, cracids, parrots, woodpeckers, some woodcreepers and jacamars, potooes) require large home ranges and at least some of them are hunted, an important point to consider in conservation politics. The Nouragues area has still a 'complete' bird community in that all the large species of the Guianan Shield forest block are present with viable populations and species richness is high; this supports those (see e.g. Thiollay 1989, Robinson *et al.* 1990) who claim for the conservation of large areas including all main forest types and their natural disturbance regimes.

The explanation of why so many tropical forest bird species are rare, or why so few are common, is still debated (e.g. Karr 1977, Terborgh *et al.* 1990, Thiollay 1994). The answer is complex in that various factors interact. Low detectability and/or seasonal movements and yearly fluctuations may introduce serious sampling biases. Naturally low density is the most widespread form of rarity where low prey availability, high predation pressures and interspecific competition may constitute important limiting factors. However patchy distribution is another form of rarity phenomenon in relation to marked habitat specialization's with subtle microhabitat requirements, particularly for foraging and breeding. Natural disturbance, edaphic variability, treefall gaps and ensuing successions are classical factors of high heterogeneity of both habitat and resources distribution in tropical forests as well as their four-dimensional mosaic pattern, highly variable and heterogeneous is space and time (Richards 1952, Sutton *et al.* 1983, Brokaw 1985, Karr & Brawn 1990). This would allow species-specific specialization's and promote species richness but because of the small size of the components of this habitat mosaic this would also entail large home ranges or territories particularly for species with life-long pair-bond and settlement (which may be the case for most forest species), and low densities as a result.

8.2 PATTERN DIFFERENCES BETWEEN NONPASSERINES AND PASSERINES

Throughout this chapter we have stressed differences between Nonpasserines and Passerines. In fact such disparities may exist in other tropical forests (cf. Erard 1989, Erard & Brosset in press) let alone in other bird communities but they are rarely dealt with. However, Nonpasserines are phylogenetically older than Passerines (cf. Olson 1985, Sibley & Ahlquist 1990) so their comparison may help to understand the role historical factors play in bird community structuring. They are larger species and distributed along a wider diet range than Passerines. Their adaptive zone includes specialization's like preying on large animals (invertebrates and even vertebrates) with 'sit-and-wait' behaviour, feeding on aquatic organisms, nectar-feeding (favoured by small body size and specific hovering flight in hummingbirds, see Norberg 1990), aerial feeding (with high performance flight devices in swifts and falcons), frugivory (including pulp, pulp-and-seed and seed eating of larger fruits than those consumed by Passerines). It looks as if Passerines could exploit forest resources only in niches left empty or not saturated by Nonpasserines; their small size and low densities would enhance this opportunism and would also favour more species richness than abundance. These ecological constraints and organization processes are not exclusive of those usually recognized, e.g. low food availability, habitat heterogeneity, interspecific competition, predation and diseases pressure (see Thiollay 1991a). Of course this remains very speculative and requires much further research.

8.3 LIMITS OF FIELD DATA

As in many other surveys, data used for the present study are still fragmentary and their analyses are based on still too broad and oversimplified categories. Habitat use should be much better quantified and habitats themselves should be better defined. Forest mosaic is surely more complex for a bird than for a human observer. We need also better measures of density in all habitats, in number of individuals per unit area, i.e. taking into account all individuals present, not solely breeding units. Species diets and their variations along the annual cycle both within and between habitats are still poorly known. Though sometimes considered outdated, thorough autecological studies are a prerequisite before processes responsible of the structuration of tropical forest bird communities can be correctly evaluated. Furthermore, other vertebrates and invertebrates of the study site should also be considered, because they can be either prey, predators or competitors of birds. This should involve multidisciplinary teams working on particular sites in the long term and on a scale sufficiently large to include all resident species and particularly the largest ones (see also Robinson *et al.* 1990, Terborgh *et al.* 1990, Levey & Stiles 1994). The Nouragues should be one of these sites.

Another question is why there are so many monospecific genera and conversely 9.5% of Nonpasserine genera and 6.3% of Passerine genera represented on the site by at least 4 species, and two Passerine genera (*Myrmotherula* and *Tangara*) including up to 7 species. The presence in tropical forests of almost 'sibling species' is an interesting point which may help to understand how regional factors participate to the dynamics and the structuring of communities (see Cracraft & Prum 1988, Brosset 1996).

A final comment will be that too many studies of bird communities 'select' the species they consider in their analyses. Usually only breeding or resident species are considered in depth, vagrants are neglected or listed separately. Do really vagrants play no role in the community? Every community is a dynamic assemblage of species. To-day vagrants can be to-morrow residents. Even if they do not settle for long in the area, their study may provide insights into the resistance to invasion of the community, as well as on the potentialities of colonization after disturbance.

Acknowledgements

We owe a great debt to A. Brosset, P. Charles-Dominique, the late J.-L. Dujardin, P. Gaucher, B.Goguillon, the late P. de Mercey, P. Raynaud, F. Thiollay, O. Tostain, P. Unitt and C. Vansteenwegen who collected data on the birds of the Nouragues area, their status and their life histories. O. Claessens commented on a draft of this paper.

This white-tailed trogon (*Trogon viridis*) is not afraid of the canopy walkway. *Photo Patrick Jansen.*

CHAPTER 13

THE ADAPTIVE SIGNIFICANCE OF FLOCKING IN TROPICAL UNDERSTOREY FOREST BIRDS: THE FIELD EVIDENCE

Mathilde Jullien and Jean-Marc Thiollay

1. Introduction

Flocking behaviour is very widespread among animals (Barnard 1983, Morse 1985, Faaborg 1988) which has given rise to an abundant literature. Multispecific associations of birds occur both in temperate, subtropical and tropical areas. Birds display a wide array of social and territorial behaviours, which play a major role in shaping their density, spatial distribution, breeding and foraging strategies and inter- or intraspecific relationships (Jullien & Thiollay 1998). A given species may harbour different social behaviours in different stages of its life history. We focus here on birds consorting together exclusively when they are actively foraging. Even when it is restricted to foraging birds, flocking behaviour may have different origins, forms and functions. This is probably why the two major and non-mutually exclusive hypotheses proposed to explain how birds may benefit from feeding in flocks are still debated. The first hypothesis claims that flocking may enhance predator avoidance, and thus decrease predation risk, either through an early detection of predators resulting from vigilance conferred by 'many eyes and ears' (Powell 1974, Hoogland 1979, Lazarus 1979) or by confusion and dilution effects during attacks (Pulliam 1973, Bertram 1978). At the same time, individuals foraging in flocks may spend less time scanning for predators and devote more time searching for food (Lazarus 1972, Powell 1974, Popp 1988). The alternative hypothesis postulates that flock members can increase their foraging efficiency using information transferred actively or passively by the more successful (Krebs *et al.* 1972, Krebs 1973, Greig-Smith 1978, Herrera 1979, Sasvari 1979, Giraldeau 1984, Giraldeau & Lefebvre 1986), the more experienced and/or the elder flockmates they imitate or follow to food patches (Barnard 1983, Krebs & Davies 1987).

These two hypotheses are usually tested either on dense, monospecific feeding aggregations found in open marine and terrestrial temperate habitats (e.g., in gulls, geese, waders and granivores) or on winter flocks of passerines in temperate forests (Morse 1970, Ekman 1986, Smith & Buskirk 1988, Rollfinke & Yahner 1991, Suhonen 1993a and b). Little information is available in the Neotropics (but see Munn & Terborgh 1979, Powell 1979, 1985, Gradwohl & Greenberg 1980, Wiley 1980, Munn 1985, Greenberg & Gradwohl 1985, 1986) although foraging flocks are a prominent feature of bird communities in the tropical forests (Buskirk 1976, Morse 1977, Powell 1985, Thiollay & Jullien 1998).

Bongers *et al.* (eds.), Nouragues. Dynamics and plant-animal interactions in a neotropical rainforest, 143-159.
2001 Kluwer Academic Publishers. Printed in the Netherlands.

The aim of this paper is first to summarize the principal characteristics of the different types of associations observed in French Guiana (PART I). This paper also compares the natural history and selective values of these different types of flocking behaviour (PART I) with particular emphasis on the most complex of all, the mixed-species flocks (PART II). At Nouragues, birds were observed consorting in seven distinct groups; they include opportunistic associations around temporary and localized food sources (fruiting trees, army ants), monospecific flocks of group-living or colonial breeding species and multi-species canopy flocks of omnivores where mating systems or other life history traits may overshadow foraging requirements as flocking determinants (Jullien 1997, Thiollay & Jullien 1998). As written by Powell (1985), 'to effectively evaluate (the validity of the two hypotheses previously explained), it is necessary to have a working knowledge of the structure, dynamics, and spatial organization of mixed-species flocks'. To this end, we focus only on a typical and widespread mixed-flock type, maybe the most highly evolved, stable and socially structured: the year-long association of insectivores in the mid-level understorey of neotropical lowland forests which have been studied in details by one of us (MJ) from 1991 to 1999. We propose to assess and discuss the adaptive significance of this flocking type from results already published (Jullien 1997, Jullien & Thiollay 1998, Thiollay & Jullien 1998). The study site and the methods used to collect and analyse the data have been explained elsewhere (Jullien 1997, Jullien & Thiollay 1998, Thiollay & Jullien 1998).

1.1 BEHIND THE SCENE: THE TROPICAL FOREST ENVIRONMENT

Understanding the function of mixed-flocking needs a clear appreciation of the major constraints in primary rain forest ecosystems (Gentry 1990, Thiollay 1991a). Low insect availability is suggested by low foraging success of tropical forest foliage gleaners compared to similar temperate species (Jullien 1997, Thiollay 1988). Many invertebrates appear irregularly and impredictably distributed in space and time, and relatively inaccessible, because of high diversity, low density and productivity, seasonal variations (Leigh *et al.* 1983), habitat heterogeneity and multiple antipredator defences. Foraging activity is further limited by short daylight period (at best 10h/day in the understorey), frequency of rains and midday heat. There is also much circumstantial evidence of a high predation pressure (emphasized by studies on breeding success), maybe more because of a wide variety of predators than high specific densities (Greene 1988, Thiollay 1989).

From the point of view of bird detectability and predation risk, tropical forest habitats may be divided into: (a) litter and low above ground vegetation, where birds, when they are cryptically coloured and slow moving, are difficult to see and may be attacked only from the above; (b) relatively open undestorey of mature forest with dim light, where foraging birds are rather conspicuous but where stalking predators are not easy to detect; (c) dense vegetation (second growth of vine-tangles) at any level where birds are both difficult to detect and to reach unobtrusively; (d) upper canopy, gaps and edges where birds are conspicuous but can be approached at close range only from one side (cover). Obviously (b) and (d) are likely to be the most risky habitats, whereas (a) and (c) are more protective, although Brawn *et al.* (1995) suggested that terrestrial birds might be subject to a high predation rate. The occurrence of mixed-flocks matches this decreasing order of safeness (Fig. 1).

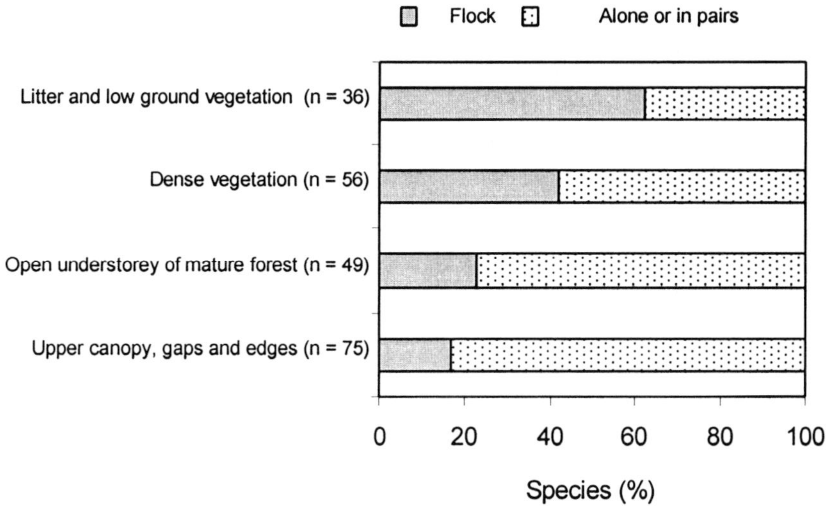

Figure 1: Number of bird species foraging in flocks and alone or in pairs, in the dangerous (open understorey of mature forest; upper canopy, gaps and edges) and safe habitats (litter and low above ground vegetation; dense vegetation), expressed in percentage of the total number of species recorded in each habitat (From Jullien 1997 and Thiollay & Jullien 1998).

2. General review of social groupings

2.1 SOCIAL FORAGING PATTERNS: THEIR RELATIVE IMPORTANCE AND CHARACTERISTICS

Thiollay & Jullien (1998) attempted to document the relative frequency and characteristics of social foraging behaviour among 216 species in the avifauna of French Guiana (see Appendix after Thiollay & Jullien 1998). A total of 13,550 records of different encountered foraging individuals was collected at the Nouragues field station from 1991 to 1994 (see Appendix in Thiollay & Jullien 1998). Overall, 133 bird species were detected foraging either in pairs (19.0%, Table 1) or solitarily (42.6%, Table 1). Feeding in pairs is a common behaviour among insectivores (Fig. 2) whereas foraging solitarily is widespread among carnivores (raptors), nectarivores (hummingbirds) and lek-mating frugivores (manakins, some cotingas, Fig. 2). The remaining 83 species (38.4% of the total community) were mostly or exclusively birds foraging in flocks of different composition, including: (1) multi-species upper canopy flocks, (2) understorey mixed-species bird flocks, (3) monospecific associations, (4) opportunistic gatherings of frugivores, and followers of (5) army ants, (6) Red-Throated Caracaras (*Daptrius americanus*) and (7) tamarin or capuchin monkeys (Table 1).

TABLE 1. Distribution of foraging behaviours in the bird community of a primary neotropical rainforest in French Guiana (N = 216 species; from Jullien 1997, Thiollay & Jullien 1998).

Social foraging behaviour	Total number of species *	% of 216 species
Solitary	92	42.6
Pair	41	19.0
Multispecies canopy flocks	31	14.4
Monospecific groups	21	9.7
Understorey mixed flocks	20	9.3
Gatherings of frugivores	4	1.9
Army ant followers	4	1.9
Caracara followers	2	0.8
Monkey followers	1	0.4

* Each species was confined in a single class of social foraging behaviour, i.e., where > 50% of records occurred.

Multi-species upper canopy flocks

Multi-species canopy flocks are highly mobile associations of small insectivores, frugivores, and omnivores (Fig. 2). Less common than the understorey flocks (Munn 1985), the canopy groups are larger (Terborgh 1990) and apparently the most abundant flock type in terms of the number of species observed at least once foraging within. In all, 81 species were identified in these canopy flocks (Fig. 3), 31 of which were regular participants (50 - 95% of all specific records, Table 1). Canopy flocks were dominated by tanagers (*Chlorospiza*, *Coereba*, *Cyanerpes*, *Dacnis*, *Euphonia*, *Hemithraupis*, *Tachyphonus*, *Tangara*), flycatchers (*Myiopagis*, *Tolmomyias*), as well as the Ash-winged Antwren *Terenura spodioptila* (Thamnophilidae), the Pink-throated Becard *Pachyramphus minor* (Tyrannidae), the Tropical Gnatcatcher *Terenura spodioptila* (Sylviidae) and the Slaty-capped Shrike-Vireo *Vireolanius leucotis* (Vireonidae). Each canopy flock was apparently led by the Fulvous Shrike-Tanager *Lanio fulvus* (Thraupinae), a noisy and sallying species. This kleptoparasite may increase its foraging success at obtaining prey through the use of false alarm calls that distract the bird that flushed it, a behaviour similar to *Lanio versicolor* in Amazonian Peru (Munn 1986). Each multi-species canopy flock was always composed of a permanent core of mated pairs or family groups of three obligate flock participants. They were the Pink-throated Becard, the Ash-winged Antwren, as well as the Fulvous Shrike-Tanager, which switched frequently from canopy to understorey flocks when the two flocking types happened to meet (Jullien & Thiollay 1998).

Understorey mixed-species bird flocks

The mixed-species and yearlong associations of medium size insectivorous birds occur at mid-level of relatively open understorey in tall, closed-canopy mature forest stands (Fig. 2). Eighty-six species were identified in these understorey flocks (Fig. 3), 20 of which were regular flock members (Table 1). Details on the composition, organization, structure and dynamic of this flocking type are summarized in next section PART II.

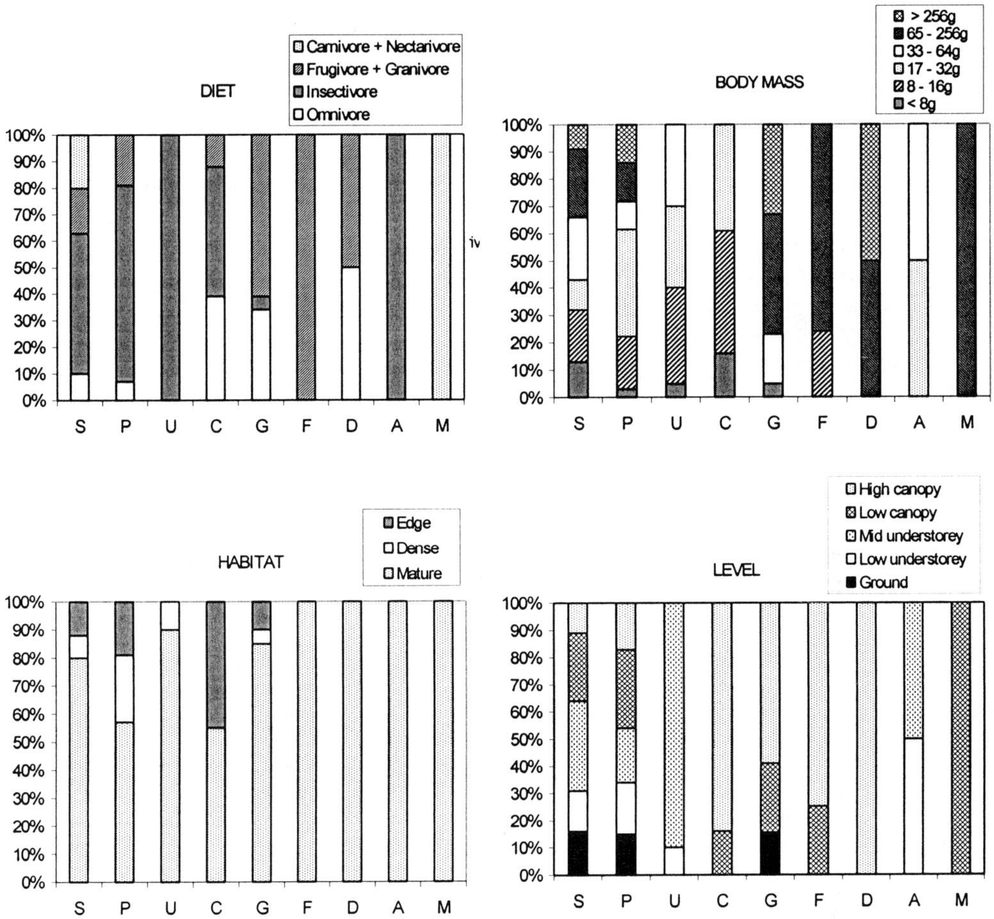

Figure 2: Distribution of species, in percentage of the total community, among categories of diet, size, habitat and foraging level. S = Solitary; P = Pairs; C = Multi-species upper canopy flocks; U = Understory mixed-species bird flocks; G = Monospecific associations; F = Opportunistic gatherings of frugivores; A = Army-ant followers; D = Followers of Red-Throated Caracaras; E = Monkey followers (After Jullien 1997 and Thiollay & Jullien 1998).

Monospecific groups

Most of the monospecific groups are permanent and stable associations of large frugivores and granivores foraging at every level in the mature forest (Fig. 2). They were composed of family groups of 4 to 15 individuals (mated pairs, unmated adults, youngs and immatures). Thirty-three species were identified in these groups (Fig. 3), 21 of them were regular participants (Table 1) belonging to different social systems. A first social system was represented by permanently group-living co-operative breeders. This behaviour was documented in a variety of species from a large terrestrial omnivore, the Gray-winged

Trumpeter *Psophia crepitans* (de Mercey 1994), to a raptor, the Red-throated Caracara *Daptrius americanus* (Thiollay 1991b), a small understorey insectivore, the Musician Wren *Cyphorhinus arada* and an arboreal frugivore, the Purple-throated Fruitcrow *Querula purpurata*. The second social system included gregarious frugivores, with solitary (most Psittacidae and the Yellow-Green Grosbeak *Caryothraustes canadensis*) or colonial breeding (the Golden-winged Parakeet *Brotogerus chrysopterus*, the Green Oropendola *Psarocolius viridis*, the Yellow and Red-rumped Caciques *Cacicus cela* and *C. haemorrhous*).

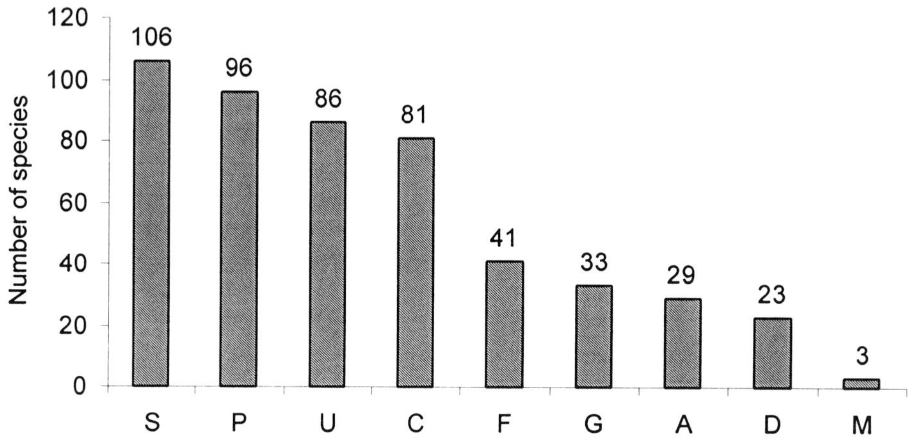

Figure 3. Total number of bird species observed at least once foraging in one of the 9 social categories. This number is independent of the flocking propensity of each species. S = Solitary; P = Pairs; C = Multi-species upper canopy flocks; U = Understorey mixed-species bird flocks; G = Monospecific associations; F = Opportunistic gatherings of frugivores; A = Army-ant followers; D = Followers of Red-Throated Caracaras; E = Monkey followers (From Jullien 1997 and Thiollay & Jullien 1998).

Opportunistic gatherings of frugivores

Mono- or multispecific frugivorous flocks correspond to short-term and opportunistic gatherings of small (Pipridae, Thraupidae), medium size (Cotingidae) and larger birds (Psittacidae, Ramphastidae, Cracidae) around fruiting trees (Fig. 2). They included 41 species (Fig. 3) but only four of them were regular flock members (Tab. 1). They were the Golden-headed Manakin (*Pipra erythrocephala*), the Pompadour Cotinga (*Xipholena punicea*), the Purple-breasted Cotinga (*Cotinga cotinga*) and the Spangled Cotinga (*Cotinga cayana*). Because of a constantly changing social and environmental (temporary and localized food sources) surrounding, these associations have not evolved any complex social behaviours and interspecific relationships nor developed adapted defence strategies, in contrast to the three previous forms of flocking.

Army-ant followers
Willis and Oniki (1978) have described this common and typical foraging association of medium size insectivorous birds following swarm-raiding army-ants in the understorey of the Neotropics (Fig. 2). In French Guiana, 29 species were observed foraging occasionally around army-ants (Fig. 3). Only 4 of them (1.9 % of the total community, Tab.1) foraged exclusively at swarms snatching fleeing insects flushed by the foraging ants (Tab. 1). They were the Plain-brown Woodcreeper *Dendrocincla fuliginosa*, the White-chinned Woodcreeper *Dendrocincla merula*, the White-plumed Antbird *Pithys albifrons* and the Rufous-throated Antbird *Gymnopithys rufigula*.

Caracara followers
Caracara followers are an intriguing foraging association of canopy birds, originally reported by Thiollay (1991b). The Red-throated Caracara *Daptrius americanus* is a predator of wasp nests and the only co-operative group-living raptor in French Guiana. This species constitutes permanently groupings of 4 to 5 individuals, including generally one breeding pair accompanied by non-readers, youngs and immatures. Permanent groups of Caracaras were almost followed by other birds foraging independently from them, on different food sources, in different ways and even at different levels. Fig. 2 reveals that Caracara followers are mostly medium size to large canopy frugivores. This particular gregarious system occurred in a variety of species, including some parrots (*Amazona*, *Pionites*, *Pionopsitta*, *Pionus*), parrolets (*Touit*), toucans (*Ramphastos*), aracaris (*Pteroglossus*), toucanets (*Selenidera*) and Cotingas (*Perissocephalus*, *Phoenicircus*, *Querula*, *Rupicola*). Twenty-three species were seen foraging with Caracaras (Fig. 3). Most of these species tended to stop following the Caracaras when the raptors moved out of their territory. Nevertheless, two regular attendants, the Red-fan Parrot *Deroptyius accipitrinus* and the Green Oropendola *Psarocolius viridis*, were observed following the Caracaras for several hours.

Monkey followers
Monkey followers constitute another interesting association where two raptors, the Double-toothed Kite *Harpagus bidentatus* and the White Hawk *Leucopternis albicollis*, were observed staying below active troops of Capuchin (*Cebus apella*) and Tamarin (*Saguinus midas*) monkeys to take advantage of cryptic prey (e.g. large insects, lizards, small snakes) they disturbed or flushed when moving (Tostain *et al.* 1993, Kessler unpublished data, Zhang & Wang 2000).

2.2 FUNCTIONS OF FLOCKING

In French Guiana, flocking takes many forms, which probably have different origins and functions. All the flocking types described above have been well documented throughout the Neotropics, although the question of their adaptive significance still remains open. Nevertheless, we can draw some conclusions as discussed below (Table 2).

(1) Most canopy flock members are omnivores and insectivores foraging actively on cryptic food sources that are difficult to detect or to be dislodged and chased. Thus, they all share a unique foraging syndrome characterized by high mobility, conspicuousness and active searching methods, exclusive of an efficient vigilance (Jullien 1997). As a consequence, they are submitted to the highest risk of detection and attack by predators. These foragers derive a little foraging enhancement (Powell 1985) but a substantial

reduction of their own predation risk by their association with vigilant species. The Fulvous Shrike-tanager (*Lanio fulvus*) acts as a sentinel because its flycatching and vigilant behaviour allows it to notice first the presence of predators. Only this species gains an important feeding benefit from this association either through the 'beater effect' or the emission of false alarm calls to kleptoparasite flockmates (Munn 1986, Jullien 1997).

TABLE 2. Adaptive significance of flocking among the bird community in French Guiana (After Jullien 1997)

			Adaptive significance of flocking				
Flocking types	Degree of association	Flock members	Food	Foraging behaviour	Degree of vigilance	Feeding enhancement	Predator avoidance
Canopy mixed-species bird flocks	Permanent	1 kleptoparasite & sentinel (*Lanio*) +	Flying insects or flushed	Sit-and-wait	High	+++	- - -
		Non vigilant species	Cryptic, non-flying insects	Scrutinize	Low	- - -	+++
Monospecific groups	Permanent	Vigilant	Fruits or Insects	Sit-and-wait	High	Complex	Complex
Opportunistic gatherings of frugivores	Short-term	Vigilant	Localized and temporary	Sit-and-wait	High	+	-
Army-ant followers	Permanent during foraging	Vigilant	Insects flushed by ants	Sit-and-wait	High	+++	+
Caracaras followers	Short-term	Vigilant	Fruits	Sit-and-wait	High	- - -	+++
Monkey followers	Short-term	Vigilant raptors	Insects, lizards, snakes flushed by monkeys	Sit-and-wait	High	+++	- - -

(2) An evaluation of the adaptive significance of monospecific groups seems complex and is still precluded by the absence of sufficient data. For the co-operatively group-living raptor *Daptrius americanus*, Thiollay (1991b) emphasizes that both their "signalling of prey location, partial food sharing and allofeeding", and their well-developed anti-predator vigilance are "likely to promote their maintenance within the group" and thus to facilitate learning and survival of less experienced birds.

(3) Many studies emphasize that the short-term and opportunistic associations of frugivores have more attributes consistent with the foraging enhancement hypotheses, probably because the formation of this flocking type is imposed by the exploitation of localized and temporary food sources (Powell 1985). Thus, they correspond to casual associations in which birds consort together when the opportunity presents itself.

(4) Army-ant and monkey followers, by snatching prey flushed by foraging ants and monkeys, especially derive a feeding benefit from this association even if they are constrained to adopt the habitat, level and course imposed by these animals. Here, the presence of other birds is, at best, a secondary benefit, not one sought after intentionally. Predator avoidance resulting from vigilance conferred by "many eyes and ears" also appears as a secondary benefit, i.e. as a consequence of this flocking behaviour.

(5) As underlined by Thiollay (1991b) and Jullien and Thiollay (1998), no direct enhancement of food finding or feeding rates and no benefit from each other's foraging are apparent among the Caracara followers (frugivores), except that the Caracaras might provide protection because of their high vigilance, early warning system and loud mobbing of predators. These frugivorous species enter in the open understorey (dangerous habitat) almost only with the Caracaras. This would be an example of habitat selection influenced by predation risk and of flocking behaviour evolving as an anti-predator defence (Thiollay 1991b).

Powell (1985 p.726) emphasizes that the determinants of flocking "are a combination of spatial, temporal, climatic and behavioural factors". Thus, evaluating the adaptive significance of flocking is complex and implies that we have a complete knowledge of the composition, structure, social organization and dynamics of each organization. To date, these information are only available for the multi-species flocking system of understorey birds in Amazonian-type forests. Thus, we end this chapter by drawing a specific review of this particular flocking type. We first summarize the principal characteristics of the mixed-species understorey bird flocks before discussing their function.

3. The adaptive significance of mixed-species understorey bird flocks

3.1 COMPOSITION, ORGANIZATION AND DYNAMICS OF THE MIXED-SPECIES UNDERSTOREY BIRD FLOCKS: A SUMMARY

This particular flocking behaviour was studied in detail at the Nouragues field station from 1991 to 1994 (Jullien 1997, Jullien & Thiollay 1998). According to their flocking propensity and occurrence (Jullien & Thiollay 1998), understorey flock members can be classified into six categories from the constant and obligate flock members to the accidental participants (Table 3). The most typical feature of this flocking type is a set of 10 permanent insectivorous members, or core species. Three species were sit-and-wait birds (2 *Thamnomanes*, 1 *Hylophilus*) whereas the seven remaining species were active foragers (4 *Myrmotherula*, 1 *Xiphorhynchus*, 1 *Automolus*, 1 *Philydor*) exploiting the understorey of the mature forests (Table 3). All accidental, occasional and frequent but not permanent flock members were species intermediate between highly vigilant and slow-moving birds in open understorey, or birds foraging in a protected environment (near the ground or in dense vegetation), or species with a different food diet, or species maintaining a smaller or a larger home-range than the flock territory, or species with a number of pairs in a given area lower or higher than the number of flocks on the same surface (Table 3).

The 10 constant and obligate flock members never foraged alone. Obligate flock members shared a common territory, which was defended year-round by each species against conspecifics of neighbouring flocks. Flock membership was limited to a single breeding adult pair of every core species (breeding population), often accompanied by one or more extra birds. These additional individuals were either immatures born in the same flock or unpaired birds in adult plumage coming from different groups (floating population). Flock number, size, and species composition, as well as territorial boundaries were highly stable between seasons and years (Munn & Terborgh 1979,

TABLE 3. Characteristics on the different categories of flock membership I = Insectivorous; O = Omnivorous; F = Frugivorous

Categories of flock menbership	Number of species	Diet	Habitat	Strata	Flocking propensity (%)	Occurrence (%)	Species home-range / Flocks home-range *	Number of pairs / Number of flocks **
Permanent flock members	8	I	Mature forest	Mid	100	100	=	=
Permanent flock species of some groups only	2	I	Mature forest	Mid	94-100	53-60	=	<
Permanent flock species ranging over 2-3 flocks	33	I	Mature forest Dense forest	Mid	70-90	40-67	= or>	< or =
Frequent but not permanent flock members	5	I	Mature forest	Mid Canopy	40-65	27-100	<, = or >	<, = or >
Occasional flock members	24	I O	Mature forest Dense forest Forest streams	Ground Mid Canopy	6-29	7-53	<, = or >	<, = or >
Accidental flock members	24	I O F	Mature forest Dense forest Treefall gaps	Ground Mid Canopy	5-19	0	<, = or >	< or =

* Species home-range can be smaller (<), equal (=) or larger (>) than flock home-range.
** Number of pairs per species in a given surface can be lower (<), equal (=) or higher than the number of flocks in the same area.

Greenberg & Gradwohl 1986, Jullien & Thiollay 1998). The spatial and temporal stability of these mixed flocks suggests that the populations of at least the 10 core species may be highly stable and that the permanent flocking behaviour may strongly limit population levels. For every core species, the general pattern may be summarized as follows (Fig. 4): flock membership was limited to a single breeding adult pair, which tended to remain for life in the same flock. This pair usually did not attempt to breed each year (irregular breeding attempt) and produced no more than one fledgling per year on average (low productivity). This young first remained with its parents for an unusually long period (10 months on average). Then it either stayed in its natal flock or dispersed in successive groups (21months on average and sometimes up to 3 years) searching for a mate and an opportunity to settle. Delayed reproduction may be caused by a high survival rate of breeding adults (Jullien 1997), a limitation of flock membership to one breeding adult pair, a saturation of all suitable forest habitats with contiguous flock territories. Such a floating population may buffer stochastic demographic events by providing new breeders to offset losses (Jullien & Thiollay 1998). Delayed reproduction of new recruits imposed partially by the dominance of a single pair per flock was similar to that in group-living species (Emlen & Vehrencamp 1985). This flocking behaviour is one of the many constraints that maintain the density of most tropical primary forest birds at a low level and thus enhance the coexistence of so many species (Powell 1989, Thiollay 1991a, 1994, Jullien & Thiollay 1998).

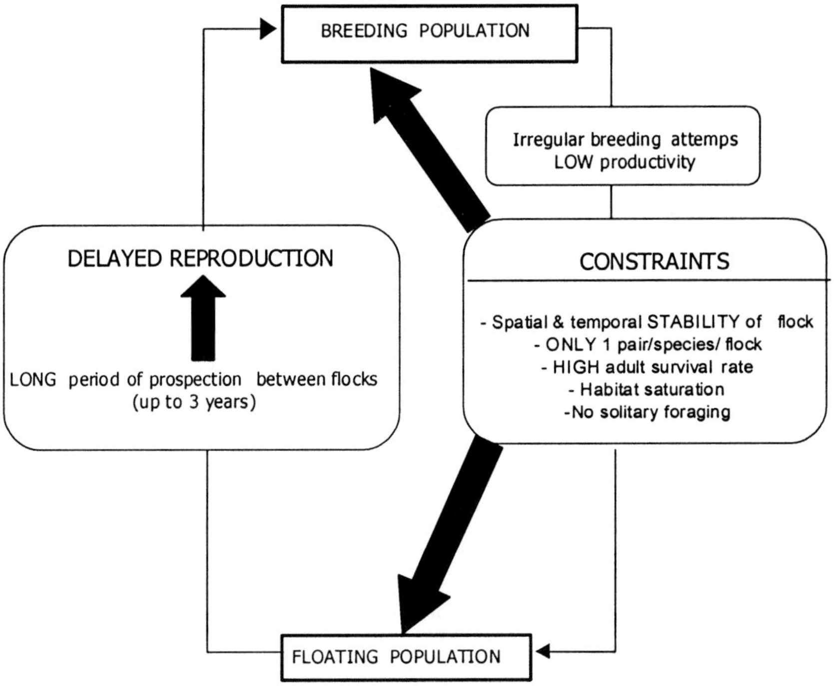

Figure 4. Population dynamic of the 10 constant and obligate members of the mixed-species understorey bird flocks.

This social structure was also maintained during the breeding season through special behavioural adaptations, at least for the obligate and permanent flock members. The main features adjusting nesting activities and daily movements of the flock and allowing the species involved to stay permanently in the flock can be summarized as follows (Jullien 1997): nest building occurred only when the flock foraged close to the nest. Male and female shared all the breeding duties (incubation and nestling care) and spent in turn 2-hours periods on the nest while their mate was foraging in the flock. An always single young was produced and joined the flock within one day of fledging (see also Brosset 1969, Munn & Terborgh 1979). These behavioural and ecological adaptations allowed the parents to keep the feeding rate ($< 2\ h^{-1}$), prey carrying movement, and predation risk, to a minimum. Equal share of nesting duties by both sexes, low feeding rates and early ability of nestlings to follow adults eased the maintenance of exclusive foraging in flocks by obligate flock members during all nesting stages.

To summarize, this flocking type is the only known case of territorial defence of a single hunting area by several species together, one of the most stable, permanent, highly evolved and socially structured associations, and an example of the complex interspecific interactions governing the dynamic and organization of tropical forest communities (Wiley 1971, Powell 1977, Munn & Terborgh 1979, Gradwohl and Greenberg 1980, Powell 1985, Greenberg and Gradwohl 1986, Powell 1989, Jullien 1997, Jullien & Thiollay 1998).

3.2 BENEFITS AND COSTS OF UNDERSTOREY MIXED-SPECIES FLOCKING

The benefits and costs of the typical mixed-species flocking described just above are examined, first in terms of improved foraging efficiency, and secondly in terms of predation risk.

Benefits and costs associated with foraging

The possible feeding benefit of flocking, through local enhancement of foraging success by information transfer between flock members has given rise to various hypotheses and experiments (Sasvari 1992, Master *et al.* 1993, Giraldeau *et al.* 1994). It has been suggested that birds might monitor each other, either to learn the most rewarding foraging sites and techniques from more experienced individuals, or to avoid searching sites previously investigated by another individual, or to join successful birds on patchy food sources, or to kleptoparasitize flockmates, or to catch cryptic prey flushed by neighbours ('Beater effect').

The obligate flock members vary in their capacities to derive feeding benefits from their association with other flock participants (Fig. 5).

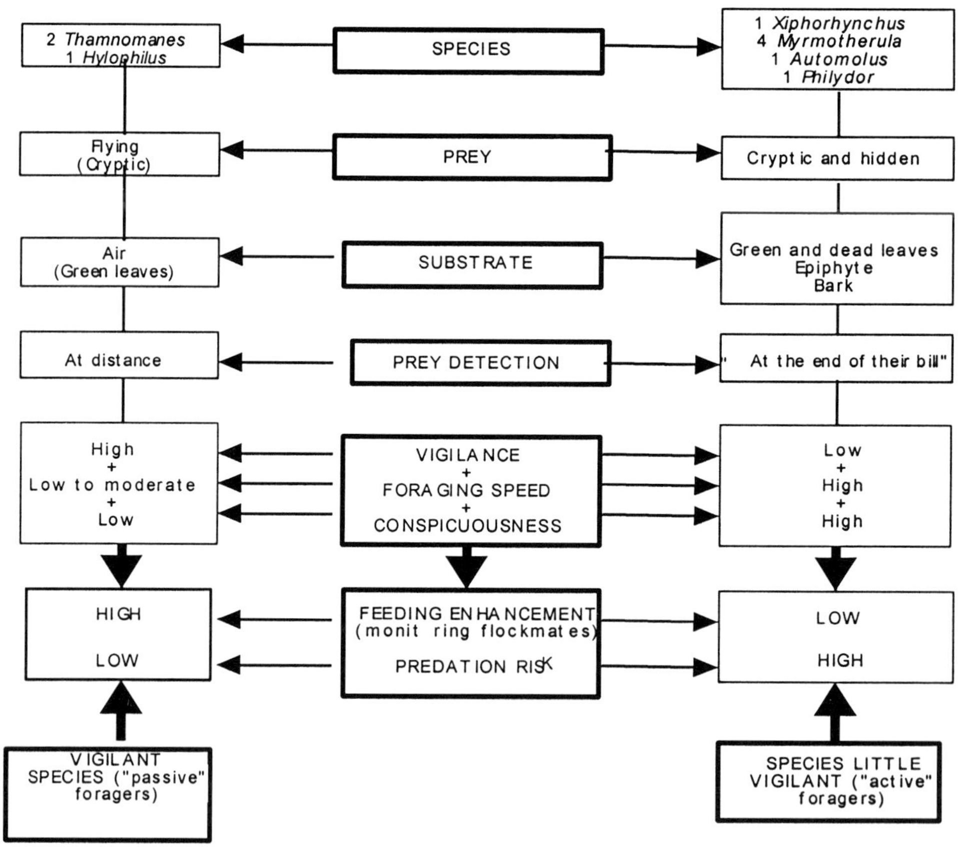

Figure 5. Consequences of the foraging behaviour of the 10 core species on their capacities to derive feeding and antipredator benefit from their association with other flockmates (After Jullien 1997).

Three obligate flock members (the Dusky-throated Antshrike *Thamnomanes ardesiacus*, the Cinereous Antshrike *Thamnomanes-caesius*, the Tawny-crowned Greenlet *Hylophilus ochraceiceps*) are sit-and-wait (or "passive") foragers. Techniques used to capture prey include (a) sallying, i.e. catching an item from air or the undersurface of a leaf, and, less frequently (b) gleaning, i.e. plucking an insect from the undersurface of various substrates. Being vigilant improves their chances of locating food sources at distance as well as noticing nearby or approaching predators (Fig. 5).

Seven other obligate flock members are (a) green foliage and twig gleaners (the White-flanked Antwren *Myrmotherula axillaris*, the Longed-winged Antwren *Myrmotherula longipennis*, the Gray Antwren *Myrmotherula menetriesii*), (b) dead leave probers and gleaners (the Brown-bellied Antwren *Myrmotherula gutturalis*, the Rufous-rumped Foliage-gleaner *Philydor erythrocercus*, the Cinamon-rumped Foliage-gleaner *Automolus infuscatus*), and (c) epiphyte (*Phylidor erythrocercus*) or (d) bark probers (the Chestnut-rumped Woodcreeper *Xiphorhynchus pardalotus*) of cryptic and hidden invertebrates (Fig. 5, Jullien 1997). Thus, these 'active' foragers are constrained to scrutinize, search on, under and deeply inside various substrates, with quick and ceaseless movements. As a consequence, they are highly conspicuous when foraging and little vigilant (Fig. 5, Jullien 1997). Being little vigilant decreases their chances to monitor visually the foraging behaviour of the other flock members. No evidence was ever found that these 'intensive' foragers intentionally and closely follow or join flockmates (Jullien 1997). Each species uses its own foraging method, substrate and level, with no apparent co-ordination with other species (Jullien 1997). As a result, these species gain low or moderate feeding benefits to be associated with other birds: 4.1% to 12.4% of their captures correspond to prey flushed exclusively by the other active foragers they also kleptoparasitized (Fig. 6).

Being the only sit-and-wait foragers, the two Antshrikes and the Greenlet are the only specialists regularly benefiting from such opportunities: *Hylophilus ochraceiceps*, *Thamnomanes ardesiacus* and *Thamnomanes caesius* derived respectively 27.1%, 42.9% and 59.8% of their captures from insects made available by the beater effect of active flock members they obviously watched, followed and kleptoparasitized (Fig. 6).

Flocking may also incur feeding costs through depressed food availability by exploitative and interference competition. When preys are scarce, resource depletion by neighbours is likely to occur. Disturbance by flock members also makes some insects escape, hide or freeze and become inaccessible to followers. However many prey are so cryptic that, on average, a first foraging attempt may not reduce significantly the chance for a second bird to be successful on the same patch. Agonistic interactions and displacements may disturb foraging activities, but they are relatively rare in Guianan flocks. The kleptoparasitic behaviour of *Hylophilus ochraceiceps*, *Thamnomanes ardesiacus* and *Thamnomanes caesius* was responsible for the loss of 7.1% to 9.1% of prey caught by the "active" foragers. Due to the dilution effect resulting from flocking life, only 8.2% to 13.6% of the total number of hostile interactions were observed in the active foragers (0.4% to 13.6% in the vigilant species). Although aggressive interactions were rare, at least one subordinate species (*Thamnomanes ardesiacus*) was restricted to lower levels and periphery of flocks by a dominant congener (*Thamnomanes caesius*, Jullien 1997). As similar cases among ant followers (Willis 1981) a subordinate status may be a stable strategy where competitive costs are presumably balanced by flocking benefits (Cimprich & Grubb 1994).

When expressed as the product of foraging height, foraging manoeuvre and prey substrate, niche overlap between small species (average niche foraging overlap = 0.821 ± 0.095 for *Myrmotherula axillaris*, *longipennis*, *menetriesii*) is greater than between the 6 large flocking birds (average niche foraging overlap between all other flocking species = 0.153 ± 0.267). The relatively large flock territory size (mean = 9.76 ha, range = 3.21 ha - 14.31 ha, Jullien & Thiollay 1998), supposed to be enforced by the larger core species (Powell 1989), may be more beneficial to smaller species (insurance against over exploitation of food sources) than costly (energy lost in extensive movements). Powell (1989) pinpoints that the smallest species 'can coexist with greater niche overlap because their density is flock dependent rather than resource dependent'. As a result, we found no suggestion of within flock interspecific competition arising from foraging niche similarities (Jullien 1997).

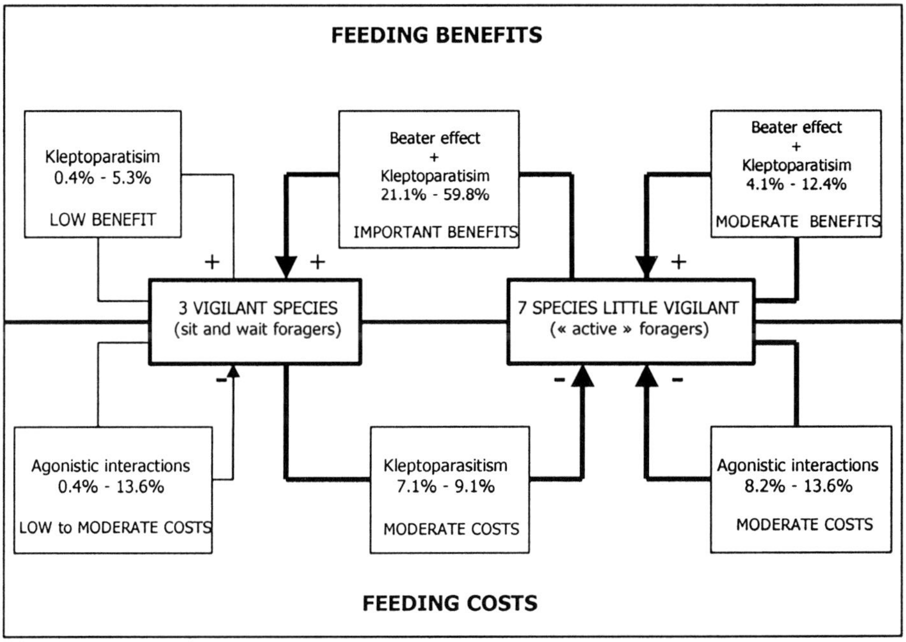

Figure 6. Benefits and costs of understorey mixed-species flocking associated with foraging.

Within flock competition may be another cost of flocking. Intraspecific competition has been demonstrated in temperate winter flocks of tits where dominant individuals survive better and displace subordinates from the reachest food patches or safest foraging sites (Koivula *et al.* 1995). Any intraspecific competition is kept to a minimum in tropical flocks where a single pair of each species usually occurs (Munn & Terborgh 1979, Jullien & Thiollay 1998).

Benefits associated with predation risk

The antipredator benefits of flocking include an early warning of predators through the accumulated vigilance of several individuals. The protection thus gained is important because predators rely mostly on surprise to strike at close range. The confusion effect derived from birds scattering in different directions may also mislead the attacking predator, helping vulnerable prey to escape. The obligate flock members vary in their capacities to maintain awareness while foraging (Fig. 5). Not being vigilant decreases the chances of the seven active foragers to locate possible predators whereas being conspicuous facilitates their detection by raptors and snakes (Fig. 5). These active foragers derive a substantial reduction of their own predation risk by their association with the three vigilant species. As *Lanio fulvus* for the canopy flocks, the greenlet and the two antshrikes function as sentinels by emitting early alarm calls to signal the presence of predators to which all the other flockmates respond (Jullien 1997), an advantage that these active foragers may find higher than the cost of being robbed.

Alarmed birds in flocks, however, not only flew, usually diving to lower levels, but also froze and became inconspicuous. When the predator was not an immediate danger for birds aware of its presence (e.g. snakes, owls), flock members indulged in-group mobbing. Such behavioural defences apparently deter predators from attacking the yet highly detectable flocks. During 2132 hours spent following flocks in French Guiana a mere 13 attacks by raptors were witnessed, only one of them successful (M.J., unpublished data). This would translate into an average of 1 attack/16 days and 1 capture/7 months. Similar rates, much lower than on other flock types, were also observed in various study sites (Table 4). The potential cost of flocking, i.e., increasing conspicuousness of individuals, is probably far outweighed by its deterrent effect.

Our obligate flock species had the highest mean annual survival rate (mean = 0.751, range = 0.669-0.861; see also Greenberg & Gradwohl 1986 and 1997) ever calculated for tropical passerines (Karr *et al.* 1990, Faaborg & Arendt 1995, Jullien & Thiollay 1998), higher than local solitary species or facultative flock members (Jullien & Clobert 2000). This might be related to a lower predation rate, which has not yet been estimated, or to a better individual fitness of flocking birds (Jullien & Clobert 2000).

4. Antipredator advantages of multispecies territoriality

We have already noticed that (Jullien & Thiollay 1998), besides the immediate benefits of flocking (early warning system, confusion effect) the daily and year-long association of several species within a permanent home-range is likely to have potential advantages in the defence against predators. First, the interspecific long-term familiarity may help them to have a better knowledge and understanding of their respective behaviours and vocalizations, both in routine foraging sessions and in case of threat by predators. An evolutionary consequence may be a finely tuned, better co-ordinated and more efficient response to alarm calls. This was suggested by the widely different reactions of flock members to alarm calls given by *Thamnomanes sp.*, the usual sentinels and most vocal species in these mixed flocks. Almost identical alarm calls (to the human ear) were given in different situations both to harass and kleptoparasitize another flock member (false alarm, see Munn 1986; Terborgh 1990), and to warn of the presence of a non lethal threat (e.g., a human observer) and to signal a potentially dangerous predator. In the first case, only the individual attacked seemed to be lured for a few seconds (enough to make it drop the prey). In the second case, most birds responded, some of them gave alarm calls toward the intruder but they did not stop and shortly resumed normal foraging. In the last case, all birds immediately dove into safer sites and froze to become inconspicuous.

Moreover, as in any territorial and sedentary species, a continuous and intimate knowledge of an area, its food resources, dangers and refuges, may increase an individual's fitness compared to a bird using an unfamiliar area. This is especially important in such a complex environment with so many cryptic prey and inconspicuous predators. The daily use of highly specific gathering and bathing sites, at times when birds were particularly vulnerable, also suggested that the core members may have learnt by experience that these sites were safer than others apparently similar areas that were never used.

5. Ecological correlates of flocking

The regular flock attendants shared a set of characteristics, unique in the bird community. They were typically small (< 50 g), actively foraging insectivores in the mid-understorey of mature forest stands. Their rapid foraging behaviour, i.e. quick and ceaseless foliage or bark gleaning or deep searching and probing in dead leaves, epiphytes or bark, involved unusually low vigilance rates. At least obligate species tended to avoid edges or gaps and this may be why mixed flocks declined or disappeared in forest fragments (Stouffer & Bierregaard 1995) and logged forests (Thiollay 1992).

The apparent plumage mimicry of birds in mixed flocks, noticed by Moynihan (1962) in the Andes may not be related to flocking. Tropical forest birds tend to share convergent plumage patterns to minimize detectability against their particular background and flock members were not significantly more similar between them than were all bird species at the same level.

In striking contrast, non-flocking birds foraged either in dense vegetation, and/or very low, or on trunks or with long scanning periods (Jullien & Thiollay unpublished data). Yet, a high proportion of them foraged in pairs, a compromise between the costs of flocking and the risk of solitariness.

6. Differences with other mixed flocks

In French Guiana, as elsewhere, the permanent and socially structured understorey flocks seem to be an adaptation to a specific set of constraints: widely scattered food distribution pattern, diverse prey, cryptic or hidden, mobile and elusive, necessarily conspicuous and non-vigilant foraging methods, unprotective surrounding vegetation. When one or more of these constraints are relaxed, flock types may have evolved weaker and less elaborate social interactions and antipredator behaviours. They have also permanent species, but they are more irregular in space, time, size and composition. Groups of frugivores and army-ant followers are opportunic gatherings that may be more vigilant (fruit eaters) or less exposed (near ground). The more conspicuous canopy flocks (noisy, mobile, colourful, large, wide-ranging) are in a more open environment, presumably safer for surprise attacks.

In shorter and more tropical open forests (mountain, monsoon, semi-deciduous, secondary or disturbed), the distinction between canopy and understorey flocks tends to disappear and a single flock type ranges across mid and upper forest levels. Winter flocks of temperate forests have some characteristics in common with tropical understorey flocks (moderate size, stable membership, dominance relationships, permanent home range), but they do not persist during the breeding season. In winter, the conditions are quite similar to those found in tropical forests: arthropod food is sparsely distributed, cryptic, hidden and available during a short daylight period often shortened by rain, hence the necessity to search large areas, actively and continuously, leaving little time for vigilance and increasing the relative benefit of foraging in a flock.

Conversely, during the temperate breeding season, food supply is more abundant, richer (caterpillars), easily accessible and during a longer daylight period. Therefore it may be more efficient to hold an exclusive territory, small and well known to limit movements and detection by predators.

7. Conclusions

None of the behavioural traits of understorey mixed flocks suggest a significant enhancement to feeding success or efficiency, due to flocking but independent of the effect of flocking on vigilance. Only predictions derived from the antipredator hypothesis were supported. To reduce predation risk, birds have the following alternatives. (1) Evolve colour patterns matching their usual background and efficient escape tactics or flight manoeuvres: flocking species in general have no poorer adaptations than other species in this field. (2) Remain hidden in dense vegetation to be inaccessible or well in the open to avoid stalkers: mixed flocks usually exploit intermediate habitats. (3) Adopt an inconspicuous behaviour, reduce movements and increase individual vigilance, as do solitary birds: flocking species exhibit opposite trends. Therefore, improving collective vigilance by foraging close to other birds may be the only remaining strategy to minimize predation risk when exploiting an unsafe habitat and prey types that require a risky searching behaviour. The species following Red-throated Caracaras (*Daptrius americanus*) in the understorey, where they do not forage alone (Thiollay 1991b) offer an even more striking example of an interspecific association allowing the use of a habitat that would be avoided otherwise from fear of predators. On a broad geographic scale, the apparent correlation between the number of bird predators and the mixed flocking propensity in tropical forest is also currently investigated.

Mixed flocking is one of the many complex interspecific relationships that play a critical role in maintaining the high diversity of tropical forest bird communities. Permanent flock territories limit the density of obligate flock members (Greenberg & Gradwohl 1986, Jullien & Thiollay 1998) and thus contribute to lower their abundance, a mechanism crucial to species coexistence in tropical forest (Powell 1989, Thiollay 1991a).

Acknowledgements

This manuscript has benefited from insightful comments by Drs. R.O. Bierregaard and R. Greenberg. The project was funded by a grant from the National Centre of Scientific Research (CNRS.) within the National Program on Biodiversity (Director: Prof. R. Barbault). M.J. was supported by a grant from the Ministère Français pour l'Enseignement et la Recherche Scientifique during her field work, and by the University of Montana (Montana Cooperative Wildlife Research Unit, Avian Studies Program) during the redaction of this manuscript.

Toon Rijkers climbs a tree with his portable IRGA to measure photosynthesis. *Photo Frans Bongers.*

CHAPTER 14

HABITAT SELECTION, AMBIENT LIGHT AND COLOUR PATTERNS IN SOME LEK-DISPLAYING BIRDS

Marc Théry and John A. Endler

1. Introduction

Ambient light is probably the most variable physical factor in tropical rain forests. High light intensity above forest canopy is reduced through tall and dense vegetation to about 0.1 - 1 % in the understorey. Moreover, whitish ambient light spectra of direct sunlight is drastically changed to yellow-green in forest shade, to bluish in woodland shade, or to yellow-orange in small gaps depending upon the geometry of vegetation and the fraction of the sky covered by cloud (Endler 1993). In addition, in these habitats ambient light is purple at dawn and dusk, and converges to white with cloudy weather during the day. These changes in light quantity and quality have profound effects on seed germination, photosynthesis, and direction or timing of plant morphogenesis, including flowering and fruiting phenologies (reviews by Endler 1993, 1997, Vázquez-Yanes & Orozco-Segovia 1993, Smith 1995, Fankhauser & Chory 1997, Théry 2000). Because the conspicuousness of colour patterns depends upon the interaction between ambient light and reflectance spectra of colour pattern elements (Lythgoe 1979, Endler 1986, 1990, 1991, 1992, Endler & Théry 1996), ambient light quality will also interact with animal vision systems, and therefore affect visual communication between animals. This can profoundly affect survival and reproduction, as well as communication between animals and plants, crucial to pollination, seed dispersal, and herbivory (Endler 1993, 1997).

Leks are sites where groups of males congregate and display to females, females choose and mate with a small subset of available males, and males only provide gametes but no parental care or resource to females (Bradbury 1981). These sites are long lasting and used by successive generations of males and females, although particular display territories can be abandoned and others newly occupied (M. Théry unpublished data). Each species shows some constancy in the degree of male clustering, preferred lek habitat, vegetation structure, and nature and dimension of display perches (Théry 1990b, 1992). Therefore, it was reasonable to ask whether ambient light and display height affect the nature and distribution of colour signals on males, mate choice, or spatial distribution of lekking males. We addressed these questions in 5 sympatric lekking manakins (Passeriformes: Pipridae; nomenclature according to Prum 1992): *Manacus manacus*, *Dixiphia pipra*, *Pipra erythrocephala*, *Lepidothrix serena* and *Corapipo gutturalis*, and in the Guianan cock-of-the-rock *Rupicola rupicola*, a closely related member of the Cotingidae.

Bongers *et al.* (eds.), Nouragues. Dynamics and plant-animal interactions in a neotropical rainforest, 161–165.
2001 Kluwer Academic Publishers. Printed in the Netherlands.

Reflectance and ambient light spectra were measured with a Li-Cor LI-1800 or an Ocean Optics PS-1000 portable spectrometer. Three colour variables, brightness, hue and chroma, were computed using the segment classification method of Endler (1990) between 400 nm and 700 nm. The generated colour space is independent of assumptions of the vision of any specific species, but does not include the UV (300-400nm). Consequently we may be underestimating differences visible to birds but not humans in this study. Nevertheless, most of what is happening in the manakins and *R. rupicola* is in the long wavelengths, so leaving out the UV has a minimal effect on the results (Endler & Théry 1996). Hue angle is a function of the wavelengths at which the spectrum changes most rapidly. Chroma is function of the slope of a spectrum at the wavelength that defines the hue and represents colour saturation or purity. Brightness is total reflectance, total irradiance, or total radiance depending on the source of the spectrum.

2. Nature and disposition of colour signals

Females of all lekking manakins are mainly green, whereas adult males are very colourful or boldly patterned (Théry 1990b, Ridgely & Tudor 1994). *P. erythrocephala* and *D. pipra* are coloured on the top of the head, the first of golden-yellow and the second of white. *P. erythrocephala* also has scarlet thigh feathers. On the other hand, *C. gutturalis* and *M. manacus* are mainly coloured in the underparts: *C. gutturalis* of a snowy white throat and *M. manacus* of a white belly and orange legs. Black upperparts of *C. gutturalis* and *P. erythrocephala* have shiny blue reflection in direct sunlight. Males *L. serena* have yellow belly, blue rump, white front, and yellow-oran22/05/00ge chest patch. The brilliant white front of *L. serena* is the brightest colour signal we have ever measured in an animal (J. Endler and M. Théry unpublished data).

Manakin leks are distributed along a specific height gradient (Théry 1997), which determines light incidence and influences disposition of colour patches on males (Théry 1990b). Head-coloured males have the highest display sites (*D. pipra*, *P. erythrocephala*), whereas low displaying males receiving light reflected from lek sites are coloured in the underparts (*M. manacus*, *C. gutturalis*). However, the blue iridescence of *C. gutturalis* is located in the upperparts and only appears when adult males are flying in open canopy or display in a sunfleck on a mossy log. On the other hand *L. serena*, the only species to display at mid-level of dark understorey has colour patches both in the under- and upperparts.

Females *R. rupicola* are dark olive grey, the most coloured female part being the interior of the beak, bright yellow-orange. Males are bright orange with spectacular ornamental feathers, with some white and dark brown on wings (Snow 1982, Endler & Théry 1996).

3. Conspicuousness and crypsis of female and male colour patterns

The simplest way for an animal to be cryptic is to have colour patches matching a random sample of the visual background in time and place of maximum predation risk (Endler 1978, 1984). In a colour space (Endler 1990), colour specifications derived from reflectance spectra of females and juveniles largely overlap with those of visual backgrounds, whereas males are largely outside the range of visual backgrounds, indicating that they contrast markedly with backgrounds (Fig. 1, Endler & Théry 1996). A similar result is found when a colour space including the UV and based upon Passerine vision is constructed (J. Endler and M. Théry unpublished data). Manakin females are highly cryptic on small cup nests made of leaf and/or green moss and seen against green foliage. Female *R. rupicola* are hardly visible on a bracket of mud and rootlets against the granite or laterite of nesting caves, but the bright interior of their beak should be conspicuous to nestlings. Males and females are usually brighter than backgrounds, although males are considerably brighter than females and juveniles. Among males, *R. rupicola* averages only slightly brighter than backgrounds. Although males have higher chroma than females, male colour patches either have high chroma or high brightness, but not both. Only males of *L. serena* have some patches with both bright and saturated colours (Endler & Théry 1996).

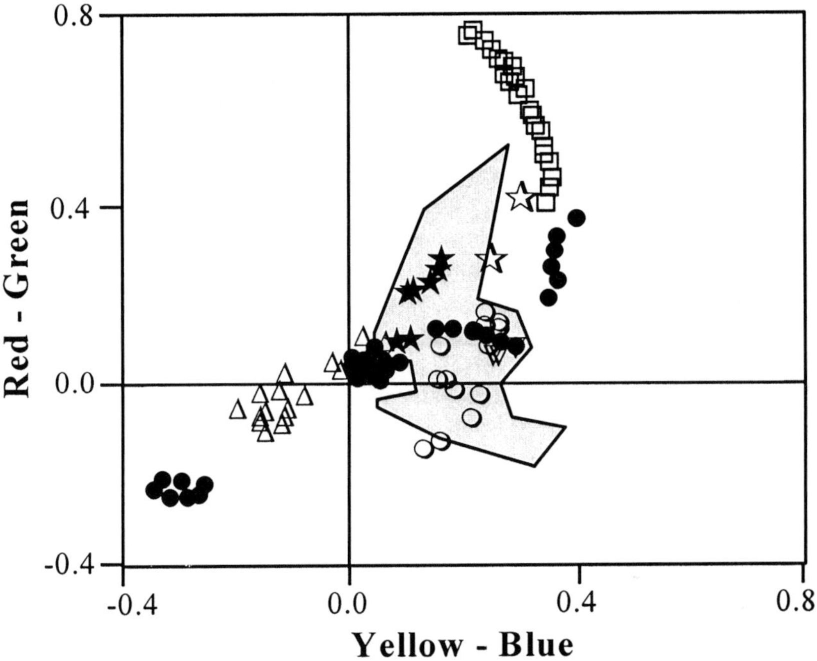

Figure 1. Colour specifications derived from reflectance spectra of male, female and juvenile birds (symbols) relative to the range of variation of visual backgrounds spectra (shaded) plotted in the colour space of Endler (1990). Each point represents a single reflectance spectrum measured by a Li-Cor LI-1800 or an Ocean Optics PS-1000 portable spectroradiometer, from which hue angle and chroma have been extracted using the segment classification method of Endler (1990) between 400 nm and 700 nm. The ordinate represents relative brightness of red and green spectral segments, and the abscissa is the difference between yellow and blue segments. The distance from the origin is chroma (colour saturation) and the angle clockwise from the top of the ordinate is hue. The direction of vectors approximately corresponding to Munsell hues are indicated by arrowheads (R = red, Y = yellow, G = green, B = blue, P = purple).
Bird symbols: *R. rupicola* males: closed stars: dark and pale wing bars, open stars: wing strings, squares: rest of body; *Corapipo gutturalis*: upward-pointing triangles: males, downward-pointing triangles: females and juveniles; *Lepidothrix serena*: closed circles: males, open circles: females and juveniles.
The black and white colour pattern elements of males, and the low chroma patches of juveniles and females cluster tightly near the origin and so are difficult to distinguish.

The heterogeneity of light habitats (Bongers *et al.* this volume) affects the conspicuousness of colour patterns. In the three species we have analysed, *R. rupicola, L. serena* and *C. gutturalis*, the colour and/or brightness contrast of birds is greater at times and places of display, and reduced when birds are not displaying or leave the lek (Endler & Théry 1996).

R. rupicola males display in the dry season when a beam of small gap light passes over their horizontal perch, and display simultaneously partly in a small gap (most often on wings and crest) and partly in forest shade. Males are relatively cryptic in forest shade because their orange plumage weakly reflects greenish ambient light. On the other hand, males are extraordinarily conspicuous with high intensity and chroma from the part of the body in small gap and low intensity in the adjacent forest shade (Endler & Théry 1996). This arises because the male reflectance spectrum mismatches the forest light spectrum and matches that of small gaps, making it an efficient reflector in small gaps but a very inefficient one in forest shade. Males also minimize visual contrast when not displaying. Forest shade of the lek forest has higher chroma than the average for the surrounding forests, and therefore males are more cryptic in the lek area when not displaying than they are in foraging areas.

L. serena males mate during the rainy season, and use the alternation of clouds and unobstructed sunlight to optimize conspicuousness (Endler & Théry 1996). Groups of males move downhill as the sun rises above the hill at the opposite side of the valley, staying in the shadow in front of the advancing zone of forest shade and small gaps (Théry 1990c). Particularly, the turquoise rump patch strongly contrasts with the back and tail in cloudy weather because it is a better match to the whitish light of cloudy weather than to the yellowish light of forest shade, and the yellow-orange chest adds complementary colour contrast to the pattern. The white front is highly conspicuous in the unusually dark understorey, but is also conspicuous is other light environments.

C. gutturalis displays and mates during the dry season on sections of mossy logs fallen in primary forest understorey. Males display both in sunflecks of spectra intermediate between those of small gaps and large gaps, and in forest shade (Endler & Théry 1996). They only use the brightest section of logs characterized by high light intensity contrast with the surrounding (Théry 1987, 1990b, Théry & Vehrencamp 1995), and stop displaying with natural or experimental obstruction of sunflecks (M. Théry unpublished data). Black and white males have highest visual contrast when rapidly flying over sun patches or when displaying simultaneously in both sun and shade. Additionally, the blue iridescence of the back, appearing with direct sunlight, increases the difference in effects of the light environments when displaying closely to the female. In these three species, colour patterns and display behaviour, including the choice of where and when to display, may coevolve together to optimize communication efficiency in different ambient light spectra, as in guppies (Endler 1991).

4. Colour signals, male dominance and mate choice

In a study of mate choice in *C. gutturalis*, light intensity measurements were taken to analyse the interaction between site characteristics, male displays and female visitation (Théry & Vehrencamp 1995). Males displaying at arenas with highest light intensity and brightness contrast with the surrounding received significantly greater numbers of female visits. Light properties of the display site, i.e. timing, intensity and contrast of sun patches, determine with lek attendance the attractiveness of male display to females, whereas male interactions subsequently mediate copulatory success. Attendance and conditions of visual communication appear as primary cues for female choice in *C. gutturalis*, but this is not used by females as an indicator of the males' relative dominance because dominance status is not correlated with light characteristics of display sites.

We found no significant variation of male colour (area, brightness, hue, chroma) among adult males *C. gutturalis* (M. Théry unpublished data). On the contrary, in *R. rupicola*, a two-fold variation of colour saturation was measured among different males of the same lek during two successive years. There is also a strong variation in the saturation of orange-red on the beak, around the eyes, and on male legs. The most colourful male was dominant among other males (M. Théry & C. Lehoucq unpublished data), and therefore likely to have a higher reproductive success (Trail 1985). Displays and threat of competing males are more frequent than female visits in *R. rupicola*, and determine social dominance relationships and relative disposition of display sites, largely influencing the distribution of female visits and matings (Trail 1985, Trail & Adams 1989).

5. Conclusion

In lekking birds, specific habitat requirement and courtship display suggest that lek location may be influenced by simple physical factors that make display and communication more efficient (Théry 1990b; Théry 1997, Westcott 1993; Endler & Théry 1996). We have shown that light quality is essential in habitat preference of three forest lekking birds of French Guiana: *R. rupicola*, *L. serena* and *C. gutturalis* (Endler & Théry 1996). Light patterns and the spatio-temporal distribution of sunflecks and shade also influence lekking behaviour, mate choice, and location of colour patches in manakins (Théry 1987, 1990b, Théry & Vehrencamp 1995).

Plumage colour appears as a bird adaptation to optimize crypsis and conspicuousness in specific courtship ambient light, under constraints of both natural and sexual selection. Sensory drive can set the initial direction of the Fisher process and profoundly affects the direction of sexual selection (Endler 1992, 1995). The evolution of sensory systems, signals and behaviour is coupled and predictable from the knowledge of environmental physics, and the biophysics and neurobiology of signal transmission and reception. For example, visual conditions and spectral sensitivity may favour the use of blue and red colours in signaling, but carotenoid-based colours of birds (red, orange, yellow) may indicate feeding success and immune resistance of males, so red would be favoured over blue. Similarly, colour vision of frugivores may result from coevolution with fruit colour, but particular environmental conditions will favour particular sensory characteristics and signals. For that reason, visual systems, display signals and behaviour can result from adaptation to environmental conditions constraining colour signal design of both food items and conspecifics.

Acknowledgements

JAE thanks the US National Science Foundation and MT the CNRS for financial support.

Eggs of the poison arrow frog *Dendrobates ventrimaculatus* in the axil of a bromeliad. Nouragues, 1991. *Photo Marga Born.*

CHAPTER 15

DISTRIBUTION AND LIFE HISTORIES OF AMPHIBIANS AND REPTILES

Marga Born and Philippe Gaucher

1. Introduction

Amphibians and reptiles are significant components of many tropical communities. They usually occur in relatively high population densities, which often exceed the biomass of other vertebrates (Vitt 1990). This makes them important predators and thus regulators of the arthropod fauna. Their high biomass provides a prey base in many habitats as well. Besides this central position in the food web of many ecosystems, amphibians and reptiles are functionally important as biological indicators of environmental degradation owing to their specific habitat requirements and their limited ways to disperse. With the ever-increasing devastation of tropical rainforest, the selection of protected areas, which represent pristine ecosystems, becomes more and more urgent. Active conservation of such ecosystems with their specific species and their interactions is exclusively possible if this conservation is based on an extensive knowledge of the ecosystem concerned.

In this chapter the herpetofauna of two sites within the Nouragues Nature Reserve, a protected area of about 1,000 km² of pristine undisturbed rainforest, which has recently (1995) secured the status of nature reserve, is reported. The Nouragues forest is characterised by seasonal rainfall, with a long dry season from September to November and a short dry season in February-March. The major forest communities - highland, riverine and inselberg - are described elsewhere in this book.

2. Field studies

Within the Nouragues Nature Reserve two sites have been studied with respect to amphibians and reptiles, the Nouragues station and the Arataye station, approximately 17 km from each other.

2.1 NOURAGUES STATION

The first site within the Nouragues Nature Reserve, the Nouragues Station (4°5'N, 52°41'W), is situated on a plateau at the foothills of an inselberg. Major forest communities and the climate are described elsewhere in this book (Grimaldi & Riéra Chapter 2, Poncy *et al.* Chapter 4, de Granville Appendix 2). The area includes a hilly zone, a creek valley and swamp forest (a low area with slow running water). Apart from the numerous streams in this area, there is only one isolated temporary forest pond (about 15 x 20 m, depth up to 0.5

Bongers *et al.* (eds.), Nouragues. Dynamics and plant-animal interactions in a neotropical rainforest, 167–184.
2001 Kluwer Academic Publishers. Printed in the Netherlands.

m), which is filled for about two months during the long rainy season (April-May-June). Rainfall at Nouragues varies between years so it happens that in years without sufficient rainfall the pond dries up completely between two rain showers. Permanent forest ponds do not occur at this site. Fieldwork was conducted at irregular time spans between April and October in 1991, except for the month July, from mid-December 1994 to mid-January 1995 and from 1-12 June 1998.

Fieldwork was done in addition to a field study on the ecology of Dendrobatid frogs. Species richness was determined by visual encounter surveys in combination with surveys at amphibian breeding sites. In visual encounter surveys a person walks through a habitat for a described time period, systematically searching for animals. Little time was spent during the night, except for the period between April 23rd and the end of May 1991, when the temporary forest pond was visited two times a week. In 1998, the visual encounter surveys were supplemented with aural surveys of calling anurans. The species survey was performed almost entirely in the forest surrounding the field station within an area of 3 km^2. The swamp forest was investigated only 4 times. Observations were made from 0-2 m above ground. Hiding places were not studied for logistic reasons. When a species was seen, its scientific name, date, time, location, vertical position, sex (if recognisable), microhabitat and behaviour were recorded.

The presence of one arboreal frog was determined by its call only. This canopy frog, which had been caught for the first time in 1997 by one of us (P.G.) at the Arataye site, is relatively common at Nouragues. This species seems not to be associated with streams. Calling males were recorded throughout the year – except for the driest months - at Nouragues, Arataye and in primary forest at the Piste de St. Elie. Its very characteristic, regularly repeated call ('tonk, tonk, tonk'), that is often associated with a 'metronome', is most commonly heard from the early evening until the early morning. This frog, which was eventually referred to as a new species, turned out to be *Hyla hadroceps.* The description of this species by Duellman & Hoogmoed (1992) was based on a specimen from the extreme southern Guyana. It distinguishes itself from *Phrynohyas* and *Osteocephalus* by having a single vocal sac.

One species, *Rana palmipes*, is included on the Nouragues list although it was not found within the searching area of 3 km^2 surrounding the camp, but within 5 km of the campsite.

2.2 ARATAYE STATION

The second site, the Arataye Field Station (3°59'N, 52°35'W), is bordered by a flat area in the North, by forested hills in the East and by the Arataye River in the South and West. This study site contains small creeks, a swamp forest and a large (about 3 ha) forest pond, which fills during the rainy season between December and July. After the rainy season there remains a small water body with a surface of 4 m^2. The pond is situated in forest, characterised by large gaps with dense bushes. The canopy is low (15-20 m) except from some emergent trees (*Caryocar* sp.). Fieldwork at Arataye was conducted between September 1995 and September 1998, in the forest surrounding the field station within a radius of one km. Surveys were performed in the early morning and late afternoon, night-time fieldwork was carried out from dusk to mid-night. Transects were walked weekly.

At either of the two study sites, species richness was determined by visual encounter surveys initially. However, for amphibian species richness this method was replaced by aural encounter surveys of calling frogs, once the different calls were known (Marty &

Gaucher 1999). Six species could not be located by their call only (*Dendrobates tinctorius, D. ventrimaculatus, Lithodytes lineatus, Rana palmipes, Osteocephalus cabrerai and Scinax cf. x-signatus.*

Species, which were hard to identify, were preserved in formalin and identified by either M.S. Hoogmoed, curator of the National Museum of Natural History in Leiden, the Netherlands (Nouragues specimens) or J. Lescure, Muséum National d'Histoire Naturelle in Paris, France (Arataye specimens).

3. Results

3.1 SPECIES NUMBERS

We recorded 54 species of amphibians (21 genera, 8 families) and 58 species of reptiles (44 genera, 14 families) at Nouragues (Table 1 and Appendix 5). At Arataye 65 species of amphibians (25 genera, 9 families) and 65 species of reptiles (53 genera, 18 families) were

TABLE 1. Number of genera and species within each family of amphibians and reptiles. Data collected in 1991, 1994 and 1998 at the Nouragues Field Station and between September1995 and September 1998 at the Arataye Field Station, French Guiana. N = Nouragues, A = Arataye.

FAMILY	GENERA		SPECIES		FAMILY	GENERA		SPECIES	
SITE	N	A	N	A		N	A	N	A
Amphibians					Reptiles				
Frogs & Toads					Snakes				
Allophrynidae	1	1	1	1	Boidae	2	4	2	4
Bufonidae	2	2	6	5	Colubridae	17	18	22	21
Centrolenidae	2	2	2	3	Elapidae	1	1	2	3
Dendrobatidae	3	3	6	7	Viperidae	2	2	4	2
Hylidae	5	5	18	27	TOTAL	22	25	30	30
Leptodactylidae	5	7	18	17					
Microhylidae	2	3	2	3					
Pipidae	0	1	0	1	Turtles				
Ranidae	1	1	1	1	Chelidae	2	2	2	3
TOTAL	21	25	54	65	Emydidae	1	1	1	1
					Kinosternidae	0	1	0	1
Total amphibians	21	25	54	65	Pelomedusidae	0	1	0	1
					Testudinidae	1	1	1	1
					TOTAL	4	6	4	7
Reptiles									
Lizards & Amphisbaenians					Crocodilians				
Amphisbaenidae	0	1	0	2	Alligatoridae	1	1	1	1
Gekkonidae	3	4	4	5	TOTAL	1	1	1	1
Gymnophthalmidae	7	7	8	7					
Iguanidae	0	1	0	1	Total reptiles	44	53	58	65
Polychrotidae	1	1	4	4					
Scincidae	1	1	1	1					
Teiidae	3	3	3	3					
Tropiduridae	2	3	3	4					
TOTAL	17	21	23	27					

encountered. The most important anuran families at both sites are the Hylidae and the Leptodactylidae; among the reptiles most snake species belong to the Colubridae and most lizard species to the Gymnophthalmidae and the Gekkonidae.

3.2 CHANGING SPECIES DISTRIBUTIONS

Documented occurrences of several herpetofaunal species in French Guiana remain sparse. Species distribution patterns are indubitably a reflection of this lack of material. Therefore it must be emphasised that such patterns often give more insight in the occurrence and search behaviour of herpetologists.

Lescure (1975, 1976, 1981a, 1982, 1986), Lescure *et al.* (1995, 1996), Hoogmoed & Avila-Pires (1991) and Born (1996a, 1996b and unpublished data) reported already the majority of amphibian species found at Nouragues and Arataye from other localities in French Guiana. Most of the lizard species in this area have been reported in literature from throughout the country (Hoogmoed & Lescure 1975, Gasc 1976, 1981, 1990, Hoogmoed & Avila-Pires 1991, Born 1996b and unpublished data, Avila-Pires 1995).

During the progress of this book, herpetofaunal studies at Saül and Petit Saut were conducted (Kok 2000, Vidal unpublished data, de Massary 1999, de Massary & Lescure 1998). At Petit Saut, the building of the Petit Saut dam flooded 365 km^2 pristine rainforest. The herpetofauna of the created forest islands was studied in order to examine the effect of forest fragmentation on the herpetofauna in French Guiana. The area under study at Petit Saut partly covers the same area, which was inventoried by Hoogmoed & Avila-Pires (1991). In this way we could add one frog species, seven lizard species and fifty-five snake species to the Petit Saut species inventory conducted by Hoogmoed and Avila-Pires.

Nevertheless, the herpetofauna of Nouragues and Arataye does include noteworthy records of species, for which additional data for extending distribution ranges were recorded. Some of them were recently reported from Saül as well (Kok 2000).

Amphibians

For three amphibian species, present at Nouragues but absent at Arataye, distribution ranges extended. These species are: *Atelopus flavescens*, *Leptodactylus longirostris*, *Leptodactylus myersi.* Extended ranges were found for twelve amphibian species present at both sites: *Allophryne ruthveni*, *Atelopus franciscus*, *Epipedobates hahneli*, *Hyla* sp. 1 (one other record from de Piste de St. Elie), *Leptodactylus rhodomystax*, *Leptodactylus stenodema*, *Physalaemus petersi*, *Hyla hadroceps* (first record in French Guiana), *Osteocephalus oophagus*, *Scinax proboscideus*, *Phrynohyas resinifictrix* and *Phrynohyas coriacea.* Lescure et al. reported the latter for the first time in 1996, although at Nouragues calling males were recorded in May 1991 at the isolated forest pond, where they congregate for only two nights.

Range extensions were found for nine species due to their presence at Arataye: *Colostethus baeobatrachus*, *Hyla brevifrons*, *Hyla ornatissima*, *Hyla melanargyrea* (first and only other record in French Guiana from Hoogmoed & Avila-Pires, 1991), *Scinax cruentommus* (no other records in French Guiana), *Lithodytes lineatus*, *Leptodactylus petersii*, *Pipa pipa* (no other records from the interior) and *Hamptophryne boliviana.*

Lizards and Amphisbaenians

The Nouragues and Arataye data expanded the distribution area for one amphisbaenian: *Amphisbaena alba* (present at Arataye), a semi-fossorial species, which has not been reported before from the central part of the country. Distribution ranges extended for four lizard species: *Bachia flavescens*, *Prionodactylus argulus*, *Mabuya nigropunctata* and *Alopoglossus angulatus*. *Alopoglossus angulatus* (present at Arataye) inhabits the leaf-litter layer near creeks and is considered as rare (Gasc 1990, de Massary 1999); it was only mentioned from Village Bienvenue in the south eastern part of the country (Hoogmoed 1973), from Petit Saut by Hoogmoed & Avila-Pires (1991) and de Massary (1999).

Snakes

Initially, few studies in French Guiana focussed on the inventory of snakes. Gasc and Rodrigues (1980) listed several species mainly from the south eastern part of the country and some from other localities. Chippaux (1986) conducted a species inventory in the coastal area and collected additional data in the south eastern part, in the Northwest and in the central part of the country. Chippaux' plots in the central part (Saül, Maripasoula and Camopi) were visited only occasionally.

Recently, it seems that snakes excited the interest in an increasing degree, considering the number of publications on this subject. Starace (1998), Vidal (1994-1995, unpublished data) and Vidal *et al.* (1999) focussed on this group of reptiles. Starace reported for several species additional locations throughout French Guiana. Vidal reported 55 additional snake species for the Petit Saut area. Among them were two species, which were recorded for the first time in French Guiana. One of those two species (*Xenopholis scalaris*) could be recorded at Arataye as well. However, most knowledge on the distribution of snake species in French Guiana is based on the work of Chippaux. Besides Chippaux' herpetological species inventories, there are three other studies at the coastal area, conducted at Petit Saut (Hoogmoed & Avila-Pires 1991, Vidal unpublished data) and at the Piste de St. Elie (Born 1991 unpublished data).

Thirty-one snake species present at Nouragues and/or Arataye (see also Appendix 5) have not been reported before from the interior part of French Guiana, obviously due to the lack of material. These species are: *Atractus badius*, *A. Schach*, *Chironius fuscus*, *C. carinatus*, *Drymarchon corais*, *Drymobius rhombifer*, *Helicops angulatus*, *Imantodes lentiferus*, *Leptodeira annulata*, *Liophis breviceps*, *Oxybelis fulgidus*, *Pseustes sulphureus*, *Rhinobotryum lentiginosum*, *Siphlophis cervinus*, *Taeniophallus brevirostris*, *Tripanurgos compressus*, *Xenodon rhabdocephalus*, *Xenopholis scalaris*, *Micrurus collaris*, *M. psyches*, *M. surinamensis*, *Bothrops brazili*, *Bothrops bilineatus**, *Corallus caninus**, *C. hortulanus**, *Chironius scurrulus**, *Erythrolamprus aesculapii**, *Leptophis ahaetulla**, *Philodryas viridissimus**, *Tantilla melanocephala** and *Xenoxybelis argenteus**. Marked species (*) were recently mentioned from the centre of the country by Starace (1998) as well.

Atractus schach was reported from the northwestern part of the country only.*Imantodes lentiferus* and *Taeniophallus brevirostris* were mentioned from one locality in the Southeast and only recently from the northern part as well. The presence of *Micrurus collaris*, *M. surinamensis* and *Rhinobotryum lentiginosum* extends their distribution ranges southwards to an important degree because they are known from the northern part of the country only. The only other citation of *Micrurus collaris* seems to be that of Chippaux (1986) from Cacao. At Nouragues it was found at night in August 1991, at the campsite on a rock, close to the electric generator. *Xenopholis scalaris* have been reported only recently from two other locations in the northern part of the country (Starace 1998, Vidal et al. 1999).

Turtles and crocodilians

From the turtle species recorded at Nouragues (four) and Arataye (seven), *Geochelone denticulata* is the most common species in French Guiana with distribution records from throughout the forested parts of the country by Fretey (1987) and from the Piste de St. Elie (Born unpublished data). *Phrynops nasutus* was already known from the western part of the country and from Oyapock in the southeast (Fretey 1987) so its discovery at Nouragues and Arataye extends this distribution range to the Northeast. *Platemys platycephala* (present at both sites), *Rhinoclemmys punctularia* (present at both sites), *Phrynops gibbus* (Arataye), *Kinosternon scorpioides* (Arataye) were already reported from the coastal area or from the Southeast (Fretey 1987, Metrailler & Le Gratiet 1996) but not yet from the central part of the country. *Podocnemis cayennensis*, present at Arataye along the Arataye River, was known from the Northeast to Oyapock, but was not reported from the interior, so its distribution range extends more westwards.

Medem (1983), Hoogmoed and Avila-Pires (1991) and Born (unpublished data) mentioned *Paleosuchus trigonatus* from the mid-coastal area, from Cayenne and further eastwards, so its presence at Nouragues and Arataye, extends this range southwards.

3.3 LIFE HISTORIES

Amphibians

The ecological summary (Appendix 5) of amphibian species at Nouragues and Arataye represents several reproduction strategies (reproduction strategies derived from Duellman & Trueb 1986, Hödl 1990 and Crump 1974). Frequencies of the different breeding systems of the species at both sites are tabulated in Table 2.

The majority of species is dependent on water in some way: for egg deposition or for further development of their larvae. Therefore, amphibians are most conspicuous at breeding ponds or streams and surveys at those sites are most effective. At Nouragues 15 species were encountered during the long rainy season in the isolated temporary forest pond, usually in large congregations of explosive breeding. Outside the breeding season most of them were found only very occasionally. Ten of the pond breeders lay their eggs unprotected in stagnant water (*Allophryne ruthveni*, *Ceratophrys cornuta*, *Bufo margaritifera*, *Bufo* sp. only at the edge of the pond, *Chiasmocleis shudikarensis*, *Phrynohyas coriacea*, *P. venulosa*, *Osteocephalus taurinus*, *Scinax proboscideus*, *S. ruber*). Five other species, which were found at the same forest pond (*Phyllomedusa vaillanti*, *P. tomopterna*, *P. bicolor*, *Hyla leucophyllata* and *Hyla* sp. 1), lay eggs on leaves overhanging water, thereby reducing egg predation. Hatching tadpoles fall into the water to feed and develop. Another group (three species), which was found at the forest pond, reduces egg predation by laying eggs in a foam nest on the surface of (*Physalaemus petersi*) or near the forest pond (*Leptodactylus pentadactylus*, *L. knudseni*). Five other foam nest producers are *Leptodactylus longirostris*, *L. myersi*, *L. stenodema*, *L. rhodomystax* and *L. mystaceus*.

Except for *L. longirostris*, which was found breeding in water-filled depressions on the inselberg and near the cascades (a stream current embedded in granite substrate), the other Leptodactylids were not found at their breeding sites but throughout the season, usually at forest edges and along forest trails. Besides the tadpoles of *Leptodactylus longirostris* one other tadpole species, most probably of *L. myersi*, was recorded in the depressions between the *Pitcairnia* vegetation on the Inselberg. Duellman & Trueb (1986) reported tadpoles of this frog from water filled cracks on granitic rock. They moved from one crack to another by flipping the tail and moving themselves across the rock. *Leptodactylus cf. wagneri* was found breeding in depressions near the swamp forest or in the inundated edge of the crique Nouragues.

TABLE 2. Different anuran breeding strategies at the Nouragues Research Station (54 species) and the Arataye Field Station (65 species) in French Guiana, expressed as proportion of the amphibian fauna. Number of species is given in parenthesis. Data collected in 1991, 1994 and 1998 at Nouragues and between September 1995 and September 1998 at Arataye. Notes: (1) = includes *O. cabrerai*, which is reported to breed in tree holes, although we saw it on vegetation along small streams. (2) = In *Colostethus*, except *C. degravillei*. (3) = This category does include *Leptodactylus stenodema* and *L. pentadactylus* (both species were categorized in 2.g.), although they were reported to construct foam nests away from water as well. Categories derived from Duellman, & Trueb (1986).

	Reproductive system	Site			
		Nouragues		Arataye	
		%	N	%	N
1	Eggs and tadpoles in water				
	a. aquatic eggs & tadpoles free in streams	0.15	(8)	0.12	(8)
	b. aquatic eggs & tadpoles free in stagnant water	0.20	(11)	0.31	(20)
	c. aquatic eggs & aquatic tadpoles in constructed basin near streams	0.02	(1)	0.02	(1)
	d. aquatic eggs & tadpoles in water-filled tree-hole or bromeliad (1)	0.07	(4)	0.06	(4)
2	Eggs terrestrial or arboreal and tadpoles in water				
	e. eggs on leaves overhanging stagnant water & tadpoles free in water	0.09	(5)	0.08	(5)
	f. eggs on leaves overhanging streams & tadpoles free in water	0.04	(2)	0.05	(3)
	g. eggs in foam nest near or on stagnant water & tadpoles free in water	0.17	(9)	0.12	(8)
	h. arboreal or terrestrial eggs & tadpoles carried to small water body or inundated stream edge (2) on back of parent	0.09	(5)	0.09	(6)
3	Eggs and tadpoles out of water				
	i. eggs & tadpoles in a terrestrial clay built foam nest	0.04	(2)	0.03	(2)
	j. terrestrial eggs, direct development into a frog, no free-living tadpole	0.11	(6)	0.06	(5)
	k. eggs carried in cells on back of female, development into young frog, no free-living tadpole	0.00	(0)	0.02	(1)
	l. eggs terrestrial & tadpoles carried on parents' back until metamorphosis	0.02	(1)	0.02	(1)
4	Unknown	0.00	(0)	0.02	(1)
	Summary:				
1	Total species dependent on water for their reproduction (3)	0.85	(45)	0.84	(55)
	-breeders using a constraint water body: water-filled tree-hole, bromeliad axil (d,h)	0.17	(9)	0.15	(10)
	-breeders using a temporary or permanent forest pond (b,e,g)	0.46	(25)	0.51	(33)
	-stream breeders (a,c,f)	0.20	(11)	0.18	(12)
2	Total species independent on water for their reproduction (i,j,k,l)	0.17	(9)	0.14	(9)
3	Unknown	0.00	(0)	0.02	(1)

Among the stream breeders (Table 2) 9 species were present along the crique Nouragues (*Atelopus flavescens, A. franciscus, Bufo guttatus, Otophryne pyburni, Hyla fasciata, H. geographica, H. granosa,* the recently described *Hyalinobatrachium nouragensis*, (Lescure & Marty, 2000) and *Cochranella oyampiensis*). The latter lays eggs on leaves overhanging a stream or creek. *Bufo guttatus* was occasionally found at night, sitting on the large rock boulders bordering the creek. Another stream breeder, *Hyla boans*, scoops out a small pool at the edge of a stream or uses natural depressions. Eggs and tadpoles of this species were found in natural depressions at the inundated edges of the cascades at Nouragues. Although not associated with stream sides, the opportunistic breeder *Bufo* sp. was regularly found in small breeding congregations at the inundated edges of the creek Nouragues. This species seems to breed throughout the year whenever it finds a small body of standing water. Tadpoles and calling males were found in a puddle on top of a fallen tree trunk within the forest at the Nouragues Station.

Probably due to their reproduction strategy, three species (*Osteocephalus oophagus, Phrynohyas resinifictrix, Hyla hadroceps)* occurred throughout the forest without being obviously associated with any microhabitat. These species lay their eggs in arboreal waters: water-filled tree cavities (or bromeliads in *O. oophagus*) sometimes high up in the canopy. Tadpoles of *Phrynohyas resinifictrix* or *Hyla hadroceps* were found only once at Nouragues: in a water-filled tree-hole 1.5 m above the ground in mature forest. Although *Osteocephalus cabrerai* is reported to breed in water-filled tree-holes as well, this species was present near small creeks at both sites and at the Piste de St. Elie. Considering its presence along streams, we question whether this frog uses tree holes for its reproduction.

Another group of eight species, which were present throughout the forest, includes *Adenomera* spp. and *Eleutherodactylus* spp. They belong to a group of frogs, which are independent of water for their reproduction. Thereby they eliminate competition with other aquatic animals, predation by aquatic predators or unfavourable circumstances such as a breeding site that runs dry. *Adenomera* spp. produces a clay-built foam nest between roots in the ground. Eggs are laid in this nest subsequently. Advanced tadpoles hatch from each egg. They stay in the nest subsequently, drawing on their yolk reserves until they are fully metamorphosed. Two such foam nests were found at Nouragues by accident, during intensive searchings of the litter herpetofauna. One contained 5 tadpoles, the other only eggs. Within two weeks, two of the five tadpoles hatched into miniature frogs, the other tadpoles died. *Eleutherodactylus* deposits eggs on land in moist situations. A tough membrane provides protection against desiccation. Development and metamorphosis take place within the egg-capsule so there is no free-living stage of a tadpole. Noteworthy is the occurrence of many young *Eleutherodactylus chiastonotus* at the end of the dry season in the drier parts of the swamp forest and Palm swamp at Nouragues.

In *Pipa pipa*, eggs are imbedded in special brooding pouches on the dorsum of the female. Upon completion of development the froglets push their way out and swim away. Although this frog - which was recorded at Arataye in 1998 - is an aquatic species, it was assigned to the category of species that have their eggs and tadpoles out of water (Table 2).

Seven other species lay their eggs on land in moist situations, but they are still dependent on water for further development of their larvae. *Dendrobates tinctorius, Epipedobates femoralis, Colostethus beebei* and *C. baeobatrachus* (only present at Arataye) lay their eggs on the ground or on bromeliads (*Dendrobates ventrimaculatus* and *Epipedobates hahneli*). One, or both parents, return to the nest regularly. Hatching tadpoles wriggle on the parents back and are subsequently transported to small water bodies like water-filled bromeliad axils or water-filled cavities in a tree or log. Tadpoles of *Colostethus degranvillei* are carried on the back of their parents until metamorphosis (Lescure 1984). *Colostethus beebei* releases its tadpoles at the inundated edge of stream. Tadpole transport and their subsequent release in suitable water bodies could be observed in all Dendrobatid species at Nouragues. *Dendrobates tinctorius* occurred most commonly in canopy gaps, but not on the inselberg while *D. ventrimaculatus* seemed to prefer both canopy gaps and the open, sunny vegetation of bromeliads on the inselberg. *Epipedobates femoralis, E. hahneli* and *Colostethus beebei* were found throughout the forest while *C. degranvillei* seemed to be more associated with stream sides.

Examination of the Arataye data reveals similar trends for reproductive modes. Of the amphibians encountered at Arataye 84.6% is dependent on water (Nouragues 83.3%), while 13.8% (Nouragues 16.7%) is independent on water for their reproduction (Table 2). The Arataye area contained 33 stagnant water breeders. Most of them (24) were found almost exclusively in the large temporary forest pond, which in fact is an inundated part of the forest. Among them were explosive breeders like *Ceratophrys cornuta, Osteocephalus leprieurii, Phrynohyas coriacea, P. venulosa, Hamptophryne boliviana, Scinax cf. x-signatus, Scinax* sp. 1, *Chiasmocleis shudikarensis* and *Allophryne ruthveni*. These species migrate to the pond after heavy rains in December, breed for only one or two nights to disappear again until the next rains in May.

Five species within the group of stagnant water breeders (*Hyla fasciata, H. granosa, H. multifasciata, L. mystaceus* and *L. rhodomystax)* seemed to be strongly associated with the swamp forest at Arataye. Conversely, at Nouragues these frogs occurred along the slow running creek. The forest pond at Arataye dries up only occasionally. This enables the other inhabitants of the pond, electric eels of the genus *Electrophorus* to survive the dry season. Noteworthy is that these same electric eels, besides the aquatic turtles *Phrynops gibbus*, *P. nasutus* and *Kinosternon scorpioides*, form the chief predators on eggs and tadpoles of all pond breeders, but on *Osteocephalus leprieurii* in particular. The majority of the stream breeders at Arataye inhabited both the creeks and the Arataye River. This group included almost the same species as were found in Nouragues. At this site they occurred only along the creeks. Among them was *Bufo marinus*, usually encountered in open areas near streams or rivers. In Arataye it was present along the Arataye River but in Nouragues it was found only sporadically at the campsite, probably because its preferred habitat is scarce at Nouragues.

Reptiles

Reptile species recorded at the two sites under study are predominantly egg-layers with a few exceptions. The lizard *Mabuya nigropunctata* gives birth to live young. Most lizard species (*Gonatodes humeralis, G. annularis*, *Anolis fuscoauratus, A. nitens, A. ortonii, A. punctatus*, *Ameiva ameiva*, *Kentropyx calcarata*, *Tropidurus plica*, *T. umbra*, *Iphisa elegans*) lay their eggs between loose bark on a tree trunk, among roots, in holes in rotten trees or logs, at tree bases among leaves in the litter layer (*Anolis* spp., *Bachia flavescens*, *Coleodactylus amazonicus*) and between stones (*Gonatodes* spp). *Tupinambis teguixin* and *Gonatodes* are also reported to use termite nests for egg deposition. Sometimes these nests are high up in trees and sometimes they are used as a communal nest (in *Tupinambis teguixin*). Eggs of *Kentropyx calcarata* and *Leposoma guianense* were found once at the Nouragues Field Station. The first clutch contained 7 eggs, which were glued together on a leaf; they hatched within two weeks. The second clutch contained two eggs, glued on a leaf, which hatched within three weeks. See also data in Avila Pires (1995) on this subject.

Among the snakes encountered at Nouragues and Arataye, *Boa constrictor*, *Corallus hortulanus*, *C. caninus*, *Eunectes murinus*, *Epicrates cenchria*, *Bothrops atrox*, *B. bilineatus* and *B. brazili* are viviparous as well. *Helicops angulatus*, an aquatic snake, was reported to be facultative oviparous or viviparous (Mattison 1995). Viviparity eliminates the necessity of leaving the water to lay eggs. Next to nothing is known about the egg-laying sites of snakes. They hide their eggs so well that they are rarely found. However, a clutch of four eggs of *Drymobius rhombifer* was found at the Arataye site, under a piece of fallen bark on the ground in May 1996. After three weeks, three eggs hatched.

3.4 COMPARISON TO OTHER SITES

Species occurrence at the five sites in French Guiana (Fig. 1) and at the site in adjacent Brazil is tabulated in Appendix 5. Data are taken from Born 1996a, 1996b (amphibians and

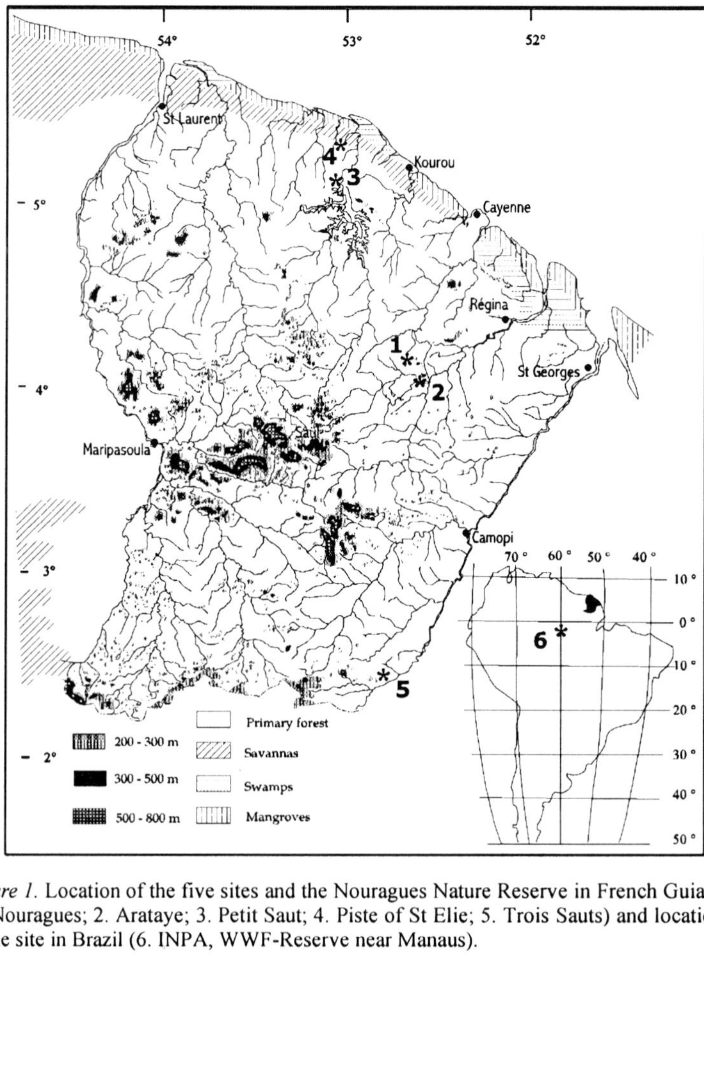

Figure 1. Location of the five sites and the Nouragues Nature Reserve in French Guiana (1. Nouragues; 2. Arataye; 3. Petit Saut; 4. Piste of St Elie; 5. Trois Sauts) and location of the site in Brazil (6. INPA, WWF-Reserve near Manaus).

reptiles, Nouragues), unpublished data fieldwork 1991, 1994, 1998 (Nouragues) and 1991 (amphibians and reptiles, Piste de St. Elie), Hoogmoed & Avila Pires 1991 (amphibians & reptiles Petit Saut), de Massary 1999 (lizards Petit Saut), Vidal unpublished data fieldwork 1994-1995 (snakes Petit Saut), Starace 1998 (snakes Petit Saut), Lescure 1982 (amphibians Trois Saut), Chippaux 1986 (snakes Trois Sauts and Petit Saut), Gaucher unpublished data 1995-1998 (amphibians and reptiles Arataye), Zimmerman & Rodrigues 1990 (amphibians & reptiles INPA/WWF Reserves, 90km NW of Manaus, Brazil). Taxonomical names on the different species lists are updated as far as possible. For example the current genus name of some treefrog species changed from *Ololygon* into *Scinax*. The *Bufo typhonius* group was found to consist of *Bufo margaritifera* and *Bufo* sp. The current proper names of some other species mentioned in the different species lists are (old names in parenthesis): amphibians: *Atelopus spumarius* (*A. pulcher*), *Cochranella oyampiensis* (*Centronella oyampiensis*), *Hyalinobatrachium taylori* (*Centronella taylori*), *Dendrobates ventrimaculatus (D. quinquevittatus)*, *Epipedobates hahneli* (*E. pictus*), *Leptodactylus myersi (L. rugosus)*, *Osteocephalus cabrerai* (*O. buckleyi*), *Otophryne pyburni* (*O. robusta*), lizards: *Alopoglossus angulatus* (*A. carinicaudatus*), *Anolis nitens (A. chrysolepis)*, *Bachia flavescens* (*B. cophias*), *Mabuya nigropunctata* (*M. mabuya)*, *Tropidurus plica (Plica plica)*, *Tropidurus umbra (Plica umbra)*, *Tupinambis teguixin* (*T. nigropunctatus*), snakes: *Corallus hortulanus* (*C. enydris*), *Liophis typhlus* (*Leimadophis typhlus*), *Liophis lineatus* (*Lygophis lineatus*), *Liophis reginae* (*Leimadophis reginae*), *Taeniophallus brevirostris (Rhadinaea brevirostris)*, *Xenoxybelis argenteus (Oxybelis argenteus)*. Subspecies names are avoided. We used Frost (2000) for current names of amphibian species, and Avila-Pires' (1995) taxonomical classification of lizard species.

The Trois Sauts records included 10 species (1 Hylidae, 3 Microhylidae, 6 Leptodactylidae), which were not actually captured at Trois Sauts, but east or westwards of it. Those species have been added to the Trois Sauts list because they are almost certainly present at this site, according to Lescure (1982). These species are: *Adenomera hylaedactyla*, *Lithodytes lineatus*, *Hamptophryne boliviana*, *Eleutherodactylus inguinalis*, *Eleutherodactylus gutturalis*, *Adelophryne gutturosa* and *Chiasmocleis hudsoni*. Three other species: *Osteocephalus cabrerai*, *Synapturanus mirandaribeiroi* and *Leptodactylus knudseni* have been added to the list for the same reasons but were actually collected between Camopi and Village Pina. Lescure (1982) listed two other species, (*Phrynohyas venulosa* and *Hyla multifasciata*, Hylidae) which were not reported from Trois Sauts by herpetologists but by local Indians.

For similar reasons, *Phyllomedusa vaillanti*, *Micrurus lemniscatus* and *Bothrops bilineatus* were included on the list of the INPA-WWF Reserve in Brazil (Zimmerman & Rodrigues 1990) although they were never collected at the INPA-WWF Reserve.

4. Discussion

4.1 SPECIES RICHNESS

Amphibians

If we compare the number of amphibian species encountered at Nouragues with that for Trois Sauts and Arataye, the inventory of frogs at Nouragues probably is not yet complete. Compared to Arataye and Trois Sauts, Nouragues has 11 species less, although the Trois

Sauts list includes 10 species, which were not collected there actually (Table 3 and 5). This lower species richness is partly due to differences in sampling effort and methods.

Table 4 tabulates a summary of the sampling regimes at the different sites under comparison. Examination of this Table reveals that Trois Sauts and Arataye are relatively thoroughly sampled sites. Both sites were sampled for many months; moreover, sampling effort in Trois Sauts was increased considerably by the co-operation with local Indians (the Wayãpi), who helped with the collection of frogs.

TABLE 3. Species richness of amphibians and reptiles at five sites in French Guiana and one in Brazil. Data taken from Born 1996a and 1996b, unpublished data fieldwork 1991, 1994, 1998 (Nouragues) and 1991 (Piste de St. Elie), Hoogmoed & Avila Pires 1991 (Petit Saut), de Massary 1999 (Petit Saut), Vidal unpublished data fieldwork 1994-1995 (Petit Saut), Starace 1998 (Petit Saut), Lescure 1982 (Trois Saut), Chippaux 1986 (Trois Sauts & Petit Saut), Gaucher unpublished data 1995-1998 (Arataye), Zimmerman & Rodrigues 1990 (Brazil, INPA/WWF Reserves near Manaus). Category (Reserve) includes Nouragues and Arataye. • = Total reptiles at Trois Sauts does not include lizards, turtles and crocodilians. •• = Total reptiles Brazil does not include turtles and crocodilians. ND = No data available.

	Nouragues	Arataye	(Reserve)	Petit Saut	Piste St. Elie	Trois Sauts	Brazil
Total amphibians	54	65	(70)	38	34	65	42
Total reptiles	58	65	(82)	93	54	• 44	•• 86
Lizards & amphisbaenians	23	27	(29)	24	19	ND	24
Snakes	30	30	(45)	67	31	44	62
Turtles	4	7	(7)	1	3	ND	ND
Crocodilians	1	1	(1)	1	1	ND	ND
Total number of species	112	130	(152)	131	88	•109	••128

The amphibian species inventory at Arataye was mainly based on aural surveys of calling amphibians. Because many amphibians - especially small species that call from a hidden position - can be extremely difficult to find, this sampling method is very efficient if species composition rather than abundance is required. Additional fieldwork at Nouragues (1998) focused on aural nighttime surveys. Due to this fieldwork, together with some identified frog calls that were registered in 1991 at the temporary forest pond at the same site, we could add some species to the Nouragues list. In contrast to Arataye and Trois Sauts, the total number of night surveys at Nouragues was low, due to a larger proportion of daytime fieldwork on the ecology of Dendrobatid frogs.

The lower species richness in Nouragues versus Trois Sauts and Arataye is most pronounced in the Hylidae (Table 5 and Appendix 5), whereas species richness in the other families is approximately equal. Compared to Arataye and Trois Sauts, Nouragues has, respectively, 9 and 7 species of Hylid frogs less. If compared to Petit Saut (species richness of amphibians based on only two weeks of intensive sampling), the Piste de St. Elie and the INPA-WWF Reserve, Nouragues has respectively 3, 3 and 2 more species of Hylid frogs. In contrast, Arataye has 12, 12 and 11 more species of Hylid frogs if this site is compared to Petit Saut, the Piste de St. Elie, and the Brazilian site respectively. Arataye has one Leptodactylid species less than Nouragues, *Leptodactylus longirostris* that occurred only on the Inselberg. Petit Saut, the Piste de St. Elie and the Brazilian site have a lower number of Leptodactylid frogs than Nouragues and Arataye.

TABLE 4. Summary of the sampling regimes at the different sites and their results, expressed in total number of species. Aural encounter survey = survey of calling amphibians in an area. Transect = Transect sampling. Pit-fall = Pit-fall trap sampling. Clear cut = In Brazil (INPA/WWF-Reserves near Manaus) part of the snake species were brought in by workers who were clear cutting several hundred hectares in the reserve area. Island sampling = The lizard inventory at Petit Saut (b) was conducted on several small islands of 2 ha tropical rainforest, which were created when a 350 km² area was flooded by the building of a dam on the Sinnamary river. Notes: At the Piste de St. Elie a roadside transect (10 km) was surveyed almost every day in order to search for live and dead snakes (run over by passing cars). The sites Trois Sauts, Petit Saut and Brazil include data from different species inventories concerning amphibians, lizards/amphisbaenians and snakes, which are expressed by different letters (a), (b) and (c) respectively. Data based on the authors mentioned in legend of Table 3. ND = No Data; M = Months; W = Weeks.

Site	Sampling time	Visual encounter survey	Aural encounter survey	Breeding site survey	Transect	Pit-fall traps	Clear-cut	Island sampling	Sampling area	Total species found
Nouragues	4 M	•	•	•					3 km²	112
Arataye	30 M	•	•	•	•				3 km²	130
Petit Saut (a,b,c)	2 W	•		•					ND	67
Petit Saut (b)	ND	•				•		•	ND	23
Petit Saut (c)	ND	•			•				ND	60
Trois Sauts (a)	Many M				•	•			10 km²	65
Trois Sauts (c)	18 M				•	•			ND	44
P. St. Elie (a,b)	2 W	•	•	•					5 km²	67
P. St. Elie (c)	2 W	•			•				10 km trail	21
Brazil (a)	36 M		•	•	•				65 km trail	42
Brazil (b)	35 days				•				ND	24
Brazil (c)	40 M				•		•		1661 km trail	62

By far the majority of the amphibians, which were absent at Nouragues but present at Arataye, are nocturnal species. At Arataye they were found breeding either in stagnant water of the forest pond, or in small pools in the inundated riverbanks of the Arataye. The study site Nouragues has only one shallow, isolated forest pond, which is filled up with water during periods with very heavy rainfall. However, a few days without rain would cause this pond to dry up and entire cohorts of tadpoles to be lost. This was observed twice during the long rainy season in 1991, and thus resulted in a negative effect on the breeding success of several species in that year. Although the lower number of nocturnal amphibians at Nouragues may in part reflect sampling error due to lower nocturnal sampling effort, it is this lack for appropriate breeding sites that definitively limits amphibian species composition. The numbers of foam nest builders like the Leptodactylid species, are almost equal at both sites. Probably they are less dependent on the life span of forest ponds because tadpole development takes place in the foam nest. Similar trends are found in the streamside breeders (Centrolenidae, Bufonidae) and species that have their eggs and tadpoles out of water (*Eleutherodactylus* spp., *Adenomera* spp.) or that are only dependent on very small water bodies for tadpole development (Dendrobatidae).

TABLE 5. Species richness within the different anuran families at five sites in French Guiana and one in Brazil. Data based on the authors mentioned in legend of Table 3. Category (Reserve) includes total of Nouragues and Arataye.

	Total number of species within each family						
SITE	Nouragues	Arataye	(Reserve)	Petit Saut	Piste St Elie	Trois Sauts	Brazil
Family							
Allophrynidae	1	1	(1)	0	0	1	0
Bufonidae	6	5	(6)	5	4	5	3
Centrolenidae	2	3	(3)	2	0	2	1
Dendrobatidae	6	7	(7)	3	4	6	3
Hylidae	18	27	(28)	15	15	25	16
Leptodactylidae	18	17	(20)	10	11	19	13
Microhylidae	2	3	(3)	3	0	5	5
Pipidae	0	1	(1)	0	0	1	1
Ranidae	1	1	(1)	0	0	1	0
Total number of species	54	65	(70)	38	34	65	42

Reptiles

Lizards and amphisbaenians. The lizard species composition at Nouragues is almost similar to that of Arataye, except for some species. Among this group are species which are usually extremely difficult to find due to their hidden way of life, like the fossorial amphisbaenians *Amphisbaena fuliginosa, A. alba* and the lizard *Alopoglossus angulatus,* all present at Arataye but absent at Nouragues. In contrast to Nouragues, the Arataye study site includes an open, riverine habitat - the Arataye River - where *Iguana iguana* was commonly found. Unfortunately one lizard species, of the streamside associated genus *Neusticurus* could not be identified, but this species is almost certainly either *Neusticurus rudis* or *N. bicarinatus*, both present at Nouragues. This lizard was not listed in the Appendix and Tables. One other lizard species, which occurred at Arataye, is *Pseudogonatodes guianensis*. This lizard appears only briefly on the surface and is strongly associated with *Astrocaryum* palms, where it lives among the accumulated humid leaves at the tree base (Gasc 1986). The semi-aquatic species *Uranoscodon superciliosus,* most commonly found on vegetation along creeks and rivers, did only occur at Arataye.

Unfortunately there is no site-study available on the lizard fauna at the Trois Sauts region, but some species were registered in literature: *Anolis nitens*, *A. fuscoauratus*, *Gonatodes humeralis*, *Arthrosaura kockii*, *Neusticurus bicarinatus*, *Tupinambis teguixin*,

Tropidurus plica, *Uracentron azureum*, *Thecadactylus rapicauda*, *Leposoma guianense*, *Ameiva ameiva*, *Kentropyx calcarata*, *Prionodactylus argulus*, *Polychrus marmoratus*, *Uranoscodon superciliosus* (Hoogmoed & Lescure 1975, Gasc 1981). If compared to Petit Saut and the Piste de St. Elie, Nouragues has 1 species less and 4 species more respectively. The study site at the Piste de St. Elie includes secondary forest and other open habitats bordering a forest road where sun-loving species like *Cnemidophorus lemniscatus* (Teiidae) and *Polychrus marmoratus* (Polychrotidae) were active during the day. Nouragues has 1 species less if compared to the INPA-WWF Reserve in Brazil (Table 3). The lizard inventory at this site is considered almost complete according to the authors Zimmerman and Rodrigues (1990). It is noteworthy that species considered as rare or inconspicuous at the INPA-WWF Reserve (like *Polychrus marmoratus*, *Iphisa elegans*, *Neusticurus rudis*, *N. bicarinatus*) are common or at least frequently seen in one or more of the sites in French Guiana.

Within the study sites under comparison, there are only three species exclusively found in the INPA-WWF Reserve (*Leposoma percarinatum*, *Bachia panoplia* and *Anolis philopunctatus*). They were missing in all sites in French Guiana. Strangely the first species is distributed in northern South America east of the Andes but has not yet been reported from French Guiana. Of the second species no other records are known than Manaus and a location in Colombia (Avila-Pires, 1995). Of *Anolis philopunctatus* no other records are known than the INPA-WWF Reserve.

The Arataye site is richer in species than the other sites in French Guiana and Brazil. Arataye has respectively 3, 8 and 3 species more than Petit Saut, the Piste de St. Elie and the INPA-WWF Reserve. The different sampling regimes, most probably, account for these differences in species richness, especially for the Piste de St. Elie.

Snakes. Within herpetofaunal communities, snakes seem to be the animals with the most unpredictable microhabitat. Snakes are always found by accident. In other words, there is no likely place where we could look for them. Several species live outside our direct range of observation due to their arboreal way of life. Ground dwelling species with a cryptic coloration or colour pattern, do not flee when someone approaches. They stay where they are and remain immobile, depending on their camouflage. Other, fossorial species are often extremely difficult to find. This might explain the fact that even sites that are close to each other, like Arataye and Nouragues, have few species in common (15 snakes or 33.3 % of the total number of 45 snakes species occurring at both sites). *Boa constrictor*, *Chironius fuscus*, *Bothrops atrox*, *Xenoxybelis (Oxybelis) argenteus*, *Leptodeira annulata* are widespread species; they are present at all sites. The semi-aquatic snake *Helicops angulatus* (distributed throughout northern South America) and *Taeniophallus (Rhadinaea) brevirostris* (distributed in the Amazon Basin and the Guianas) were present at all sites as well.

At Nouragues most snakes were found in the transition from the dry to the wet season. Four species (the arboreal *Imantodes lentiferus* and *Oxybelis fulgidus*, the rare and mainly arboreal *Siphlophis cervinus* and the terrestrial species *Micrurus collaris*) were found only once in the cabins at the campsite while the rare species *Rhinobotryum lentiginosum*, was found twice at the campsite. Three other common, terrestrial species (*Bothrops brazili*, *B. atrox* and its mimicking species *Xenodon rhabdocepahlus)* were active at the campsite several times. Most other snake species occurring at Nouragues were conducted in the forest surrounding the camp except for *Drymarchon corais*. This large snake was found in the *Clusia* vegetation on the inselberg and at the same time, the only snake species that we could record in this habitat. This large snake is not exclusively associated with the inselberg habitat. At Nouragues it was seen in the forest as well, once while it was struggling to prey a large opossum.

The lower proportion of nighttime sampling effort at Nouragues in comparison to Arataye seems to be of no effect to the total number of strictly nocturnal species, which were encountered at Nouragues (9 species or 30.0 % at Nouragues versus 5 species or 16.7 % at the Arataye study site, see also Appendix 5). Certain species that occur in primary forest may be easier to find in other, open habitats. The relative large number of species recorded at the Piste de St. Elie illustrates this. Here, a roadside transect, which dissected the forest surrounding the camp, was investigated regularly. Both living snakes (which were crossing the road), as well as dead snakes (which had been run over by passing cars) could be easily collected in this way. Although little or no information on microhabitat results from this method - we can only draw some macro habitat information from the forest bordering the road – it turned out to be very efficient to determine species numbers. In only two weeks sampling, 21 of the 31 species were encountered at the roadside transect.

Similar results were obtained at the INPA-WWF Reserves in Brazil where almost 50 % of the total number of species were encountered during a clear-cut survey, and at the roadside transects at Petit Saut.

4.2 FAUNAL RESEMBLANCE FACTORS

Similarity or dissimilarity with the other sites in French Guiana and in Brazil, is expressed by means of the faunal resemblance factors (FRF). In Table 6 we show the FRF of both sites and of both sites combined (Total Nature Reserve). The herpetofauna of Nouragues is most similar to the nearby site Arataye (approximately 17 km from each other, FRF 0.74). The FRF's for amphibians, and lizards & amphisbaenians, are almost equal and relatively high (0.82 and 0.84 respectively), while the FRF for snakes is strikingly low. This low FRF reflects the difficulty to encounter snakes in the forest and thus the under-sampling of this group. Inventorying lizards is much easier, because there are many specific habitats to look for lizards. Furthermore, lizards are not prone to long periods of inactivity, thereby reducing the chance to be found. The inventory of frogs at both sites is probably nearly complete because we have sampled long enough to find rare or infrequent breeders.

Besides these remarks, we discuss the FRF of the whole Nouragues Nature Reserve with other localities since both sites were by far inventories at the smallest scale and adjacent. The FRF between Nouragues Nature Reserve and two other well-sampled sites, Petit Saut and the INPA/WWF Reserve in Brazil is lowest for amphibians (0.63 and 0.43 respectively) and highest for lizards and amphisbaenians (0.87 and 0.75 respectively). This might be explained by the fact that lizards are less limited in their ways to disperse or less restricted to special (breeding) habitats than amphibians do. The FRF for snakes are 0.75 and 0.62 respectively. The FRF for snakes between Brazil and the Nouragues Nature Reserve (0.62) is rather high, considering the distance between the two sites (approximately 1100 km). This value is even higher than the FRF for snakes between the two sites within the Reserve (0.62 versus 0.50). This difference is indubitably the effect of sampling methods.

TABLE 6. Faunal resemblance factors of the different taxonomic groups between Nouragues/Arataye and five other sites in French Guiana and Brazil. Faunal resemblance factor = FRF = 2C / (N1+N2), where C is the number of species in common of two sites, N1 = number of species in site 1 and N2 = number of species in site 2 (Duellman 1965). Distances are measured in a straight line. Notes: Turtles and crocodilians were not included because data on these were not available for two sites. Moreover, available data on these species at the other sites hardly contribute to the FRF. * = FRF for Trois Sauts does not include lizards & amphisbaenians. Data based on the authors mentioned in legend of Table 3. Total Nature Reserve includes Nouragues and Arataye.

SITE	Arataye	Petit Saut	Piste St. Elie	Trois Sauts	Brazil
NOURAGUES					
Distance from Nouragues in km	17	120	140	205	1100
1. Total reptiles & amphibians					
Species in common	85	78	60	65*	63
FRF	0.74	0.66	0.63	0.60*	0.54
2. Total amphibians					
Species in common	49	29	31	43	21
FRF	0.82	0.63	0.70	0.72	0.44
3. Total reptiles					
Species in common	36	49	29	22*	42
FRF	0.65	0.68	0.56	0.45*	0.60
4. Lizards & amphisbaenians					
Species in common	21	21	15	-	17
FRF	0.84	0.89	0.71	-	0.72
5. Snakes					
Species in common	15	28	14	22	25
FRF	0.50	0.58	0.46	0.59	0.54
ARATAYE					
Distance from Arataye in km		137	157	188	1083
1. Total reptiles & amphibians					
Species in common		84	67	71*	65
FRF		0.67	0.65	0.61*	0.52
2. Total amphibians					
Species in common		33	31	50	24
FRF		0.64	0.63	0.77	0.45
3. Total reptiles					
Species in common		50	35	21*	41
FRF		0.68	0.65	0.42*	0.57
4. Lizards & amphisbaenians					
Species in common		21	16	-	19
FRF		0.82	0.70	-	0.75
5. Snakes					
Species in common		29	19	21	22
FRF		0.60	0.62	0.57	0.48
TOTAL NATURE RESERVE					
(Nouragues and Arataye)					
1. Total reptiles & amphibians					
Species in common		100	73	84*	77
FRF		0.73	0.64	0.66*	0.57
2. Total amphibians					
Species in common		34	33	53	24
FRF		0.63	0.63	0.79	0.43
3. Total reptiles					
Species in common		66	40	31*	53
FRF		0.80	0.65	0.53*	0.66
4. Lizards & amphisbaenians					
Species in common		23	17	-	20
FRF		0.87	0.71	-	0.75
5. Snakes					
Species in common		42	23	31	33
FRF		0.75	0.61	0.70	0.62

The small differences in FRF's among the sites in French Guiana can probably be attributed mostly to differences in sampling effort as well. Thoroughly sampled sites are showing greater similarity with the Nouragues Nature Reserve.

4.3 RECOMMENDATIONS FOR INVENTORY TECHNIQUES

In view of the different sampling techniques, aural sampling of amphibian breeding sites, is a relatively simple but very effective way to determine presence or absence of anuran species. To utilise such techniques we need tape recordings of frog calls, from which the calls can be learned, and from which different calls can be verified. Once the different frog calls are known, it is in most cases very easy to identify the different species that are calling. An aid to identify vocalisations is to obtain suitable recordings of calling activity in the field and to analyse them afterwards. Frogs can easily be stimulated to start calling, if they are confronted with their own type of call in the field. However, even when aural surveys are used to survey a habitat as completely as possible, voiceless species would be missed. On the other hand, species that are never seen but commonly heard because they spend most of the time in leaf-litter, would be missed if only visual surveys are used. Therefore, a combination of techniques (aural and visual) is recommendable.

Noteworthy is that most breeding site surveys focus on adults. Nevertheless, larvae are present at the breeding sites for longer periods than adults are. Therefore we recommend sampling both adults and larvae. Larvae are often less simple to identify than adults. If there is any doubt about larval identification, they should be reared through metamorphosis.

We already noted that there is no special place to look for snakes. Moreover, snakes are prone to long periods of inactivity; they feed infrequently, reducing our chance to spot them. Roadside surveys seems an appropriate technique to sample snakes. However, this method favours maximising of species numbers; ecological information on (sub) habitat is not obtained in this way for example. Roadside surveys require infrastructure thus, habitats in primary forest cannot be surveyed with this technique. During our fieldwork most snakes were encountered during rainfall after extended periods of drought, in the early morning hours and late afternoon.

Lizards are much easier to find than snakes. There is far more information on their ecology and species-specific habitat preferences. Although they do not congregate at special breeding sites, we know for most of the species in which habitat they occur in particular.

Acknowledgements

We thank M.S. Hoogmoed, I. Ineich, F.J. Bongers, P. Charles-Dominique and F.J. Sterck for their helpful comments on the manuscript and the Dutch Foundation for the Advancement of Herpetology for financial support.

CHAPTER 16

FISHES OF THE ARATAYE RIVER AND THEIR SPATIO-TEMPORAL ORGANIZATION

François J. Meunier and Thierry Boujard

1. Introduction

Half a century ago, the scarcity of ichtyological studies in French Guiana was contrasting with the situation of the other countries of the Guyana shield. Indeed, in 1949 Puyo registered only 76 species of freshwater fishes in French Guiana though one century sooner Schomburgk (1841) already described 84 species in the rivers of Guyana and at the beginning of the twentieth century this number was raised to more than three hundred (Eigenmann 1912). It is also early in this century that the Surinam ichtyology spread out (Outboter 1993). So, when the various French institutes, INRA and National Museum of Natural History, and later IRD (ex ORSTOM) planned intensive ichtyological studies at the end of seventies, French Guiana ichtyology was far backward to that of the neighbouring countries. But two decades later, owing to several extensive works in the main rivers and coastal swamp (Keith 1997) and to many intensive inventories like the study on the Arataye (Boujard *et al.* 1990a,b), we can consider that delay was recovered and may be overtaken (Meunier *et al.* 1999). Effectively, checklists and dichotomic keys (Le Bail *et al.* 1984, 2000, Rojas-Beltran 1984, Planquette *et al.* 1996, Boujard *et al.* 1997, Keith *et al.* 2000) are now available and various biological studies have been conducted (Boujard *et al.* 1988, Loir *et al.* 1989a, b, Renno *et al.* 1989, 1990, Pascal *et al.* 1994, Hostache *et al.* 1993, 1995, and others). Moreover, ecological studies on fish communities of coastal streams (Mérigoux *et al.* 1998), tributaries or 'criques' (Rojas-Beltran 1986) have been published. It should also be pointed out that several species have been identified as good models for the study of various biological phenomenons or for their high reactivity to water quality. Among other examples, can be cited *Leporinus friderici* used to study the role of neural crests in the ontogeny of scales (Zylberberg & Meunier 1996) and *Apteronotus albifrons* and its significant changes of electric signal characteristics in response to water pollution (Thomas *et al.* 1996a,b).

The specific richness of French Guiana rivers is very high with more than 480[1] species (Planquette *et al.* 1996, Keith *et al.* 2000, Le Bail *et al.* 2000) and it shows a high degree of endemism (Planquette *et al.* 1996, Meunier *et al.* 1999). The Arataye that borders the Nouragues Forest Reserve is a tributary of the Approuague river, the fourth river of French

[1] With the marine species that live in brackish water

Bongers *et al.* (eds.), Nouragues. Dynamics and plant-animal interactions in a neotropical rainforest, 185–189.
2001 Kluwer Academic Publishers. Printed in the Netherlands.

Guiana (Boyé *et al.* 1979). The ichtyological fauna of the Approuague basin and its main tributaries are now relatively well known (Planquette *et al.* 1996, Boujard *et al.* 1997, Keith *et al.* 2000, Le Bail *et al.* 2000) and it is at the present time possible to give an overview of the richness and specificity of the freshwater fish community of the Arataye river and of the environmental variables that may explain the assemblage composition of its ichtyofauna.

2. Richness and diversity of fish in the Arataye river

The checklist of fishes living in the Arataye river is mainly the result of two intensive assessments of specimens living in an area near the Nouragues site, between the waterfalls Saut Japigny and Saut Pararé (Boujard *et al.* 1990a,b). The first inventory took place during the dry season (November 1988), with low water level, the second during the wet season (July 1989), with high water level. These two inventories were completed by here and there selective fishing in the same region and between Saut Japigny and the confluence of the Arataye with Approuague river during the last twenty years (see Planquette *et al.* 1996; Keith *et al.* 2000, Le Bail *et al.* 2000). The systematic status of some species mentioned by Boujard *et al.* (1990a,b) have been corrected in order to take into account the recent revisions of the Tetragonopterinae (Zanata 1997) and the Cichlidae (Kullander & Nijssen 1989), and several new species have been recently described (Géry *et al.* 1991, 1996, Huber 1991, Géry 1992, Muller & Isbrücker 1993). The checklist that we give in this book is the present state of the knowledge for the considered geographical area. We think that future studies should probably bring only minor complements to the current data.

The Arataye river flows in Approuague at about 135 km of its mouth. The prospected zone is located in the lower part of Arataye and belongs to the upper part of the Approuague basin. The checklist, presented in Appendix, with its 120 species shows the great specific richness of the river, as for numerous other neotropical rivers especially in French Guiana (see Meunier *et al.* 1998). Among other examples, Boujard and Rojas-Beltran (1988a) have observed very similar species diversity in the upper part of the Sinnamary river with 102 species identified. In the Arataye river two orders of Teleosts are prevailing as in other French Guiana rivers: Characiformes (nine families) and Siluriformes (nine families) with respectively 46.6% and 31.6% of the known species. Gymnotiformes, Cyprinodontiformes, Synbranchiformes and Perciformes (Cichlidae, Sciaenidae) are the four other characteristic groups of this river. The distance between the mouth of Approuague and the Arataye rivers, and the numerous rocky ridges that cross the banks of the Approuague could be responsible for the absence of Clupeiformes, this latter order being generally restricted to the lower part of the rivers (see, for Sinnamary river, Boujard & Rojas-Beltran 1988a).

The ichtyological fauna of the Arataye river confirms its belonging to the eastern zone of French Guiana. Indeed, various studies, especially electrophoretic studies on the Anostomid *Leporinus friderici* (Renno *et al.* 1989), have emphasised the distribution of two forms of this fish in two distinct geographical zones that extend widely on the bordering countries in the East as in the West. *L. friderici acutidens* is an eastern marker, *L. friderici friderici* being known in the western rivers as Sinnamary, Mana and Maroni (Renno *et al.* 1990). Other species such as *L. melanostictus*, *Bryconops aff. caudomaculatus* (probably a sub-species of *B. caudomaculatus*; see Planquette *et al.* 1996), *Astyanax leopoldi*, *Ancistrus cf leucostictus*, *Hepapterus tenuis*, *Crenicichla johanna* and

Geophagus camopiensis are living only in Approuague and Oyapock rivers, whereas *Rhamdella leptosoma* is known in very few samples in Rupununi (Guyana) and Arataye rivers (Boujard *et al.* 1990b). In fact, this species is probably more abundant than expected (Le Bail *et al.* 2000). Lastly, among the 13 endemic species of the Approuague basin (Meunier *et al.* 1999), 6 inhabit the Arataye river; they represent specific richness of the Arataye that call for special attention with respect to their protection: *Creagrutus planquettei*, *Moenkhausia aff. grandisquamis*, *Corydoras spilurus*, *Corydoras approuaguensis*, *Pimelodella megalops* and *Ituglanis* sp. The endemic species *Moenkhausia aff. grandisquamis* is probably in the Arataye-Approuague system a vicariant of *M. grandisquamis* that inhabits all the other rivers of French Guiana (Planquette *et al.* 1996). Another remark is for *Thayeria ifati*, a very scarce fish in French Guiana and only known in three sites, one in the Maroni river (Planquette *et al.* 1996) , the second being the Arataye river (Boujard *et al.* 1997), and the third in the Approuague river at Saut Grand Mathias (Géry *et al.* 1991).

3. Spatial heterogeneity and fish communities

French Guyana rivers show a typical longitudinal zonation. For example, the Sinnamary river can be divided in three zones (the upper part of the river, the middle part with its large meanders draining the flat old coastal plain and the estuary which represent the lower part) according to diverse geological, pedological, hydrological and phytosociological indices (Boujard & Rojas-Beltran 1988a); it is the upper basin that looks the most complex (Boujard & Rojas-Beltran 1988a,b). The Nouragues zone of the Arataye river between Saut Japigny and Saut Pararé, a stretch of 6 km and their secondary tributaries, belong to this 'upper basin' category of French Guiana rivers and is a good illustration of the spatial heterogeneity of fish communities. This area is a highly structured patchwork of biotopes arranged as a mosaic.

In the Arataye river that is 50 meters wide in the selected area, Boujard *et al.* (1990c) described five main typical biotopes. Many secondary tributaries draining the dense rain forest into the main river constitute the first biotope; these are 1 to 5 m wide, less than 1 m deep during the dry season and during the wet season they overflow into the surrounding forest. The rapids at the level of the rocky ridges constitute the second biotope; they are characterised by the abundance of rocks in dry season and turbulences of water at rainy season. Two other biotopes are formed by river meanders. The eroded banks of the river (concave) are high, steep, cluttered with dead trees and shaded by rain forest canopy. The sedimented banks of the river (convex bars) are low, with low riverside vegetation specially the mocomoka arum (*Montrichardia arborescens*); these banks are flooded during the wet season. The fifth biotope corresponds to the straight reaches of the rivers between the meanders and/or the rapids, bordered by dense and shrubby vegetation flooded during the wet season.

Seasonal cycles play an important role on the temporal organisation of the fish community. Species diversity index is high and differs between biotopes during wet and dry seasons. During dry season, the concave banks exhibit low species diversity while rapids exhibit high species diversity. Conversely, during the flooded periods, the highest species diversity is observed in the secondary tributaries but larger individuals are caught mainly at the river edges of the straight parts of the main river. Moreover, the floodplain is inhabited by juvenile communities specially at the end of the wet season, but the fish fauna of these inundated areas is also composed of adults of small species (< 15 cm) (Tito de Morais *et al.*, 1995).

Some fishes can characterise a particular biotope. During the dry season, the secondary tributaries are inhabited mainly by small species as *Bryconops affinis*, *Corydoras aeneus*, *Glanidium leopardum,* and with overflowing *Myleus ternetzi* and *Hoplias aimara* become common, whereas they are less abundant in the main river. Three species in high density characterise the rapids: *Hypostomus gymnorhynchus*, *Moenkausia aff. intermedia* and *Geophagus camopiensis*. The convex regions show a high species diversity of small species. On the contrary, large species are common in the straight sections with domination of *Myleus rhomboidalis* during the dry season, while the fish community is dominated by *Cyphocharax sp. (gr. spilurus)* during the wet season. The concave region is depleted of small species and *Ageneiosus brevifilis* is abundant during the wet season.

Physical characteristics of water, like transparency, conductance or depth could also be factors of the structuration of fish communities. For example, turbid waters with low light (little tributaries in the floodplain under the canopy) may favour fishes with sensory adaptation alike Gymnotiformes (Roriguez & Lewis 1997) whereas visually oriented fishes prefer clear waters. This sensory adaptation plays an important role for detection of preys (Tejerina-Garro *et al.* 1998).

During our studies massive longitudinal migrations were not observed but lateral or short-distance migrations to the tributaries or to the overflooded areas were recorded. Fish movements from the rivers to the flooded areas, give the opportunity to have access to abundant food sources for phytophagous and detritivorous fishes (Goulding *et al.* 1988, Boujard *et al.* 1990d, Lowe-McConnell 1998) as for predators (Fernandes 1997) (accumulation of fat reserves) and to breeding zones (Ponton & Copp 1997). They result in a significant decrease of catches in the main rivers (Boujard *et al.* 1990c). For example, *Myleus ternetzi* that eats mostly leaves, ingests drifting leaves with poor nutritional value during dry season (it lives in the main river). Conversely during the rainy season it invades the flooded forest and its stomach contains leaves of riverine vegetation such as *Inga meissneriana* (Mimosaceae) and *Cydista aequinoctialis* (Bignoniaceae) with higher nutritional value. Similar results have been obtained for *Myleus rhomboidalis* that feeds on seeds during the rainy season and on leaves (low nutritional value) of Podostemonaceae (= 'coumarou grass') during the dry season (Boujard *et al.* 1990d) as well as for *Leporinus friderici* an omnivorous fish that essentially feeds on seeds, fruit and terrestrial insects during the wet season and seems to be starving during the dry season (see also Lowe-McConnell 1964, Carvalho 1980, Goulding 1980, Junk 1985, Roubach & Saint-Paul 1994)

The seasonality of feeding is synchronised by rains and causes annual variations in growth rate. There are two dry seasons and two rainy seasons per year. Both dry seasons seems to be sufficiently marked to induce a slow growth *annulus* (Lecomte *et al.* 1985). As a result, several species show two annual growth cycles per year: Ariidae (Lecomte *et al.* 1989), Anostomidae (Boujard *et al.* 1991), Serrasalmidae (Lecomte *et al.* 1993) and the atipa (*Hoplosternum littorale*) also show slow growth during the dry season of March (Boujard & Meunier 1991). Armored catfish (Loricaridae) reduce also their growth performance during the dry seasons (Powers 1983). Such growth cycles are also known for some African tropical fish. In every case the periods of fast growth are in phase with the rainy season when food is abundant and rich. Furthermore, growth performance of

Leporinus friderici was correlated to the total duration of the flooded periods (Boujard *et al.* 1991).

Spawning is another vital function that is tightly related to external factors but field data on spawning in French Guiana rivers are very scarce (Ponton & de Mérona 1998). Nevertheless, Le Bail *et al.* (1989) showed that *Myleus ternetzi* could reproduce throughout the year but percentage of mature animals was the highest during February-April, i.e. just prior to the main rainy season (May-July). *Leporinus friderici* seems to spawn only during the rainy season (Boujard 1992). Moreover, spawning of Atipa, *Hoplosternum littorale*, have been well studied recently. It appears that in this species, gonadic maturation starts before intensive raining, the GSI increasing in October-November (Hostache *et al.* 1993) and spawning appears at the rainy season in December-January (Pascal *et al.* 1994). The role of the floodplains as nursery is another factor of temporal organisation of fish species, at least for the juvenile communities (Ponton & Copp 1997, Ponton & Tito de Morais 1997).

4. Conclusion

The Arataye river shows a specific richness similar to the other French Guiana basins and as the main tributary of the Approuague river, a certain degree of endemism. The structure of the fish communities is the result of joint action of various factors, leading to a balance between several aquatic compartments. So the whole system must be protected against the anthropic activities that frequently upset this balance: see for example the consequences of the Petit-Saut's dam (on the Sinnamary river) on the biology of the fish in general (Tito de Morais & Lauzane 1997) and specially on the young Characiformes (Ponton & Copp 1997). Severe conservation programs must be enacted to protect and to manage this original patrimony.

Acknowledgements

We are grateful to P. Keith and P.-Y. Le Bail who gave us data before the published the second issue of the '*Atlas des poissons d'eau douce de Guyane*'. This study was supported by the AllocationSpécifique 1988 of the French Ministry of National Education.

Pithys albifrons. Photo Pierre Charles-Dominique.

PART 4

FRUGIVORY AND SEED DISPERSAL

CHAPTER 17

RELATIONSHIPS BETWEEN SEED DISPERSAL AND BEHAVIOURAL ECOLOGY

Pierre Charles-Dominique

1. Introduction

Seed dispersal can be achieved by very different means, such as wind (anemochory), water (hydrochory), gravity (barochory), autopropulsion at dehiscence (autochory), or active transport by animals (zoochory). Like elsewhere in tropical rain forests (Van der Pijl 1969, Hladik & Hladik 1969, McKey 1975, Howe & Smallwood 1982, Janson 1983, Gautier-Hion *et al.* 1985), zoochory is the most common means of dispersal at Nouragues. At this site, 121 species of fruit-eating vertebrates have been observed, which, by their behaviour, contribute to the dispersal of seeds: - 188 species of birds (out of 436 bird species present on the site), - 55 species of mammals (including 23 species of bats and 32 species of non-flying mammals, out of 166 mammals species found at the site and other vertebrates that eat fruit, e.g. turtles). 'True' frugivores the species that only digest the pulp of the fruit and discard most of the seeds, whether these actually transit through the digestive tract (endozoochory) or not (synzoochory). The species that forage for fruit in order to eat their seeds (frugivorous-granivorous species, Charles-Dominique *et al.* 1981) are not considered as true frugivores but rather as seed predators.

However, distinctions are not always easy to establish because the definition of the diet is based on the predominant feeding habits of the animal species and does not consider occasional particular behaviours. In some cases, typical seed predators may behave as true dispersers for certain plant species. For example, the usually seed destructive squirrel *Sciurus aestuans* contributes to the dispersal of the palm *Maximiliana maripa* as it only eats the pulp of its fruit and leaves the seed (Charles-Dominique *et al.* 1981). Similarly, although major seed predators otherwise, parrots ingest only the pulp of the palm *Jessenia bataua* and disperse its seeds (Sist 1989). On the other hand, typical frugivorous species may destroy the seeds of some plant species: during a brief period of the year, the monkey *Cebus apella* eats the immature fruit of different species and in so doing spoils the seeds (Zhang 1994).

A number of plant species producing dry pulpless fruit are termed 'rodontochorous' because they are only dispersed through the particular behaviour of large-size terrestrial rodents (agoutis *Dasiprocta aguti*, acouchies *Myoprocta exilis*). These seed predators bury seed surplus in order to build up reserves, but the fact that a number of these caches are not

Bongers *et al.* (eds.), Nouragues. Dynamics and plant-animal interactions in a neotropical rainforest, 191–196.
2001 Kluwer Academic Publishers. Printed in the Netherlands.

retrieved constitutes the key to the plants dispersal and regeneration (Forget *et al.* Chapter 25) Rodontochory is found in many plant species, including species which are primarily zoochorous or anemochorous. This secondary transport (Janzen 1986, Smythe 1970, Vander Wall 1992) can have either a positive or a negative effect on the reproductive success of the plants involved.

The general principle of zoochory is fairly simple. The seed is associated, on the one hand, to the nutritious elements (pulp, arilla) actively sought by the frugivores which eat them, and, on the other hand, to attractive signals (colours, contrasts of brilliance, movements, smells, etc.) on which the frugivores base themselves to locate the ripe fruit. The zoochorous fruit is thus a composite organ, the different parts of which meet different functional needs. 1. The seeds, their physiological characteristics, their size, and their number and disposition (number by fruit, number by infructescence) contribute to the reproductive 'strategy' of the plant. 2. The nutritious appendices (pulp, arilla), in which the frugivores are interested, display extremely variable structures, origins and chemical compositions which are the result of many convergence phenomena (Sabatier 1983). The structure and the chemical composition of the pulp are correlated with morphological, behavioural and digestive traits of the animal dispersers associated to the plants. 3. The signals composed of different anatomical and biochemical structures (pigments, volatile substances) ensure the location of the fruit and the identification of their state of maturity. These traits are correlated with the sensorial characteristics of the associated dispersers. 4. The characteristics of the protective structures of the fruit (secondary compounds in the seed, structure and thickness of the pericarp, dehiscence mechanisms, position within the tree, structure of the peduncle, etc.) are linked to defence mechanisms against seed predators or against inefficient frugivores which disperse the seeds in unfavourable environments.

The differences of accessibility of zoochorous fruits are based on combinations of characters: sizes, shapes and attachment structures of the fruit and seed, chemical composition of the pulp, presence of protective organs (tough teguments), types of dehiscence, position of the fruit on the branches, nature of the maturity signals, etc. To each type of zoochorous fruit corresponds a more or less diversified group of frugivorous animals whose size, anatomy, sensorial physiology and behaviour are compatible with the fruit's characteristic features. The wide range of life habits displayed by the different coexisting frugivorous species determine numerous specialisation's which orient them towards particular fruit types. By convergence, although they may be phylogenetically distant, different plant species belonging to a same functional fruit type category produce similar fruit. These categories of fruit correspond to different syndromes that allow attributing them to the types of frugivorous animals that disperse them (Van der Pijl 1957, 1972, Gautier *et al.* 1985). Syndromes such as small-seeded bat-dispersed fruit, large-seeded bat-dispersed fruit, fruit dispersed by small birds, fruit dispersed by large birds, primate-dispersed fruit, fruit dispersed by large terrestrial mammals, ant-dispersed fruit, rodent-dispersed fruit, etc. are easily identifiable. In this chapter, which introduces the chapters 18 to 22, we will evaluate the importance of zoochory in the plant communities from Nouragues and the effects of the behaviour of different types of frugivores on seed dispersal.

2. Relative importance of zoochory at Nouragues

In order to estimate the relative importance of zoochory within the plant communities found in Nouragues, the following studies were considered: - for birds: Erard et al. 1991; Théry and Larpin (1993); - for primates: Julliot & Sabatier (1993) and Zhang (1995); - for marsupials: Charles-Dominique *et al.* (1981), Atramentowicz (1988) and Julien-Laferrière (1991); - for the kinkajou (Procyonidae): Julien-Laferrière (1993); - for bats: Charles-Dominique (1986, 1991), Brosset & Charles-Dominique (1990), Brosset *et al.* (1996) and Cockle (1997); - for rodents: Guillotin (1982) and Forget (1994, 1996); - for the anatomy of fruits in the Guianas: Sabatier (1983) and Roosmalen (1985); - for the floristic inventories: Poncy *et al.* (1998). Fruits have been examined in all but 152 of the 1254 plant species found at Nouragues so far. Regarding that these 152 species are relatively poorly represented locally, we have a fairly good overall picture of the area. A fruit is considered zoochorous or anemochorous when biological observations and its external characteristics are sufficiently typical. Other fruits (non-zoochorous and non-anemochorous) are placed in a third category. These fruits are dry and do not display wind-dispersal traits. Information remains scarce for most of them although some are dispersed by the rodents that also feed on them. Consequently, zoochorous and anemochorous fruits only will be discussed here.

Overall, 70.9% of the phanerogams collected at Nouragues produce animal-dispersed fruits, and 24.4% produce wind-dispersed fruits (N = 910). However, these data must be corrected in function of the main vegetation formation types found on the site. High forest covers 90% of the area, and the remaining surface concerns small inclusions of particular formation types such as rock-savannas, low forests, bamboo thickets and liana forests (Poncy *et al.* Chapter 4, Larpin Chapter 5). Very few botanical data are available on the two latter vegetation types and comparisons will therefore only involve high forest (Poncy*et al.* 1998), low forest (Larpin 1993) and rock-savannas (Sarthou 1992) (Table 1).

TABLE 1. Comparison of fruit types in the different vegetation formation types

	Animal-dispersed (%)	Wind-dispersed (%)	Neither (%)	All examined fruits (N)	Fruit unknown (N)
All vegetation types	70.9	24.4	4.7	910	136
Rock-savannas	31.9	64.0	4.2	119	22
Low forests	69.5	29.0	1.4	348	35
Low forests & rock-savannas	61.4	32.9	5.7	386	36
High forest	77.9	18.1	4.0	524	100

It is clear that rock-savanna is the vegetation type which harbours the highest proportion of wind-dispersed species and the lowest proportion of animal-dispersed species. This habitat is characterised by severe constraints of water deficit during the dry season and has favoured species resistant to dry conditions (de Granville 1982, Sarthou 1992). Wind-dispersal is compatible with the physical and microclimatic conditions that prevail there. Low forests generally border rock-savannas and a number of wind-dispersed species from the savannas are also found there, which explains why this forest type displays a higher proportion of wind-dispersed species than the high forest.

The analysis of the dispersal mode in relation to the main biological types (tree, liana, epiphyte, terrestrial herbaceous species) yields further interesting information (Table 2).

The retained types have been voluntarily simplified in order to roughly group plants which share certain biological characteristics: - trees (trees, shrubs), - lianas (woody vines which germinate on the ground), - epiphytes (plants which establish on another plant, including true epiphytes, hemi-epiphytes such as many *Clusia* and *Ficus*, and parasitic plants such as the Loranthaceae), - herbs (terrestrial herbaceous plants which establish on the ground).

TABLE 2. Comparison of the different biological types and their dispersal modes

	Animal-dispersed (%)	Wind-dispersed (%)	Neither (%)	N	Number of families
Trees	87.8	8.1	4.12	655	69
Lianas	76.8	20.5	2.68	112	32
Epiphytes	55.5	39.9	4.62	173	16
Herbs	43.0	42.5	14.36	181	24

The progressive regression of zoochory from trees, to lianas, epiphytes and finally herbs is immediately visible. However, the epiphyte and herb categories are both dominated by the essentially wind-dispersed families Orchidaceae, Poaceae and Cyperaceae. When these families are left aside, the proportion of zoochorous species rises to 78.7% (14.8% of anemochorous species) for epiphytes (15 families, 122 species) and to 61.4% (18.1% of anemochorous species) for herbs (21 families, 127 species). Orchidaceae, Poaceae and Cyperaceae still aside, the 117 remaining families, corresponding to 1016 documented species, are respectively represented for the animal dispersed species by 84% in the high forest, 83% in the low forest and 50% in the rock savannas. We can see from these results that zoochory is an important component of these moist tropical ecosystems, even when relatively drought-determined habitats such as low forests and rock-savannas (Sarthou 1992) are also taken into account.

3. The seed dispersal modes and their consequences on dispersal

The combined action of a variety of fruit-eating vertebrates produces a very heterogeneous transport of seeds, comparable to a series of movements directing seeds towards particular sites. For the plant species involved, the redistribution of seeds on new establishment sites depends on the diet of the 'transporter' animals, the routes they most frequently use, their resting places, the way they defecate or regurgitate the seeds, and their social behaviour, which often regulates their movements. As the following examples will show, the biological characteristics of fruit-eating animals differ from one species to the next and lead to diverging dispersal patterns.

- size of the disperser: overall, various anatomical and phenological traits of small-seeded plant species favour small sized frugivorous vertebrates. As small sized animals are more abundant, both in number of species and in number of individuals, the resulting seed shadow is much more homogeneous than that generated by larger vertebrates.
- mode of locomotion: terrestrial, arboreal or flying mammals use very different routes and supports, and correspondingly leave the seeds in different locations. Fruit-eating birds usually defecate from a perch whereas bats do so in flight (Charles-Dominique 1986). The result of this is that the seed rain produced by small sized birds is denser on the perches, or the ground under the perches, than elsewhere, whereas that produced by bats is more evenly distributed. The expulsion of seeds by bats of the genus *Artibeus* during flight constantly

feeds the soil seed bank of the pioneer tree *Cecropia obtusa*, omnipresent in proportions varying from 32 to 400 by m² (Prévost 1981). In contrast, the extremely small seeds of Cyclanthaceae and Araceae epiphytes dispersed by understorey Carolliinae bats reach the base of tree trunks and other inclined surfaces where the development of the seedlings is optimal (Cockle 1997 and Chapter 24).

- social structure: solitary life or life in a group have consequences on the degree of spatial heterogeneity of the dispersal pattern. Moreover, social characteristics may force certain species to make greater use of particular sites and thus to disperse their food-plant species there in greater densities than anywhere else ('gardening' of Liebermanet *al.* 1979). This is the case of lekking birds such as manakins and cock-of-the-rocks (Théry 1990a, 1990b, Théry & Larpin 1993, Krijger *et al.* 1997). Similarly, howler monkeys defecate predominantly at their sleeping quarters and accumulate below them, year after year, great quantities of seeds of Sapotaceae and other plant families on which they feed, producing strikingly higher seedling and treelet densities of the corresponding species on these restricted areas than elsewhere (Julliot 1997).
- anti-predatory defences: the concentration of frugivorous animals on fruiting trees attracts predators, especially as foraging for fruit often involves movements in the open. Many feeding strategies have appeared which cut down time spent on exposed sites. Very often, frugivores quickly gather their food in the fruiting tree and retire to more a sheltered place to process what they have collected. This is the case of many bird species that accumulate fruit in their crop and digest a little way away, regurgitating or passing the seeds that fall under their shelter. Phyllostomidae bats generally pick fruits on the wing, without perching, and fly to a feeding roost located nearby, to eat it out of predators' sight (Morton 1973, Charles-Dominique 1986). Bats' feeding roosts are characterised by overhead cover and open space underneath to allow a quick escape by plunging and flying away in zigzags in the undergrowth. A number of large-seeded bat-dispersed fruits (*Symphonia globulifera, Andira spp, Dipteryx odorata, Parinari spp,* etc.) often accumulate under such feeding roosts, where they can be the object of secondary dispersal by rodents, which eat them, store the surplus in caches, and 'forget' some of them (Forget 1993).
- secondary dispersal: a certain number of plant species, such as *Vouacapoua americana* (Forget 1994, see Forget *et al.* Chapter 25), rely on rodents for their dispersal, but rodents can also disperse many other species which may have been previously wind- or animal-dispersed. This secondary dispersal may have a positive action when the seeds are buried out of reach of seed-eating insects (Janzen 1986, Forget 1994, 1996). Ants also play a role in dispersal, and may transport either true myrmecochorous species or small seeds found in bird or bat splats (Cockle 1997). Dung beetles which bury the faeces of fruit-eating vertebrates (Estrada & Coates-Estrada 1986) constitute yet another group of secondary dispersers of small-size seeds, which are thus deposited out of reach of seed-eaters and more evenly distributed in space (Feer 1999).
- foraging strategies: 'trapliner' fruit-eating species disperse seeds at a longer distance than species that focus on localised resources. Toucans (Thiollay 1987) and the bats *Artibeus lituratus* and *A. jamaicensis* (Morton 1973, Charles-Dominique 1986) can visit trees more than 10 km apart in the course of a single day or night. This foraging behaviour allows dispersal over great distances and, consequently, the colonisation of new areas ('Colonisation Hypothesis', Janzen 1970, Connell 1971). In contrast, territorial species generally disperse seeds within 100 m of the parent plant. It would appear that this range

still permits the maintenance of species in the environment as long as the microclimatic conditions are favourable to their establishment ('Escape Hypothesis', Janzen 1970, Connell 1971).

Unlike wind-dispersal, zoochory is oriented and does not operate at random (or in relation to the dominant wind directions). This allows a greater efficiency of dispersal towards particular sites. Behavioural differences observed between frugivore species appear to play a key role in the structuration of the vegetation. Some behavioural traits result in higher heterogeneity (Fleming & Heithaus 1981, Howe & van Kerckhove 1979, McKey 1975, Hartshorn 1980) while others, in contrast, contribute to distribute more evenly the seeds of particular plant species (Garber 1986). The resulting seed shadow, associated to features of the forest, ecophysiological characteristics of the seedlings and predation, maintains biodiversity by creating new plant assemblages in a mosaic structure, of which each unit corresponds to a phase of maturity. This results in the spatial redistribution of new plants, always different from the pre-existing situation.

Because of zoochory, tropical forests are characterised by important dispersal distances, at least for part of the seed production, and by a relatively selective orientation of the transport. In the past, these ecosystems have undergone periodical variations in extent, retreating to refuge zones when the climate became drier and spreading rapidly during the more humid periods. Zoochorous species are the first to recolonise new areas. Indeed, 98% of the newly established plant species in a French Guianan regrowth are animal-dispersed (Prévost 1983). Autochorous, barochorous and rodontochorous species can only disperse over small distances and only reach these newly forested areas long after the zoochorous and anemochorous species. Therefore, their presence in high proportions on a site could be correlated with the age of the forest block or with the proximity of ancient refuge areas (Charles-Dominique *et al.* Chapter 7).

CHAPTER 18

FRUGIVORY AND SEED DISPERSAL BY THREE NEOTROPICAL PRIMATES: IMPACT ON PLANT REGENERATION

Catherine Julliot, Bruno Simmen and Shuyi Zhang

1. Introduction

The tropical rainforest is a mosaic both in the spatial distribution of plant species, which are often clumped, and in the forest dynamics with the juxtaposition of areas of different age (e.g. Hubbell 1979, Alexandre 1989, Whitmore 1991). The role of frugivorous animals is central to the mechanisms of natural regeneration of tropical forests, since 60 to 95% of tropical plant species mainly depend on them for their seed dispersal (Charles-Dominique Chapter 17). Though monkeys represent the largest biomass of frugivores in tropical forests (25 to 40%; according to Eisenberg & Thorington 1973), as yet, except for some particular studies (Sourd & Gautier-Hion 1986, Janson *et al.* 1986), most of the studies of fruit choice by frugivorous vertebrates mainly concern birds (Erard & Théry Chapter 22). Other studies analysed the interactions between fruits and a community of vertebrate frugivores including several primates species (Janson 1983, Knight & Siegfried 1983, Gautier-Hion *et al.* 1985, Dowsett-Lemaire 1988).

During the two last decades, several authors (e.g. McKey 1975, Hartshorn 1980, Howe & Vande Kerckhove 1981) have stressed the importance of frugivorous animals in the preservation of the floristic heterogeneity of tropical forests. Paradoxically, there has been very little research on the relation between seed dispersal by frugivorous animals and seedling distribution of exploited plant species. Only a few recent studies provide information on this topic concerning single animal species (Dinerstein 1991, for rhinoceros; Julliot 1992, for red howler monkey; Théry & Larpin 1993, for cock-of-the-rock; Voysey, 1996, for gorilla), or single plant-animal interaction (Tutin *et al.* 1991, for gorilla and *Cola lizae*). Lieberman *et al.* (1979) also suggested that baboons (*Papio anubis*) increase the concentration of edible plants in the most exploited area of their territory, and thus performs 'inadvertent gardening'. Milton (1980) has made Panama similar observations for the mantled howler monkeys (*Alouatta palliata*) at Barro Colorado Island, Panama. She suggested that seed dispersal by howler monkeys could be the origin of the patchy distribution of food resources in the howler home range.

Six Primate species are present at Nouragues. In this chapter, we focus on the three most abundant species, *Alouatta seniculus* (Red howler monkey), *Cebus apella* (Tufted capuchin monkey) and *Ateles paniscus* (Black spider monkey), whose frugivore diets and

Bongers *et al.* (eds.), Nouragues. Dynamics and plant-animal interactions in a neotropical rainforest, 197–205.

seed dispersal patterns were studied during several field periods between 1988 and 1996. According to the importance of the different food categories, the diet of each primate species can be classified as insectivore-frugivore for the capuchin monkey, folivore-frugivore for howler monkey, and mainly frugivore for spider monkey (Guillotin*et al.* 1994, Simmen *et al.* Chapter 8). Although leaves or prey represent more than half of the total diet of howler and capuchin monkeys respectively (Julliot & Sabatier 1994, Zhang 1994), fruits account for a large part of the diet, especially during the period of peak fruit production. Moreover, the feeding strategies and ranging patterns of both species are principally related to fruit availability and distribution (Julliot 1992, Zhang & Wang 1995). Indeed, for the three primate species, the proportions of the different food categories in the diet showed more or less important seasonal variation in relation to fruit availability. Seed dispersal patterns by those species will be presented and discussed, particularly for howler monkeys, with regard to their impact on forest regeneration.

2. Frugivory and fruit selectivity

Single troops of howler and capuchin monkeys exploited at least 97 and 135 plant species as fruit or seed resources in two annual feeding-cycles at our study site, representing 27% and 26% of the annual diet respectively (Julliot 1992, Julliot & Sabatier 1994, Zhang 1994). Spider monkeys were observed to exploit fruits and seeds of 73 plant species during a 6-month study (Simmen & Sabatier 1996), and Roosmalen (1985) reported that at least 171 fruit species were exploited by this primate species in a similar habitat in Surinam, representing about 83% of the annual diet. Including all data collected between 1988 and 1997, howler monkeys were observed to exploit at least 149 species as fruit resources at the Nouragues Station. Similarly, 103 and 180 fruit species were recorded overall for spider and capuchin monkeys respectively. These figures are probably underestimated, especially for spider monkeys that have not been studied over comparable periods of time at Nouragues.

Sapotaceae is the most important plant family exploited by the three primates, with 19% and 11% of the total fruit consumption for howler and capuchin monkeys respectively during two-year studies (Table 1). This could be partly due to the abundance of Sapotaceae in the study site (the first family in species number and the third one in population density; Poncy *et al.* Chapter 4). Other high-ranking fruits belong to Moraceae for howler monkeys, Arecaceae and Mimosaceae for capuchin monkeys, and Myristicaceae and Moraceae for spider monkeys. Most important overlap in the fruit diets were observed during the fruiting period (February-April, Simmen *et al.* Chapter 8) between howlers and the other species. This was principally due to the lower diversity of the frugivore diet of howler monkeys that mainly exploited fruit species with abundant crops. The overlap in the fruit choices by spider and howler monkeys, then, reached 95%, Sapotaceae being the main plant family exploited simultaneously by both primates. At this time, spider and howler monkeys generally concentrated most of their feeding activity on a few species which had either a mass fruiting phenological pattern or which were represented by a few large trees producing food over several months.

Fruit selectivity of howler and capuchin monkeys was analysed in relation to morphological characteristics of fruits and their availability in the habitat during the two-year studies (Julliot 1992, 1996a, Zhang 1994). Although both howler and capuchin monkeys eat fruits differing in morphological characteristics, they consumed preferentially medium-sized (5-50 g and/or > 5 cm) fruits with juicy pulp, yellow-orange colour, and a small number of well-protected seeds. This is in accordance with results of previous studies in African and American tropical forests (Janson 1983, Knight & Siegfried 1983, Gautier-Hion *et al.* 1985, Wheelwright & Janson 1985 ; Dowsett-Lemaire 1988). At our study site, yellow-orange colour seems to be an important criterion of fruit choice by both primates; red fruits are also preferentially consumed by howlers in comparison with their availability, and brown fruits by capuchin monkeys. Fruit size also seemed to be a determinant characteristic in the monkeys' choice. It is usually considered that 'mammal fruits' are larger than 'bird fruits' (Janson 1983, Knight & Siegfried 1983, Gautier-Hion *et al.* 1985). At our study site, 'high-ranking' species are essentially medium-sized fruit species, while most of the 'low-ranking' species had smaller fruits (< 5 g and/or < 2 cm). Several authors (Knight & Siegfried 1983, Wheelwright 1985) noted that small-fruited plants attract potentially more seed dispersers than those with large fruits, particularly for large fruits with hard pericarps. Consequently, it is possible that there is much higher plant-disperser specificity in the case of large protected fruits (see also Janson 1983).

TABLE 1. Main plant families exploited as fruit resources by howler (Julliot 1992) and capuchin (Zhang 1994) monkeys during two annual feeding cycles at the Nouragues station, and during six months of simultaneous observation (Simmen & Sabatier 1996) for both species and spider monkeys.

	Plant Family	Proportion of fruit consumption (%)	
		2-year studies	6-month study
Alouatta seniculus	Sapotaceae	18.8	58.8
	Moraceae	14.5	25.9
	Caesalpiniaceae	9.2*	
	Celastraceae	8.4	
Cebus apella	Sapotaceae	10.9	28.2
	Arecaceae	8.5	11.1
	Mimosaceae	7.4	
Ateles paniscus	Sapotaceae		34.9
	Myristicaceae		16.0

* exploited only for unripe fruits

Morphological characteristics of fruits are certainly not the only criterion by which monkeys choose fruit. Other characters, such as nutrient contents, the occurrence of secondary compounds, palatability and spatial aspects of fruit displays must also be taken into account (Glander 1978, Milton 1980, Sourd & Gautier-Hion 1986, Janson *et al.* 1986, Simmen 1991). Several of these authors noticed that actual preference of primates for a plant species may result from the relative availability of surrounding species and may decrease when the plant spectrum is modified. The analysis of the howler and capuchin monkeys' selectivity also showed that a particular plant species, preferentially chosen at one period of the year, could be of low preference during another period according to the availability of other food species. Fruit choices were also examined in relation to the biochemical composition of fruit and sugar perception of the three monkey species

(Simmen 1991). Although there are no data on the discrimination of soluble sugars by howlers, they might have a sensitivity comparable to that of spider monkeys, since taste thresholds for sugar decrease with increasing body size across primates (Simmen *et al.* 1995). Among nineteen fruit species tested for pulp sugar content (Simmen & Sabatier 1996), concentrations as high as 90% of the dry weight were found in the pods of *Inga* (Mimosaceae), a plant genus highly selected by capuchins. Fruits of Sapotaceae, the main family exploited by the three primates, also have high sugar content (from 54.5% to 90.6%) and show similar combinations of sucrose, fructose, and glucose across species. Simmen and Sabatier (1996) noted that Sapotaceae fruits have sugar contents 15 to 75 times higher than taste thresholds, and, probably, are very attractive in terms of the gustatory sensation.

Although the seed size and the protection of fruit were not important in the monkeys' choice, these two characteristics were important with regard to seed dispersal. Except for some parrot species that are seed predators, only mammals can open fruits with hard and indehiscent pericarps with their teeth; and the size of seeds limits the number of seed dispersers that could ingest them. In accordance with the observations of Janson (1983) for a primate community in Peru, primates in our study could be considered to be the principal seed dispersers of plant species that have fruits with a hard and indehiscent pericarp and/or large seeds. Consequently, it is possible to determine a 'primate syndrome' ('syndrome'; in the sense of Van der Pijl 1972) whose plant species mainly depend on primates for their seed dispersal: middle- or large-sized pulpy fruits with yellow or orange colour, hard and indehiscent pericarp, and few large seeds usually well protected (Julliot 1994, 1996a). That is the case of the Sapotaceae family whose fruits, in South America, often show all of these characteristics simultaneously (Julliot 1994).

3. Patterns of seed dispersal

The three primate species disperse by endozoochory most of the plant species they used as fruit resources (Table 2). Howlers ingested intact nearly all seeds of the fruit species they exploited (Julliot 1996b). Capuchin and spider monkeys exhibit lower proportions because they both destroy or drop the seeds of several fruit species.

Seed size of fruit species exploited by the three primate species fluctuated from 0.1mm diameter (*Ficus* sp., Moraceae) to 4 cm x 2.5 cm (*Pouteria laevigata*, Sapotaceae), which was the upper limit of seed size that howler monkeys were observed to ingest (Julliot 1996b). Spider and howler monkeys are the only arboreal frugivores in South America able to ingest seeds of 4 cm x 2 cm without damaging them (Roosmalen 1985, Julliot 1996b). The capuchin monkeys also eat fruits with a very hard pericarp, but they do not ingest seeds larger than 2 cm x 1cm, dropping these under the parental tree (Zhang 1994).

Germination tests were carried out for several species dispersed by howler or capuchin monkeys (Julliot 1996b, Zhang 1994). Results showed no significant differences of germination rates and germination latency between ingested and fresh seeds for most plant species tested, even though both primates exhibit important differences in transit times. Indeed, seed passage through the digestive tract of capuchin and spider monkeys are relatively short (3 and 4 hours on average, respectively). In contrast, the transit time of howler monkey is very slow, 20h40min on average (Julliot 1996b). Consequently, although howler monkeys have long resting period and do not spend much time travelling, the mean distances of seed dispersal are quite similar to capuchin and spider monkeys (see Table 2).

TABLE 2. Characteristics of seed dispersal by howler and capuchin monkeys at the Nouragues Station during two-year flied studies (Julliot 1992, Zhang 1994), and by spider monkeys in Surinam during a full year study (Roosmalen 1985).

	Alouatta seniculus[b]	*Ateles paniscus*	*Cebus apella*[a]
Number of exploited fruit species	97	171[c]	135
Percentage dispersed by endozoochory	89	80[c]	84
Transit time (first appearance of seeds)	20h40"	4h[a]	3h
Mean distance of seed dispersal	290 m	250 m[a]	390 m

Data from: (a) Zhang (1994), (b) Julliot (1992), (c) Roosmalen (1985)

In addition, some aspects of howler monkey behaviour lead to a special pattern of seed dispersal (Julliot 1996b). First, howlers used small stable home ranges in comparison with the two other species (Simmen *et al.* Chapter 8). Second, among the several sleeping sites used by a troop, some of them, called 'main sleeping sites', were used regularly all year long. Lastly, their low transit time induces a bimodal rhythm of defecation with a first peak in the morning when they left their sleeping tree, and sometimes a second one in the early afternoon, at the end of a resting period. All the group members defecate simultaneously, 60% of the defecations are localised under the sleeping sites, and about 35% under the main sleeping sites. All these points induce a seed concentration of particular plant species at highly localised sites in the forest. In contrast, individual capuchin or spider monkeys defecate independently during their daily travels. Thus, they disperse seeds more randomly and homogeneously over their home ranges. Capuchin defecations are very small and clumped; however, the larger defecations produced by spider monkeys are usually scattered by the understorey trees and branches and dispersed over a large area, each seed being isolated from each other (Andresen 1999). The mean number of large seeds (2 cm long) in the clumped defecations of a single howler troop may reach up to 200 seeds according to the fruit species, with a mean of 15 ± 38 seeds / faeces (n = 440 faeces) (Julliot unpublished data).

4. Impact of seed dispersal by howler monkeys on forest regeneration

Seed concentration produced by the particular pattern of seed dispersal by howler monkeys can modify locally the regeneration potential of the forest.

The dynamics of forest regeneration, after a natural gap or a slight thinning, depends on three sources: i. fresh seed rain, ii. seedling bank, and iii. seed bank (Alexandre 1989, Garwood 1989). The first one, 'fresh seed rain', occurs after the gap formation and is composed of seeds brought from the surrounding forest by dispersers, principally birds and bats. The two other sources constitute the regeneration potential of the forest before the gap:

- the 'seedling bank' in the understorey corresponds to seeds of plant species able to germinate with little light, but needing much humidity. Seedlings grow quickly with the help of a small opening in the forest canopy, but they cannot survive if the sunlight is too bright and would disappear if the opening were too large.

- the 'seed bank' is composed of latent seeds in the soil. These plants are generally pioneer species whose seeds need a lot of light to germinate. These species occur in large gaps where previous species cannot remain.

To determine the impact of howler monkeys on these two last parameters, the seedling population and the composition of the soil seed bank were analysed around the main sleeping sites of a howler monkey troop at the Nouragues Station (Julliot 1997, Julliot unpublished data). The seedling surveys were made on five plots 32m wide and 26m long including a monkeys' sleeping site, and compared with five control plots. The location of control plots was chosen in order to be in the same floristic habitat as the sleeping site plot and to be as far as possible from any sleeping sites. The survey was done according to a 2 x 2m grid map of the plot. The survey concentrated on seven plant taxa (*Parahancornia fasciculata* (Hub.) Ducke, *Virola kwatae* R. Benoist, *Minquartia guianensis* Aublet, *Quiina obovata* Tulasne, *Chrysophyllum lucentifolium* Cronguist, *Pouteria torta* subsp. *glabra* Pennington and *Pourouma* spp.) that were among the most consumed by howler monkeys and whose seeds were dispersed without damage. All the seedlings of these plant species under 1 m high were counted every two quadrats.

The results show that the total number of seedlings is 2.5 times higher for sleeping site plots than for control plots (1552 versus 592 seedlings; Table 3). Significant differences

TABLE 3. Total number of seedlings per plant species for the five sleeping sites of howler monkeys and control plots. Each batch included a sleeping site plot and the corresponding control plot.

Batch	P.f.	V.k.	M.g.	Q.o.	C.l.	P.t.	P.spp	Total
A. SLEEPING SITE PLOTS								
1	3	86	65	64	89	8	105	420
2	8	37	24	2	266	2	82	421
3	1	70	23	2	10	5	68	285
4	11	9	114	0	10	5	61	210
5	5	89	9	3	82	1	27	216
Total	28	291	235	71	563	21	343	1552
B. CONTROL PLOTS								
1	3	21	5	0	0	1	15	45
2	0	12	0	5	28	1	123	169
3	3	10	0	3	151	0	46	213
4	0	1	6	4	2	1	6	20
5	2	8	32	1	12	0	90	145
Total	8	52	43	13	193	3	280	592

P.f.: *Parahancornia fasciculata* (Apocynaceae) - P.f.: *Virola kwatae* (Myristicaceae) -
M.g.: *Minquartia guianensis* (Olacaceae) – Q.o.: *Quiina obovata* (Quiinaceae) -
C.l.: *Chrysophyllum lucentifolium* (Sapotaceae) – P.t.: *Pouteria torta* subsp. *glabra* (Sapotaceae) –
P.spp.: *Pourouma* spp. (Cecropiaceae).

were found in favour of sleeping site plots, for five species (*P. fasciculata, V. kwatae, M. guianensis, C. lucentifolium* and *P. torta* subsp. *glabra*). The particular spatial distribution of seedlings on the sleeping site plots in comparison with control plots (see Fig. 1, for the example of batch 1) was tested by the Geary's index (Julliot 1997). The results show a non-random distribution of seedlings for most of the species on sleeping site plots. On control plots, this distribution is non-random in only 9 cases, and in 6 of them seedlings were localised around a fruiting tree. A more complex analysis of seedling distribution performed for the sleeping site plot of batch 1 reveals a clumped distribution of seedlings for 4 of the 7 plant species with clump size of 64 m^2 diameter (Julliot 1997).

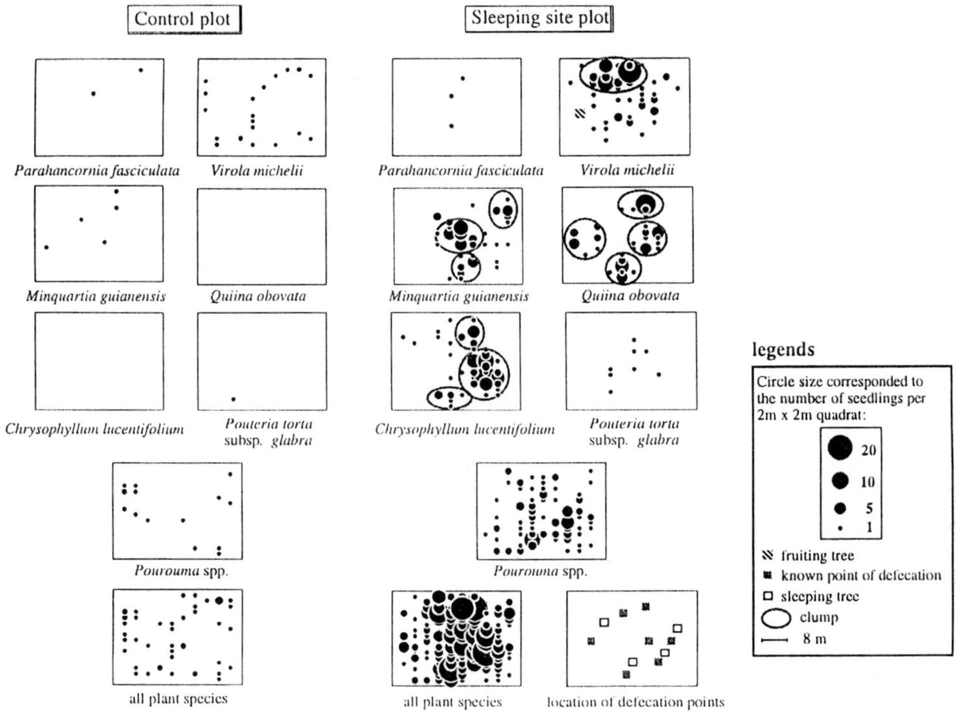

Figure 1. Spatial distribution of seedlings for each of the seven plant species individually, and for all the seven species, on the sleeping site plot and on the corresponding control plot of batch 1. A sleeping site plot together with its matched control plot made up a batch.

Seed dispersal by red howler monkeys seems to induce two scales of heterogeneity in seedling distribution: i. a large-scale of heterogeneity, corresponding to a higher density on the sleeping site plots of plant species dispersed by howler monkeys, and ii. a small-scale of heterogeneity, corresponding to the clumped distribution of these species inside the sleeping site plots.

The composition of the soil seed bank was carried out for two of the five sleeping site plots and their corresponding control plots (Julliot unpublished data). Soil samples were exposed to full sunlight and every germination was noted. After six months of experiment, the number of germinations was significantly higher for sleeping site samples, with a mean of 243 seeds/m² versus 181 for control samples. Identified germinations belonged to 19 taxa from which seven taxa were disseminated by howler monkeys. These taxa amounted to 58% of the soil seed bank for sleeping site plots versus only 24% for control plots. Thus, it appears that the concentration of howler monkey defecations around their sleeping sites modifies the floristic composition of the soil seed bank to a large extent.

These results seem to indicate that seed dispersal by howler monkeys modifies locally the regeneration potential of the forest through the seedling bank and the seed bank, by increasing, under their sleeping sites, the numbers of the plant species they eat and disperse. Other frugivorous animals have a similar impact on forest regeneration, such as the cock-of-the-rock (*Rupicola rupicola*) whose the lek social structure induced a seedling concentration of plant species exploited by this bird on their courtship sites as showed by Théry and Larpin (1993) at Nouragues. These studies indicate that frugivorous animals, including some primate species that are considered as mainly folivores, may have an important role in the floristic structure of tropical forest, either by increasing the spatial heterogeneity of the forest, like howler monkeys, or, on the contrary, in reducing this heterogeneity, like probably spider or capuchin monkeys, through different patterns of seed dispersal linked to the social behaviour and physiological characteristics of the dispersal agent.

5. Post-dispersal survival of seeds and seedlings dispersed by monkeys

It may not be possible to predict patterns of seedling distribution from the activity patterns of frugivorous animals without considering other factors that act later on dispersed seeds and seedlings, such as predation and secondary seed dispersal by rodents, or mortality induced by pathogens and fungi (Forget *et al.* Chapter 25).

It was often considered that clumped seed dispersal by large frugivorous vertebrates, such as howler monkeys, has a negative impact on seed and seedling survival (Howe 1980; Chapman 1989). According to Howe (1980), seed dispersal by howler monkeys is indirectly the origin of most seed and seedling mortality of *Tetragastris panamensis* (Burseraceae) for which they are the principal dispersers on Barro Colorado Island (Panama). He estimated that only 11% of seeds left parent trees with some chance of survival. According to Howe (1980) and Chapman (1989), clumped seed dispersal by howler monkeys does not correspond to the 'Escape Hypothesis' of Janzen (1970) and Connell (1971), because the same density-dependant mortality factors of seeds and seedlings, especially predation of seeds by rodents and competition between seedlings, exist on the defecation sites of howler monkeys and under the parent trees. However, several studies have shown that mortality, due to pathogens (Augspurger 1984) or predators (review by Howe & Smallwood 1982), is not necessarily correlated with density of seeds and seedlings. Howe (1989) also suggests that plant species with a 'clump-dispersal model' must have developed chemical or mechanical defences against predators or pathogens. The seedling surveys under howler monkey sleeping sites at the Nouragues Station included several stages of young plant development and reflected several years of dissemination, as well as seed and seedling mortality. The results of these surveys indicate that, even if clumping results in high mortality due to the concentration of seeds, the accumulation of seeds over time and their continuous replacement modifies the seedling population under howler sleeping sites, by increasing the seedling density of plant species they eat, and overwhelms the effects of predators and pathogens.

In a comparative experiment, Zhang and Wang (1995) analysed the impact of different seed dispersal patterns by spider and capuchin monkeys on seed survival of one plant species, *Ziziphus cinnamomum* (Rhamnaceae), dispersed by both primates. Results seem to indicate that, although spider monkeys dispersed more seeds of this species and may be considered as an important seed disperser, capuchin monkeys are more efficient since a seed dispersed by capuchins had a survival rate 2.6 times higher. Several experimental studies have been started at the Nouragues Station to determine whether clumped seed dispersal by large frugivores, such as primates, may induce a higher mortality of seeds and/or seedlings. Preliminary results indicate that many different factors interfere in seed and seedling survival, such as seed morphology and physiology, seedling development, seasonal variations of food available to terrestrial mammals, local habitat diversity, etc. (Forget *et al.*, Jansen & Forget Chapter 26). All these factors may have an effect upon the foraging and hoarding activities of seed and seedling predators, and, thus, modify the post-dispersal survival rate of seeds and seedlings, independently from the patterns of seed dispersal.

At Pararé site, Forget and Sabatier (1997) made a survey of *Virola kwatae* (Myristicaceae) seedlings at 3-month old over 0.45 ha (180 m x 25 m) including two fruiting trees of this species. At this site, spider monkeys removed on average 52% of the seed production of *Virola* trees. In total, 306 seedlings were inventoried and marked over the survey area, and their survival was followed during 4 years. The spatial distribution of young seedlings showed higher density around fruiting trees (< 20 m from tree), and under the main arboreal pathway used by spider monkeys. Thus, it seems that seedling distribution of *V. kwatae* at a distance from parent trees is consistent with foraging and defecating behaviour of spider monkeys. However, after 4 years, the seedlings still alive were located at the edge of treefall and branchfall gaps which seem to indicate that juveniles of *V. kwatae* required several successive gaps in their neighbourhood to persist in the forest understorey. The recruitment of plant species is also related to habitat characteristics, such as the light environment, and these habitat factors may overrule the effects of seed dispersal by frugivorous animals.

Fungi cover dead wood on the forest floor. *Photo Pierre Charles-Dominique.*

CHAPTER 19

FRUGIVORY AND SEED DISPERSAL BY BATS

Pierre Charles-Dominique and Anya Cockle

1. Introduction

Bat communities have been studied at several places in French Guiana. Investigations were initiated on the St Elie road (ECEREX), where they focused on the seed dispersal of pioneer plant species (de Foresta *et al.* 1984, Charles-Dominique 1986, Brosset *et al.* 1996), and from there spread to a few other sites. Among them, studies undertaken at Nouragues examined in particular the role of understorey bat species in the dispersal and regeneration of Cyclanthaceae and Araceae hemi-epiphytes of the genus*Philodendron* and *Rhodospatha* (Cockle 1997, and Chapter 24). Over 2500 bats were caught at Nouragues, revealing a relative homogeneity in relation to other bat communities in French Guiana, in particular concerning feeding behaviours. The determination of the different species' diets was made possible by the systematic collection of faeces upon capture, complemented by observations made under known feeding and day roosts.

A total of 101 bat species have been reported from French Guiana to this date, including 56 species that were recorded at Nouragues (Brosset & Charles-Dominique 1990, Brosset *et al.* 1996, Simmons & Voss 1998, Brosset *et al.* Chapter 10). These figures are likely to change, and about ten more species are expected to be found in the region. At Nouragues, further investigations will undoubtedly yield several additional rare species that have only been recorded once or twice in French Guiana so far. As detailed by Brosset *et al.* 1996 and Chapter 10), the Phyllostomidea family dominates the bat community in terms of both number of species and number of individuals (95.6% of the bats captured and 77% of the species). This strictly Neotropical family comprises several subfamilies, each one being characterised by particular feeding habits, in French Guiana (Charles-Dominique unpublished data) as well as in other countries where they have been studied (Fleming 1986, Voss & Emmons 1996, Kalko *et al.* 1996).

2. Frugivorous bats

The Phyllostominae include 28 species in French Guiana and 16 at Nouragues in particular. All are insectivorous or predators of small vertebrates. However, *Phyllostomus hastatus*, which usually feeds on large insects and small vertebrates (Bradbury 1977), may occasionally take fruit (1 observation out of 47). It appears that foraging and hunting strategies, as well as the modes of prey localisation and selection, vary considerably from

Bongers *et al.* (eds.), Nouragues. Dynamics and plant-animal interactions in a neotropical rainforest, 207–215.
2001 Kluwer Academic Publishers. Printed in the Netherlands.

one species to the next, thus contributing to the segregation of ecological niches (Kalko *et al.* 1996).

The Stenodermatinae are represented by 18 species in French Guiana and by 16 at Nouragues alone. All are frugivorous, but some species such as the *Artibeus* spp. may supplement their diet with pollen or nectar.

The Carolliinae comprise only three species in the region, all of which are also present at Nouragues. Like the Stenodermatinae, they are frugivorous but may also take some pollen and nectar (Charles-Dominique unpublished data).

The Glossophaginae comprise seven species in French Guiana, including four at Nouragues. This group specialises in the exploitation of nectar and pollen, which deeply influenced their anatomy (elongation of the muzzle and tongue, presence of horny papillae at the tip of the tongue, reduction of the size of the ear and nasal leaf, reduction of the size of teeth, or even their disappearance, in particular atrophy of the lower incisors, etc.). *Glossophaga soricina* is the least specialised of the group and may occasionally include some fruit in its diet. This species which is abundant in disturbed area is absent from Nouragues.

The Desmodontinae are constituted by the true vampire bats, which feed exclusively on the blood of vertebrates (one species, *Desmodus rotundus*, collected at Nouragues).

Only 19 species of true frugivores, belonging to the subfamilies Carolliinae and Stenodermatinae, will be examined here. They alone account for 63.8% of the netted bats at Nouragues.

3. Methods

The results that will be exposed below are part of several studies undertaken between 1987 and 1996 at Nouragues. Much of it has not been published under a synthetic form. Only data from netted bats will be examined, as it is only at this moment that faeces can be collected, either while the bat is being eased out of the net or a few minutes later, inside the bag (one hour at minimum in the bag). Unluckily, frugivorous Phyllostomidae bats have a short gut transit time and most individuals have already emptied their bowels when they are netted. The small seeds collected this way can be identified to species or genus level by comparison with a reference collection made with plant samples of the area. Complementary information was also gathered by stretching collectors under day or feeding roosts known to be used by a particular species followed with radio-tracking. This second method made it possible to collect the larger seeds which are not ingested but transported in the mouth, and which cannot be found in the nets (bat faeces often contain only seedless pulp and juices from these large-seeded fruits). In this case, the collectors were visited every morning

The nets of generally 9 m long and 12 m in rare occasions were placed along the ground, between 0 and 2.5 m high, taking care to frequently change their location in order to prevent the bats from memorising them. Sixty-five different netting sites were sampled, some of them several times (in this case leaving at least one month between netting sessions). Overall, 1805 net.hours were put in. The yield (0.44 netted individual per hour and per net) was rather low, as is usually the case in primary forest because of the 'dilution' of the bats in an extensive three-dimensional volume. Captures took place within 3 km of the campsite.

Moreover, an important netting effort was deployed in the higher strata of the forest. Canopy nets were stretched between two thin ropes looped taut between the ground and two conveniently placed branches. In order to allow the free vertical translation of the net from ground to lower canopy, the understorey must be locally cleared of obstacles: ancient tree fall or windbreak gaps can be suitable in this respect. The small number of convenient sites and the time needed to install the ropes properly considerably restricted the sampling in relation to ground netting. The bats more easily detect canopy nets than understorey nets and, moreover, they are often repeatedly opened night after night for several weeks, which undoubtedly made possible the memorisation of their location by sedentary species. Two situations were sampled: on the one hand, 'mid-canopy' nets between 10 and 25 m below the canopy's upper limit (7 sites); and on the other, 'upper canopy' nets placed 5 to 15 m from the canopy's upper limit but on the border of a large opening (4 sites). The position along larger clearings allows the interception of high-flying bats, which forage in the open space above the forest, as they dip along the forest borders. Net heights were measured from the local upper limit of the canopy rather than from the forest floor because the actual canopy height is very variable and the upper limit is more determinant in the spatial localisation of the bats' movements. In the mid-canopy, 2673 net.hours yielded 0.12 bat/hour/net, whereas in the upper canopy, 2585 net.hours yielded 0.02 bat/hour/net only. Despite its low yields, this method allows the capture of species which would otherwise go undetected (these data only take into account frugivorous Phyllostomidae). Such a heterogeneous sampling makes statistical analyses difficult to apply, but differences between understorey and canopy results are so obvious that conclusions can still be drawn (the differences are even more obvious when other families are examined, such as the Molossidae, Vespertilionidae and Emballonuridae).

4. Results

4.1. CHIROPTEROCHORY

The specialisation of frugivorous animals on certain types of fruit is very variable, but it is particularly obvious in Neotropical bats. This lead van der Pijl (1957) to describe the 'syndrome' of chiropterochory. This phenomenon is much more marked in southern America than for the Old World's Megachiropteran bats, which appear less specialised (Morton 1973). In Phyllostomidae bats, the location, identification and picking of the fruit takes place in flight, this last operation lasting but a mere fraction of a second during which the fruit is detached and taken away. The only exceptions are the infructescences of *Philodendron* (Araceae), *Evodianthus, Asplundia* and *Thoracocarpus* (Cyclanthaceae), on which the bats must land in order to eat the berries (Cockle 1997, and Chapter 24). The fruit is thus generally picked on the wing, minimising in this way the predation risks. However the bats must afterwards reach a suitable perch to eat it out of predators' reach (Morton 1973, Fleming 1986, Charles-Dominique 1986). Bats are particularly vulnerable when perched, and the choice of the roost - day roost for sleeping, or night roost for feeding and resting - is determinant for security. Feeding roosts generally display a wide-open space below them and a dense screen above (leaves, liana tangles, etc.). When disturbed, the bat lets itself drop and plunges in the undergrowth, flying away in zigzags. This habit distinguishes Phyllostomidae bats from frugivorous birds, which are very mobile when perched and can easily hop around inside dense vegetation, often picking the fruit and eating it on the spot. Even though Pipridae and Trogonidae birds may pick fruit in flight (Moermond & Denslow 1983), they generally perch again very close by. This behavioural difference is essential and has important repercussions on seed dispersal: frugivorous birds tend to defecate from a perch, whereas frugivorous Phyllostomidae bats defecate more often during flight (Charles-Dominique 1986, Thomas *et al.* 1988).

Phyllostomidae bats detect fruit and identify their state of maturation in flight, by smell. It follows that not all fruits can be noticed and selected by the bats (van der Pijl 1957). Most 'bat' fruits do not change colour when ripe, remaining greenish or whitish, like the surrounding leaves: 'cryptic' fruit (Cooper *et al.* 1986); instead, they start to emit particular scents which guide the bats. However some species can be also detected by echolocation (Kalko & Condom 1998). Such fruit are placed outside the foliage, above it on erect peduncles or hanging below it, so as to be freely accessible to oncoming flying animals. Chemical analyses show that their pulp contains little lipids, but simple C_6 sugars directly absorbed (Charles-Dominique 1986). Some of them (*Cecropia obtusa*, Cecropiaceae; *Solanum* spp., Solanaceae; *Vismia* spp., Clusiaceae) also contain small quantities of non-polymerized amino-acids, which are also directly absorbed and constitute the only source of nitrogen for the bats which feed on them (Charles-Dominique 1991). These fruits, quantitatively poor in nutritive elements but qualitatively rich, are regarded by some authors such as McKey (1975) as 'poor and non specialised', as opposed to the 'rich and specialised' fruits with which taxa of frugivorous animals have been able to evolve (for example, the lipid-rich fruit of the Lauraceae, associated to the Cotingidae birds, (McKey 1975, Snow 1981, Erard *et al.* 1989). However, bat fruits can also be considered specialised because their characteristics make them much more accessible to bats than to any other animals. To compensate for the relatively poor nutrient content of these fruits, Phyllostomidae bats such as *Artibeus* spp. or *Carollia perspicillata* feed 30 to 40 times in a single night, ingesting up to twice their body weight in fruit pulp (Charles-Dominique 1986, 1991). This feeding habit also involves short gut transit times of 5 to 10 min, which limit the accumulation of large quantities of food in the digestive tract and thus keep the body light enough for efficient flight (Fleming 1986, Charles-Dominique 1986, Cockle 1997).

The question of the origin of nitrogen compounds in the diet of Phyllostomidae bats is still under debate. Fleming (1986) reports the presence of insect remains in faeces of *Carollia perspicillata* in Costa Rica, whereas none have been found in French Guiana in several hundred faeces collected, whether at the nets or under roosts. Moreover, chemical analyses of bat fruits suggest that, at least in French Guiana, the quantities ingested should be sufficient to provide an appropriate amount of nitrogen (soluble amino acids, Charles-Dominique 1986).

Two types of chiropterochorous fruit can be distinguished. The first includes fruits containing a great number of small to minute seeds that are swallowed along with the pulp and therefore pass through the digestive tract. The second comprises fruits with large seeds that are not ingested but discarded after feeding. Large seeds are dropped under feeding roosts, along with the small pellets of fibrous matter which remain after the bat has absorbed the fruit juices out of the pulp (van der Pijl 1957, Vasquez-Yanes *et al.* 1975, Heithaus *et al.* 1975, Janzen *et al.* 1976, Morrison 1978, 1980, Bonaccorso 1979, Fleming & Heithaus 1981, Dos Reis & Guillaumet 1983). These oval pellets, pressed against the palate, reflect the size of the bat (length equal to the distance separating the upper canines).

4.2. FRUGIVORY AND FLYING ZONATION

Understorey captures involve much higher numbers than canopy captures (68.5%), which makes comparisons difficult. However, despite this sampling unevenness, *Carollia perspicillata, C. brevicauda, Sturnira tildae, S. lillium, Uroderma bilobatum*, and, up to a point, *Rhinophylla pumilio*, are all found more often in the understorey; see Bonaccorso 1979) and Handley (1967) for comparisons in Central America. In contrast, many Stenodermatinae such as *Ametrida centurio, Artibeus gnomus, A. concolor, Chiroderma trinitatus, C. villosum*, are more common in the canopy. The larger *Artibeus* (*A. lituratus, A. jamaicensis, A. obscurus*) appear to use both canopy and understorey strata, as they forage high in the trees but commute regularly to understorey feeding roosts through gaps (Charles-Dominique 1986), Table 1.

TABLE 1. Number of netted bats in the different sampled strata of the forest

	Understorey	Mid-canopy	Upper canopy
Carollia brevicauda	31	0	0
Carollia perspicillata	174	2	0
Rhinophylla pumilio	196	17	0
Sturnira lillium	4	0	0
Sturnira tildae	11	1	0
Ametrida centurio	2	12	8
Artibeus concolor	2	6	3
Artibeus obscurus	125	76	15
Artibeus gnomus	16	54	1
Artibeus jamaicensis	71	47	9
Artibeus lituratus	89	37	4
Chiroderma trinitatum	5	12	0
Chiroderma villosum	14	18	0
Mesophylla macconnelli	9	4	0
Platyrrhinus helleri	27	31	1
Uroderma bilobatum	21	3	0
Vampyressa brocki	1	4	0
Vampyrodes caraccioli	0	1	0
Total	798	326	41

Two main categories of frugivorous bats can therefore be distinguished:

1) Understorey bats: *Carollia perspicillata, C. brevicauda, Rhinophylla pumilio* (Carolliinae), *Sturnira lillium, S. tildae* and *U. bilobatum*. At least in the case of the Carolliinae, these bats have relatively short and broad wings that allow them to near-hover and to make their way through very dense vegetation. In French Guiana, *C. perspicillata* is the only one of the three Carolliinae species to survive in areas degraded by human activities, such as agricultural land and villages, where, as in the case of the *Sturnira*, its numbers rise five to ten-fold (Cosson 1994, Brosset & Charles-Dominique 1990).

2) Canopy bats: all belong to the subfamily Sternodermatinae and can be divided into two further groups: the small to medium Stenodermatinae capable of flying in the cluttered

space of the mid-canopy stratum (*Artibeus gnomus, Vampyressa* spp., *Mesophylla macconnelli, Ametrida centurio, Chiroderma trinitatum, Platyrrhinus helleri*); and the larger Stenodermatinae which are relatively fast fliers because of their weight and which are therefore restrained to the more open spaces of the upper canopy or treefall gaps (*Artibeus lituratus, A. jamaicensis, A. obscurus, A. concolor, Chiroderma villosum*).

The Stenodermatinae community appears to display a regular gradient in body size (Fig. 1), which certainly contributes to the separation of the ecological niches. These bats transport the fruits they have picked in flight to a feeding roost. As they cannot manage weights above 40 to 60% of their own body weight (Morton 1973, Charles-Dominique 1986), the size of the fruit is determinant for the smaller bat species. On the other hand, these species can exploit the more cluttered areas that the larger species cannot reach.

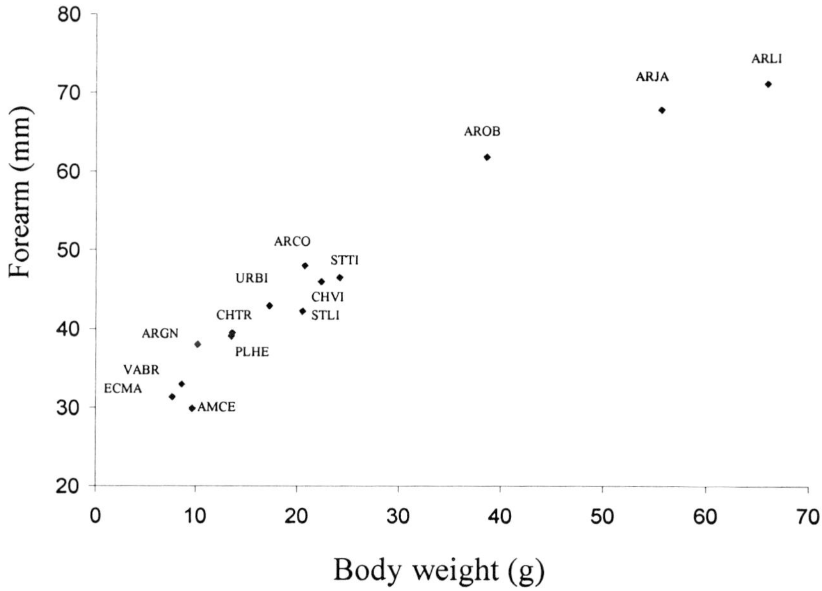

Figure 1. Relation between forearm length and body weight among frugivorous bats of the canopy: ARLI = *Artibeus lituratus* (n=149), ARJA = *A. jamaicensis* (n=207), AROB = *A. obscurus* (n=159), ARCO = *A. concolor* (n=28), STTI = *Sturnira tildae* (n=64), CHVI = *Chiroderma villosum* (n=42), URBI = *Uroderma bilobatum* (n=24), STLI = *Sturnira lilium* (n=23), CHTR = *Chiroderma trinitatum* (n=26), PLHE = *Platyrrhinus helleri* (n=49), ARGN = *Artibeus gnomus* (n=73), VABR = *Vampyressa brocki* (n=18), ECMA = *Ectophylla macconelli* (n=24), AMCE = *Ametrida centurio* (n=34).

4.3. DIETARY CHOICES

The examination of faeces collected upon capture (Table 2), completed by observations of food remains under roosts, displays a clear separation between the diets of the Carolliinae (understorey) and the Stenodermatinae (canopy). The first group concentrates on the fruit of understorey lianas such as *Philodendron* spp., *Rhodospatha venosa* (Araceae), *Evodianthus funifer, Asplundia heteranthera, A. brachyphylla, Thoracocarpus bissectus* (Cyclanthaceae), *Schlegelia paraensis* (Bignoniaceae), *Marcgravia coriacea*

(Marcgraviaceae) and understorey shrubs such as *Solanum* spp. (Solanaceae) and *Piper* spp. (Piperaceae) (Cockle 1997).

TABLE 2. Occurrences of seeds of different plant taxa found in the faeces of captured frugivorous Phyllostomidae bats at Nouragues.

	Rp	Cp	Cb	Sl	St	Ao	Aj	Al	Ph	Ub
Evodianthus funifer	10	1	1		1					
Asplundia spp.	5	2	1		1					
Thoracocarpus bissectus	3				1					
Philodendron spp.	15	6	1	1	1					
Schlegelia paraensis	2	1								
Piper spp.		2	1							
Solanum spp.		3			1					
Vismia sp.		1								
Ficus spp.							9		2	
Cecropia obtusa						7	4	5		2

Rp: *Rhinophylla pumilio*, Cp: *Carollia perspicillata*, Cb: *Carollia brevicauda*, Sl :*Sturnira lillium*, St: *Sturnira tildae*, Ao: *Artibeus obscurus*, Aj: *Artibeus jamaicensi*, Al: *Artibeus lituratus*, Ph: *Platyrrhinus helleri*, *Ub: Uroderma bilobatum*

In the secondary forests of French Guiana, the diet seems to shift slightly in order to include a greater proportion of *Solanum, Piper* and *Vismia* (Clusiaceae) and less of the Araceae and Cyclanthaceae (Cosson 1994, Charles-Dominique 1986), thus reflecting the differences in plant composition. Most investigations carried out in various South American countries show that *C. perspicillata* feeds mainly on *Solanum, Piper, Vismia* and several other genus not found in French Guiana such as *Muntjaca* (Fleming & Heithaus 1981, Fleming 1986, for Costa Rica, Dos Reis & Guillaumet 1983, for Brazil). These differences are more attributable to local plant compositions than to actual dietary preferences. *Solanum, Piper* and *Vismia* are also eaten at Nouragues, but in much smaller proportions as these species are relatively scarcer in primary forests. Data collected under the roosts of *Rhinophylla pumilio* (Cockle 1997) confirm the capture data, and complete it with somewhat rarer observations of *Rhodospatha* spp. (Araceae), *Aechmaea sp* (Bromeliaceae), *Coussapoa* spp., *Cecropia sciadophylla, C. obtusa* (Cecropiaceae), *Ficus* spp. (Moraceae), all small-seeded, and of the large-seeded *Licania canescens, L. membranacea* (Chrysobalanaceae), *Symphonia globulifera* (Clusiaceae), *Byrsonima sp* (Malpighiaceae) and *Siparuna sp* (Monimiaceae). *C. perspicillata* has also been observed transporting the large-seeded fruits of *Protium araguense* (Burseraceae) and *Symphonia globulifera* (Clusiaceae).

The diet of the Stenodermatinae is mainly based on *Ficus* spp. (Moraceae) and *Cecropia obtusa* (Cecropiaceae), which have small seeds, and, for the larger *Artibeus* bat species, the large-seeded *Andira* spp., *Dipteryx odorata* (Fabaceae), *Symphonia globulifera* (Clusiaceae), as well as many Chrysobalanaceae of the genus *Parinari, Couepia, Licania*, difficult to identify to species level. Under a *Uroderma* roost, fruit remains of *Licania* spp. and *Ficus* spp. have been found. There is therefore little dietary overlap between Carolliinae and Stenodermatinae, at least in the case of the well-documented genera *Artibeus, Chiroderma* and *Platyrrhinus*. For example, the seeds of *Cecropia obtusa* and

Ficus spp., which constitute the base of the *Artibeus* diet, are seldom found in the faeces of Carolliinae bats.

A few exceptions can however be reported. *Uroderma bilobatum* sometimes includes *Solanum* spp. in its diet (observations from other parts of French Guiana), and the *Sturnira* appear to have similar feeding habits to that of the Carolliinae, with *Solanum, Piper, Philodendron* and some Cyclanthaceae. The genus *Sturnira* has long been placed in a subfamily of its own, the Sturnirinae, on account of anatomical traits: dietary characteristics seem to support such a distinction.

Many Stenodermatinae remain poorly documented, either because they are rarely captured (*Vampyressa* spp., *Vampyrodes caraccioli*) or because the faeces yield very little exploitable matter (*Ametrida centurio*, for example, only seems to produce greenish or yellowish 'fruit juices' with no solid particle to help identification).

4.4. CONSEQUENCES FOR SEED DISPERSAL

Because of their dietary specialisations, frugivorous Phyllostomidae bats are often associated with plant species which have evolved in accordance with this means of dispersal. In consequence, these plants are largely dependent on bats for the dispersal of their seeds. We showed that bats remove 80% of *Cecropia obtusa* fruit, and 95% of both *Solanum rugosum* and *Vismia guianensis* fruit (Charles-Dominique 1986). Similarly, by indirect estimations, it appears that the small understorey bat *Rhinophylla pumilio* is responsible for the dispersal of the great majority of the Cyclanthaceae *Evodianthus* and *Asplundia* seeds (Cockle 1997).

Another consequence of chiropterochory is the mode of dissemination of the smaller seeds. After ingestion, seeds are mostly defecated during flight (Charles-Dominique 1986) because defecation at roost, where the bat hangs head down, is a rather complicated process involving repositioning of the pelvis and projection of the faeces, with the risk of dirtying the pelage. The evacuation of faeces during flight makes it possible for the seeds of *Cecropia obtusa* defecated by the larger *Artibeus* to reach treefall mounds in the centre of gaps (Charles-Dominique 1986, 1993). In the same way, the very small seeds of the *Philodendron* and various Cyclanthaceae hemi-epiphytic lianas are able to colonize the near-vertical tree bases where the best conditions for their subsequent development are met (Cockle 1997).

Dispersal in flight also contributes to a greater 'evenness' of the seed rain, thus indirectly acting on the structure of the soil seed bank: all 25 x 25 cm soil samples collected at Nouragues for the study of the seed bank contained seeds of *Cecropia obtusa*.

5. Conclusion

Nineteen species of frugivorous bats have been reported from Nouragues, all of which display similar biological traits (fruit removal in flight, use of a particular feeding roost, short gut transit, defecations mostly in flight, etc.). The separation of the ecological niches of the different species is based on combinations of several factors which minimise interspecific competition: 1- the forest stratum exploited (understorey *versus* canopy), 2- the fruit species chosen (Carolliinae and *Sturnira* diets, *versus* the diet of other Stenodermatinae, with very little overlap), 3- the size of the bat species, which determines the maximum size of the fruit chosen in relation to flight capacities, 4- the anatomy of the wing and shape of the patagium, in relation to the body weight, which allow different flight possibilities (manoeuvrability, hovering, speed, etc.) and may oblige some species to avoid the cluttered spaces open to other species.

These specialisation's allow the coexistence of many different sympatric species, which together constitute a community characteristic of the habitat. The species richness found in Neotropical rain forests is unique: nowhere in Africa, Asia or Australia more than 3 to 5 coexistent frugivorous bat species can be found (Brosset 1986, Thomas 1982, Fleming 1993). We have also shown that the entire bat community reacts to modifications of the environment, many specialised species disappearing and generalists tending to explode demographically (Brosset *et al.* 1996). Similar analyses may be undertaken on the frugivorous species alone, which are easier to capture and whose diets can be easily determined by the examination of the faeces. The frugivorous bat community of Nouragues is very similar to that of other primary forest sites in French Guiana, such as St Elie road (Brosset & Charles Dominique 1990), Paracou (Simmons & Voss 1998), St Eugène near the Petit Saut dam (J.F. Cosson unpublished data) and the Kaw Mountain (A. Cockle unpublished data). These studies will therefore serve as a basis for comparative investigations undertaken in various artificially disturbed habitats in French Guiana.

A group of *Ectophylla macconnelli* bats in a « tent » made of a leaf of a young *Astrocaryum sciophilum* palm. *Photo Pierre Charles-Dominique.*

CHAPTER 20

FRUGIVORY AND SEED DISPERSAL BY KINKAJOUS

Didier Julien-Laferrière

1. Introduction

The kinkajou, *Potos flavus* (Carnivora, Procyonidae), is one of the most abundant and most active seed dispersers among nocturnal arboreal frugivores in Neotropical forests (Walker and Cant 1977, Bonaccorso *et al.* 1980, Charles-Dominique *et al.* 1981, Estrada & Coates-Estrada 1985, Howe 1986, Robinson & Redford 1986, Janson & Emmons 1990, Julien-Laferrière 1999). However, the ecology of this species surprisingly has received little attention and very few field studies focusing on kinkajous have been published (Walker & Cant 1977, Kays & Gittelman 1995, Kays 1999). To know the place of this mammal in the community of neotropical frugivores and its role in forest dynamics, it is necessary to investigate direct parameters such as diet or foraging behaviour as well as patterns of seed dispersal and fate of seeds, but also indirect ones such as space use patterns, activity, or social organisation (Rowell & Mitchell 1991, Charles-Dominique 1995, Zhang & Wang 1995a). One focal point is to evaluate the efficiency of the frugivore in seed dispersal, by analysing the fate of seeds dispersed and, ultimately, the recruitment of new reproductive plants (Houle 1995, Schupp & Fuentes 1995).

During a field study at the Nouragues station (7 months dispatched between 1991 and 1993), I radio-tracked several kinkajous at three different seasons, in order to analyse adaptations in the kinkajou's foraging behaviour to seasonal variations in food availability. Radio-tracking allowed recording quantitative data on use of space, activity, individual relationships, diet, and seed dispersal in this species. I also surveyed the fate of dispersed seeds. I will present here some data on diet, fruit selection, seed dispersal, and fate of dispersed seeds, in an attempt to find the 'missing link' between fruit eating and seedling recruitment.

2. Methods

Seven adult kinkajous (two males and five females) were radio-tracked during three different seasons: 1) January-March, the beginning of the rainy season, is characterised by an abundance of fruits; 2) July-August, the end of the rainy season, by a paucity of both flowers and fruits, and 3) September-November, the dry season, by a paucity of fruits and an abundance of flowers (Sabatier 1985, Julien-Laferrière & Atramentowicz 1990, Zhang & Wang 1995b). Two of the five females were followed during two different seasons.

Kinkajous were trapped in double-door wire traps ('Tomahawk Live Trap' 107 x 38 x

Bongers *et al.* (eds.), Nouragues. Dynamics and plant-animal interactions in a neotropical rainforest, 217–225.
2001 Kluwer Academic Publishers. Printed in the Netherlands.

38 cm), baited with banana and honey. The traps were hauled up trees, then tied on or under a big horizontal branch in the lower canopy (20-25 m high). Selected trees were flowering or fruiting and regularly visited by kinkajous, according to preliminary observations. Once captured, kinkajous were anaesthetised, checked and equipped with radio transmitters (Biotrack TW2 and SR1 in 1991, SS2 in 1993), then released at nightfall. They were then followed alternatively during the first or the second half of the night. The seven individuals were followed for a total of 87 half-nights (two individuals during 19 half-nights in July-August 1991, two during 31 half-nights in January-March 1993, and five during 37 half-nights in September-November 1993).

Diet, patterns of fruit and seed use, and characteristics of plants used (biological type, height) were determined by direct observation, completed with the collection and identification of the pieces of fruits coats and seeds dropped during feeding. As animals were followed at close proximity, it was possible to record most faeces at the moment of defecation, and to identify the seeds defecated.

Two experiments were carried on in January-March 1993 to examine the fate of seeds dispersed by kinkajous.

- Germination trials. To test for possible effects of gut transit on seed germination and seedling growth, 332 seeds of five plant species (41 to 93 seeds per species) were collected and planted on similar conditions. Among them, four species (*Pourouma bicolor*, *Virola michelii*, *V. surinamensis*, and *Inga huberi*) were among the most heavily used by kinkajous during this period. The fifth, *Cecropia obtusa*, was added because this was the only species used for fruits during the three seasons. To mimic kinkajou gut transit, several seeds from each species (except the minute seeds of *C. obtusa*) were gently cleaned to clear them of their aril or surrounding pulp; some seeds of *I. huberi* were also cleaned of their external smooth coat. Seeds were then periodically checked to determine their germination rate and germination time (time between seed planting and germination) (Hladik & Hladik 1969, Estrada & Coates-Estrada 1984, Julliot 1996), as well as the rate and time of appearance of seedlings. Ninety-six seeds (5 to 36 per species) were collected from kinkajou defecations during radio-tracking sessions, the others were taken from ripe fruits freshly fallen from the trees exploited by the focal animals. I used tins 10 cm in diameter by 15 cm in height, drilled on the bottom, placed at one meter in height on a shadow place at the edge of the forest, and protected from rodents and insects by a wire mesh covered with a mosquito net. Tins were filled with original, presumably abiotic, soil collected under one building of the camp. According to their size, two (*P. bicolor*, *Virola spp*, *I. huberi*) or 30 (*C. obtusa*) seeds were deposited on the soil of each tin, then provided with water.
- Survey of seeds defecated in situ. 175 seeds of *P. bicolor* defecated by kinkajous were marked with threads *in situ* (Forget 1990) and checked periodically thereafter. Seeds not found within a 10 m radius around the defecation place were considered to have been secondarily dispersed by rodents (Forget 1990) or ants (Levey & Byrne 1993). Various defecation sites allowed to compare the fate of seeds placed in different natural situations: without dispersal, *i.e.* under the source tree or another fruiting tree from the same species ('food tree species': 85 seeds), or dispersed under different patterns: under a fruiting tree from another species ('other fruiting species': 36 seeds), under an adjacent tree (dispersal distance < 10 m: 'neighbouring tree': 34 seeds), or at any place in the forest (greater dispersal distance: 'understorey': 20 seeds). *P. bicolor* was chosen as a focal species because it was the most heavily used for fruits by kinkajous during the study period.

Similar surveys were performed at the same time on seeds of *V. michelii* (N = 7), *V. surinamensis* (N = 15), and *I. huberi* (N = 17) defecated by kinkajous. I used χ^2 tests, or Fisher exact probability tests for the smaller samples (Siegel 1956), to test through time the effects of treatments on the fate of seeds, i.e. gut transit on the time and rate of germination and the survival rate of seedlings in experiment 1 (germination trials), situation of defecation on the rate of seeds present, eaten, or secondarily dispersed, and seedling survival, in experiment 2 (defecated seeds).

3. Results

3.1 DIET AND FOOD CHOICE

Kinkajous are mainly frugivorous, although they also may feed on nectar (Charles-Dominique *et al.* 1981, Julien-Laferrière 1993, 1999; Kays 1999), and occasionally on insects (Bisbal 1986, Redford *et al.* 1989). At the Nouragues station, 17 out of 23 faeces I collected and carefully dissected contained only seeds and fruit pulp, from one to 5 different species; only 1 to 6 whole insects or arthropod fragments were present in the other six faeces, and were probably swallowed incidentally.

Forty-eight plant species were exploited by kinkajous: 41 species from 22 families for fruits, and 7 species from 6 families for flowers (Table 1 and 2). Direct observations allowed determining the biological type of plants visited in 43 species. Nearly 80% were large canopy trees (height > 25 m, 15 species), or plants held up by these trees: vines (13 species) and epiphytes (6 species). Smaller trees (12 to 25 m in height, 9 species) were visited less often, and were always included in the upper vegetation stratum.

All species were not used with the same intensity, but a small number was predominant in the diet. For example, in January-March 1993, 70% of the feeding visits occurred in the trees of 5 out of 18 species, and represented 84% of the feeding time; the first two more visited species included nearly half (49%) of the feeding time (Table 3). The fruits of *Strychnos erichsoni*, *Coussapoa angustifolia*, and *Cecropia obtusa* were predominantly used in July-August 1991 (Julien-Laferrière 1993), and the flowers of *Eperua falcata* in September-November 1993 (Julien-Laferrière 2000).

Direct observations allowed to precisely identifying the fruit part eaten in 32 species. The fruits of 11 species were swallowed whole, whereas in 20 species only the medium part of the fruit was swallowed together with the seeds (mesocarp in 11 species, aril in 9 species). In one species, *Symphonia globulifera*, the seeds were not swallowed but dropped under the feeding tree, after kinkajous had eaten the fruit pulp by biting off several pieces.

Fruits eaten whole were small to minute fleshy fruits (maximum length x diameter 1.5 x 1 cm, *Schlegelia paraensis*) with a soft pericarp and seeds strongly attached to the mesocarp. These were berry-like fruits, with numerous minute seeds (*S. paraensis*, *Brosimum acutifolium*, *Ficus* sp., young fruits of *Cayaponia sp*), or pseudo-fruits (*i.e.* infrutescences: *Cecropia*, *Coussapoa*). One larger berry-like fruit (*Chrysophyllum lucentifolium*, 3.5 cm in diameter) was also eaten whole but was picked in small pieces instead of being swallowed in one time.

In fruits partly eaten, the part eaten was always strongly attached to the seed but easy to detach from the epicarp. These fruits were indehiscent, moderately hard, drupe-like fleshy fruits with one (*e.g. Pourouma bicolor*) to few (*e.g. Strychnos erichsonii*) fairly small seeds, or dehiscent, exhibiting an arillate seed at maturity (*e.g. Virola, Protium, Inga*).

TABLE 1. Plant species used for fruits by kinkajous

Family	Species	Biological type	Fruit condition	Part eaten	Season
Bignoniaceae	*Schlegelia paraensis*	V	m	whole	III
Burseraceae	*Protium* sp. (93-55)[a]	t	m	aril	I
Cecropiaceae	*Cecropia obtusa*	t	m	whole	I-II-III
	Cecropia sciadophylla	t	m	whole	III
	Pourouma bicolor	t	m	mesoc.	I
Clusiaceae	*Clusia* sp. (93-93)[b]	E	m	mesoc.?	III
	Symphonia globulifera	T	m	peric.	I
Cucurbitaceae	*Cayaponia* sp. (91-32)	V	i	whole	II
Dilleniaceae	*Doliocarpus* sp. (93-15)	V	i	aril	I
Flacourtiaceae	*Laetia procera*	T	m	mesoc.	I
Hippocrateaceae	*Cheiloclinium* sp. (93-33)	V	i	aril	I
	Salacia sp. (93-59)	V	m	mesoc.	III
Loganiaceae	*Strychnos erichsonii*	V	m, i	aril	II
Melastomataceae	*Mouriri huberi*	T	m	aril	III
Menispermaceae	*Abuta imene*[b]	V			I
Monimiaceae	*Mollinedia* sp. (91-43)	t	i		II
	Siparuna sp. (91-31)[b]				II
Moraceae	*Brosimum acutifolium*	T	i	whole	III
	Coussapoa angustifolia	E	m	whole	II-III
	Coussapoa latifolia	E	m	whole	III
	Ficus guianensis	E	m	whole	III
	Ficus trigona	E	m	whole	II
	Ficus sp. (93-86)	E	m	whole	III
Myristicaceae	*Virola michelii*	T	m	aril	I
	Virola surinamensis	T	m	aril	I
	Virola venosa	T	m	aril	I-II
Papilionaceae	*Inga* cf. *brachystachys* (93-16)[b]				I
	Inga cinnamomea[b]				I
	Inga huberi	t	m	mesoc.	I
	Inga sp. (93-18)[b]				I
Passifloraceae	*Passiflora garckei*	V	m	mesoc.	II
Polygalaceae	*Moutabea* cf. *guyanensis* (93-23)[b]		m		I
Rhamnaceae	*Ziziphus cinammonum*	T	m	mesoc.	I-II
Rubiaceae	*Psychotria* cf. *anceps* (91-31)[b]	t			II
Sapotaceae	*Chrysophilum lucentifolium*	T	i	whole	I
	Manilkara huberi	T	i	mesoc.	I
	Pouteria hispida	T	i	mesoc.	I
Sapindaceae	*Matayba* sp. (93-65)	T	m	whole	III
	Paullinia capreolata	V	i	mesoc.	I
Tiliaceae	*Apeiba echinata*	t	m	mesoc.	III
	Apeiba glabra	t	m	mesoc.	III

[a]: numbers beginning by 91 or 93 are personal collection numbers, collection available atBrunoy Laboratory; [b]: seeds collected in faeces; Key: Biological type: V = vine; E = epiphyte; T = large canopy tree (h ≥ 25 m); t = small canopy tree (h = 12-25 m). Fruit condition: m = mature; i = immature. Part eaten: mesoc. = mesocarp; peric. = pericarp. Season: I = January-March 1993; II = July-August 1991; III = September-November 1993.

TABLE 2. Plant species of which flowers are exploited by kinkajous (key as in Table 1).

Family	Species	Biological type	Fruit condition	Part eaten	Season
Bignoniaceae	*Schlegelia* sp. (91-37)	V	m	nectar	II
Caesalpiniaceae	*Eperua falcata*	T	m	nectar	III
Cariocaraceae	*Cariocar glabrum*	T	m	nectar	III
Cucurbitaceae	*Cayaponia* sp.2 (91-33)	V	m	whole	I
	Cayaponia sp.3 (93-19)	V	m	whole	I
Marcgraviaceae	*Marcgravia coriacea*	V	i	nectar	II
Papilionaceae	*Inga rubiginosa*	T	m	nectar	II

3.2 DISPERSAL AND FATE OF SEEDS

All but one plant species, *S. globulifera*, had their seeds swallowed by kinkajous. However, the fruits of 9 out of the 42 species sampled were eaten unripe. Consequently, 79 % of the species used for their fruits were actually dispersed by endozoochory. The largest seed swallowed (found in one faeces) was this of *Inga cinnamomea* (3.3 x 1.7 x 0.9 cm, 2.8 cc in volume). *S. globulifera* was the only species for which the seed was too large (2.5 cm in diameter, 8cc in volume) to be swallowed, and was dropped at the base of the tree.

TABLE.3. Plant species visited for fruit used in January - March 1993: number of visits, time of visit, number of trees visited and occurrence in defecations.

Species	Nb of visits	% of the total time of visit (3290 min)	Nb of trees	% occurrence in defecations (N = 64)
P. bicolor	17	25.4	5	75.0
V. surinamensis	6	23.9	1	7.8
V. michelii	10	21.3	3	1.6
C. lucentifolium	6	9.9	2	0
I. huberi	3	3.3	2	7.8
Others	mean = 1.4 (range 1-3, N = 13 spp.)	mean = 1.9 (range 0.5-9.9)	15	mean = 2.3 (range 1.6-4.7, N = 9 spp.)

Seed dispersal distance was measured from the two kinkajous monitored in 1991. Three out of the 16 defecations recorded occurred under the tree where the animal was feeding or under a fruiting tree from the same species. The other seed samples were dispersed on average at 200m (SD = 75, range 70-340 m) from the source tree. Germination trials pointed out that, according to the plant species, transit through the kinkajous' gut differently affected germination time and survival rate of seeds (Table 4).

Pourouma bicolor. The first germinations were recorded at 1.5 month. Germination rate at this time was higher in defecated seeds compared to whole fruits (X^2=41.59, 1df, $P<0.001$) and intact seeds ($X^2 = 27.45$, 1 df, $P < 0.001$), but was not different between whole fruits and intact seeds (Fisher exact probability test, 1 df, $P > 0.05$). Seedlings were not recorded before 9 months. Seedling survival rate at 9 months was lower in whole fruits compared to intact seeds ($\chi^2 = 9.27$, 1 df, $P < 0.01$) and defecated seeds (χ^2=10.92, 2 df, $P < 0.001$), but was not different between intact and defecated seeds (Fisher test, 2 df, $P > 0.05$).

TABLE 4. Results of germination trials according to species and seed treatment. Germination delay and delay of seedling appearance, with the corresponding rate and time (columns « first »). Maximum rate of germination and maximum number of seedlings with the corresponding time (columns 'max').

Species	Treatment	# seeds	germination				seedling			
			first		max		first		max	
			rate	time	rate	time	rate	time	rate	time
P. bicolor	W	29	0	-	0	-	48	9	48**	9
	I	28	14	1.5	14	1.5	89	9	89	9
	D	36	83	1.5	83***	1.5	89	9	89	9
V. michelii	I	19	0	-	0*	-	32	9	32	9
	C	15	27	2	27	2	47	9	47	9
	D	12	42	1	50	1.5	67	9	67*	9
V. surinamensis	I	25	0	-	0	-	88	9	88	9
	N	26	0	-	0	-	73	9	73	9
	D	11	0	-	0	-	100	9	100	9
I. huberi	I	12	58	0.5	58	0.5	25	0.5	92*	9
	C	12	100	0.5	100	0.5	75	1	75	1, 9
	N	12	33	0.5	33	0.5	33	0.5	50	9
	D	5	20	0.5	20	0.5	80	0.5	100	1, 9
C. obtusa	W	30	0	-	0	-	0	-	0	-
	I	30	0	-	0	-	83	0.5	87*	1
	D	30	0	-	0	-	57	1	57	1

Rates are in %, times are in months. Records have been performed at 0.5, 1, 1.5, 2 and 9 months after sowing, except in *P. bicolor* (no data at 2 months), *I. huberi* and *C. obtusa* (no data at 1.5 and 2 months). W = whole fruit ; I = intact seed ; D = defecated seed ; C = cleaned seed ; N = naked seed. Statistical difference: * $p < 0.05$, ** $p < 0.01$, *** $p < 0.001$ (Fisher exact probability test)

Virola michelii. Germination time for defecated seeds (0.5 to 1 month) was shorter than those recorded for cleaned seeds (1.5-2 months) and intact seeds (>2 months). Maximum germination rate was lower for intact seeds compared to cleaned seeds (Fisher test, 1df, $P < 0.03$) and defecated seeds (Fisher test, 1 df, $P < 0.002$), but was not different between cleaned and defecated seeds (Fisher test, 1 df, $P > 0.05$). As in *P. bicolor*, seedlings appeared between 2 and 9 months. Rate of seedling survival at 9 months was higher in defecated than in intact seeds ($X^2 = 5.95$, 1 df, $P < 0.02$), but not higher in cleaned *vs* intact seeds ($X^2 = 0.30$, 1 df, $P > 0.05$), nor in defecated *vs* cleaned seeds (Fisher test, 1 df, $P > 0.05$).

Virola surinamensis. Germination time was longer than 2 months in defecated, cleaned as well as intact seeds, and the rate of seedling survival at 9 months was not different between the three treatments ($X^2 = 4.67$, 3 df, $P > 0.05$).

Inga huberi. Germination delay was short, the maximum rate was recorded at 0.5 months. The first seedlings also were recorded at 0.5 month, except in cleaned seeds where they appeared at one month. Maximum seedling rate as survival rate of seedlings at 9 months was higher in intact seeds compared to naked seeds (Fisher test, 1 df, $P = 0.04$), but was not

different between the other samples (Fisher test, 1 df, P > 0.05 in each case).
Cecropia obtusa. Seedling appearance was faster for intact seeds (<0.5 month) compared to defecated seeds (0.5-1 month) and whole fruits (>1 month). Maximum seedling rate (1 month) was higher in intact seeds compared to defecated seeds (X^2=5.25, 1df, P<0.02), and in defecated seeds compared to whole fruits (X^2= 21.01, 1 df, P < 0.001). No germinations were visible, and no seedling was found alive after 9 months in any sample.

Sixty-four defecations were recorded during January-March 1993. Among these, the situation of 53 was determined. Twelve (23%) occurred under the source tree or another tree from the same species ('food tree species' case no dispersal), 4 (8%) under an adjacent tree ('neighbouring tree': short distance dispersal), 17 (32%) under a fruiting tree from a different species ('other fruiting species': large dispersal distance) and 20 (38%) at any place in the forest, during movements of the disperser ('understorey': large dispersal distance).

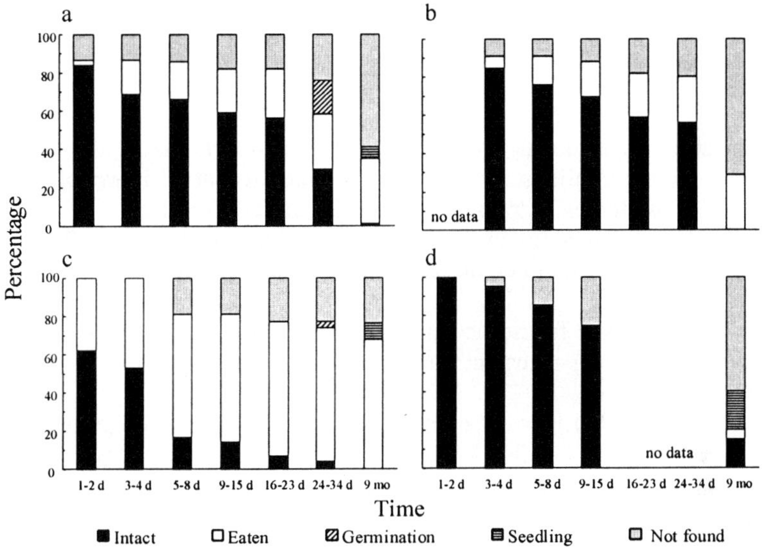

Figure 1. Fate of *Pourouma bicolor* seeds defecated *in situ* by kinkajous. Percent of seeds found intact, eaten by rodents, germinated, or established as seedlings, and seeds not found i.e. secondarily dispersed, from 1 to 34 days and after 9 months in four patterns of defecation: a = 'food tree species' (N = 100 seeds), b = 'neighbouring tree' (N = 34), c = 'other fruiting species' (N = 43), d = 'understorey' (N = 37; see text for the definitions of patterns).

Pourouma bicolor seeds were by far the more frequently defecated: seeds of this species were present in three-quarters of the 64 defecations recorded. This species also involved the higher number of visits to fruiting trees (Table 3). The fate of seeds defecated differed according to the situation (Fig. 1). The distribution of seeds was significantly different among the four situations at all time intervals between 3-4 and 16-23d (3-4d: χ^2=30.19; 5-8d: χ^2 = 44.24; 9-15 d: χ^2 = 42.21; 6ddl and P < 0.001 for each; 16-23 d: χ^2 = 26.66, 4 ddl;

24-34 d: $\chi^2 = 35.43$, 6 ddl; $P < 0.001$ in both cases). The sample 'same tree species' did not obviously differ from a uniform distribution until 16-23 d, where eaten seeds were less numerous and intact seeds more numerous than expected. At 24-34 d germinations were more numerous and eaten seeds again less numerous than expected. The overrepresentation of intact seeds and underrepresentation of eaten seeds constantly increased from 3-4 d until 24-34 d in the sample 'neighbouring tree', whereas the sample 'different species' displayed an opposite tendency. The 'understorey sample' had an increasing excess of intact seeds and deficit of eaten seeds from 3-4d to 9-15 d (next checks have not been available on this sample).

On a longer period of time (9 months), the distribution of seeds or seedlings again differed according to the situation ($\chi^2 = 25.45$, 6 df, $P < 0.001$). The proportions in the 'same tree species' and 'neighbouring tree' samples did not differ from a uniform distribution. Conversely, the 'different species' sample presented an excess of non-dispersed (intact or eaten) seeds and a deficit of dispersed seeds, whereas the 'understorey' sample presented a deficit of non-dispersed seeds and an excess of germinations or seedlings.

4. Discussion

Kinkajous appear as opportunistic frugivores and flower-eaters that feed on numerous plant species from numerous families. Fruits eaten belonged to varied morphological types. As revealed by previous studies (Atramentowicz 1988, Julien-Laferrière 1999), fruit selection by kinkajous, like some sympatric didelphid marsupials, is not influenced by fruit morphology, size, colour, number of seeds, or nutritional value, but depends mainly on their abundance and the accessibility of the edible part: the diet shifts seasonally, following the abundance in flowers or fruits; the fruits have soft, easy to open, husks, or are dehiscent, giving direct access to the part eaten; the part eaten, mesocarp or aril, is generally poor in nutrients but rich in water.

However, the part eaten is in all cases strongly tied to the seeds (Sabatier 1983). This morphological trait favours swallowing of seeds, and kinkajous swallowed the seeds for all but one plant species. However, kinkajous ate unripe fruits from a fairly large number of species, independently of the seasonal food supply. This impact, comparable to that of seed predators, was never observed before in this species (Julien-Laferrière 1993, 1999). Thus, 79% of plant species used for fruits by kinkajous (N = 43) were actually dispersed. By comparison, on the same site, 89% of the species used for fruits by howling monkeys *Alouatta seniculus* (N = 97) were dispersed (Julliot 1996), and 81% (N = 135) by capuchin monkeys *Cebus apella* (Zhang 1994).

More than one-third of kinkajou defecations occurred in the understorey. In *P. bicolor,* the most heavily species used for fruits, the lack of dispersal does not seem detrimental to the seed survival on the short term (until one month): predation was lower on seeds deposited under the fruiting tree or under trees from the same species, and germinations were relatively numerous. Predation also was lower on seeds dispersed at short distance, as on seeds deposited on the understorey. Conversely, seeds deposited under other fruiting trees suffered higher predation. However, seed post-dispersal intensity was not significantly different among the four situations. This suggests that predator satiation (Janzen 1971), but not secondary disperser satiation (Forget 1992, 1993) occurred. On a longer term (9 months), *P. bicolor* seedlings were more numerous and eaten seeds less numerous only for seeds dispersed in the understory. In *I. huberi*, 5 out of 10 seeds dispersed in the understorey had given viable seedlings after 9 months, opposed to 1 out of 7 seeds dispersed under another fruiting tree.

These results suggest that the pattern of long-distance undirected dispersal is advantageous as a way for seeds or seedlings to escape predation by rodents or competition with other seedlings, as postulated by the 'escape hypothesis' (Janzen 1970, Howe & Smallwood 1982). Transit by the gut or cleaning increased the germination rate of the seeds in two of the five species tested (*P. bicolor*, *V. michelii*), had no influence in two others (*V. surinamensis*, *I. huberi*), and decreased the germination rate in the last one (*C. obtusa*). Transit by the gut reduced germination time in one species (*V. michelii*), had no obvious effect in three others (*P. bicolor*, *V. surinamensis*, *I. huberi*), and increased germination time, compared to intact seeds, in another (*C. obtusa*). Seedling survival was enhanced by gut transit in two species (*P. bicolor*, *V. michelii*), was not influenced in two others (*V. surinamensis*, *C. obtusa*), and was reduced, compared to intact seeds, in another (*I. huberi*). The varying influence of gut transit of seeds on their subsequent germination, according to the plant species and the dispersing animal, seems the rule in plant-frugivore interactions (*e.g.* Hladik & Hladik 1969, Lieberman & Lieberman 1986, Julliot 1996). The terms of the disperser-dispersed interaction may be very different between plants that are closely related phylogenetically but have different ecological constraints (Barnea *et al.* 1990). Positive, neutral, or negative influence of seed transit on germination depends on the compatibility between the digestive physiology of the consumer and the physiology of seed germination (Vasquez-Yanes & Orozco-Segovia 1984). For the plant, a possibly detrimental effect of seed gut transit may be negligible compared to the decisive advantage of dispersal.

Acknowledgments

This study was supported by the Centre National de la Recherche Scientifique (URA 1183, presently UMR 8571). I also have to thank especially the Centre IRD of Cayenne for their logistic help in the field, D.Sabatier who determined most of the vegetable samples, P. van der Meer who recorded a part of the germination results, D. Hammond, J. den Ouden and P.-M. Forget who usefully commented on earlier drafts of this paper.

Toon Rijkers discusses the uses of forest fruits with Desmo and Wemo Betian. Nouragues, 1994. *Photo Gerrie Offerman.*

CHAPTER 21

FRUGIVORY AND SEED DISPERSAL BY TERRESTRIAL MAMMALS

François Feer, Olivier Henry, Pierre-Michel Forget and Marc Gayot

1. Introduction

Fruit is a major food resource for vertebrates inhabiting Neotropical rainforests (Terborgh 1986, Dubost 1987, Janson & Emmons 1990). Once on the ground after natural fall, or having been dropped by flying or arboreal vertebrates beneath fruiting trees as well as away, fruits and seeds are also dispersed and eaten by ground-dwelling vertebrates. Terrestrial mammalian frugivores comprise some of the most species-rich vertebrate groups (e.g. rodents) and one of the most important groups in term of biomass (e.g. ungulates) in this bioregion. These mammalian groups contain many species typical of the tropical lowland forest of South America.

Different levels of frugivory are related to dietary needs, body size, specific habitat and resource availability. This paper reviews published studies on the diet composition of some of the commonest mammalian species exploiting the fruit production in French Guiana. When data are available we analyse the dietary responses to temporal variations in fruit abundance, which is the resource factor most likely to affect the feeding strategies of frugivores. The terrestrial frugivores play different roles toward seed fate according to their feeding behaviour or masticatory abilities. Ingestion of intact seeds and granivory associated with caching behaviour promotes seed dispersal and may influence early recruitment of seedlings. The objective of this paper is therefore to highlight the role of different mammal species of French Guiana as seed-dispersal agents. Morphological characteristics of seeds were considered when they are likely to affect the role of terrestrial mammals as seed predator or disperser.

The diet data of small rodents is based on samples collected at Saut Pararé for *Neacomys guianae* (Guillotin 1981), and at Paracou field station for *Oryzomys capito* and *Proechimys cuvieri* (Henry 1994). The samples of larger species like *Dasyprocta leporina*, *Tayassu tajacu*, *Mazama gouazoubira*, *M. americana* and *Tapirus terrestris* were collected throughout French Guiana (Gayot 2000, Henry 1994, Henry *et al.* 2000). Stomach contents were washed and filtered through sieves of different mesh sizes: 1 mm^2 for small rodents, 1 and 5 mm^2 for agouti, collared peccary and deers, and 0.4 and 36 mm^2 for lowland tapir. For small rodents, all the contents were analysed and categorised as seeds, pulp, animal matter and leaves and fibre. For the larger species, all large particles were separated into the different food types. The rest was mixed and homogenised, then 100 g for tapir and 10% for other species were randomly removed and categorised. Each food category was dried

Bongers *et al.* (eds.), Nouragues. Dynamics and plant-animal interactions in a neotropical rainforest, 227–232.

and weighed and the proportion extrapolated to the entire sample of small particles. The total percentage of the different food types was completed by adding the corresponding analysis of the small particles and large particles. Fruit and seed samples were measured and identified by D. Sabatier when it was possible. No data from French Guiana were available on the species of carnivores that are partly frugivorous.

2. Diversity of frugivory

The fruit consumption by terrestrial mammals of the neotropical forest varies between the studied taxonomic groups (Table 1). This variation is related to differences in body weight, metabolic requirements, digestive systems, foraging behaviour and the availability of fruits on the forest floor (Henry 1997). In French Guiana, the rate of fruit consumption increases with body weight amongst rodents from the small murid species to the large dasyproctid species and echimyid. These species balanced their diet by ingestion of arthropods (Guillotin 1981, 1982, Henry 1997). *Neacomys guianae*, the smallest species studied, consume insects in greater proportion than fruits whereas *O. capito* consume these two food categories in similar proportions. In French Guiana, the abundant and large spiny rats (*Proechimys* spp.) are the most frugivorous, as in Peru and Central America (Smythe 1970, Emmons 1982, Adler 1995). Because small rodents have small absolute trophic needs but relative high-energy requirements, they should focus on insect prey, which are rich in energy and protein. Henry (1997) showed that pregnant females of*O. capito* and *P. cuvieri* which require the highest quality food are the most insectivorous compared to males and nonpregnant females.

TABLE 1. Diet composition of 8 species (Nc: *Neacomys guianae*; Oc :*Oryzomys capito*; Pc: *Proechimys cuvieri*; Dl: *Dasyprocta leporina*; Mg: *Mazama gouazoubira*; Tj: *Tayassu tajacu*; Ma : *Mazama americana*; Tt: *Tapirus terrestris*) of frugivores from French Guiana in percent of dry weight in the total sample of stomach content.

	Species							
	Ng [a]	*Oc* [b]	*Pc* [b]	*Dl* [b]	*Mg* [c]	*Tt* [b]	*Ma* [c]	*Tt* [d]
Body weight	13-22 g	20-50 g	200-500 g	4-6 kg	11-18 kg	18-26 kg	20-48 kg	225-300 kg
Sample size	N = 10	N = 36	N = 36	N = 36	N = 34	N = 36	N = 28	N = 27
Diet items								
Seed		7.0	13.0	46.7		44.5		17.6
Pulp		40.5	53.2	37.3		17.1		10.8
Total fruit	35.0	47.5	66.2	84.0	68.3	61.6	56.0	28.4
Leaves/fiber	6.5	0.7	2.2	7.1	25.0	28.4	36.5	71.6
Animal matter	58.5	51.8	31.6	8.9	0.3	10.0	0.5	0.0

[a] after Guillotin 1981, [b] after Henry 1994, [c] after Gayot 2000 consumption of flowers not included,[d] after Henry *et al.* 2000

Agoutis consume much less animal matter than smaller rodents and more fruit than any of other studied ground dwelling vertebrates in French Guiana (84% of dry weight). Unlike small rodents, agouti harvest more seeds than pulp. These differences are related to its high absolute energy demand and better ability to forage on large seeds, which often have a thick, hard endocarp. Observations of acouchi in French Guiana let assume that its diet is very similar to agoutis. The diet of the paca (*Agouti paca*) is not documented in French

Guiana; however it is considered as less of a frugivore and granivore than the agouti in Panama (Smythe *et al.* 1982). Thus it may be less affected by fruit shortages. Nevertheless Beck-King *et al.* (1999) recently showed in a Costa Rican secondary forest that its diet mainly comprised fruits and large seeds.

The diet of collared peccary (*T. tajacu*) in French Guiana is dominated by fruits, followed in importance by leaves, fibers and animal matter. A similar diet profile was observed for this species in Peru (Bodmer 1989) whereas white-lipped peccaries (*T. pecari*) consume more fruits than does the collared peccary (Bodmer 1989, Kiltie 1981b) in relation to the exploitation of flooded forest (Bodmer 1990b). Like agouti, the collared peccary of French Guiana consume more seed than pulp. This dietary trait should be related to nutritional requirements as it is the most accentuated among pregnant females (Henry 1997).

The diet of red and grey brocket deers collected in French Guiana showed that fruit (pulp and especially seed) constitute the major part of their diet. The rest of their diet is essentially composed of leaves, flowers and occasionally fungi (Gayot 2000). Similarly, in Peru, the diet of brocket deers consist of more than 80% of fruits (Bodmer 1989). In Surinam, reproductive plant parts accounted for over 50% of rumen volume of red brocket (though the sample was collected outside the peak fruitfall period; Branan *et al.* 1985).

The lowland tapir contrasts with other ungulates of French Guiana by consuming leaf and fiber in greater proportion than fruit. Nevertheless fruit parts were rarely absent and reached a maximum proportion of 70% of dry weight. A similar proportion of fruit was observed in the diet of lowland tapir in Peru (Bodmer 1990a). In Brazil, Fragoso (1997) signalled that tapirs consumed many fruits. A lower consumption of fruit was reported for Baird's tapir (*T. bairdii*) in Costa Rica (C. Foerster, unpublished data) and in Panama (Terwilliger 1978) which may be associated to the richness of undergrowth vegetation in these sites of Central America. Conversely, the presence of large monotypic clumps of fruiting palms (e.g. *Mauritia flexuosa* in Peru, Bodmer 1990a; *Maximiliana maripa* in Brazil, Fragoso 1997) may be proposed to explain the relatively high frugivory for a large non-ruminant ungulate. Nonetheless, this habitat heterogeneity seems not to be present in the more homogenous forest of French Guiana where the tapir consumed a high diversity of fruit.

3. Seasonal variation in diet

Fruit production in the forest of French Guiana reaches a maximum between February and May and influences the seasonal diet of terrestrial frugivores. Henry (1997) showed that for small rodents the maximum intake of fruit occurred during the fruit-rich season. During this period the consumption of pulp was especially high. Like *O. capito*, *P. cuvieri* ate significantly more insects than seeds during the fruit-poor season. Agouti appeared also to be opportunistic, harvesting more pulp than seeds when ripe, pulpy fruits were plentiful. During the season of fruit shortage (June-September), the consumption of seeds for this species was maximal, what could be related to exploitation of buried large seeds (Forget 1990, Henry 1999). Seed storage by caching may also be seasonally important for acouchi (*Myoprocta acouchi*) whose diet is probably similar to that of agoutis. Smythe *et al.* (1982) showed the role of seasonal contribution of seed reserves for large rodents in Panama.

Throughout the annual cycle, collared peccary appears truly granivorous, eating more seeds than pulp, maximum consumption of seeds occurs during the season of fruit scarcity.

During this period peccary may most likely search for buried seeds including the seeds stored by scatterhoarding rodents (see Kiltie & Terborgh 1983) as was observed for white-lipped peccaries (Kiltie 1981a). Both deer species presented a maximum consumption of fruit during the fruit-rich season but the proportions of seeds did not significantly change (Gayot 2000).

The fruit consumption of the lowland tapir in French Guiana tracked fruit availability as with Baird's tapir in Costa Rica (C. Foerster, unpublished data.). In Peru Bodmer (1990b) did not find seasonal variation in diet of the lowland tapir, what he related to exploitation of extended exploitable flooded areas during dry season, a temporary habitat absent from French Guiana.

4. Seed dispersal

The fate of seeds harvested by terrestrial mammals depends primarily on their food handling behaviour and their masticatory and digestive systems. Rodents handle fruits, clean of pulp and tegument, and gnaw or cache seeds using their forepaws (Smythe 1978). Ungulates swallow, chew fruits and soft seeds, eventually regurgitating (ruminants) or defecating intact seeds (Janzen 1983). Both groups of animals participate at different levels to seed dispersal but their role as seed predators must be taken into account in defining their potential effect on plant recruitment.

4.1. SCATTERHOARDING BY RODENTS

Food hoarding has been particularly investigated for tropical species in the last decade (Forget 1990, 1991, 1996, Hallwachs 1994). Two types of hoarding behaviour are defined, each with contrasting consequences for plant recruitment. On the one hand, Myomorph rodents generally concentrate (larder hoarding) food items in a small part of their home range, such as in their nesting burrow. In contrast Dasyproctids and Sciurids scatter (scatterhoarding) food resources through their individual's home range (Morris 1962). In the latter rodent species, a certain fraction of the cached seeds escape post-dispersal predation (Janzen 1970) suggesting that rodents are potential seed dispersers of plant species upon which they rely for food (Price & Jenkins 1986, Vander Wall 1990, see Jansen & Forget this volume).

Small rodents. Spiny rats carry and larderhoard seeds into their burrow where they presumably are either consumed or placed in a situation unfavourable for germination (Guillotin 1982, Forget 1991). Forget (1991) first documented scatterhoarding of seeds from *Astrocaryum paramaca* (Arecaceae) by spiny rats in French Guiana, demonstrating that this species primarily functions as seeds predator but also functions as seed disperser. Evidence of seed dispersal was also documented for *P. semispinosus* in Panama (Adler & Kestell 1998). In this respect, terrestrial Echimyidae may be similar to Heteromyidae (*Heteromys* spp.) (Brewer & Rejmánek 1999).

Experiments were conducted at Nouragues using thread-marked seeds of *Chrysophyllum lucentifolium* (Sapotaceae), *Manilkara huberi* (Sapotaceae), and *Osteophloeum platyspermum* (Myristicaceae), following Forget (1990). Ten to 15 seeds were placed on two kind of plots on each of 5-19 sites. One plot was covered by an exclosure allowing only small rodents to enter. The second plot was not protected, open to all terrestrial vertebrates. Seed fate was examined after 20 days and cached seeds were searched inside a circle of 5m radius. Thread marked Sapotaceae seeds were occasionally retrieved in holes of spiny rats. Experiments showed that a majority of seeds of *C. lucentifolium* were eaten but 5% were cached under the litter at a distance not exceeding 5 m, presumably by spiny rat. Caching rates of 8 and 47% were recorded in the same conditions for *M. huberi* and *O. platyspermum*, respectively.

Large rodents. Following Sabatier's (1983) observations on the hoarding behaviour of captive acouchi, Forget (1990, 1996) analysed the fate of seeds of *Vouacapoua americana* (Caesalpiniaceae) and *Carapa procera* (Meliaceae), two abundant tree species at Nouragues forest (see also Jansen & Forget this volume). During the 1987 mast fruiting of *V. americana*, using thread-marked seeds, Forget (1990) showed that almost 100% of *V. americana* seeds were removed from three experimental sites, nearly 70% of them being buried by rodents and the rest being eaten by mammals in the late wet season. Similarly, approximately 75-100% removal of *C. procera* seeds was associated with scatterhoarding in Paracou (Forget 1996). Despite retrieval by caching rodents and pilferage by other mammals, many seeds escape consumption and establish seedlings, explaining why there is a dense seedling distribution around parent trees (see for an example Forget 1994) or around defecation sites of howler monkeys (Julliot 1997).

Sabatier (1983) listed 22 species of fruits and seeds consumed by agouti or acouchi including the above cited large seed species, the small seeds of *Virola kwatae* (Myristicaceae) and these of *Tetragastris panamensis* (Burseraceae). While the role of agouti-acouchi as seed dispersers of *Carapa* spp. or *V. americana* is clear (Sabatier 1983, Forget 1990, 1996, 1997), it remains less important for the smaller seeded plant species in French Guiana. Overall caching rates by rodents observed on open plots at Nouragues were 72% for *O. platyspermum* (weight 3.5 g), 18% for *M. huberi* (1.6 g) and 13% for *C. lucentifolium* (1.4 g). A similar increase of caching rate with seed weight was observed in Panama (Forget *et al.* 1998) and was experimentally demonstrated in Costa Rica (Hallwachs 1994).

4.2 SPIT DISPERSAL BY UNGULATES

The presence of *Astrocaryum sciophyllum* and *Mauritia flexuosa* pulp but not the seed in lowland tapir stomach contents from French Guiana (Henry *et al.* 2000) suggested that seeds were spat out during mastication. Captive collared peccaries were observed rejecting the seeds of Arecaceae they cannot break with molars. The seed spitting behaviour was also recorded with peccaries and lowland tapir in Peru (Bodmer 1991). Intact seeds of *Spondias mombin* (Anacardiaceae) were recorded in rumens of brocket deers of French Guiana. According to the observations of seeds of *S. mombin* regurgitated during rumination by white-tailed deer (*Odocoileus virginianus*) (Janzen 1985), it may be assumed that this plant species, as well as other hard seeded species (e.g. Anacardiaceae, Combretaceae, Malpighiaceae, Sapotaceae, etc.), are potentially dispersed by brocket deers.

4.3 GUT DISPERSAL

Intact seeds in abomasum contents (i.e. ruminated food) of red and gray brocket deers and in stomachs of peccaries are few and small (Bodmer 1989, 1991, Gayot 2000, Henry 1997) whereas they frequently occurred in stomach, cecal and fecal samples of tapirs (Bodmer 1991, Janzen 1982, Fragoso 1997, Rodrigues *et al.* 1993). For instance, in French Guiana, intact seeds occurred in 81% (N = 27) of the stomachs of lowland tapirs (Henry *et al.*

2000). A majority of intact seed speoies (67% of 36) were smaller than 1 cm in diameter (of short axis for elongated seeds) but the maximum size reached 2.5 cm (*Elephantomene eburnea*, Menispermaceae). Most of these seeds (81%) are smooth and not protected by a lignified endocarp. The most frequently encountered species came from fleshy pulped fruits, such as *S. mombin*, *Bagassa guianensis* (Moraceae) and the largest, *Landolphia guianensis* (Apocynaceae) (6.0-8.5 cm diameter). Data from French Guiana showed a higher occurrence of intact seeds (60% of 68 seed samples) and smaller seeds than samples from Peru (Bodmer 1989) which may correspond to differences in habitat (see Hammond & Brown 1995).

Lowland tapirs may be considered a good potential dispersal agent for a wide array of seeds although its powerful jaws suggest it may also be a seed predator (Janzen 1982, Janzen & Martin 1982, F. Feer unpublished data). Tapir ingested seeds may be sufficiently scarified to begin germination into the gut and to be killed by digestion (Janzen & Martin 1982, Janzen 1981). Additionally tapirs prefer to defecate in water which may not facilitate establishment unless the seeds are adapted to inundated habitats (e.g. *S. mombin*, *Saccoglotis* sp., Humiriaceae). At Nouragues area, many tapir faeces are also deposited in permanent dry sites (see also Fragoso 1997).

5. Conclusion

The dispersal agents of a tree species generate an initial seed shadow which is commonly reshaped by post-dispersal seed mortality or secondary dispersal. The final seed distribution generates the seedling shadow, an important starting point in the plant recruitment, hence the necessity to understand the overall action patterns of the terrestrial frugivores and granivores. Rodents are known to consume or disperse the seeds contained in primate dung (Estrada & Coates-Estrada 1991, F. Feer unpublished data) and also those in large herbivore dung (Janzen 1982). This was also observed at the Nouragues site, with spiny rat, acouchy and agouti removing a variety of seeds (e.g. Burseraceae, Moraceae, Myristicaceae, Sapotaceae) defecated by howler monkeys (P.M. Forget unpublished data, F. Feer unpublished data) and it also may occur in the same manner with tapir dung (Fragoso 1997). Rodents interact during a short time with large dung beetles burying or relocating seeds from primate or tapir dung (Estrada & Coates-Estrada 1991, Andresen 1999, Feer 1999). Competing terrestrial granivores are involved in the fate of cached seeds. Dasyproctid rodents are likely to exploit the superficial caches of spiny rats, and peccaries are known to search the forest soil for buried seeds (Kiltie 1981a). Ecological interactions implying many partners such as seed eating insects, scatterhoarding rodents and tapirs are often involved in the recruitment pattern of seedlings (see Fragoso 1997). The result is that seed dispersal processes take place on a more complex spatial and temporal scales than previously described. The study of interaction of small and large-scale dispersers with predators open large research perspectives on key processes involved in forest structure.

CHAPTER 22

VEGETARIAN SPECIES IN THE BIRD COMMUNITY WITH AN EMPHASIS ON FRUGIVORY AND SEED DISPERSAL

Christian Erard and Marc Théry

1. Introduction

It is quite a common place to say that Neotropical forest bird communities are among if not the richest ones in the world. Though recognised since a long time (MacArthur*et al.* 1966, Pianka 1966, Orians 1969), and as well as global patterns of diversity, their high species diversity remains the subject of many debates (Thiollay 1991, and papers in Keast 1990 and Ricklefs & Schluter 1993). Guianese forest bird communities have already been presented and discussed (Erard 1986, 1989; Thiollay 1986, 1988, 1994; Tostain *et al.* 1992; Thiollay *et al.* Chapter 12); though still somewhat preliminary, these studies constitute good bases for future research. Compared to those found in African tropical forests and particularly to those of temperate forests, these Guianese bird communities include a high proportion of species feeding on vegetable material, particularly on fruits (Erard 1989) though temperate bird communities may display in autumn and winter an important increase of vegetarian species. An old-world observer is also struck by the high diversity of fruiting plants, a number of them (e.g. palms, figs, Melastomataceae, Rubiaceae) being key species as food resources for birds, not to say vertebrates.

An increasing literature stresses the importance of plant-bird interactions in the functioning of tropical forests, particularly in the forest dynamics through pollination and seed dispersal mechanisms (e.g. Stiles 1985, Howe & Westley 1988, Leigh 1999). Vegetarian species are also fundamentally interesting for their social behaviour, particularly for the link between their social organisation systems and spatio-temporal patterns of plant-resources distribution.

Rather than an overview of the problems posed by frugivory and plant-bird interactions and thus duplicate at least large parts of other texts (Snow & Snow 1988, Jordano 1992, Fleming & Estrada 1993, Levey *et al.* 1994), we opted for this chapter to expand the analysis of the bird-list of the Nouragues area (see Thiollay*et al.* Chapter 12 and Appendix 4), focusing on frugivores, nectarivores and omnivores in order to examine what kind of patterns can be detected in the structuration of the avian community in relation to these particular food categories. We will also make a general presentation of the main results of the studies conducted in the area on the frugivorous birds (Erard & Sabatier 1989, Erard*et al.* 1989, 1991; Erard & Théry 1994, Krijger *et al.* 1997, Sabatier 1983, Théry 1990a,b,c,

Bongers *et al.* (eds.), Nouragues. Dynamics and plant-animal interactions in a neotropical rainforest, 233–241.

1992, 1997; Théry *et al.* 1992, Théry & Larpin 1993, Théry & Vehrencamp 1995, Endler & Théry 1996) and suggest some avenues for future research.

2. Site and methods

We deal here with the avifauna of the lowland rainforest extending from the inselbergs of the Nouragues to the Arataye river, a tributary of the Approuague river, where is installed a camp for field studies at Saut Pararé and another one downstreams where programmes on canopy study will develop. Summing the most intensely surveyed areas at the three sites, the study area extends roughly over 30 km².

This forest area is a mosaic of various habitat types. Those found around the field station (Thiollay 1994, Jullien 1997) and important for birds consist mainly in (1) high mature forest with high closed canopy and open understorey, (2) lower, denser and more irregular forest, (3) low and dense stands in naturally disturbed areas with an extreme profusion of vine tangles, (4) patches of swamp palm forest and seasonally flooded banks along streams, (5) dry and dense forest rich in Myrtaceae and Clusiaceae on the lower slope of rocky outcrops, (6) top of inselbergs with bare rocks, grasses, clumps of shrubs and treelets. They are found everywhere between Nouragues and Arataye but their proportions in the forest mosaic are not constant and still unmeasured at this scale. Likewise, both small and large treefall gaps of various ages exist in varied numbers and proportions in every habitat and are a very important component of the spatial heterogeneity. Furthermore, in this area some openings occur (locally called 'cambrouzes') with very dense bamboo-like vegetation, their presence being most probably at least in part due to ancient human activities though local effects of soil peculiarities cannot always be excluded. Along the river, riverine and naturally disturbed forests exist as well as secondary forests due to old human settlements at Saut Pararé (for details on this site, see Sabatier 1983, Maury-Lechon & Poncy 1986).

All bird species recorded in this area are listed in the Appendix 4 which summarises data on their body mass, abundance, foraging location and behaviour, diet, social organisation and habitat, i.e. a set of information that helps the reader to get a broad but rather coarse picture of the niche of every bird species. This information is still preliminary for many species. Abundances were evaluated only in the forest mosaic of the area gridded with permanent 100 m x 100 m quadrats at Nouragues site. An extrapolation of the densities to the entire area may underestimate some species particularly because these densities are not the number of pairs per sample area of accurate specific habitat but the number of pairs per area of pooled habitats in the 100-ha sample of the Nouragues forest mosaic. Furthermore, the term pairs refers in reality to social units, so the abundance index does not reflect the real number of individuals actually present on the site. Nevertheless it is quite useful and better than 'common', 'fairly common' or 'rare'. Various observers gathered these data on the birds. The observations were standardised only after 1990. On the quadrats, the classical spot-mapping method was used, proving quite satisfactory for territorial species (see also Terborgh *et al.* 1990). Mist-netting produced also many data on secretive species and allowed morphometric measurements as well as capture-recapture data on ringed birds. Colour ringing and radio-tracking were used to follow individuals of certain species in space and time (Erard *et al.* 1989, Théry 1990a, c, 1992, 1997, and many still unpublished data by various observers).

Diets were studied mainly by visual observations of feeding birds at fruiting trees and also of birds followed during their foraging movements or feeding activities at nest. Nevertheless, data were also obtained in nearby sites from digestive tract analyses of game birds hunted by local people (Erard *et al.* 1991, Théry *et al.* 1992).

It is not always easy to classify birds according to their diet because the reality is that of a continuum from completely animalivorous to completely vegetarian species. Things are also complicated by the fact that qualitative and quantitative analyses do not necessarily lead to the same conclusions. For instance the trumpeter *Psophia crepitans* might be qualified as an omnivore by visual observation of foraging troops but is clearly a frugivore when digestive tracts are examined (Erard *et al.* 1991, see also Erard *et al.* 1989 for other differences between qualitative and quantitative data). Furthermore and probably for many species seasonal diet variations are important in relation to food availability and breeding cycles.

With all these limitations in mind we considered to be vegetarian a species which was assumed to base its diet at least for two thirds on vegetable matter. Likewise an animalivore is a bird with at the very most one third of its diet consisting of vegetable matter. Species that base between one third and two thirds of their diet on vegetable matter were called omnivores. These proportions were similarly used to define frugivores, nectarivores and insectivores (arthropod- or invertebrate-eaters would be better). Among these animalivores, particularly among the insectivores, we took count for the present analysis of those species that occasionally include a significant proportion of vegetable matter in their diet. The number of these occasional frugivores is most probably still underestimated; further field observations of foraging birds will almost certainly add a lot of other species to the present list. In reality most nectarivores do not strictly rely on nectar for feeding, more or less regularly their diet includes small insects but the exact proportions of these are not known. Some Passerines (in particular Emberizidae, e.g. *Coereba, Cyanerpes, Chlorospiza*) are known to eat small arthropods and are also partly nectarivores but as the real proportion of nectar in their diet is still undetermined, and as they have been observed ingesting also small fruits and flowers, they have been classified with the omnivores.

3. Species richness and taxonomic composition

Overall, 433 species make up the Nouragues avifauna (see Appendix 4), among them 73 are true frugivores, 63 qualify as omnivores and 23 are nectarivores. Among the 222 insectivores, 29 species are occasional frugivores. Thus we can consider for this chapter that 188 species ingest vegetable matter, particularly fruits.

Table 1 gives the taxonomic composition of the vegetarian part of the avifauna. It is worthwhile to note that half of the bird families and almost half of the genera present in the Nouragues area eat fruits. Though only 38% of the total number of species present in the bird fauna (43% when nectarivores are included) eat fruits, these proportions stress once more the importance of fruits in the structuration and dynamics of tropical forest bird communities.

This is true for both Passerines and Nonpasserines, and points to a long evolutionary history of these vegetarian communities. Nevertheless it is interesting to note that a larger proportion of Nonpasserines are more strictly frugivores than Passerines: Nonpasserine frugivores account for 23.3% of all Nonpasserine species and for 56.8% of vegetarian Nonpasserines whereas these figures are respectively 12.3% and 27.2% for Passerines.

Indeed there are a lot of fruit-specialist Nonpasserine families, i.e. Tinamidae, Cracidae, Psophiidae, Columbidae, Psittacidae and Ramphastidae whereas in the Passerines we found only the sub-oscines Cotingidae and Pipridae.

TABLE 1. Taxonomic composition of the vegetarian part of the Nouragues avifauna. Particularly in Passerines, the same family or genus can include species of different diet categories and be listed several times accordingly.

		Frugivore	Omnivore	Nectarivore	Insectivore-frugivore	Total	Avifauna
Non-	Families	8	4	1	1	14	33
Passerines	Genera	25	4	18	1	48	116
	Species	42	8	23	1	74	180
Passerines	Families	3	9	-	5	11	18
	Genera	23	37	-	23	79	159
	Species	31	55	-	28	114	253
Total	Families	11	13	1	6	25	51
	Genera	48	41	18	24	127	275
	Species	73	63	23	29	188	433

Table 2 accounts for the distribution of vegetarian bird species according to their abundance index. These 188 vegetarian species do not statistically differ in abundance from the 245 non vegetarian species ($\chi^2 = 5.52$, df = 2, P > 0.05). Nevertheless the 74 vegetarian Nonpasserine species tend to be statistically more abundant than the 106 non vegetarian ones (a 2 x 2 contingency table for indices 1 *vs.* 2 + 3 gives ($\chi^2 = 24.8$, df = 1, P < 0.001), but there is no difference in abundance between vegetarian and non vegetarian Passerines.

TABLE 2. Distribution of the vegetarian bird species according to their abundance index in the Nouragues area. Each entry gives the total number of species and, in parentheses, numbers respectively of Nonpasserines + Passerines. Total = total number of vegetarian species; Avifauna = total number of species in the bird community. Insectivores are occasionally frugivorous invertebrate-eaters

	< 1 pair / km²	1-9 pairs / km²	> 10 pairs / km²
Frugivores	33 (20+13)	37 (21+16)	3 (1+2)
Omnivores	31 (2+29)	27 (5+22)	5 (1+4)
Nectarivores	8 (8+0)	12 (12+0)	3 (3+0)
Insectivores	17 (1+16)	12 (0+12)	-
Total	89 (31+58)	88 (38+50)	11 (5+6)
Avifauna	233 (115+118)	177 (59+118)	23 (6+17)

Since Nonpasserines are larger than Passerines, and since birds are gape-limited for fruit-size selection (Wheelwright 1985), this suggests that there would be more large fruits either in species number or available quantities or both than small ones and that they constitute a food resource sufficiently abundant and available throughout the year to sustain important populations of specialised Nonpasserine frugivores. However, comparisons of strict frugivores *vs.* other vegetarian species and of strict frugivores + omnivores *vs.* other vegetarian species do not reveal any statistically significant difference.

This pattern would thence simply imply that Nonpasserines have more greatly benefited than Passerines from tropical forest plant resources during evolutionary times and that coevolutionary or at least mutualistic patterns could be more easily detected among Nonpasserines than among Passerines. This suggests also that Nonpasserine selection pressure on fruits would be stronger than Passerine one and points to the need for more phylogenetically-based comparisons when looking for coevolutionary patterns between plants and their fruit-consumers.

Table 3 details the distribution of the vegetarian species in the various habitats found in the Nouragues area (see Appendix 4). As well as totally (χ^2 =15.44, df=6, P<0.02) as separately for Nonpasserines (χ^2 = 16.20, df = 4, P < 0.01) and Passerines (χ^2 =12.19, df = 5, P < 0.05) the distribution of the vegetarian species is statistically different from that of non vegetarian species.

TABLE 3. Distribution of the vegetarian bird species among the various habitats in the Nouragues area. Total = total number of vegetarian species; Avifauna = total number of species in the bird community. I = insectivores (invertebrate-eaters) that are occasional frugivores.

	Primary forest	Gap	Disturbed forest	Stream, swamp	Large opening	River	Inselberg	Total
Frugivores	52	1	8	1	7	2	2	73
Omnivores	37	4	2	2	15	3	-	63
Nectarivores	11	2	1	3	4	2	-	23
Insectivores	9	1	5	3	6	3	2	29
Total	109	8	16	9	32	10	4	188
Avifauna	228	21	41	29	58	42	14	433

Vegetarian species are relatively less numerous in swamp forest and along forest swamps and rivers than non vegetarian species (10.1% of vegetarian species *vs.* 20.7% of non vegetarian ones found in these habitats). Also, vegetarian species tend to be a little more diverse in mature forest (58.0%) and openings (17.0%) than non vegetarian ones (respectively 48.6% and 10.6%).

These patterns again point to the importance of fruits in the evolution of tropical forest Nonpasserines. They also indicate that swamp forests and river banks are less favourable for vegetarian species than primary forest. Openings with their grasses and shrubs producing small fruits and seeds are much more suitable as a bird habitat for Passerine than Nonpasserine vegetarian species.

Nevertheless here also a difference between Nonpasserines and Passerines can be detected. Vegetarian Nonpasserines have proportionately more species in mature forest (67.6 *vs.* 32.8%) and fewer in swamp forest and along rivers (9.5 *vs.* 15.5%) and in openings (8.1 *vs.* 8.9%) than non vegetarian Nonpasserines. On the other hand, vegetarian Passerines relative richness is more important than non vegetarian one in openings (22.8*vs.* 3.9%) but is almost equal in swamp forest and along forest streams and rivers (10.5 *vs.* 9.5%). Species numbers are however proportionately similar in mature primary forest for both categories.

When abundance index and habitat (Table 4) are crossed, it appears that vegetarian species have higher abundance indices in mature primary forest than their non vegetarian counterparts; this is particularly true for Nonpasserines (χ^2= 14.28, df=1, P < 0.001). That forest swamps, riversides and inselbergs are bad habitats for vegetarian birds is confirmed.

TABLE 4. Distribution of bird species in the different habitats of the Nouragues area according to their abundance index.

Habitat	Vegetarian species			Whole avifauna		
	<1 pair/km²	1-9 pairs/km²	>10 pairs/km²	<1 pair/km²	1-9 pairs/km²	>10 pairs/km²
Primary forest	25	73	11	72	137	19
Gap	4	4	-	10	8	3
Disturbed forest	12	4	-	26	15	-
Stream, swamp	5	4	-	19	10	-
Large opening	30	2	-	56	2	-
River	9	1	-	39	3	-
Inselberg	4	-	-	11	2	1
Total	89	88	11	233	177	23

It is worth to look at the foraging location of the vegetarian species in mature primary forest and contrast it to that of non vegetarian species (Table 5). Globally vegetarian species show statistically significant differences in their locations relatively to non vegetarian species (χ^2 = 66.34; df = 5; P < 0.001), being evidently non aerial, not specialised on the exploitation of trunks and epiphytes, and forage less by pouncing from a perch though 'gulpers' (*sensu* Levey *et al.* 1994) like trogons, cotingids and manakins, as well as occasional frugivorous Passerine insectivores can snatch fruits in flight either directly or while hovering. Furthermore vegetarian species forage proportionately much more in the forest canopy (where fruits are concentrated) than non vegetarian ones which appear more homogeneously located from ground to canopy.

TABLE 5. Location in mature primary forest of vegetarian bird species in the Nouragues area. Total = total number of vegetarian species. Avifauna = total number of species in the bird community. For more detail see Appendix 4.

	Aerial	Ground	Low foliage	Middle foliage	Canopy foliage	Middle perch	Canopy perch	Low bark	Middle bark	Canopy bark
Frugivores	-	7	2	4	39	-	-	-	-	-
Nectarivores	-	-	2	2	7	-	-	-	-	-
Omnivores	-	2	2	4	27	1	1	-	-	-
Insectivores	-	-	-	1	7	-	1	-	-	-
Total	-	9	6	11	80	1	2	-	-	-
Avifauna	7	25	9	32	112	6	8	2	17	10

When species are grouped into low-level foragers, middle-level foragers and canopy foragers, the 109 vegetarian species and the 112 non vegetarian ones (aerial are excluded) addicted to primary forest differ significantly in their foraging location (χ^2 = 28.76, df = 2, P < 0.001). Then, respectively for low, middle and canopy foragers, we have the following figures: 14, 11 and 75% for vegetarians and 19, 39 and 42% for non vegetarians.

Both vegetarian Nonpasserines and Passerines differ significantly from non vegetarian (χ^2 = 45.72, df = 5, P < 0.001 for Nonpasserines; χ^2 = 38.25, df = 4, P < 0.001 for Passerines) in their location in primary forest. The most striking difference is that 7

Nonpasserine frugivores, one omnivore and 2 nectarivores but no non vegetarian Nonpasserines are terrestrial nor associated with lower understorey. Conversely, only 2 Passerine frugivores and 3 omnivores against 23 non vegetarian Passerines forage on the ground and/or in lower understorey.

This suggests that fruits on the ground are large and available in sufficient quantity and diversity throughout the year to support a guild of specialist frugivores (including not only Nonpasserine birds but also mammals such as rodents, cervids and pecaries). Low flowering plants would permit 2 nectarivores to stay in the lower understorey where they can complement their diet with small insects. The relative paucity of vegetarian Passerines in lowest understorey would suggest rather poor small fruit resources that prevent frugivores even small species from specialising in this forest level. That the question requires a more ecosystemic approach (accounting in particular of the interactions between the various vertebrate faunas present) is apparent from the fact that among non vegetarian species there are 17 terrestrial Passerines but no Nonpasserines, a situation very different from that in Gabon (Erard 1986, 1989).

Another point to underline is the difference in proportion of seed-dispersers *vs.* seed-predators between vegetarian Nonpasserines and Passerines: 64% of the former (*Opisthocomus* and Nectarivores being excluded) are seed-predators whereas 95.6% of the latter are seed-dispersers. However, as already pointed out by Levey *et al.* (1994), the dichotomy between seed-dispersers and predators is not so clear-cut. Indeed, seed-predators like curassows (here *Crax alector*) may disperse small seeds (Erard & Sabatier 1989, Erard *et al.* 1991, Erard & Théry 1994). The action of terrestrial seed-predators may be beneficial for the plant through limitation of seed density under the fruiting tree. On the other hand the beneficial role of dispersers may not always be prominent: see the case of 'marshers' (*sensu* Levey *et al.* 1994) that waste seeds and drop them below the fruiting tree or in its very close vicinity.

4. Bird-plant mutualism

Our knowledge of bird-plant mutualism in the Nouragues area is still scanty. We just now begin to be able to list bird species that mainly or scarcely, permanently or sporadically feed on fruits. Indeed as pointed out above, the present list is most probably still incomplete, particularly as regard the numerous insectivores that may occasionally ingest fruits. We still have a poor knowledge of the seasonal variability of bird species diets. Even for those species clearly recognised as frugivores, we lack adequate detailed data on foraging behaviour, fruit selection and seed-handling processes, relations between gustative and digestive apparatus and physiology and fruit-processing according to its morphology and chemistry.

Nectarivores remain to be studied. Data on them are still very incomplete and almost anecdotal. Hummingbirds and some Emberizidae (e.g. *Cyanerpes* and *Chlorophanes*) are active pollination agents. Observations exist of *Coeroba flaveola, Cyanerpes caeruleus, C. cyaneus, Chlorophanes spiza* and particularly *Topaza pella* visiting regularly flowers of *Eperua* trees (plants particularly visited by nocturnal mammals, mainly nectarivorous bats) in late afternoon, even at dusk, when they open and at dawn before they close. Likewise, numerous nectarivorous birds concentrate on flowering *Symphonia globulifera*, going from tree to tree. Various hummingbirds share red, yellow or white inflorescences, small-billed species visiting flowers with short corolla whereas long-billed species though preferring long corollas take also the nectar of smaller flowers. Though obvious the role of these nectarivores as pollinators remain to be studied in term of fitness for the plants.

Just a few groups of frugivores have been studied in detail, either because they have particular social behaviour: lekking Pipridae (Théry 1990a,b,c, 1992, 1997; Théry & Vehrencamp 1995; Endler & Théry 1996; Krijger *et al.* 1997) and *Rupicola rupicola* (Erard *et al.* 1989, Théry & Larpin 1993, Endler & Théry 1996), or because they are large, mainly terrestrial frugivores and are game birds suffering almost anarchic hunting pressures: terrestrial *Tinamus major, Crax alector* and *Psophia crepitans*, and arboreal *Penelope marail* (Erard & Sabatier 1989, Erard *et al.* 1991, Théry *et al.* 1992, Erard & Théry 1994).

All species studied based their diet on the fruits of a large range of plant species without sharp specialisation on particular fruit types (i.e. drupe, berry, etc.). Nevertheless, Pipridae consume mainly fruits of Melastomatacea and Rubiaceae, two plant families that appear as keystone food resources for them. Diurnal fruit removal by manakins ranged from 61 to 89% for 20 species of Melastomataceae (Théry 1990a). Thus manakins are among if not the most important seed-dispersers for this plant family. Indeed the overlapping succession along the year of the fructifications of the various species of this plant family sustain populations of manakins. However, seed dispersal efficiency varies between sexes. Female manakins spend much time in gaps and other forest microhabitats favourable to the germination of Melastome seeds whereas males concentrate seed dispersal on their leks in less favourable sites, where they spend most of their time (Krijger *et al.* 1997).

From the detailed study of the diet of the large terrestrial and arboreal frugivores it appeared that though *Crax alector* eats the fruits of 80 plant species, only 12 account for 76% of its diet; likewise, the other seed-predator *Tinamus major* ingests the fruit of 38 plant species but only 9 make up 79% of its diet. For seed-dispersers, the terrestrial, partly invertebrate-eater *Psophia crepitans* takes the fruits of 55 plant species, 7 of them making up 78% of its diet, but the arboreal *Penelope marail* appears still more selective, eating the fruits of only 24 plant species, 75% of its diet being accounted for by 4 plant species, 2 of which accounting for 60% (Erard *et al.* 1991, Théry *et al.* 1992).

Key fruit species are thus apparent for these large frugivores and presently some fruit species can even be identified as important for and specifically consumed by one or the other of these large frugivorous bird species suggesting tight mutualistic relationships or even coevolutionary processes. However, the possibility that these high species-specific bird-plant interactions are magnified by an incomplete knowledge of the animal-plant interactions in the ecosystem cannot completely be ruled out as long as other terrestrial and arboreal frugivorous vertebrates have not been thoroughly studied.

Fruits eaten by birds in the Nouragues area fit in with the classical bird fruit selection syndrome (Gautier-Hion *et al.* 1985, Willson 1986, Jordano 1992). However studies of large frugivores showed that this general syndrome could be subdivided into more specific syndromes which allow defining *Tinamus*-type, *Crax*-type, *Psophia*-type and *Penelope*-type fruits. Furthermore, though these syndromes are rather well clear-cut, they can be linked in different ways depending on whether qualitative (i.e. mere list of fruiting plants) or quantitative (fruit quantity ingested per plant species) data are used in the analyses. Qualitative analyses oppose arboreal to terrestrial frugivores whereas quantitative analyses contrast seed dispersers and predators (Erard & Théry 1994, Erard, Théry and Sabatier unpublished data).

Bird-eaten fruit plants seem to make a trade-off between several strategies: e.g. attraction of arboreal seed-dispersers, defence against terrestrial seed-predators and rewarding of terrestrial seed-dispersers.

5. Some avenues for future research

We do not want to duplicate here a lot of lines for future research developed by others (e.g. Willson 1991, Jordano 1992, Levey *et al.* 1994, Leigh 1999) but simply underline some of the too numerous gaps that remain in the background necessary to place correctly bird-plant interactions in an evolutionary perspective.

These mutualistic relations between plants and the birds that consume their fruits are thus to be considered in the general frame of both coevolution and predator-prey interactions; in this latter case, not only between plants and their fruit-consumers but also between plants and all their predators as well as between frugivorous birds and their own predators (Sallabanks & Courtney 1992). Indeed, this frame should be enlarged to include also parasites and even insectivores as the former may contribute to the regulation of the populations of frugivores and the latter may regulate plant predators and parasites.

We need also to evaluate the benefits in terms of fitness that plants receive from frugivores (Fischer & Chapman 1993, Jordano & Herrera 1995). Seed dispersal is a necessary but not sufficient one for real mutualism. Which bird-dispersed seeds have the best chance of germination and to develop into a fruiting plant? What are in these conditions the best avian seed-dispersers: permanent or occasional frugivores? We do not know yet what role do these latter play in terms of forest dynamics.

Among the other important things that remain to be studied are the chemical processes developed by plants that may stimulate avian frugivory and promote seed dispersal but obstruct seed predation (Mack 1990, Horner *et al.* 1992, Simms 1992, Steel *et al.* 1993). This requires ecophysiological studies of both plants and birds.

Another point of interest also directly linked to coevolution is that of the adaptations of birds to frugivory. These may be physiological as pointed out above but they must also be morpho-anatomical (Moermond *et al.* 1986, Jordano 1987, Karasov & Levey 1990, Martinez del Rio & Restrepo 1993, Théry 1997). Indeed they can integrate both physiology and morpho-anatomy. Thus chemical contents of fruits (e.g. carotenoids and secondary metabolites) might be very important both for the plant and for the bird, and make the basis of a strong and tight evolutionary interaction between plant signalling and bird vision as well as metabolism (Willson & Whelan 1990, Endler & Théry 1996, Levey & Cipollini 1998). These studies should be comparative and in an evolutionary perspective, i.e. based on very robust phylogenies of both plants and birds.

As a final comment, we will emphasise once more the need and importance of teams for long-term, we would even say very long, studies in tropical forest ecology before a proper understanding of processes is attained, and at least sometimes before real patterns are even detected.

The epiphytic orchid species *Scaphyglottis amethystina* (orchidaceae). Nouragues, 1994. *Photo Gerrie Offerman.*

PART 5

FOREST DYNAMICS AND RECRUITMENT

CHAPTER 23

TREE-FALLS AND CANOPY GAPS: PATTERNS OF NATURAL DISTURBANCE

Peter J. van der Meer and Frans Bongers

1. Introduction

In most tropical rain forests, between 0.5 and 3.6% of the trees (>10 cm diameter) die per year (Phillips & Gentry 1994). A large proportion of these trees fall over and create canopy gaps ('hole in the forest canopy existing through all height levels' [Brokaw 1982]). Estimates of the forest area annually affected by canopy gaps (sensu Brokaw) range between 1-2% (e.g. Clark 1990, Hartshorn 1990, Jans *et al.* 1993). This process of canopy gap formation and the ensuing forest development ('patch dynamics'; White & Pickett 1985) plays an important role in the population dynamics of most plant species (for reviews see Bongers & Popma 1988, Denslow 1987, Pickett & White 1985, Platt & Strong 1989).

The spatial variation in the number, size and distribution of new canopy gaps may be very high. Canopy gap formation is not a completely random process: in many forests there seems to be an increased chance of canopy gap formation around old canopy gaps (Tanaka & Nakashizuka 1997). Increased turbulence around gaps (Brokaw 1985, Hubbell & Foster 1986), variation in local topography (Poorter *et al.* 1994), or asymmetrical crown shapes of trees around gaps (Young & Hubbell 1991) could cause this. Clustering of gaps is likely to have important implications for the population dynamics of many species, and may be an important factor shaping the structure of the forest.

The Nouragues forest is one of the few tropical rain forest sites where gap formation has been systematically monitored over a relatively long period (5 years) on a large area (12 ha). Van der Meer & Bongers (1996a) found that tree-falls (but not canopy gaps) were clumped away from both old and large gaps. In another study at the same site Van der Meer & Bongers (1996b) suggest that certain areas in Nouragues are more prone to canopy disturbance than others, which was thought to be related to the locally shallow soils. These two latter studies were based on 3 years of monitoring in the 12 ha study area.

In this chapter we will present additional data on tree-falls, branch-falls, and the formation of canopy gaps at the above mentioned 12 ha study area for a period of five years. Also, we investigated the spatial relationships between tree-fall and branch-fall events and canopy gaps, and try to answer the question whether the location of a canopy gap does increase the chance of a tree-fall or branch-fall in its immediate neighbourhood.

Bongers *et al.* (eds.), Nouragues. Dynamics and plant-animal interactions in a neotropical rainforest, 243–250.
2001 Kluwer Academic Publishers. Printed in the Netherlands.

2. Methods

The study was conducted on a permanent sample plot of 12 ha (300 m x 400 m) of closed rain forest at Nouragues, French Guiana (for detailed site description see elsewhere in this book). We monitored formation of canopy gaps larger than 4 m^2 (cf. Brokaw 1982) between the beginning of 1991 and the end of 1995 at intervals of varying length (3 months-2 years). We determined location and date of gap formation, and marked the centre with a plastic picket. Gap size was measured as soon as possible after gap formation using three methods:

1. cf. Brokaw (1982): the area enclosed by the eight points (eight directions from the gap centre, 45° intervals) where the first vegetation above 2 metres is encountered.
2. cf. Runkle (1981): the area bordered by the stembases of the canopy trees>20m tall surrounding the canopy opening.
3. cf. Van der Meer (Van der Meer & Bongers 1996a); the area enclosed by the eight points (eight directions from the gap centre; 45° intervals) where the first vegetation above 20 metres is encountered.

The height of the vegetation was checked using an optical range finder (Ranging Optimeter 120; range 2-30 m); the exact location of the vertical projection of the vegetation was determined using a pentagon prism.

Monitoring of tree- and branch-falls did not start until the end of 1992. By December 1992 all trees>10 cm in diameter at breast height (dbh; 130 cm) in the 12 ha plot were permanently tagged, numbered, measured (dbh) and mapped to the nearest meter. At the end of 1993 and 1995, the plot was checked again for all fallen and damaged trees. Damaged trees were trees which had lost at least one major branch with a base>10cm diameter. All fallen and damaged trees were grouped in *events*. A distinction was made between an *initiating* tree- or branch-fall (fall of a tree or branch which was not caused by the fall of another tree or branch) and a *domino* tree-fall or branch-fall (trees or branches which are brought down by the fall of another tree or branch). The initiator of the event was determined on the basis of the position of the fallen stems and/or branches, and other damaged and/or fallen trees in such events were marked as domino tree-falls or domino branch-falls. We considered 3 types of falls: tree-falls, branch-falls, and dead stem-falls, and we recorded 2 modes of tree-fall: uprooted and snapped.

Spatial patterns were analysed by a refined nearest neighbour analysis (Upton & Fingleton 1985). A Monte Carlo simulation technique (Besag & Diggle 1977) was used to test the random or non-random nature of the cumulative nearest neighbour distances between gaps and/or tree- and branch-falls. The minimum distance between two points was set at 4m, as the minimum gap radius is 2 m. For further details see Van der Meer (1995) and Van der Meer & Bongers (1996a, b).

3. Results

Canopy gap formation 1991-1995

Between 1991 and 1995 a total of 41 canopy gaps (>4 m^2) were created. The area covered by gaps varied considerably per year and by gap estimation method, ranging between 47-341 m^2 (Brokaw), 162-1591 m^2 (Van der Meer), and 576-4,442 m^2 (Runkle) (Fig. 1).

The average gap size varied per year and per method, but not significantly (Kruskal-Wallis $P > 0.05$), with the 1991 gaps being largest. 1993 was the most dynamic year with the highest total area in newly formed gaps (for all three methods), and also with the highest number of gaps formed. During the following year 1994 the lowest number and area of gaps of the five-year study period was observed. Over the five years, a total of 824, 3922 and 12663 m^2 (Brokaw, Van der Meer, Runkle respectively) was covered by new canopy gaps, being 0.69, 3.27 and 10.55% of the study area respectively. The average annual forest turnover rate (annual percentage forest area covered by newly created canopy gaps) over the five years was 0.14% (Brokaw), 0.65% (Van der Meer) and 2.11% (Runkle).

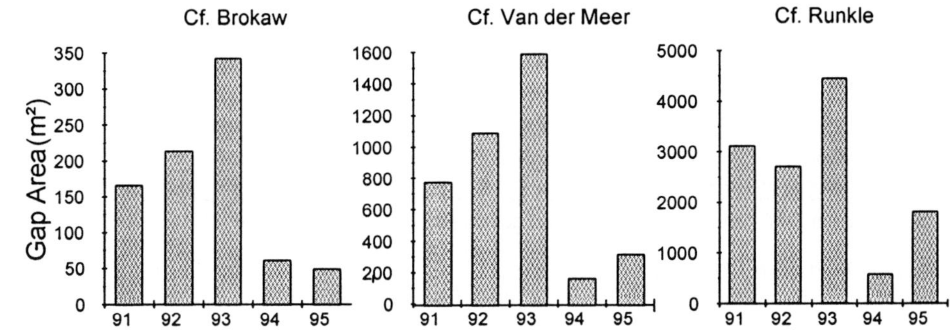

Figure 1. Annual area (m^2) covered by canopy gaps in a 12 ha forest study area from 1991 till 1995, using 3 different gap definitions (Brokaw, Van der Meer, and Runkle; see text). Note different scale for axes.

The frequency distribution of gap sizes over the 5 years showed different patterns for the different gap measuring methods (Fig. 2). Using Brokaw's method, small gaps were most abundant, and their number decreased gradually with doubling gap sizes, although gaps between 18-34 m^2 in size are relatively rare.

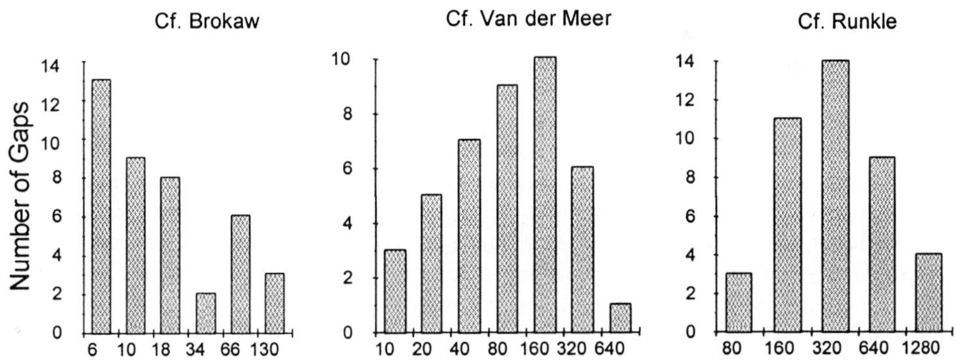

Figure 2. Size frequency distribution for canopy gaps formed between 1991-1995 in a 12ha forest study area, using 3 different gap definitions (Brokaw, Van der Meer, and Runkle; see text). Note different scale for axes.

Most gaps (63%) created between 1991 and 1995 were formed by a tree-fall: of these 26 tree-falls, 16 had snapped and 10 were uprooted. Twelve gaps (29%) were created by a fallen limb, and 7% of the gaps were initiated by a fallen dead stem.

TABLE 1. Number of natural tree- and branch-fall events during 1993-95, and the number and diameter of involved fallen and damaged trees.

	Gap creating events	Non-gap creating events	All events
Events			
no. of tree-fall events	13	31	44
no. of branch-fall events	8	22	30
no. of dead stem-fall events	4	4	8
TOTAL	25	57	82
Fallen trees			
Number	99	59	158
Mean number/event	4	1	1.9
dbh (cm)	22.5[a]	23.9[ab]	23.0[a]
Damaged trees			
Number	62	47	109
Mean number/event	2.5	0.8	1.3

Different letters indicate significant difference in dbh (one-way ANOVA; Tukey-HSD test).

Tree- and branch-fall events 1993-1995

Between 1993 and 1995 some 158 of the initial 6314 trees > 10 cm dbh fell down (= 0.8% annually) and 109 trees were damaged in 82 events (Table 1). Roughly one third (30.5%) of the events resulted in canopy gaps > 4 m^2. Most events (54%) were initiated by a fallen tree, 37% were initiated by a fallen branch, and 10% by a fallen dead stem. There was little difference between the type of initiators between gap creating and non-gap creating events (data not shown). Furthermore, 30% of the tree-fall events, 27% of the branch-fall events, and 50% of the dead stem-fall events resulted in a canopy gap.

Of the trees which fell between 1993-95, 108 (68%) had snapped, and the remaining 50 trees (32%) had been uprooted. These percentages varied little when sub-dividing the fallen trees in gap and non-gap creating initiators and domino trees, although gap initiators have slightly higher percentage of uprooted trees than non-gap initiators (data not shown).

The average diameter of fallen and damaged trees which initiated an event (initiators) was significantly higher than that of trees which had been felled or damaged by other fallen trees or branches (domino trees) (one-way ANOVA, $P < 0.05$; data not shown). Of the initiators, the trees that created a canopy gap (gap creator) had a significantly greater diameter than the trees that had not created a canopy gap (non-gap creators). Of the gap creating initiators, branch-fall trees (trees that dropped a branch) had a significantly greater diameter than snapped and uprooted trees. Also, branch-fall trees that had created a canopy gap were fatter than branch-fall trees that had not created a canopy gap. In both the group of gap creators and the group of non-gap creators, there was no difference in diameter between snapped and uprooted fallen trees (one-way ANOVA).

Spatial patterns

The spatial distribution of the 41 gaps formed between 1991 and 1995 (Fig. 3) was clustered (Monte Carlo simulation; $P < 0.05$); nearest distances of around 30 meters occurred significantly more often than they would occur in a random distribution (Fig. 4; Table 2). Also very large gaps (> 160 m cf. Van der Meer) were significantly clustered around 55 meters (Monte Carlo simulation; $P < 0.05$).

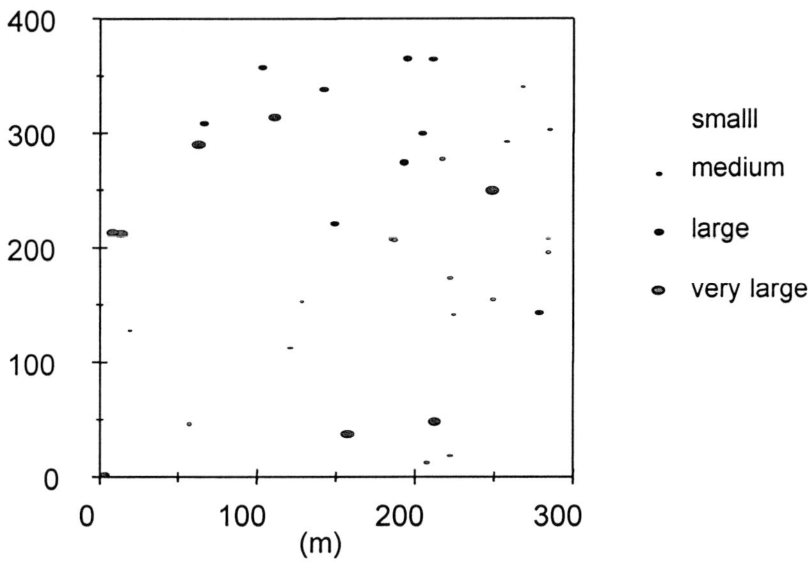

Figure 3. Location of canopy gaps formed between 1991-1995 in a 12ha forest study area at Nouragues. Gaps are grouped in four size classes (using Van der Meer gap definition): small (< 20 m^2), medium (20-80 m^2), large (80-160 m^2), very large (>160 m^2), indicated by increasing dot sizes.

TABLE 2. Spatial patterns of canopy gaps and tree- and branch-fall events in a 12 ha study site during 1991-1995 and 1993-1995. Monte Carlo simulations (200 simulations) was used to test the random or non-random nature of the cumulative nearest neighbour distances between gaps and/or events ($P < 0.05$). Trend indicates whether patterns are clustered or uniform. Distances that were outside the 95% confidence envelope of a random distribution are given.

	Distribution Pattern	Trend	Significant distance (m)
all gaps '91-'95 *(n=41)*	Non-random	clustered	32
very large gaps '91-'95 *(n=7)*	Non-random	clustered	56
all events '93-'95 *(n=82)*	Random	-	-
gap events '93-'95 *(n=25)*	Non-random	clustered / uniform	24 / 56
non-gap events '93-'95 *(n=57)*	Non-random	uniform	28
uprooted initiators '93-'95 *(n=11)*	Random	-	-
snapped initiators '93-'95 *(n=33)*	Random	-	-
branch-fall initiators '93-'95 *(n=30)*	Random	-	-
gaps '91-'92 *(n=16)* versus			
a. all events '93-'95 *(n=82)*	Non-random	clustered	18
b. gap events '93-'95 *(n=25)*	non-random	clustered	18
c. non-gap events '93-'95 *(n=57)*	Random	-	-

The location of the 82 tree- and branch- fall events between 1993 and 1995 was not significantly different from a random dispersion (Table 2). Interestingly, the 25 events which created a canopy opening were clustered around 24 m, but also seemed to be hyper-dispersed around 56 m. Non-gap events were significantly hyper-dispersed (around 28 m). Events created by uprooted trees were randomly distributed over the 12ha, as were the events created by snapped trees and by branch-falls respectively (Table 2).

The existence of a gap seemed to enhance the chance of an event in the following years (Fig. 5); the 1993-1995 events were significantly clustered around the 1991-1992 gaps (Fig. 4; Table 2). The same significant clustering around 18 m was found for the 1991-1992 gaps and the events which created a canopy gap, but was not detected for the non-gap events.

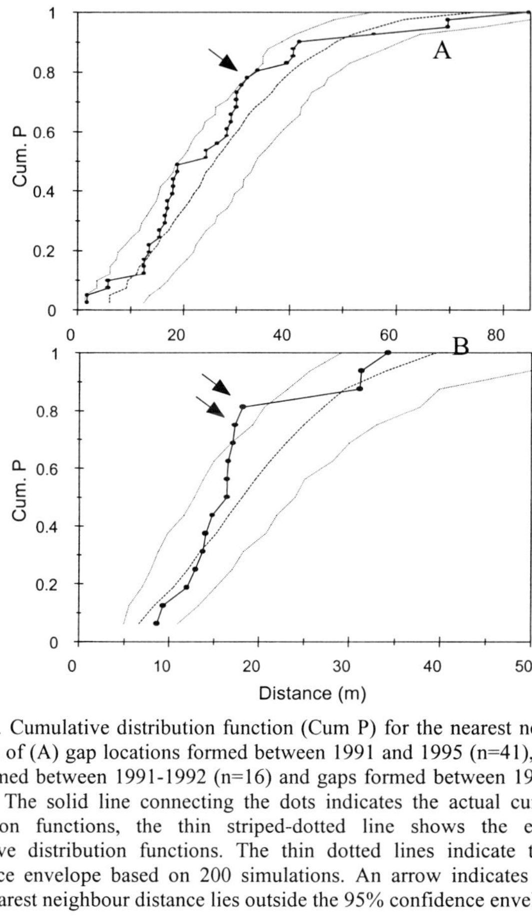

Figure 4. Cumulative distribution function (Cum P) for the nearest neighbour distances of (A) gap locations formed between 1991 and 1995 (n=41), and (B) gaps formed between 1991-1992 (n=16) and gaps formed between 1993-1995 (n=25). The solid line connecting the dots indicates the actual cumulative distribution functions, the thin striped-dotted line shows the estimated cumulative distribution functions. The thin dotted lines indicate the 95% confidence envelope based on 200 simulations. An arrow indicates that the actual nearest neighbour distance lies outside the 95% confidence envelope.

4. Discussion

As observed earlier by Van der Meer and Bongers (1996a), the main cause of tree-falls and branch-falls were other tree- and branch-falls: 72% of the fallen as well as 72% of the damaged trees were 'domino' trees (i.e. felled or damaged by other falling trees or branches). Likewise, 70% of the tree mortality in a palm forest in Puerto Rico was caused by smaller trees which were felled by falling canopy trees (Frangi & Lugo 1991). However, the causes of the fall of initiating trees or initiating branches are more difficult to determine, and is related to several external (e.g. local climate, topography, soil condition, fungi, insects, diseases) and/or internal factors (e.g. senescence, competition) (Lugo & Scatena 1996, Riéra 1983, Vooren 1986).

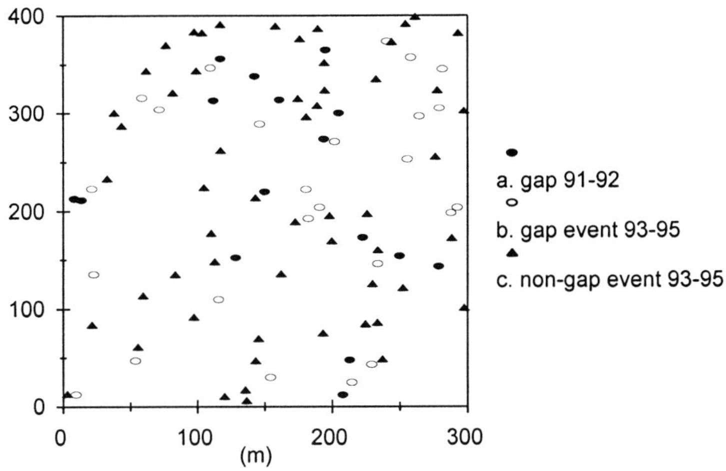

Figure 5. Location of gaps created between 1991 and 1992 (closed circles), tree-fall and branch-fall events between 1993 and 1995 which resulted in a gap (open circles), and events between 1993 and 1995 which did not result in a canopy gap (closed triangles).

The relative annual tree-fall rate over 1993-1995 (0.8%) is low compared to the reported median relative annual tree mortality of 1.6% for other tropical forests (Lugo & Scatena 1996, see also Phillips 1996 and Phillips & Gentry 1994). Over 1993 only, the relative tree-fall rate at the studied forest was 1.5% (Van der Meer & Bongers 1996a), and from 1996 till mid-1997 the relative annual tree-fall rate was 2.2% (P.J. Van der Meer and F. Bongers, unpublished data). It appears that disturbance rates in 1994 and 1995 were relatively low, which is likely to be related to the annual variation in external factors that affect tree mortality rates, like rainfall patterns and local storms. High-disturbance years in the past may also affect the annual variation in present day tree mortality rates, as the regrowth in even-aged patches starts senescing around the same time. Also, if a storm fells many trees one year there may be less susceptible trees in the following year(s).

The rate of canopy gap formation in the studied 12 ha between 1991 and 1995 was lower than the gap formation rate observed during the previous five years (1986-1990), for which Van der Meer and Bongers (1996b) report an average annual gap area of 212.6 m^2cf. Brokaw and 2970.6 m^2 cf. Runkle. This equals turnover rates of 0.18% (cf. Brokaw) and 2.48% (cf. Runkle) (van der Meer's definition was not yet used). This may seem only slightly higher than the turnover rate between 1991 and 1995 (0.14% cf. Brokaw; 2.11% cf. Runkle), but as the sizes of the gaps formed between 1986 and 1990 were measured up to 4 years after gap formation, gap sizes at the time of creation are likely to have been larger then measured. Consequently, the actual turnover rate for 1986-1990 was probably even higher than 0.18 and 2.48%. The difference in turnover rates between the two five-year periods can again be explained by the large observed variance in annual tree-fall rates as discussed above.

The locations of gaps and events were not randomly distributed over the 12 ha but tended to be clustered. Canopy gaps and tree-falls have been found to cluster before, both in Nouragues (Van der Meer & Bongers 1996a, b) as well as in a number of other forests (Hubbell & Foster 1986, Lawton & Putz 1988, Poorter *et al.* 1994, Tanaka & Nakashizuka 1997). It is clear that existing canopy gaps in Nouragues are related to the chance of the occurrence of an event in their direct neighbourhood, with more events occurring around 18 meters from gaps than expected on the basis of a random distribution. An additional nearest neighbour analysis revealed that the location of events between 1993 and 1995 ($n = 82$) was significantly related to the location of the gaps created between 1986 and 1992 ($n = 71$) (uniform; significant distances 10-15 m; $P < 0.05$). This increased tree-fall and branch-fall rate around existing gaps may be related to several factors such as asymmetric tree-crown development (Young & Hubbell 1991), exposure to wind of gap edge trees (Foster & Reiners 1986) and damaged trees in the gap edge (Putz & Chan 1986). Some of these tree-falls and branch-falls in or near gap-edges result in new canopy gaps. New gaps are only formed when the tree (or branch) falls away from the existing gap (a tree that falls into an existing gap is not seen as a new gap event). The new gap centre must be at a greater distance from the existing gap centre than the base of the initiating tree (the location of the event). This explains the smaller clustering distances of gaps and events (18 m) compared to that of gaps (24-56 m).

This clustering of gaps is likely to have important, long-lasting effects on the forest structure and the population dynamics of species. Riéra (1995) found a peculiar allotment in diameter classes and a patchy distribution of *Eperua* spp. and *Vouacapoua americana* in French Guiana, and related this to past natural disturbances. In a tropical rain forest in India, large clusters of (pioneer) species were found in areas disturbed by many tree-fall events (Pélissier 1998). Light demanding species will dominate in forest areas where large gaps are abundant, often forming a relatively open canopy. Other areas may be relatively free of (large) gaps for extended periods and be dominated by more shade tolerant species.

The spatial analysis in this study revealed that the location of all events was random, while when grouping events by initiator type (uprooted trees / snapped trees / branch-falls), locations of each sub-group of events were again random. This implies that the occurrence of events is a relatively random process, and that some tree- and branch-felling factors which could be expected to cause patchy disturbances (e.g. fungi, diseases, soil related factors) are not likely to be important causes of tree- and branch-falls in the Nouragues forest. Before, the clustering of gaps at Nouragues was thought to be related to soil factors (Van der Meer & Bongers 1996b), but these new data suggest that clustering of both gaps and events be mainly caused by pre-existing gaps. Continued observation in the existing study area as well as other forest areas at the Nouragues site is necessary for conclusive evidence.

Acknowledgements

We thank R. Beyers, D. Betian, W. Betian, J.-W. Gunnink, P. Leersnijder, A. Hofmann, M. Smeenge, R. van Heck, P.A. Jansen, F. Sterck, T. Rijkers for their help during part of the fieldwork. P.A. Jansen and N.V.L. Brokaw kindly commented on an earlier version of the manuscript.

CHAPTER 24

THE DISPERSAL AND RECRUITMENT OF CYCLANTHACEAE AND *PHILODENDRON* (ARACEAE) UNDERSTOREY ROOT-CLIMBING VINES

Anya Cockle

1. Introduction

In 1977, Grubb proposed to define a plant's ecological niche by four main components, one of which being the regeneration niche. All stages of the regeneration cycle are potentially important in the differentiation of species - or of groups of species, depending on the level of competition - and allow their coexistence. He underlined the importance, in particular, of the production of viable seeds, dispersal (in space and time), germination, establishment and development of the immature plant. The regeneration niche is one of the aspects of the ecological niche least studied, as it deals with relatively 'hidden' stages of a plant's life history. It often remains the least known part of its ecology.

Recruitment can be decomposed in recruitment of seeds (the seed rain on a site), recruitment of seedlings (after seed predation and germination), and finally recruitment of young plants (after seedling mortality). The recruitment of a population of plants is the result of a series of concatenated events which include fruit removal by the disperser, seed dispersal, seed germination and seedling establishment (Jordano & Herrera 1995). After dispersal, which involves the relationship between plant and disperser, the biotic and abiotic factors which compose the different micro-environments act as filters to determine the germination of the seeds, the establishment of the seedlings and their survival (Harper 1977). The qualitative and quantitative study of each component of the recruitment can help classify these environmental filters (Houle 1996) according to their relative importance and determine the stages to which they apply.

Dispersal determines the seed rain and hence the recruitment of the population of seeds. Secondary transport may spatially reorganise this population. Germination then selects the population of seedlings from the seed population (seedling recruitment). Seedlings, and later young plants, will have to submit to the environmental conditions of their establishment site, which will or will not allow their development to the fertile stages. The different environmental filters can not be neutral and act differentially. Factors that apply to later stages may modify or even totally obliterate the effects of a factor that had influenced earlier stages (Herrera *et al.* 1994). Their global effects may then produce spatial discordances (Jordano & Herrera 1995) between the different phases of recruitment.

Recent studies in Europe have highlighted the relationship between dispersal, colonisation and vegetative persistence (Eriksson & Fröborg 1996, Stöcklin & Bäumler

Bongers *et al.* (eds.), Nouragues. Dynamics and plant-animal interactions in a neotropical rainforest, 251–263.
2001 Kluwer Academic Publishers. Printed in the Netherlands.

1996). In tropical forests, however, the recruitment modalities of perennial non-arborescent plants have been very little studied (Horvitz & Le Corff 1993, Horvitz & Schemske 1986a 1986b 1994, Le Corff 1996, all on *Calathea*, Marantaceae). Cyclanthaceae and Araceae 'mobile' root-climbing vines constitute a poorly known and very distinctive guild in the rain forest structure, colonising the vertical tree trunks between ground level and about 10 m high. They are very common, relatively sensitive to environmental conditions, long-lived and locally persistent *via* clonality. These traits suggested that they could be used as environmental markers on a longer time scale than many trees, and this was the original idea which prompted the study of the variations of their recruitment along an environmental gradient (Cockle 1997).

2. Material and methods

2.1 THE PLANT SPECIES

Cyclanthaceae and *Philodendron* Araceae are Neotropical herbaceous monocotyledons. Eight Cyclanthaceae species are found at Nouragues, of which 3 are understorey (up to lower canopy) root-climbing vines (*Evodianthus funifer, Asplundia heteranthera* and *Thoracocarpus bissectus*). *Philodendron* species are more numerous, with five species belonging to subgenus *Pteromischum* (*P. placidum, P. guianense, P. rudgeanum*, and two yet undetermined species) and another 14 species belonging to subgenus *Philodendron* (*P. insigne, P. linnaei, P. soderstromii, P. billietae, P. grandifolium, P. ornatum, P. scandens, P. sphalerum, P. megalophyllum, P. wittianum, P. squamiferum, P. pedatum, P. melinonii, P. fragrantissimum*). These Cyclanthaceae and *Philodendron* species constitute a very distinctive guild of root-climbing vines that colonise rocks or, more frequently, tree trunks in the rain forest understorey. They tend to be more numerous in moist areas than in drier localities, both in number of individuals and number of species.

Understorey Cyclanthaceae and *Philodendron* species are long-lived and regularly produce many-seeded infructescences. Seeds are generally under 2 mm long and 0.1-0.3 mm wide, flat or elongated in shape. Germination is immediate except for some oval-seeded *Philodendron* species of subgenus *Philodendron* that may display strong dormancy.

2.2 DISPERSAL

The faeces of all fruit-eating species of bats were examined during bat-netting sessions in 1993, 1994 and 1995 for the presence of Cyclanthaceae and Araceae seeds. Seed transit through the digestive tract was timed and seed retention tested (10 experiments in captivity). Moreover, 11 roosts of the bat *Rhinophylla pumilio* were equipped with collectors and regularly checked for food remains and faeces. Four individuals of this species were equipped with radio-transmitters and followed during their nightly foraging for a total of 10 whole nights (from 6pm to 6am).

2.3 SEED RAIN

Falling material, including seeds arriving in the understorey, were collected in large funnels along an 800 m transect across the valley (9 stations, every 100 m, and 2 collectors per station) continuously during 16 months (504 days). Each collector of 0.0707m² was emptied every 2 months and the fine material was dried and sorted under a binocular microscope.

2.4 SEED DISAPPEARANCE RATES

Seeds freshly collected from bat roosts were placed on transparent plastic sheets on the leaf litter, easily accessible to Arthropods, and shielded from the rain. The seeds were left out and controlled daily until their numbers stabilised (usually after 4-8 days). Experiments were conducted in ridge-top forest, high moist forest and riverine forest, all year round, on 40 individual sites (10 in ridge-top forest, 10 in riverine forest, 20 in high moist forest). Ten additional sites, all located in high moist forest, tested the 'bat roost effect' (5 under 5 different bat roosts, where seeds accumulate, 5 controls 8 m away).

2.5 SEEDLING RECRUITMENT SITES AND SURVIVAL RATES

Three 10 x 10 m plots and one 5 x 5 m plot, located in the understorey along a mild topographical gradient (from valley bottom at 70 m a.s.l. to ridge-top at 180 m a.s.l.), were surveyed every 6 months during 18 months (April 1994, October 1994, April 1995, October 1995). Every Cyclanthaceae and Araceae individual was marked, examined (growth stage, number of leaves), and its establishment site described: height above ground (continuous variable), type of support (16 categories), slope of support (continuous variable), hardness (9 categories), rugosity (11 categories), presence of mosses (cover, 3 categories; thickness, 2 categories), presence of litter (3 categories), distance to nearest seedling of the same stage group (continuous variable). Each individual was checked for presence or absence (death) during each survey. Survival rates were calculated between October 1994 and October 1995.

2.6 THE MICROCLIMATIC GRADIENTS AND THE SPATIAL DISTRIBUTION OF CYCLANTHACEAE SEEDLINGS

Topographical and vertical microclimatic gradients were investigated with Piche evaporimeters (measure of the evaporation rate over an entire week, repeated for two weeks to eight months depending on the transects) and a Testoterm pocket thermo-hygrometer (instant measures of the temperature and relative humidity, all made under the most extreme weather conditions, i.e. during the dry season in the early afternoon). Evaporation rates were measured along six transects of variable lengths (100-800 m) at different scales (every 10 or every 100 m, or every 10 cm vertically), and also at one valley site to test small-scale horizontal and vertical variations (every metre horizontally, at 0.2 and 2 m above the ground, 20 evaporimeters over one month in the rainy season). Temperature and relative humidity were measured along four transects to study their relation to topography and to seedling establishment. Light was measured with synchronised quantum sensors in four locations along a topographical transect (over three days), and also at different heights between 0 and 3 m above the ground in valley forest (over two days).

The heights of establishment of epiphytic Cyclanthaceae seedlings (on tree trunks up to 4m, above which practically no more seedlings were noted) were recorded along 3 topographical gradients (between 60 and 180 m a.s.l.) on a total of 19 sites (of which 12 in high mature forest), each 12.6m² in surface area. The resulting vertical distribution of the seedlings was then related to local evaporation rates (measured with 19 evaporimeters at 1m above the ground).

Seedling recruitment rates were estimated experimentally. A total of 2520 seeds of different Cyclanthaceae and Araceae species were introduced at different height levels along tree trunks and followed for several months (5 to 8 checking visits depending on the date of introduction). The resulting scores were related to minimum relative humidity levels (dry season, early afternoon) in the understorey, at the corresponding locations.

3. Results

3.1 DISPERSAL

Main dispersers

A total of 959 bats were mist-netted in the understorey. Seeds of Cyclanthaceae and *Philodendron* were only found in the faeces of five understorey Phyllostomidae bat species: the Carolliinae *Rhinophylla pumilio* (7-13 g), *Carollia brevicauda* (8-14 g) and *C. perspicillata* (13-22 g), and the Stenodermatinae *Sturnira tildae* (22-25 g) and *S. lillium* (19-27 g). *R. pumilio* and *C. perspicillata* were relatively common (9.9% and 6.9% of the captures respectively), whereas the three other species together only represented 2.2% of the total catch[1].

Recapture rates of marked *R. pumilio* individuals on 44ha suggested a density of 3.1 individuals per hectare of rain forest at Nouragues (23 initially marked, then 24 caught including 4 recaptures). Although this figure needs to be taken with caution because of the low recapture rate, counts of marked bats at roost suggested similar density levels.

Diets

The primary forest bat *R. pumilio* was clearly the main disperser of Cyclanthaceae and *Philodendron* species (Table 1). Moreover, 57% of the 1432 samples of fruit (1 sample=1 fruit species collected at 1 roost control) consumed by this small bat belonged to either the Cyclanthaceae (29%, genera *Evodianthus, Asplundia* and *Thoracocarpus*) or the Araceae (28%, mainly *Philodendron*, with some *Rhodospatha*).

In captivity, an average of 75% of the seed were passed within 90 minutes after feeding, and an average of 8.7% of the seed were retained within the digestive tract until after the following meal.

Foraging movements

The radio-tracking of *R.. pumilio* during 10 whole nights showed that this bat species explored between 6 and 13 ha of rain forest each night (9.7 ± 1.8 ha, N = 10). One preferential zone was generally discernible, which was often, but not always, the same from one night to the next, and where it spent over 200 minutes. Distances covered during one

[1] It appears that other animals and birds seldom disperse these plants. Systematic ecological studies on monkeys (Julliot 1992, Zhang 1994, Simmen & Sabatier 1996), kinkajous and opossums (Jullien-Laferrière 1989) failed to find significant quantities of these fruit in their diets. Marginal dispersal occurs however (mouse-opossum *Marmosa* sp., Makin unpublished data; tanager *Tachyphonus surinamus*, Jullien unpublished data.), which is difficult to quantify.

single night ranged from 1600 to 4750 m (3185 ± 1037 m, N = 10), and the average speed, over the entire night, ranged between 152 and 506 m/h (287 ± 111, N = 10).

Activity peaked during the beginning and end of the night, with a slightly calmer period in between. General movements involved long sinuous loops of 200-500 m, over the entire prospected area, interspersed with periods of rest (up to 4 h long).

TABLE 1. Number of samples (i.e. number of times a seed species was recorded in a captured bat) and total number of individual seeds of the different botanical categories collected in the bat faeces during bat netting (1993-1995). In some cases, several species of seed were collected together from the same individual, and thus the total number of samples (98) is slightly higher than the total number of individual bats having produced seeds in their faeces.

	R. .pumilio		*C. brevicauda*		*C. perspicillata*		*S. tildae*		*S. lillium*	
No. bats caught	95		15		66		4		2	
	Samples	Seeds	Samples	Seeds	Samples	Seeds	Samples	Seeds	Samples	Seeds
Cyclanthaceae	19	4927	2	1201	3	110	3	2803		
Araceae	14	11439	1	78	6	717	1	5048	1	9200
Schlegelia Bignoniaceae	9	161	0	0	3	81				
Piper Piperaceae	2	156	5	473	10	1691				
Solanum Solanaceae	0	0	1	1	5	62				
Others	3	29	2	11	8	207				
Total	47	16712	11	1764	35	2868	4	7851	1	9200

It may be estimated from the surface area covered by the bat during foraging and from seed retention times that *R. pumilio* disperses most swallowed seeds within the 3 ha of forest around the parent plant. However, long straight flights up to 700 m have been recorded occasionally, and such movements may allow a small proportion of the ingested seeds to be dispersed to greater distances.

3.2 SEED RAIN

A total of 3039 seeds were collected, of which Cyclanthaceae and Araceae constituted respectively 39.1% and 22.1%. Overall, the Cyclanthaceae *Evodianthus funifer* produced a seed rain of 578 seeds/m²/year, *Thoracocarpus bissectus* 46 seeds/m²/year and *Asplundia* spp 8 seeds/m²/year only. The *Philodendron* subgenus *Pteromischum* produced 363 seeds/m²/year, whereas subgenus *Philodendron* only totalled 30 seeds/m²/year.

Significant spatial heterogeneity (tested using the Kruskal-Wallis procedure) was found at the topographical scale for all taxonomical categories of bat-dispersed Cyclanthaceae and Araceae species except *Asplundia* spp. Overall, about ten times less seeds of these species were deposited in the ridge-top forest than in the valley forest (Fig. 1).

3.3 SEED DISAPPEARANCE RATES

Four thousand seeds were used in 42 experiments with 10 replicates each. Most seeds of *Evodianthus funifer*, *Asplundia* spp. and *Thoracocarpus bissectus* (Cyclanthaceae) remained where they had been deposited, and average disappearance rates were of 19% (N = 1090 seeds exposed), 29% (N = 385) and 32% (N = 227) respectively. Seeds of *Philodendron* species belonging to the sub-genus *Pteromischum* were similarly little displaced (28%, N = 850), whereas the oval-seeded *Philodendron* species disappeared in great numbers (91%, N=1450).

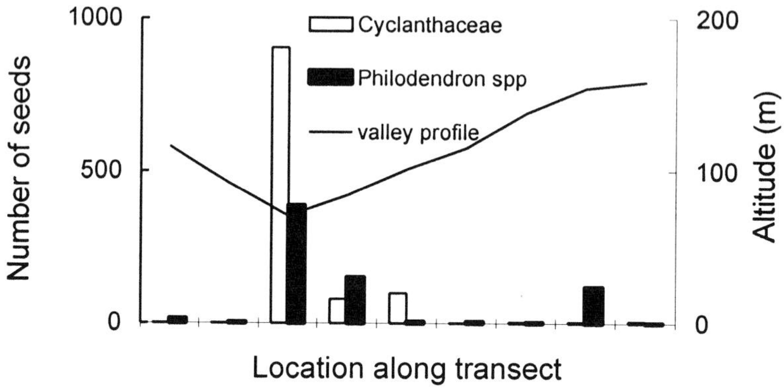

Figure 1. Topographical heterogeneity of the seed rain of the bat-dispersed Cyclanthaceae and *Philodendron* species (in black: *Evodianthus funifer, Asplundia* spp, *Thoracocarpus bissectus;* in white: *Philodendron (Pteromischum), P. melanin, P. ornate* and oval-seeded *Philodendron* species). Total number of seeds collected on each of the 9 locations, over the 504 days sampled. The black line retraces the transect profile across the valley (the horizontal and vertical scales are not the same: collecting points are 100m apart).

The main variation factors of seed disappearance rates were the seed species (Kruskal-Wallis, $P \leq 0.034$ for 9 experiments out of 12), the season (ANOVA, $P = 0.001$, for *Philodendron* subgenus *Pteromischum* only), and the interaction forest type x season (ANOVA, $P = 0.013$, for *Philodendron* subgenus *Pteromischum* only). Location under bat roosts did not produce any significant differences (Kruskal-Wallis, $P \geq 0.05$). In the case of the oval-seeded species of *Philodendron*, the carpellar membrane which sometimes remains around the seeds significantly enhanced the already high disappearance rates (Kruskal-Wallis, $P = 0.015$). Spatial heterogeneity at the scale of a few metres was very high, as well as temporal variability.

3.4 SEEDLING RECRUITMENT SITES AND SURVIVAL RATES

A total of 1754 Cyclanthaceae seedlings and 1825 Araceae seedlings were marked and monitored, on a total surface area of 325 m² in the forest understorey.

Seedling recruitment sites

Seedlings of *Evodianthus, Asplundia, Thoracocarpus* and *Philodendron* alike occurred on all types of substrates in the forest understorey. However, less than 10% were directly established on the ground, while the rest germinated on live tree trunks, various small stems, woody debris on the forest floor, tree roots running along the ground, dead leaves, etc. (Table 2). Seedlings were thus epiphytic in the broad sense.

A clear topographical gradient was noted relative to the vertical distribution of seedlings. The great majority had germinated between 0 and 3 m above the ground in the valley (99.7% of seedlings), and between 0 and 10 cm only on the ridge (97%), where establishment sites were thus reduced to a narrow band running along the ground.

TABLE 2. Numbers of Cyclanthaceae seedlings (*Evodianthus*, *Asplundia* and *Thoracocarpus*) and Araceae seedlings (*Philodendron*) found on the different types of establishment substrates. Total surface surveyed: 325m², all four plots considered together.

Family	Cyclanthaceae		Araceae	
	Number	% of total	Number	% of total
Surface roots	70	4	51	3.2
Radicels	20	1.2	13	0.8
Stems diam<1 cm	108	6.3	80	5.1
Trunks 1-5 cm	152	8.8	49	3.1
Trunks >5 cm	337	19.5	24	1.5
Erect dead stems and trunks	12	0.7	11	0.7
Dead tree buttresses	355	20.6	661	42
Woody debris diam<5 cm	94	5.4	47	3
Woody debris diam>5 cm	132	7.7	192	12.2
Litter - leaves	87	5	51	3.2
Litter - twigs	89	5.2	99	6.3
Litter - decomposed debris	93	5.4	131	8.3
Soil	155	9	149	9.5
Perched soil pockets	17	1	8	0.5
Live leaves	4	0.2	1	0.1
Others	0	0	5	0.3
Total	1725	100	1572	100

General survival rates

Survival rates varied according to the position along the topographical gradient (Fig. 2). For Cyclanthaceae seedlings, yearly survival rates ranged from 53% in the valley plot to 18% on the ridge-top plot, and from 60% to 21% for *Philodendron* seedlings. Survival was particularly low on a plot on the upper slope, where a fallen branch had created a minor opening (5% and 3%, among Cyclanthaceae and Araceae seedlings respectively). Seedlings of oval-seeded *Philodendron* species were scarce (2.7% of the Araceae seedlings) and achieved poor survival rates (27.3% overall and 31% in the valley).

Survival rates rose regularly with the size of the seedling (i.e. the number of leaves). In the valley plot, seedlings with 1 or 2 leaves displayed survival rates of 41% and 46% for Cyclanthaceae and *Philodendron* seedlings respectively, while survival of seedlings with more than 3 leaves reached 62% and 76% respectively. The correlation between survival rates and number of leaves (from 1 to more than 7) is positive and significant (Spearman, $P < 0.05$).

Influence of establishment characteristics on survival rates

On the valley plot, where the greatest number of seedlings were found, best survival rates were found on small trunks, large woody debris, large surface roots and dead buttresses (significant for both Cyclanthaceae and Araceae, χ^2, $P < 0.0001$ and $P < 0.05$). The more favourable slope values were steep near-vertical upward-facing slopes (significant effect for *Philodendron*, $P < 0.01$). Substrate hardness had a significant effect for both categories, with best survival rates on rather hard surfaces ($P < 0.05$ and $P < 0.01$ respectively). Similarly, surface coarseness of the substrate had a significant effect, with best results on rather fine-grained surfaces ($P < 0.01$ for Cyclanthaceae and *Philodendron*). Mosses had a positive effect, both in proportion of surface covered and thickness (both significant for *Philodendron*, $P = 0.002$ and $P = 0.003$ respectively). The presence of leaf litter produced a drop in survival rates (significant for Cyclanthaceae, $P < 0.05$). Establishment height was also important, with best survival rates achieved below 50 cm above the ground, and lowest rates attained at ground level (because of the leaf litter) and above 2 m high (significant for Cyclanthaceae, $P < 0.001$). The negative effect of nearest-neighbour distance was only significant when inferior to 1 cm ($P < 0.01$ and $P < 0.001$ for Cyclanthaceae and *Philodendron* respectively, distances up to 3 cm only).

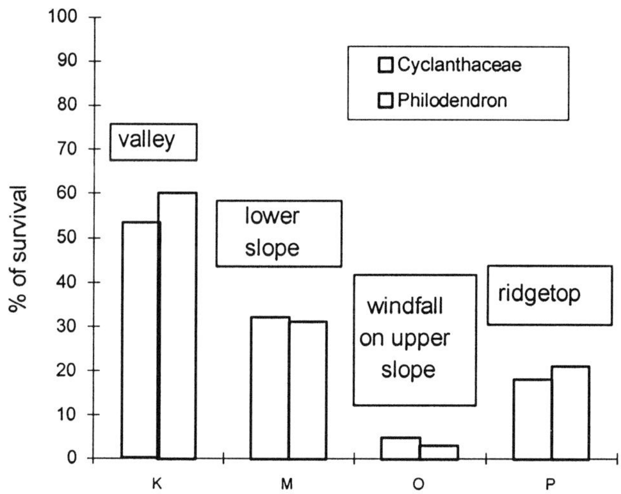

Figure 2. Yearly survival rates of Cyclanthaceae and *Philodendron* seedlings on four different understorey plots along a topographical gradient.

Net seedling growth rates

Net gain of leaves over 18 months was low but extremely variable from one individual to the next (1.4 leaves ± 1.9, min -5, max +8 for Cyclanthaceae, and 0.7 leaf ± 2.1, min -6, max +6 for *Philodendron*). In Cyclanthaceae, net growth of the seedling was negatively correlated to its size (Spearman, $P < 0.01$), and positively correlated to moss cover ($P < 0.05$) and moss thickness ($P < 0.001$). It was only correlated negatively to size in *Philodendron* ($P < 0.001$).

Older seedlings
Young individuals initialising stem elongation (stem shorter than 10 cm) displayed higher survival rates (82% and 90% for Cyclanthaceae and *Philodendron* respectively, on the valley plot). All were established between the ground and 1m high at the most.

3.5 THE MICROCLIMATIC GRADIENTS AND THE SPATIAL DISTRIBUTION OF CYCLANTHACEAE SEEDLINGS

Microclimatic gradients
Two types of microclimatic gradients were found in the forest understorey between 0 and 3 m above the ground. These involved mainly evaporation rates and relative humidity, whereas light, also tested, did not appear to follow a clear gradient, whether topographically or vertically, at this scale.

The topographical gradient
Relative humidity, temperature and particularly evaporation rates varied regularly along a slope, from thalweg to ridge. In a ridge-top situation, relative humidity was always lower and temperature and evaporation rates higher than in a valley, at a constant height above the ground.

All transects constituted by a single topographical unit (i.e. a thalweg and the slope to one of its corresponding ridges) displayed significant correlation between evaporation rates 1m above the ground and the relative elevation at the site of its measurement. Simple linear regressions were very significant, and all R^2 values were greater than 77%.

The vertical gradient
On a particular site in the forest understorey, relative humidity and evaporation rates varied vertically above the ground.

Along topographical gradients, evaporation rates were significantly lower at 20 cm above the ground than at 1m (ranks test, $P<0.01$, $N=8$ pairs), and again significantly lower 10 cm above the ground than at 20 cm or 30 cm (ranks test, $P<0.01$, $N=9$ pairs). On a same valley site, similarly, evaporation rates were significantly lower at 20 cm above the ground than at 2 m (ranks test, $P < 0.01$, $N = 10$ pairs).

Relative humidity was always higher close to the ground than at 3m. The vertical gradient was sharpest during the dry season, in the beginning of the afternoon or near an exposed topographical position (upper slope or ridge), and the least marked just after rainfall, at night or in a sheltered topographical position (valley bottom).

Both gradients in conjunction delimited sheltered valley zones where the vertical gradient was poorly marked, and exposed ridge-top zones where it became much more severe.

The vertical distribution of seedlings
The establishment heights of 4029 Cyclanthaceae seedlings were recorded on 19 sites. Seedlings established at greater heights above the ground on sites where evaporation was low, i.e. in valley bottoms (Fig. 3). In high forest, correlations of 75% (upper quartile) and 50% (median) percentiles, as well as maximum values, of seedling establishment heights with local evaporation rates at 1 m above the ground were all significant (Spearman, $P < 0.01$ and $P < 0.05$ respectively, $N = 12$). In high forest on ferralitic substrate (the most common forest type at Nouragues), the simple multiplicative regression (on the model

$y = ax^b$) of the upper quartile values of establishment height in relation to the local evaporation rates was highly significant ($P < 0.001$, $R^2 = 85.54\%$, $N = 6$).

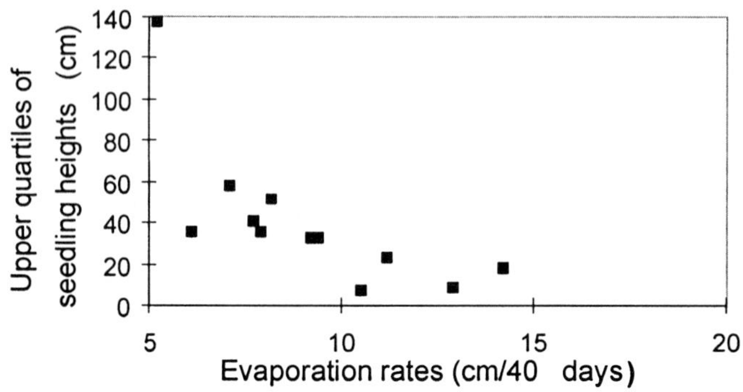

Figure 3. Upper quartile values (75% percentiles) of seedling establishment heights on the 12 high forest sites in relation to local evaporation rates at 1 m above the ground. Number of seedlings on each 12.6m² site varied between 8 and 697 (mean 227).

Seedling recruitment experiments at different heights between ground level and 2m on three sites (ridge, slope, valley) showed increasing recruitment scores with proximity to valley bottom (ranks tests, $P < 0.01$ between valley and slope, $P = 0.0109$ between slope and ridge, $N = 9$) and proximity to forest floor (Fig. 4). Scores correlated significantly and positively with minimum relative humidity values measured at each height level along tree trunks ($P < 0.0001$, $N=27$). Seedlings thus germinate and develop better closer to the ground and nearer the valley bottom, where minimum relative humidity levels are higher.

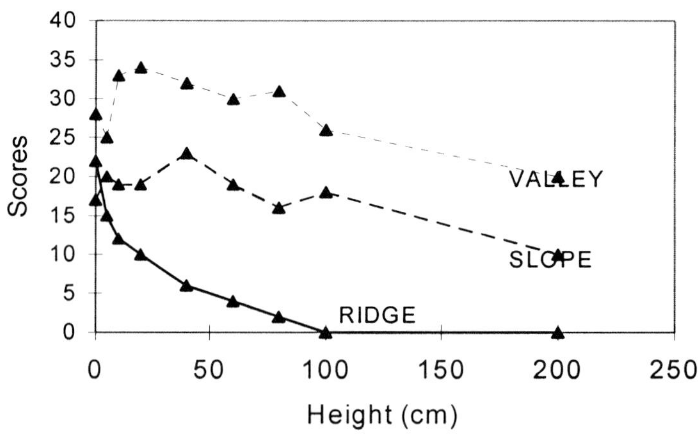

Figure 4. Variation of recruitment scores in relation to height above ground and topographical position

4. Discussion and conclusion

4.1 EPIPHYTIC ESTABLISHMENT

Tree trunks in the understorey display several characteristics: verticality, patchy distribution, constant turnover as trees die and are replaced by others, and a double microclimatic gradient (light, increasing with height, and humidity, decreasing with height). The base of tree trunks, under poor light conditions, is generally the part of the tree with the lowest density of epiphytes (Ter Steege & Cornelissen 1989, Dickinson *et al.* 1993, Wolf 1994). The initial colonisation of a tree trunk can take place *via* seed or vegetative propagation. The latter means is mostly seen in *Philodendron* subgenus *Pteromischum* and in *Thoraccocarpus bissectus*. The small bat-dispersed seeds appear nonetheless as the key to regional persistence. The small size of the seeds allows mass production and thus greater diffusion in the environment. This is a characteristic of true pioneer plant species, which cannot rely on the presence of a seed bank on their establishment site (Chambers 1995, Stöcklin & Bäumler 1996, Eriksson & Fröborg 1996), and also of a number of canopy epiphytes which compensate this way the low probability of reaching a favourable establishment site (Gentry & Dodson 1987, Ackerman *et al.* 1996). Seed rain and density of adequate establishment sites thus become the first limiting factors of recruitment (Laman 1996). Seedlings survive and develop better on the base of tree trunks and on large woody debris. It is vital for the small-sized seedlings to establish on a litter-free substrate - i.e. on an elevated, inclined substrate - as dead leaves prevent light from reaching the seedlings. This is the case for all small-seeded plants, which can only establish on naked ground (Chambers 1995) or on inclined substrates (Blanc 1989, Rousteau 1990, Grubb & Metcalfe 1996).

Recruitment bottlenecks with severe mortality rates take place during the seed rain stage because of the limitation of dispersed seeds and establishment sites, and again between the germination and subadult stages. Although yearly mortality rates are relatively low, growth and development are very slow and liable to stop or even to regress, a characteristic trait of many long-lived plants (Harper 1977). The selection occurring during the seedling stages is not randomly distributed, and characteristics of the substrate are not all neutral in relation to survival probabilities. In particular, leaf litter and lack of stability (mobile or fragile, i.e. 'soft', substrate) determine high mortality rates. In the long run, only tree bases (on which about 20% of Cyclanthaceae seedlings are established) will allow full development to the fertile stages.

4.2 CHIROPTEROCHORY

Cyclanthaceae and *Philodendron* species display two types of persistence: local persistence based on vegetative multiplication (clonality), and regional persistence based on the colonisation of new sites. Mountain refuges are rare in French Guiana, and thus long distance dispersal is the key to species survival on the long term, in particular for species sensitive to drought and climatic fluctuations (Blanc 1997). Here, dispersal distances of several hundred metres give the opportunity to reach another valley, beyond inhospitable ridge-tops. Chiropterochory thus allows both 'colonisation' (*Colonisation Hypothesis,* Howe & Smallwood 1982) and 'escape' (*Escape Hypothesis,* Janzen 1970, Connell 1971) on the topographical and regional scale.

Small understorey bats also display flight characteristics that differ from that of birds. Birds tend to fly in straight lines, for short distances, and defecate mostly when perched or at take-off (Stiles & White 1986). Bird-dispersed seeds fall on the ground, on low obstacles (dead branches, rocks) or on branches, and the colonisation of vertical tree trunks is unlikely. Bats fly almost continuously, with frequent sharp changes of direction, and often defecate during flight (Charles-Dominique & Cooper 1986). The bat-dispersed seeds, with the possibility of being projected sideways during turns, can thus more frequently reach vertical substrates such as tree trunks.

At Nouragues, it appears that most plant species that germinate on tree trunks are bat-dispersed. Not only the *Philodendron, Rhodospatha* (Araceae) and the root-climbing Cyclanthaceae, but also the lianas *Marcgravia coriacea* (Marcgraviaceae) and *Schlegelia paraensis* (Bignoniaceae). In contrast, in French Guiana, the terrestrial, saxicolous or canopy epiphytic Araceae of the genera *Dieffenbachia, Xanthosoma* and *Anthurium* tend to be dispersed by birds. The relationship between dispersal mode and establishment sites in tropical forests has been little studied. In Jamaica, epiphytes of the higher strata tend to be wind-dispersed, whereas most species establishing less than 8 m from the ground are animal-dispersed. All species which display epiphytic or saxicolous establishment between the ground and 2 m high are hemi-epiphytes or root-climbing plants, and all are animal-dispersed (Kelly 1985). In a Brazilian Atlantic forest, Fisher & Araujo (1995) show that epiphytic or terrestrial Bromeliaceae dispersed by mammals occur under 7 m above the ground, whereas bird-dispersed Bromeliaceae are heliophilous.

From an evolutionary point of view, it would be interesting to know the dispersers of the Cyclanthaceae and the *Philodendron* in other parts of their distribution: whether the relationship with one or the other of the genera *Rhinophylla/Carollia/Sturnira* is geographically stable, or varies from one region to the next. It appears stable for French Guiana at least, as the diets recorded here closely match those obtained by Cosson (1994) about 150km to the Northeast of Nouragues. In Guadeloupe, Masson *et al.* (1994) showed that the main disperser of the Cyclanthaceae *Asplundia rigida* and of some *Philodendron* is the bat *Sturnira thomasi*, both *Carollia* and *Rhinophylla* being absent from the island.

4.3 THE MICROCLIMATE AND THE TOPOGRAPHICAL RECRUITMENT GRADIENT

Two important microclimatic gradients prevail vertically in the understorey: light and humidity *sensu lato* (including both relative humidity and evaporation rates). The first apparently did not influence the seedling stages, as no constant variation was observed between the ground and 3 m high. However, it probably plays a key role in the accession to fertility of adult plants, which appear unable to flower under 2-3 m above ground level in closed forest and which tend to be more fertile in the understorey surrounding gaps. Relative humidity and evaporation rates, in contrast, appear determinant in the spatial distribution of seedlings in the understorey. Curiously, the relationship between understorey epiphytes (including cryptogams) and humidity gradients has apparently never been studied statistically before, although many authors knew intuitively of its existence.

The two gradients together determine a situation in which the seedlings' needs (humidity) and the adults' needs (light) do not occur together in space. The strategy developed by most species studied here in response to this situation is an establishment in the understorey, where humidity levels are acceptable to the seedlings, followed by elongation stages and a 'migration' upwards, where light levels allow full development and reproduction ('mobile plants', Oldeman 1974). In Jamaica, as in French Guiana, most root-climbing plants have small seeds and seedlings that establish on tree bases (Kelly 1985). In this way, the limitation of favourable sites is counterbalanced by the possibility to colonise vegetatively (by vertical elongation) sites where seedlings cannot establish (Stöcklin & Bäumler 1996), in this case higher up along the tree trunks.

The seed rain is limiting along ridge-tops, probably because the bat *Rhinophylla pumilio* is specialised on these species and because fertile adult plants are mostly concentrated along valleys. This is a conservative factor, which tends to stabilise the spatial distribution of the group of species exploited by the bat. This 'specialised' dispersal has the advantage of distributing most seeds in an overall favourable environment, where survival rates are higher. However, independently of the adults' spatial distribution, the microclimate determines the breadth of the recruitment stratum (i.e. the space between ground level and the maximum establishment height), and hence the number of favourable microsites, as well as the survival rates of seedlings. It thus constitutes the first determinant of the topographical gradient observed in these plant species. In brief, the scarcity of adults along ridge-tops appears mainly determined by two factors: the restriction of the recruitment zone by the microclimate along ridge-tops (through less establishment sites and lower survival rates), probably the most limiting factor, and the conservative dispersal of seeds by the bats in the valleys, where most fruiting adult plants occur.

The seasonal variations of the climate, with alternating dry and rainy periods, are reflected in the microclimate by alternating periods of more or less marked gradients. The more severe and the longer the dry season, the lower the establishment limits will be set, towards the ground, locally, and towards the valley, topographically, thus reducing the proportion of favourable sites. Conversely, rainy years or decades will determine an extension of the favourable establishment zones upwards and towards the ridges. On the borders of the range, better survival rates will increase recruitment rates, which will ultimately allow the extension of the range of fertile adults and, consequently, that of the seed rain.

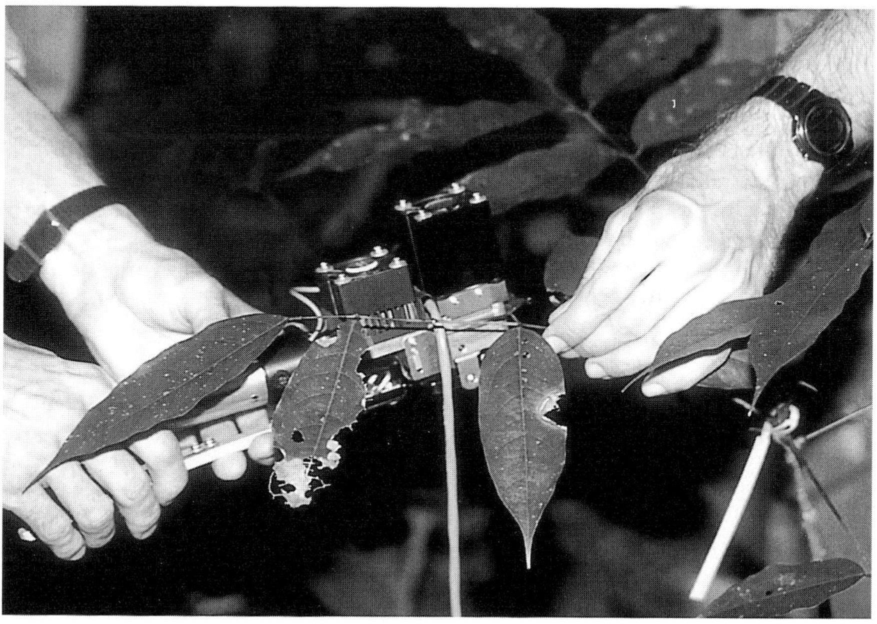

Measuring photosynthesis of the understorey tree *Vouacapoua americana*. Photo Frans Bongers.

CHAPTER 25

POST-DISPERSAL SEED REMOVAL IN FOUR FRUGIVORE-DISPERSED TREE SPECIES

Pierre-Michel Forget, François Feer, Stéphanie Chauvet, Catherine Julliot, Bruno Simmen, Françoise Bayart and Elisabeth Pagès-Feuillade

1. Introduction

In Neotropical forests, a majority of plant species relies upon arboreal and flying frugivores for the dispersal of their seeds (Howe 1986, Howe & Smallwood 1982, Charles-Dominique Chapter 17). Once the fruits are consumed and the pulp or arillote digested, seeds are either regurgitated or dropped into faeces below the parent tree or a conspecific tree, a nearby perch tree or along canopy routes frequently used by animals (Howe 1989, Howe 1993, Julliot 1996, Julliot 1997, Schupp *et al.* 1989). As animals move along pathways in the canopy or fly over trees within their home range seeking other fruit sources, they may forage into forest areas where the previously consumed tree species is at low density or absent (van Roosmalen 1985). Given the retention time in the stomach and gut, and habit of frugivores, seeds may thus be dropped below the parent tree crown, in the parent neighbourhood, or farther away from any conspecific tree (Julliot 1997). Consequently, seeds may be deposited in areas varying in the density of the dispersed tree species (Schupp 1992).

According to the predictive model of seed survival (Janzen 1970), the probability of seed survival and establishment increases with distance from conspecific fruiting trees, i.e. in forest areas having a lowest density of the conspecific trees, thus contributing to spacing between adults. Nonetheless, as Schupp (1992) suggests, the survival chance of seeds also appears to be dependent on saturation of seed predators. Theoretically, such sites occur at the proximate tail of the seed shadow, i.e. beneath parent, and in areas where tree density is overall high and where a greater proportion of seeds are likely to fall, thus saturating seed-eaters (Forget 1992). In the frugivores dispersed tree species *Faramea occidentalis*, Schupp (1992) found that the percentage of seeds surviving was positively related to tree density. Similar patterns have been reported for large-seeded barochorous (dispersed by gravity) tree species such as *Pentaclethra macroloba* in Trinidad (Rankin 1978) and *Quercus oleoides* in Costa Rica (Boucher 1981); seed removal was always higher at patches with low tree densities than at patches with higher tree density. Not only may the density of trees affect survivorship of seeds but also the edibility of their seeds. Notman *et al.* (1996) showed that the difference in seed removal was more related to species characteristics than to distance to parent, and they

Bongers *et al.* (eds.), Nouragues. Dynamics and plant-animal interactions in a neotropical rainforest, 265–274.
2001 Kluwer Academic Publishers. Printed in the Netherlands.

suggested that seed survival was dependent on the type of predators. At the large scale of a tree population, there is thus a conflict between density- and distance-dependent factors of seed mortality or survival (Schupp 1992), which may be species-dependent (Notman *et al.* 1996). Schupp (1992) suggested that seed predators contribute to spacing at low densities, but not at high densities.

This study examines the short-term effect of tree occurrence and predator saturation on tree recruitment, especially relative seed removal probabilities of four plant species that are primarily dispersed by frugivores at Nouragues. We address whether percentages in removal by vertebrates differ between species within family, and whether any interactions exist between species and forest patches as a function of tree density. Experiences were therefore carried out in a forest patch where trees are abundant and at another area where they are in low density or absent. We used seeds of four tree species that differ in their fruit and seed characteristics, and belong to two different family groups, Myristicacaeae and Sapotaceae. The latter family is more likely to be consumed and eventually secondarily dispersed by vertebrates (Brewer & Rejmanek 1999, Spironello 1999) than the former (Forget & Milleron 1991, Forget *et al.* 2000). Based on previous field observations and the literature, the first hypothesis to be tested is that less edible Myristicaceae seeds should be less attractive to vertebrates than the preferred Sapotaceae seeds. The second hypothesis is that seeds of both species should experience greater seed removal rates at deposition site in patches with low tree density than patches with high tree density. In this study, we summarise results on short-term post-dispersal seed removal, and propose a model of seed survival, providing some insights into the meaning of the disappearance and survival of seeds in light of recent studies on seed fate in Neotropical rainforests.

2. Methodology

2.1 SPECIES

The four study species are tall canopy trees with large crowns and were chosen because they are common in Guianan forests as well as at the study site, and are known to be primary food resources for frugivores, especially monkeys (Julliot & Sabatier 1993, Julliot *et al.* Chapter 18, Simmen *et al.* Chapter 8, Simmen & Sabatier 1996). Life-history traits of the species, as well as phenology and seed characteristics, are listed in Table 1.

The Myristicaceae *Virola kwatae* and *V. michelii* produce ovoid dehiscent capsules enclosing a single nutmeg-like seed covered by a red aril (Forget 1991, Forget 1997). Seeds and arils are mainly ingested by toucans (Sabatier 1983), spider monkeys (Simmen *et al.* Chapter 8) and kinkajous (Julien-Laferrière Chapter 20). Seeds are dispersed undamaged from perch trees by birds and during foraging in the canopy by arboreal mammals (Sabatier 1983, P.-M. Forget personal observations.). As a consequence of low post-dispersal predation by predators (Forget *et al.* 2000), *Virola* spp. seedling recruitment is abundant in the understorey (Forget 1997, Forget & Sabatier 1997, Julliot 1997).

TABLE 1. Comparative tree density (dbh > 30 cm/ha) and life-history traits of *Virola kwatae* and *V. michelii* (Myristicaceae) and *Chrysophyllum lucentifolium* and *Manilkara huberi* (Sapotaceae). Into brackets total number found in the plots.

	V. kwatae	*V. michelii*	*C. lucentifolium*	*M. huberi*
Density[a]				
Grand Plateau (25 ha)	0.84 (21)	0.64 (16)	1.84 (46)	0.04 (1)
Petit Plateau (12 ha)	0	1.0 (12)	0 (0)	2.17 (26)
Flowering pattern	annual	annual	annual	irregular
Fruiting	October-March	January-March	April-May	March-May
Seed size (cm)[b]	3-4 x 4-7	1.4 x 2.0	2.5 x 1.2	2.4 x 1.3
Seed fresh weight (g)[b]	4-7	2.2	1.4	1.1

[a] from Nouragues tree data base, and P.-M. Forget & C. Julliot, unpublished data
[b] from Forget (1991), Forget & Sabatier (1997), Chauvet (1997)

The Sapotaceae *Chrysophyllum lucentifolium* and *Manilkara huberi* have globose pulpy fruits, yellow-orange and orange-red, respectively, enclosing several medium-sized seeds that are swallowed and transported by arboreal mammals, principally *Ateles paniscus* and *Alouatta seniculus* (Julliot 1994, 1996, 1997, Simmen & Sabatier 1996). The pattern of dispersal is either scattered along travel routes or clumped in faeces under trees used as defecation and sleeping sites (Julliot 1994). Abundant seedling recruitment in *C. lucentifolium* occurs preferentially where seeds are defecated by monkeys (Julliot 1997, Julliot *et al.* Chapter 18). No data are available on seedling dynamics in *M. huberi*, so far, but it is known that in a related species (*M. bidentata*) dense seedling shadows may be produced in the understorey when herbivory is low (P.-M. Forget unpublished data).

2.2 SITE

The study tree species are not evenly distributed, nor do they co-occur at comparable density within the same forest patches. To test for the effects of tree density and species on seed removal, two 9 ha plots were selected based on the presence-absence and density of each focal species. Plots were 300 x 300 m, one being located at the so-called Petit Plateau (PP) area (12 ha), the other at the Grand Plateau (GP) area which extends over 25 ha. Plots were 500 m distant from each other and corresponded to different forest patches at Nouragues according to soil characteristics (Grimaldi & Riéra this volume) and tree inventory (Poncy *et al.* 1998, Poncy *et al.* Chapter 4). Specific tree densities differ between plots (Table 1). *V. kwatae* and *C. lucentifolium* are present at GP but absent at PP. On the contrary, *M. huberi* grow in dense population at PP, and is almost absent from GP. *V. michelii* was twice as abundant at PP as at GP. An underlying assumption is that frugivores dispersed seeds at each plot creating a within-forest and between-patches seed flux. For instance, it is not rare to observe *V. kwatae* seeds at PP, where adults are absent, due to long dispersal distances by monkeys such as *A. paniscus* (Roosmalen 1985). According to Schupp's (1992) hypothesis of large scale influence of specific tree density, species seed survival is expected to be similar between plots with comparable tree density in *V. michelii*, greater at GP with high density in both *V. kwatae* and *C. lucentifolium*, and greater at PP with high density in *M. huberi*.

2.3 EXPERIMENTS

Twenty sets of 5 seeds each were placed 25 m apart along two perpendicular transects 250 m long installed at the centre of each 9-ha plot. This protocol simulates loose dispersal by monkeys and birds (Howe 1989) but not large faecal clumps such as those generated by howler monkeys (Julliot 1996, 1997). Seeds were located above the ground in a depression to avoid lateral movement due to rainwashing, and their exact location was marked with a blue plastic label beside each seed location. Previous studies had shown that Sapotaceae seeds are removed more rapidly than Myristicaceae. Thus, in order to integrate early and late removal by animals, seeds were censused after 1, 7 and 42 days for the Myristicaceae, and 1, 4, 7, 14, 28 and 42 days for the Sapotaceae. Because of a staggered fruiting pattern and year-to-year variation in occurrence of fruiting between groups of species, experiments were initiated at different periods and years, i.e. 5-6 and 11-12 February 1996 for *V. michelii* and *V. kwatae*, respectively, and 14-15 April 1997 for *C. lucentifolium* and *M. huberi*. In each period, 5-seed batches of each species were set side by side about 1 m apart from each other.

2.4 DATA ANALYSES AND MODELLING

The experiment conformed to a crossed 2-factor (species and plot) design. As experiments were carried out in two different years (1996 and 1997) for the two plant families (Myristicaceae and Sapotaceae, respectively), results were analysed separately per family. Percent of seeds remaining, i.e. not removed by vertebrates, were based on the total number of seeds (N = 5) deposited per pile. Percentages of seed survival after 42 days were compared between species within family and between plots (GP and PP) using ANOVAs with sites of depots nested within plot (Forget 1992) in Systat 9.0 (SPSS 1999) after arcsin square root transformation of the data. Data were analysed after 42 days to integrate early and late removal of seeds by foraging vertebrates.

Modelling had two objectives, first to enable a comparison of failure time in species with different time census, second, to present a general model of long-term seed survivorship based on observed removal rates which are assumed to be representative of the edibility of each species. Following Boucher's (1981) equation we fitted the observed pooled data of the percent of seeds unremoved at 20 sites after 42 days to an exponential decay curve of the form: $DN/dt = -kN$ where N is the number of seeds remaining after time t and k is a constant called the instantaneous removal rate. Such equation has been found to adequately represents removal data (see also Sork & Boucher 1977). The equation is integrated as $N = N_{(0)}^{-kt}$ where $N_{(0)}$ is the initial total number of seeds (N = 100). The integrated form was used to estimate the number, which would remain and thus survive after 100 days. Note that Boucher (1981) refers to instantaneous removal rate assuming that a seed removed is entirely consumed by vertebrates, an assumption which may not be valid for our study species (see Discussion). However, to simplify, we use the term 'removal rate' considering that removed seeds were lost for regeneration at site of deposition. The constant k was then used to compare seed survivorship rates between species and to estimate the proportion of seeds that would not be removed after 100 days and would have survived to establishment. By that time, all studied plant species but *M. huberi* which has dormant seeds up to one year (P.-M. Forget unpublished data) would have established as seedlings.

3. Results

3.1 EFFECT OF PLANT FAMILY

After 42 days, Myristicaceae species were taken by terrestrial vertebrates less frequently in 1996 than Sapotaceae species in 1997 (Fig. 1). Within Myristicaceae, there was a significant interaction between the factors species and plot (ANOVA, $F_{1,\ 38} = 7.85$, $P = 0.008$) but not within the Sapotaceae family ($P = 0.208$; Table 2). In the latter, 42-day survival rate differed between plots (ANOVA, $F_{1,\ 38}=5.324$, $P=0.027$) but not between species ($P = 0.065$). Overall, average seed removal was lower at GP (Myristicaceae: range 12-28%; Sapotaceae: 12-95%) than at PP (20-74 and 98-100%, respectively; Fig. 1).

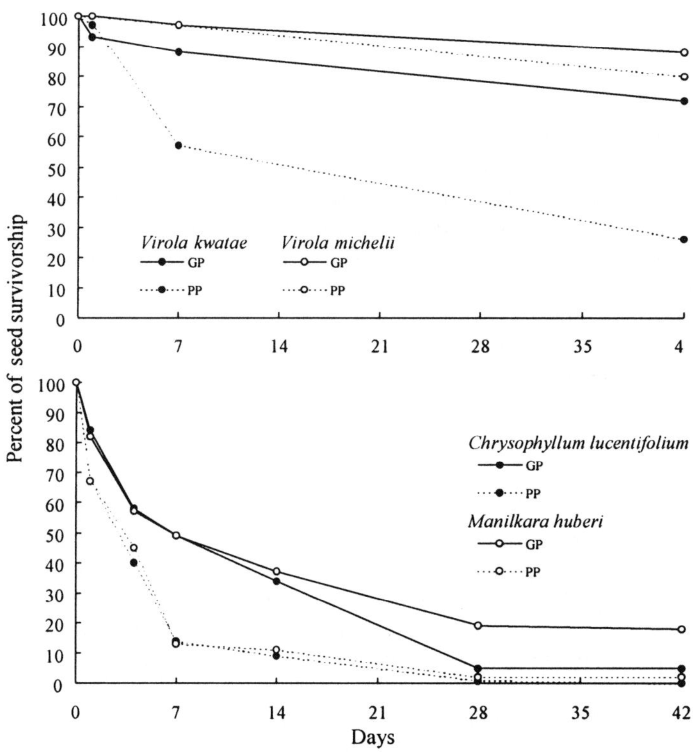

Figure 1. Average percentage of seeds of (a) *Virola kwatae* and *V. michelii* in February 1996, and (b) *Chrysophyllum lucentifolium* and *Manilkara huberi* in April 1997 remaining after 42 days at Grand Plateau (GP) and Petit Plateau (PP).

TABLE 2. ANOVA results of percent of removal of (a) Myristicaceae (*Virola kwatae* and *V. michelii*) in February 1996, and (b) Sapotaceae (*Chrysophyllum lucentifolium* and *Manilkara huberi*) in April 1997 for the effect of forest plot (Grand Plateau and Petit Plateau) and species within family at 42 days after seed placement. Data were arcsine square root transformed. Degrees of freedom indicate which error was used.

Source	Sum-of-Squares	df	Mean-Square	F-ratio	P
(a)					
Plot	10333.17	1,38a	10333.17	17.48	0.000
Species	18481.46	1,38b	18481.46	32.76	0.000
Plot*Species	4428.6	1,38b	428.594	7.85	0.008
Site(Plot)	22453.1	38a	590.872		
Species*Site(Plot)	21437.4	38b	564.141		
(b)					
Plot	1725.5	1,38a	1725.5	5.32	0.027
Species	719.7	1,38b	719.7	3.62	0.065
Plot*Species	326.0	1,38b	326.0	1.64	0.208
Site(Plot)	12316.4	38a	324.1		
Species*Site(Plot)	7560.6	38b	198.9		

3.2 EFFECT OF SPECIES AND TREE PATCHES

In accordance with the predictions of Schupp (1992), the percent of seeds removed differed both between and within species in relation with adult tree density in Myristicaceae; however this was not valid for Sapotaceae. On the one hand, the percent of *V. kwatae* seed removal was significantly lower at GP (28%) than at PP (74%, $F_{1, 38}$ = 19.82, P < 0.0001) where no tree occurs; removal of *V. michelii* seeds was not affected (P = 0.245) by forest patch (12 and 20%, respectively). On the other hand, *M. huberi* seed removal was lower but marginally not significant ($F_{1, 38}$ = 4.05, P = 0.051) at GP (82%) than at PP (98%) though the latter plot had a far greater density. Similarly, *C. lucentifolium* seed removal was not significantly different (P = 0.080) between plots (95 and 100%) despite a lack of fruiting trees at PP.

3.3 DECAY CURVES

The greatest contrast in rates of seed removal was observed among the Myristicaceae species, which have the greatest probabilities of seedling establishment after 100 days, especially in *V. michelii.* In contrast, the Sapotaceae species have the lowest probabilities (Table 3). *V. michelii*, *V. kwatae* (at PP only) and *C. lucentifolium* fit well to the exponential regression (all with R^2 > 0.915). On the contrary, in *V. kwatae* at GP and *M. huberi* at both plots, the *per-capita* value of each set of seeds seems to rapidly decline with time, i.e. within one week in the former species and within three weeks in the latter, then reach a plateau resulting in a flatter slope in the second part of the decay curve, and lower R^2 values (Table 2).

TABLE 3. Average observed 7-day removal rates and estimated 100-day survivorship f seeds based on 1-42 days data set in *Virola kwatae* and *V. michelii* (Myristicaceae) nd *Chrysophyllum lucentifolium* and *Manilkara huberi* (Sapotaceae).

		7-day observed removal rate (%)	Instantaneous Removal rate K	R^2	100-day estimated survivorship (%)
V. kwatae					
	GP	12	0.0081	0.843	44.5
	PP	43	0.0334	0.907	3.5
V.michelii					
	GP	3	0.0031	0.992	73.3
	PP	3	0.0053	0.998	58.9
C. lucentifolium					
	GP	51	0.0826	0.92	0.03
	PP	86	0.1434	0.915	0.005
M. huberi					
	GP	51	0.0498	0.78	0.7
	PP	87	0.1148	0.778	0.1

4. Discussion

4.1 WHY DO SEED REMOVAL RATES DIFFER BETWEEN PLANT FAMILIES AND SPECIES?

On the short term as well as on the long-term, removal rate is highly species-dependent. This is possibly related to three main effects: forest type, seed size and contrasting edibility between species, and year-to-year variation in survival within and between plant family.

First, regarding forest type, our results are different from what is already known from the literature. In a logged and hunted Peruvian forest, Notman *et al.* (1996) observed that seed removal rate (considered equal to predation) in a Sapotaceae (*Pouteria* sp.) was low (ca 20%) though Sapotaceae seeds are usually highly valued by animals (Brewer & Rejmanek 1999, Spironello 1999). Disturbances to forest habitat and fauna may then be the main reasons for this outcome. Indeed, likely due to hunting and rarity of larger mammals (Notman *et al.* 1996), the *Pouteria* seeds maybe not taken by terrestrial vertebrates. At Nouragues, on the other hand, due to its reserve status, independent of the density of adult trees, Sapotaceae seeds have little chance to escape from predators like peccaries, deer or rodents. All ground-dwelling vertebrates at Nouragues are present in reasonable densities today (Bongers *et al.* 2001). Perhaps, as a consequence, Sapotaceae seed survivorship is close to zero and does not seem to be affected by predator saturation in patches with contrasting adult density.

Second, seed size and nutrient quality between species may also explain the observed contrast in seed removal. On the one hand, within Myristicaceae, it is possible that the large seeds of *V. kwatae* are more visible above ground than the small seeds of *V. michelii*. As a result, they are more frequently taken by vertebrates and more likely to be cached due to their large size (Forget *et al.* 1998). Lower survival, henceforth higher removal, of *V. kwatae* seeds at the PP is possibly due here to lack of predator saturation, and intense seed removal might be explained by secondary seed dispersal (Forget &

Milleron 1991). It is also possible that change over time in removal rate is due to modification in the foraging activity of vertebrates across forest patches in relation to the increased availability of alternative desirable food resources such as *Carapa procera* (Jansen & Forget Chapter 26). On the other hand, within Sapotaceae, differences might be due to the rapid germination in *C. lucentifolium*, and leaf litter accumulation above ground which, may conceal the otherwise attractive dormant seeds of *M. huberi* (Chauvet 1997). Despite this, Sapotaceae seeds appear to be more predated and less dependent to predator saturation than Myristicaceae seeds in this year.

Finally, year-to-year variations in several characteristics of the forest habitat such as seed availability and predator population may also explain differences in seed survival between families in this study. Survival rate of *V. michelii* seeds varies between years ranging 35-73% after six weeks in 1992-1993 at Paracou (Forget *et al.* 2000), a result consistent with this study. Seed survival after six weeks was also greater in *V. michelii* (35%) than in *M. bidentata* (10%) in 1992 (P.M. Forget unpublished data). Thus, though the effect of year may not be rejected, these complementary results show that Myristicaceae produce less-preferred seeds than Sapotaceae.

4.2 A MODEL OF SEED REMOVAL

Based on the observed survival curve of four frugivores-dispersed species we propose a general model of seed survival with three theoretical survivorship curves (Fig. 2).

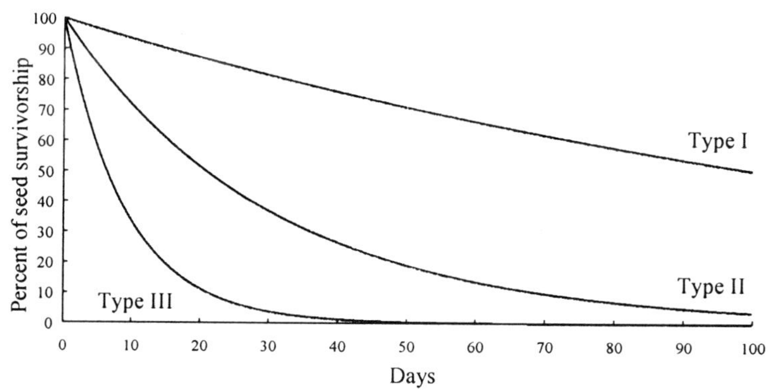

Figure 2. A model of seed survivorship showing three types of 100-day exponential decay curve. Data to build the model are simulated percentages of seeds surviving at site of deposition after 7 and 42 days, respectively, i.e. 90 % and 75 % in Type I curve ($k = -0.0071$, $R^2 = 0.9243$), 50 % and 25 % in type II curve ($k = -0.0348$, $R^2 = 0.7838$), 10 % and 1 % in type III curve ($k = -0.1156$, $R^2 = 0.7838$).

The type I curve is typical for survivorship curves of seeds which are less-preferred food items to terrestrial vertebrates, and thus less taken in a given season as observed in *Virola* spp. Such seeds are likely to be well defended against other pathogens (i.e. insects, fungus) showing high survivorship and leading to a dense population of seedlings in the understorey. On the contrary, poorly defended seeds may all die, and no seedling recruitment will occur if secondary seed dispersers do not take them.

The type II curve describes intermediate survival rate of seeds whose attractiveness decreases through time, as is the case for germinating *Vouacapoua americana* seeds (Jansen & Forget Chapter 26). Occurrence of dense seedling shadows at dispersal sites, i.e. beneath sleeping/perch trees or parent trees, will then depend on the ability of germinating seeds to escape various agents of mortality that may affect clumped seedlings (Howe 1989).

Finally, the type III curve is typical of seeds that are highly preferred food resources for vertebrates in the wet season such as small-sized seeds of *C. lucentifolium* and *M. huberi*. Such type III curve may also be observed in the dry season, as in *Dipteryx panamensis* (Papilionaceae) for which seed survival was nearly nil within four weeks in Panama (Forget 1993).

4.3 WHAT IS THE MEANING OF SURVIVORSHIP CURVES?

Survivorship curves do not adequately describe true seed survival since part of the remaining seeds may fail to establish following infestation by insects (Howe 1989, Howe *et al.* 1985) or desiccation (Forget 1990). Often linked to seed mortality, removal masks the occurrence of secondary dispersal among species with seeds large enough to be handled by rodents, especially in Myristicaceae (Forget & Milleron 1991, Forget *et al.* 1998) and Sapotaceae (Brewer & Rejmánek 1999, Chauvet 1997, Spironello 1999). In Panama, Forget and Milleron (1991) observed that removal rate of *Virola nobilis* seeds (weight 3 g) was greater where parent tree density was low and that seeds were secondarily dispersed by agoutis (*Dasyprocta punctata*). More recently, Asquith *et al.* (1997) observed that rate of *V. nobilis* seed removal and fate differed between forest types.

At Nouragues, it is not clear so far whether removed seeds in Myristicaceae are eaten or dispersed by rodents. We rarely found *Virola* spp. seed pieces above ground. It is therefore possible that removal is due to small rats (*Oryzomys* spp., *Proechimys* spp.), which may take seeds to their burrow to consume them (Brewer & Rejmanek 1999) or by deer that swallow and destroy *Virola* spp. seeds (P.-M. Forget unpublished data, in stomach content of *Mazama americana*). Nonetheless, we postulate that seeds of *V. kwatae* (weight 4-7 g) are dispersed by rodents as observed in another large-seeded Myristicaceae (*Osteophloeum platyspermum*) (F. Feer & P.-M. Forget unpublished data), as well as for other seeds in this size range in Central America (Forget *et al.* 1998, Hallwachs 1994).

Seeds of Sapotaceae that remain above ground generally germinate within two weeks and up to about a year in *Manilkara* spp. (P.M. Forget personal observation), and produce seedlings within one or two months (Chauvet 1997, Notman *et al.* 1996, P.-M. Forget personal observation). According to survivorship curves, seeds of *C. lucentifolium* and *M. huberi* that were set at both GP and PP in April, and remaining after 42 days, would have little chance (less than 1%) to establish unless part of removal corresponded to secondary dispersal with successful seedling establishment. Chauvet (1997) showed that removal rates of *M. huberi* and *C. lucentifolium* differ between forest plots (GP and PP) and throughout the fruiting season, for instance, up to 8-10 percent of seeds remained after 21 days at PP in early May. Using thread-marked seeds to relocate them and to determine their fate, Chauvet (1997) found that a greater proportion of seeds were removed and cached in *M. huberi* (24%) than in *C.*

lucentifolium (6%). Further investigation of seed fate would require taking into consideration the ratio of seeds establishing at defecating sites versus that of germinating in rodent caches in the surrounding to adequately estimate the survival probabilities of seeds dispersed by frugivorous vertebrates.

The question of the significance of seed removal is thus crucial to estimate seed survivorship. Boucher (1981) made the assumption that mammals act almost exclusively as predators of *Q. oleoides* acorn (weight 1-7 g) but Hallwachs (1994) showed that agoutis (*D. punctata*) scatterhoard them. Consequently, high instantaneous seed removal rate may correspond either to an important seed dispersal rate by rodents or to an intense seed predation depending on the availability of alternative resources. Conversely, one may postulate that low removal of seeds in dense tree patches as observed by Boucher (1981) for large-seeded species means either low dispersal or low predation of seeds by vertebrates. Because insects frequently infest unremoved seeds or desiccate, it is therefore probable that seedling establishment will be weak when seed dispersal by rodents is lacking.

Accordingly, whereas the type III decay curve predicts zero survival after 100 days, the impact of secondary seed dispersers may not be negligible to the extent that buried seeds are not recovered and eaten by vertebrates. However, because seed removal often fluctuates during the fruiting season, and because unremoved and unburied seeds may survive as seedlings, it is possible that the role of rodents as seed dispersers is anecdotal for seedling recruitment among small-sized species showing a type III curve. In that case, high removal rates might equal predation by both small and large rodents, especially when seeds are neither infested by insects after dispersal nor require burial for germination.

Acknowledgements

This study was granted by the CNRS, the Ministère de l'Environnement with a grant SOFT (Sol et Forêt Tropicaux) # 96042 to PMF and BS. We thank Christopher Baraloto for correcting a previous version of the text, and Patrick Jansen for his comments.

CHAPTER 26

SCATTERHOARDING RODENTS AND TREE REGENERATION

Patrick A. Jansen and Pierre-Michel Forget

1. Introduction

Tree seeds that drop down to the rainforest floor, either naked or embedded in fruit pulp, often disappear after a while. Many authors refer to this phenomenon as seed predation, implicitly assuming that the seed is destroyed by seedeaters such as rodents or ungulates (Forget *et al.* 1998). However, not all seed removal is by seedeaters. Many species of animals remove seeds while feeding on what is around seeds rather than on the seeds themselves. Terrestrial birds, mammals and reptiles ingest seeds while feeding on fallen fruit (e.g. Fragoso 1997, Érard & Sabatier 1988), tortoises ingest seeds while feeding on dung of frugivorous mammals (B. Josseaume unpublished data.), and dung beetles take and bury seeds with dung (Shephard & Chapman 1998, Andresen 1999, Feer 1999). All these animals may bring viable seeds into favourable conditions.

Removal even by seedeaters need not be detrimental to seeds. Some species of mammals and birds store important amounts of seed for later use in periods of food scarcity by creating numerous spatially scattered caches with few seeds each, behaviour called 'scatterhoarding' (Morris 1962). Several animals hoard seeds by hiding them in shallow caches in the soil. If such seeds, for whatever reason, happen to escape consumption by the hoarder, they may be in a good position to germinate and establish. Thus, even seedeaters may function as seed dispersers that enhance seed survival.

This chapter considers the role of scatterhoarding animals in tree regeneration in the Guianan rain forest. In this region, two important scatterhoarding animals are the Red acouchy, *Myoprocta exilis* (Wagler 1831), and the Red-rumped agouti, *Dasyprocta leporina* (L. 1758) (Dubost 1988), both caviomorph rodents. It has been hypothesised that these rodents and their food plants are mutually dependent; the rodents would not survive periods of food shortage without the food provided by the trees, while the tree species would not survive without the rodents dispersing their seed (Smythe 1978). Whether the rodents are indeed obligate mutualists for certain species of trees is highly relevant for sustained management of forest ecosystems, since interdependence could necessitate protection of animal populations (Jansen & Zuidema 2001).

Assuming that rodents bring seeds to safer sites, the fact remains that they do so in order to eat the seeds later. This paradox of seed predators functioning as seed dispersers has been the basis of much research in French Guiana. Our aim is to review evidence of scatterhoarding rodents being favourable to seeds, and to identify lacunas in the existing

Bongers *et al.* (eds.), Nouragues. Dynamics and plant-animal interactions in a neotropical rainforest, 275–288.
2001 Kluwer Academic Publishers. Printed in the Netherlands.

knowledge. First, we will discuss the question of how important scatterhoarding rodents actually are in terms of quantities of seeds removed and scatterhoarded, and what dispersal patterns they create. Secondly, we consider scatterhoarding from the viewpoint of the rodents on the one hand and the trees on the other. To evaluate whether the interaction between rodents and trees is mutualistic, we will try estimating the net contribution of scatterhoarding to regeneration by subtracting the costs of scatterhoarding. Finally, we briefly discuss some evolutionary aspects of the interaction between trees and rodents and point out directions for further research.

2. Patterns of seed removal and dispersal

Sabatier (1983) and Forget (1990) showed that the large seeds of *Vouacapoua americana* (Caesalpiniaceae), a canopy tree species that had often been assigned to the 'unassisted dispersal' group, were removed by rodents once they had been shed. Later, similar results were found in other tree and liana species that produce fruits with large edible seeds but no edible fruit pulp, such as *Carapa procera* (Meliaceae) (Forget 1996), *Astrocaryum paramaca* (Palmae) (Forget 1991) and *L. alba* (Chrysobalanaceae). Rodents also appeared to remove a great variety of other seed species. These included large seeds of bird- and monkey-dispersed species, such as the Myristicaceae (Forget & Milleron 1991) and Sapotaceae (Forget *et al.* Chapter 25), as well as large seeds of bat-dispersed species, such as the Lecythidaceae and Fabaceae (Forget 1992, 1993).

Seed removal by acouchies and agoutis is also important in terms of numbers. Jansen and co-workers laid out more than 100 batches of 10-50 large seeds and video-recorded animal activity during the following day(s). In nearly all cases, scatterhoarding rodents were the first seedeaters seen at the plots. Acouchies accounted for almost 90 per cent of cases of discovery, agoutis for less than 10 per cent. Other important seedeaters, especially in terms of seed quantities removed at a time, were collared peccaries (*Tayassu tajacu*) and white-lipped peccaris (*T. pecari*). These wild pigs eat seeds on the spot, crushing and killing all seeds but the minute (Feer *et al.* Chapter 21). Other seedeaters, such as spiny rats (*Proechimys cuvieri* and *P. guianensis*) and Guianan squirrels (*Sciurus aestuans*) rarely removed seeds. The former both scatterhoard (Forget 1991, Hoch & Adler 1997, Adler & Kestell 1998) and larderhoard seed (Guillotin 1982), while the latter scatterhoard seed in trees.

Removal experiments and video recordings have shown that acouchies and agoutis eat few of the seeds in situ. Forget (1996) found that the percentage of *C. procera* seeds in experimental batches consumed in situ was approximately 10-20 per cent, and decreased during the season. Removal rates of seed species also differ between years, depending on the absolute and relative seed availability and the relative food value of seeds (Smallwood & Peters 1986). In lean years, all consumable seed species are removed quickly, with strong competition for seeds between acouchies, agoutis and peccaries. In years of high production, the animals concentrate on preferred species (see e.g. Fig. 1). In 1996, for instance, removal of *C. procera* seeds did not set off until production of the preferred species *Licania alba* had ceased completely. Then, the animals became interested in taking even old and germinated seeds of *C. procera*.

Forget (1990) laid out thread-marked *V. americana* seeds, and searched for them after their disappearance. He discovered that a great proportion (70 %) was scatterhoarded in the direct surroundings, threadmarks protruding from the soil enabling relocation of the buried seeds. Further experiments with threadmarks and video cameras showed that the great majority of all seed species taken by acouchies or agoutis are scatterhoarded. Both species typically cache seeds individually, scattered throughout the forest surrounding a seed tree, at distances ranging from 0.5 up to 125 m (*C. procera* seed cached by an acouchy in 1999). Seeds are buried at depths from zero to 6 cm below the soil surface and are always covered with leaves. Caches are often near or in objects such as palm cones, tree logs and trunks (Smythe 1978, Forget 1990, Vander Wall 1990), and tangles of branches and lianas.

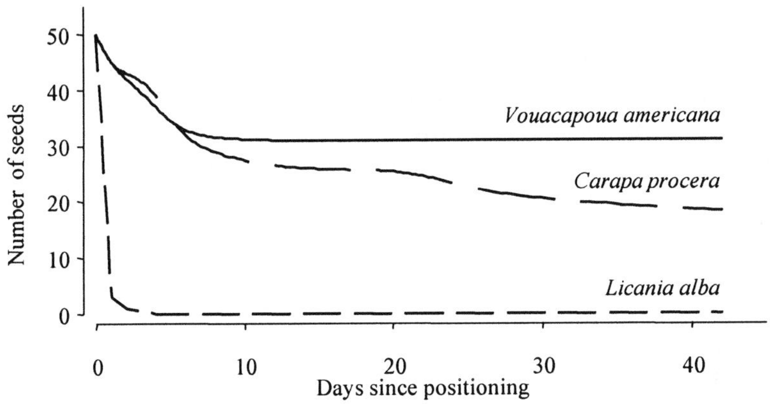

Figure 1. Removal of three seed species by scatterhoarding rodents. 50 seeds per species were placed at the forest floor along a transect with 2m inter-spacing, and 6m between conspecific seeds ($n = 50$). Nouragues, May 1996, (P.A. Jansen and A. Siepel, unpublished data)

Figure 2 shows a typical layout of caches of*Licania alba* seed after one week of video-recorded seed removal by an acouchy from an artificial feeding plot. The missing seeds may have been stored beyond the 50 m search radius.

3. The animal perspective

Hoarded seeds seem to play a vital role in the life cycle of acouchies and agoutis. Most hoarding is done during the wet season, when the majority of tree species fruit; i.e. February to April and especially May to June in French Guiana (Forget 1996). The exploitation of hoarded seeds especially occurs during the period of seed scarcity that follows. The animal's dependence on hoarded supplies during this period seems very strong. The seed proportion in the agouti diet increases when fewer seeds are available. In the leanest period of the year, seed matter even makes up almost 75% of the agouti stomach content (Henry 1999). The rodents literally live on their reserves.

Having sufficient hoarded seed available for use at the desired moment is an obligatory yet complicated task, since the rodents are not unique in their interest for the seeds. There is a constant threat of losing stored (and to-be-stored) seed reserves to food competitors, seed perishing and seed germination. Given the vital importance of reliable supplies, agoutis and acouchies face an enormous pressure to reduce these losses. Therefore, it is imperative that they adopt a profitable strategy of seed caching and cache management.

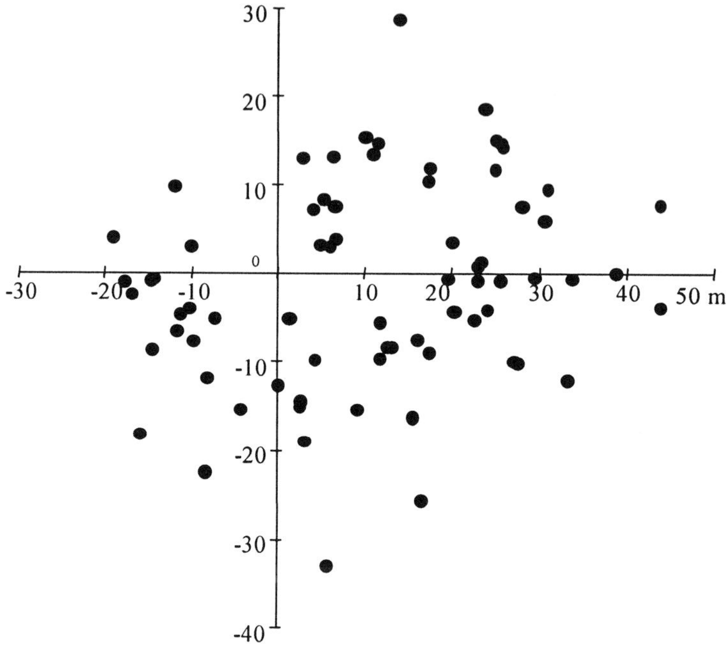

Figure 2. Spacing pattern of *Licania alba* seeds scatterhoarded by an acouchy. 52% of 139 threadmarked seeds taken from an artificial food source (origin) were relocated. Nouragues, May 1996 (P.A. Jansen and A. Siepel, unpublished data).

3.1 SELECTIVITY AND SEED VALUE

The amount of seed scatterhoarders can store is limited not only by seed availability, but also by the hoarder's time and energy budget, the availability of suitable storage sites, and the needs of competing seedeaters. In the face of these limitations, it may pay for animals to be selective in their choice of seed use. Caching efficiency might be increased by caching the most valuable seeds first. Seed value roughly depends on energy/nutrient content and storage life. For short-term hoarding, animals can simply focus on the net energy content (accounting for handling costs). Nuthatches, for example, prefer caching husked sunflower seeds to unhusked ones, since the former yield more net energy as no investment in dehusking is required (Moreno & Carrascal 1995). Long-term hoarding demands a focus on seed value at the desired moment of consumption rather than at time of hoarding. Nutritional content of cached seeds may decrease due to germination (i.e. use of seed reserves by the plant), by insect infestation, or by degradation of nutrients (Post 1992). Therefore, it may be attractive to cache slowly germinating seeds with hard, protective endocarps, or with a high content of secondary compounds that lengthen storage life (Hadj *et al.* 1996, Dearing 1997). Tannins, for instance, inhibit insect attack and, thus, help jays conserve acorn supplies (Fleck & Woolfenden 1997).

Both acouchies and agoutis are highly selective in seed caching, for which they have only their territory available. This is reflected by an enormous variation in removal speed of laid-out seed and cache spacing by acouchies for different edible seed species that are found in Guianan forests. Although these seed species differ in an infinite number of aspects, we believe that the rodents select on long seed storage life and high energetic content after storage, which conforms well to the importance of long-term supplies in their life cycles. The preference of *L. alba* and *C. procera* seeds over *V. americana* seeds of comparable fresh weight (Fig. 1), for instance, can be explained by differences in storage life and fat content (Table 1). Furthermore, acouchies clearly select uninfested seeds for caching, as do Gray squirrels (*Sciurus carolinensis*; Steele *et al.* 1996) and Blue jays (*Cyanocitta cristata*; Dixon *et al.* 1997).

TABLE 1. Properties determining nutritional value and storage life of 3 seed species scatterhoarded by acouchies and agoutis. Seeds were collected below parent trees in Nouragues in April 1996. (A. Siepel and P.A. Jansen, unpublished data).

Seed properties	Species		
	Licania alba	*Carapa procera*	*Vouacapoua americana*
Nutritional value	n = 25	n = 63	n = 50
fresh weight (g)	28.3 ± 7.5	22.9 ± 9.1	36.7 ± 10.5
dry weight (g)	9.6 ± 2.6	7.5 ± 3.0	16.3 ± 4.8
fat (g) [1]	2.4	3.7	0.2
Protein (mg) [1]	576	450	652
oligo-saccharids (mg) [1]	384	375	978
Energy (kJ) [1]	206	224	274
Storage life			
Parasitism (%)	9	6	34
seed coat type	hard	Medium	soft
time till germination	>10 mo	2-5 wks	1-2 wks

[1] Chemical analysis was done in duplicate on a mixture of uninfected seeds.

3.2 CACHE SPACING AND CACHE ROBBERY

Rather than actively defending supplies against competitors (which may be effective against other rodents, but not against peccaries and insects), scatterhoarders protect food against thieves through risk spreading or dilution. They anticipate the limited foraging perseverance of their food competitors by making the density of cached seeds so low that the average time needed for competitors to find a second cache after having found one is too long to make further searching worthwhile. The threshold is at the so-called 'giving-up density' (Bowers *et al.* 1993).

Caching a certain amount of seed in lower densities requires a greater area and, thus, a greater investment of time and energy for both transportation and cache recovery compared to caching in higher densities. Stapanian and Smith (1984) and Clarkson *et al.* (1986) reasoned that there is an optimum cache density at which the yield of remaining caches per unit investment is maximal. This optimum density should be different for seeds of different food value, high-value seeds being cached at lower densities than low-value seeds, since a greater yield merits greater investment in caching as well as in searching by food competitors (Stapanian & Smith 1984).

Larger seed species, which generally have higher food values, indeed tend to be hidden further away and in lower densities than smaller seed species (Stapanian & Smith 1984, Waite & Reeve 1993, Jokinen & Suhonen 1995, Vander Wall 1995b, Forget *et al.* 1998, Leaver & Martin 1998). Acouchies were also found to discriminate between high and low food values. When we supplied an acouchy with *C. procera* seeds covering a wide range of fresh weights, the animal hid heavier seeds further away from the source than lighter ones (Fig. 3).

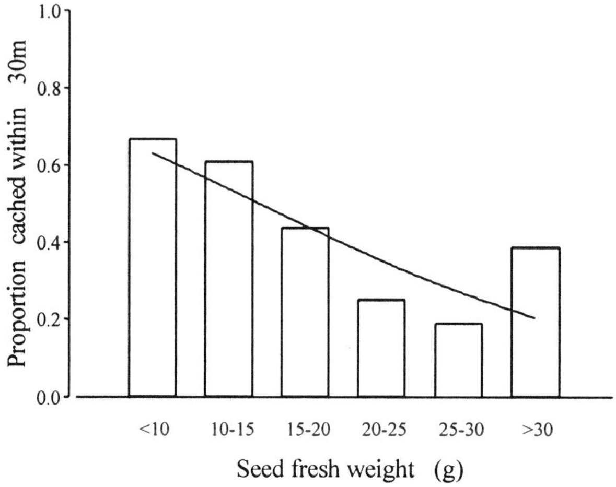

Figure 3. Proportion of *Carapa procera* seed cached within 30 m from an artificial seed source by at least one acouchy. Larger seed were taken beyond 30 m more often than smaller ones (Logistic regression on binary data: Wald=13.7, $p<0.0001$). Nouragues, May 1995, P.A. Jansen, P.M. Forget and L. van der Voort.

Following the same line of optimality reasoning, we may also predict that the optimal cache density will be lower under circumstances of fiercer competition with food competitors, since competition determines how intensively food is sought-after. Food scarcity makes a given seed more valuable and will, thus, lower the optimal cache density. Variation in local fruit availability between years (or areas) can thus lead to great differences in cache densities between years (or areas). Other factors that influence cache spacing include the characteristics of the habitat. Giving-up densities of potential cache thieves may, for instance, be lower in habitats that they can explore more easily. If rodents anticipate such factors, we may expect them to cache seeds at higher densities in habitats that are poorly accessible to pigs, such as treefall gaps with tangles of lianas and tree crowns.

3.3 CACHE PROPERTIES

Seeds are often cached near or in objects, such as logs, and seeds with hard endocarp are more common in soil near such objects than in the open (Kilty 1981). These facts have lead to the idea that the rodents use objects as visual landmarks to facilitate relocation of cached seeds. This implies the involvement of spatial memory. Although there is plenty of evidence for use of spatial memory by birds (Krebs *et al.* 1996), there is relatively little for rodents. Visual landmarks are used to remember the locations of caches by some species of chipmunks (Vander Wall 1991, Vander Wall & Peterson 1996), squirrels (Jacobs & Liman 1991, MacDonald 1997), kangaroo rats (Jacobs 1992, Barkley & Jacobs 1998), and golden hamsters (Georgakopoulis & Etienne 1994), however not by deer mice (*Peromyscus maniculatus*) (Vander Wall 1993c). Whether acouchies and agoutis use visual landmarks for the recovery of their caches remains to be studied.

An alternative explanation for preferential hiding of seeds near objects is the favourable characteristics of substrates near objects. Some substrates may be easier to dig into, or have better conservation properties. Seeds may, for instance, be less susceptible to spoilage in dryer substrates such as moulded wood along tree logs and accumulations of litter below palms (Vander Wall 1991). Dry substrates may also delay germination and, consequently, the depletion of seed reserves. Moreover, food competitors can have greater difficulty in finding seeds in dry substrates because these blur olfactory cues (Vander Wall 1991, 1993c, 1995c, Jacobs 1992). The extra investment needed for selecting these substrates would be merited by lower cache losses.

The same line of reasoning was followed by Vander Wall (1993a), who predicted that the depth at which seeds are buried increases with seed value. His idea was that food competitors are more tenacious in looking for more valuable seeds, necessitating better caching, while greater seed value at the same time permits hoarders to invest more in hiding. We think that acouchies behave according to this model, as they bury valuable seeds (e.g. *L. alba*) deeper (at up to 6 cm) than less valuable ones (e.g. *V. americana*, mostly hidden right below the surface.

3.4 CACHE MANAGEMENT

Most cached seeds disappear within a few weeks after caching. Forget (1990) found that 57 per cent of 103 cached *V. americana* seeds at three sites disappeared during one month. Likewise, we found that all of 77 *C. procera* seeds cached by (assumingly) one acouchy were gone after one month. The proportion of initial caches left untouched, however, is not a good predictor of the number of seeds surviving. We found that many of the disappeared *C. procera* seeds had been re-cached. We think that most seeds are dug up by the same animals that hid them. However, experiments with man-made caches of thread-marked seed have shown that animals also re-cache seeds stolen from supplies of other individuals (P.A. Jansen and co-workers, unpublished data). We also observed that seeds were re-cached more than once. This has also been observed in the Sierre Nevada where *Tamias amoenus* chipmunks were found to re-cache pine seeds up to four times (Vander Wall & Joyner 1998).

The continuous moving around of seeds seems part of an active management of supplies. Re-arranging caches so shortly after their creation fits to the 'rapid sequestering' strategy (Jenkins & Peters 1992). On discovery of a seed source, an animal will first hide seeds provisionally, as quickly as possible. Only in second instance, after the seed source has been depleted, the animal will optimise cache spacing. This behaviour can be understood if we consider that the risk of losing seed to food competitors will always be greater for exposed seeds than for stored seeds. Agoutis and acouchies may, thus, seek a compromise between reducing the high short-term risk of competitors taking exposed seeds and reducing the lower longer-term risk of food competitors pilfering seed caches. Kangaroo rats (Jenkins & Peters 1992) and jays (Waite & Reeve 1995) appear to behave in this way.

Another possible reason for rodents to open their caches every now and again is to inspect seed condition. This enables the rodent to judge whether a seed is still good enough to be conserved. Seeds with short remaining storage lives, such as germinated or infested seeds, are better eaten. Re-caching may also refresh the animal's memory of stored seeds.

4. The plant perspective

Scatterhoarding involves seed transportation, scattering, burial and change of the seed micro- and macro-environment. This has several potential advantages to plants. This section discusses how scatterhoarding would increase the chances of seeds surviving, germinating and establishing if the rodents recovered none of the caches they make.

4.1 ESCAPE FROM PREDATORS, PESTS AND SIBLING COMPETITION

The ultimate goal of scatterhoarding - i.e. reducing seed predation and spoilage - combines well with the interest of the parent plant. There are various ways in which the scatterhoarding may reduce seed mortality. First, transportation and scattering moves seeds beyond the direct vicinity of both the parent plant and siblings. This decreases the risk of competition between siblings. Furthermore, it decreases the risk of mortality due to pathogens and pests associated with the parent plant (Janzen 1970, Connell 1971, Hammond & Brown 1998). Most of all, however, it reduces the risk of (distance- and) density-dependent mortality due to all kinds of predators and pests (Janzen 1970, Connell 1971, Hammond & Brown 1998).

In temperate areas, experimentally scatterhoarded seeds indeed disappear more slowly at lower densities (Stapanian & Smith 1984, Clarkson *et al.* 1986), showing that the seeds suffer lower predation. Seed burial also reduces the risk of seeds being discovered by predators or being attacked by pests. Experiments have shown that buried seeds disappear at far lower rates than seeds laid out on the soil surface (Vander Wall 1990).

4.2 FACILITATION OF ESTABLISHMENT AND GROWTH

The burying of seeds by scatterhoarding rodents facilitates seedling establishment (Forget 1990, 1997a), and is even a prerequisite for germination in some rodent-dispersed species (Vander Wall 1992). The reason is probably that the availability of water for buried seed is more stable, preventing rot during moist periods and desiccation during periods of drought (Forget 1990, Kollmann & Schill 1996). The latter is especially important for recalcitrant seed species. Furthermore, roots of buried seed penetrate the ground more easily, and emerging seedlings are rooted more firmly. An unexplored idea is that burying by rodents might stimulate interaction with (endo-) mycorrhizal fungi (Pirozynski & Malloch 1988).

An interesting hypothesis is that scatterhoarding favours seedling establishment and growth by increasing the proportion of seeds in favourable microenvironments, such as treefall gaps. Both buried seeds and seedlings of *V. americana* and *C. procera*, for instance, perform better in canopy gaps than in understorey (Forget 1997a), as do other large-seeded plant species (e.g. Kazuhiko *et al.* 1997). Scatterhoarding could, for instance, be advantageous because a tendency of hiding seeds near objects renders a directed dispersal to light environments if such objects are correlated with canopy gaps. Logs in the forest, however, are mainly in the understorey since most canopy openings close within five years (Van der Meer & Bongers 1996a, b). Alternatively, scatterhoarding could be advantageous simply because a regular spacing of seeds increases the number of independent chances of seed hitting the patch-wise distributed canopy gaps.

4.3 THE FATE OF NON-SCATTERHOARDED SEED

Compared to the various potential benefits that scatterhoarded seeds face, the fate of seeds that remain below parent trees seems disastrous. Many studies at tropical sites have shown that seeds below parent trees suffer heavy density-dependent insect attack (Hammond & Brown 1998). Seeds of large-seeded species that were placed in batches below typical parent trees usually died (if not removed) due to predation by granivorous insects and ungulates (notably peccaries). For instance, moths had infested all *C. procera* seeds that had not been removed from experimental seed plots under adult trees after one month, (N= 20 adult trees, Paracou, French Guiana; Forget 1996). *V. americana* seeds that are not buried are attacked by insects (Curculionids) and lose their ability to germinate, especially in dry conditions (Forget 1990). Almost no *V. americana* seedlings establish when seeds are not dispersed, although some seedlings may survive in dry years (Forget 1997b).

Once seeds contain larvae, they have a lower chance of being removed and hoarded, because agoutis and acouchies avoid caching such seed with short storage lives. Peccaries, in contrast, do eat infested seed. Cafeteria experiments in which acouchies were observed foraging on mixtures of infested and uninfested seeds of *L. alba* showed that the animals are very skilled in distinguishing good and bad seeds by smell. All seeds were removed except for the infested ones (P.A. Jansen & A. Siepel, unpublished data). Furthermore, infested seeds have lower chances of germinating. In *C. procera*, for example, germination of infested seeds was 14% (N = 96 seeds; Forget 1996), while 97% of seeds free from infestation germinated (N = 182; P.A. Jansen, unpublished data). Moreover, infested seeds that do germinate should make smaller, less vigorous seedlings due to reserve consumption by insects.

Parent trees of *C. procera* along canopy gaps had quite high recruitment (from surface seeds) below them in some years, including seedlings that managed to emerge from infested seeds. This is in line with the idea that effects of invertebrate attack (of both seeds and seedlings) can be buffered by light availability, leading to establishment of seeds that would otherwise have died (Hammond & Brown 1998). Such buffering of infestation effects by light supports the idea that directional dispersal towards treefall gaps, all other factors being equal, could already make dispersal by scatterhoarders advantageous to trees.

5. Mutual benefit?

It is beyond doubt that rodents benefit from scatterhoarding tree seeds. Acouchies and agoutis actively choose to scatterhoard, because they need stored seeds to supplement their diet during a later period of food scarcity (Henry 1999). Whether scatterhoarding is also beneficial to the tree species that passively undergo scatterhoarding is far less clear. It is even questionable, since seedling recruitment conflicts with the interest of individual acouchies and agoutis.

The current idea that the relationship between scatterhoarders and trees must be mutualistic is largely based the logic that net survival below parent trees is often close to zero, so any survival must come from scatterhoarded seeds. The evidence, however, is merely circumstantial, and does not rule out alternative possibilities for regeneration, such as that establishment of non-scatterhoarded seeds occurs rarely (e.g. only in particular years), yet sufficiently often during a tree's life span. This section therefore focuses on the net benefit of scatterhoarding for the plants. We will discuss whether and how advantages for trees of having their seed scatterhoarded weigh up against the exploitation of seed supplies by the rodents.

5.1 THE FATE OF SCATTERHOARDED SEED

To determine whether scatterhoarding indeed contributes to the fitness of the parent plants we shall have to compare the probabilities of establishment between seeds that are scatterhoarded and seeds that are not. However, determining the fate of scatterhoarded seed is difficult. It requires that individual seeds be followed until they either die or establish. This is very much complicated by active management of cached seed by the rodents. It is meaningful that the net survival of seed scatterhoarded by any species has been estimated in so few studies (Vander Wall 1994, 1995a, 1995c, Vander Wall & Joyner 1998), only on short distances, and only for a limited part of the batch that was followed. Tracking seeds managed by acouchies and agoutis is particularly complicated because of the dense habitat and the large distances covered. The repeatedly surveying of large areas for protruding threadmarks of buried seeds is extremely labour-intensive. Tracing methods such as the spool-and-line method have been effective for tracking acorns dispersed by*Apodemus* mice over short distances (0.6 to 9.5 m; Yasuda *et al.* 1991), but appeared to be unsuitable for tracking seed dispersed by acouchies (personal observations). The use of transmitters (Sone & Kohno 1996) is also unpractical for following the fates of many seeds simultaneously. Consequently, the comparison of seed fates remains a major challenge in this field of research.

So far, attempts to follow naturally and artificially cached seed have shown that acouchies, agoutis and peccaries are very efficient in (re)locating cached seeds. Most naturally cached seeds disappear shortly after caching. Experiments with artificial scatterhoarding, which control for exploitation by the scatterhoarding animal itself, indicated that the chance of cached seeds being discovered by vertebrate cache thieves is very high (P.A. Jansen, R. Brienen & P.M. Forget, unpublished data). These findings, however, do not imply that survival is low, for two reasons. First, rodents are known to re-cache seeds, which may again escape consumption. Second, the high robbery risk for artificial caches may be an artefact of seed manipulation.

Planting experiments suggest that germinating seeds are even more vulnerable to predation by scatterhoarders than seeds. Scatterhoarding rodents attack establishing seedlings while managing or exploring their supplies, and seem to use emerging sprouts as cues (P.A. Jansen, unpublished data; N. Smythe, personal communication). Epigeal seedlings (e.g. Leguminosae, Meliaceae, Myristicaceae) can be uprooted or severely damaged when rodents remove the seed, while hypogeal seedlings (e.g. Sapotaceae) are usually killed as the cotyledons are eaten. If seed consumption continues till the seed is entirely and clearly empty, the net result of scatterhoarding must be estimated from the proportion of fully established seedlings.

5.2 SHORT MEMORIES AND OTHER ESCAPE ROUTES

Theoretically, rodents do best by hiding not a single seed more than they need, come back to eat every single one of them, and leave no benefits whatsoever for the plant. How then can scatterhoarding in any way be beneficial to plants? The most common line of reasoning is that seeds escape consumption simply because the animals forget caches, notably because memory capacity drops after scatterhoarding. Thus, the hoarder becomes a naive searcher for caches. While some scatterhoarders, such as Clarck's nutcracker (*Nucifraga columbiana*), remember caches for extremely long periods of time (Balda & Kamil 1992), diurnal rodents indeed tend to depend on short-term spatial memory for relocating caches (Jacobs 1992, MacDonald 1997).

Another possibility is that rodents play safe by always storing more seeds than they will probably ever need, because they are unable to predict what part of the hidden food will spoil or be stolen by food competitors, or how scarce food will be during the following season. Storing too few seeds is lethal and, therefore, strongly selected against. If animals habitually store food in excess, part of their supplies will normally be left and may eventually germinate and establish.

Lastly, a high population turnover of rodents may help seeds to escape consumption no matter the efficiency of rodents in depleting their caches. Both acouchies and agoutis are short-lived and are important prey for many raptorial animals. From one year to the next, Dubost (1988) recovered only five out of 16 marked agoutis at the same site (42% site fidelity), four at new sites, while the remaining seven were not recovered. Of 22 marked acouchies Dubost recovered none at all (0% site fidelity). This indicates that territories may frequently get vacant, and that the chances of a rodent not depleting its hoards could therefore be high. A logical mechanism is that seeds escape consumption because the animal dies before the depletion of its reserves. Seeds escaping consumption because the animal abandons its territory and reserves seems not logical, given the strong dependence of these rodents on stored seed.

The way in which seeds escape consumption should be visible in the spatial pattern of seedling recruitment. If seed survival and plant reproductive success were determined by loss of memory or structural excess storage, scatterhoarding would yield seedlings everywhere in every year. If, in contrast, local population fluctuations (due to death and migration) or variation in seasonal food availability determined seed survival, scatterhoarding would only yield seedlings once in several years. And if seed survival mainly depended on mortality of the individual scatterhoarder, seedling recruitment would be cohort- and territory-wise. The loss of seed crops in a number of years would then be the price paid for one successful crop.

6. Evolutionary aspects

The interactions between scatterhoarding rodents and large-seeded trees have some interesting evolutionary aspects. If the net effect of scatterhoarding for plants were negative, plants would be expected to have evolved mechanisms discouraging scatterhoarding. Instead, examples of plants exist that seem to have done the opposite; their fruit composition, seed size and phenology rather promote scatterhoarding (Vander Wall 1990). Given the large proportion of seeds affected by scatterhoarding in these plants, the reasoning that scatterhoarding must be beneficial and that plants are adapted to it is tempting. This has given rise to claims of coevolution of plants with scatterhoarders, for instance of pinion pines with chipmunks (Vander Wall 1993b), and of black walnut with fox squirrels (Stapanian & Smith 1986). However, it is impossible to prove that plants have evolved in response to interaction with scatterhoarders. What can be done, however, is measuring direction and strength of selective pressures that scatterhoarding and other factors impose on the reproductive strategies of these plants.

6.1 STIMULATING SCATTERHOARDING

The driving force behind a mutualistic interaction should be sought in the common interest of plants and rodents in preventing other granivorous insects and mammals to consume seed. The continuous threat of seeds being infested by insects or eaten by peccaries, thus becoming unattractive or unavailable for scatterhoarding, makes it important for plants to minimise the time that elapses between seed shedding and the moment at which a rodent removes the seed.

Theoretically, there are several possible ways by which plants could increase the speed and proportion of seeds being scatterhoarded. Individual plants that produce seeds in a way that is more attractive to scatterhoarding rodents could have their seeds scatterhoarded in larger numbers and more quickly than other individuals. Consequently, such individuals may be presented more heavily in future generations than others of the same species (Smith & Reichman 1984, Vander Wall 1990). Producing a greater number of seeds at a time, for example, could increase the chances of rodents finding the seed source, and starting to frequent it for exploitation. However, the high rate at which laid-out batches of seeds (simulating new seed sources) are discovered by acouchies and agoutis has convinced us that the number of seeds available at a time is of only limited importance for discovery.

In contrast, the nutritional value of individual seeds appears all the more important. Removal rates of seed species with different characteristics differ enormously. If plants producing high-valued seeds are frequented with greater fidelity than plants with less valuable seeds, producing more nutritious seeds with longer storage life may be rewarded with reduced exposure time to insects and wild pigs, i.e. reduced seed predation below the parent plant. Rodents also tend to hide valuable seeds more carefully than food of lower value (Stapanian & Smith 1986). Accordingly, larger seed tend to be transported further and spaced apart wider than smaller seeds (Stapanian & Smith 1984, Forget *et al.* 1998), which might yield survival advantages. The idea that strong competition for rodents between individuals and/or superior treatment of higher-valued seeds might have lead to frequency-dependent selection for ever-larger and more nutritious seeds is an intriguing explanation for the prevalence of large seeds among plants dispersed by scatterhoarders (Vander Wall 1990, Andersson & Frost 1996).

Plants could also influence the proportion of seeds surviving by fruiting when rodents are more likely to scatterhoard seeds. For instance, the proportion of seeds scatterhoarded increases during the wet season for both *C. procera* (Forget 1996) and *V. americana* (Forget 1990). When an individual tree sheds seed late, while seeds shed later are removed more quickly, its seeds could have short-term survival advantages. A general pattern is that plant species potentially dispersed by scatterhoarding rodents tend to mature during one specific season. Jackson (1981) found that seed fall of large-seeded species in a Brazilian lower montane moist forest peaked in the wet season, whereas small-seeded species were much less seasonal. In French Guiana, large-seeded species also tend to fruit more or less synchronically in the wet season. Could this have to do with the fact that rodents scatterhoard more in this season than in any other? Or can the synchronous fruiting be explained by reduction of predation risk because predators are satiated? Smythe (1970) has suggested such selection of scatterhoarding rodents on the fructification of large-seeded plant species.

Fruiting phenology may also influence rates of scatterhoarding. The phenologies of rodent-dispersed species vary considerably. *V. americana* sheds numerous seeds within a period of only two or three weeks, once every few years. *C. procera* seed shedding, in contrast, is modest, spreads out over two months, and occurs nearly every year (Forget 1997a). The advantage of peaked production could be that it causes satiation, and stimulates rodents to hoard the excess of seeds. The implicit idea is that the rodents will eat a greater proportion of the seed if they have the appetite. Spreading production, alternatively, may be more in pace with the rate at which animals are able to store seeds, and could reduce the duration of seeds lying waiting to be hoarded. This strategy demands a motivation of the rodent to store seeds independent of appetite, for instance, because of certain compounds that hamper instant digestion but lengthen storage life.

Given that there are limits to the amount of nutrients a plant can allocate to reproduction, plants face a trade-off between producing more nutritious seed and producing many seed. However, it might be essential to produce seed in large numbers that are nutritious at the same time. From this perspective, the phenomenon of mast fruiting may be a way for rodent-dispersed plants to escape from this limitation, by saving up energy during few years by not fruiting, in order to enable a more abundant fruiting in one subsequent year. According to the 'disperser attraction hypothesis' (Barnett 1977), the function of abundant fruiting is to attract seed dispersers, according to the 'predator satiation hypothesis' (Janzen 1971), the function is to satiate seed predators. These hypotheses are in fact complementary, with the former stressing the positive role and the latter stressing the negative role of animals. Since scatterhoarding rodents play both roles, both hypotheses may hold for the mast fruiters that have their seed scatterhoarded.

6.2 MONOPOLISING SEED RESERVES

Ultimately, provisioning and dispersal conflict. The interests of rodents and plants become completely opposite when it comes to exploitation of stored seed by the scatterhoarder. Seedlings and rodents are in competition for the seed reserves. What ways might plants have to minimise damage due to cache exploitation?

The most obvious way for plants to anticipate cache exploitation by rodents is through their timing and rate of germination. Certain plant species, such as *C. procera* and *V. americana*, start establishing before the animal actually starts drawing upon its seed supplies. The high mortality risk of establishing plants due to fungi in the wettest months and due to drought in the driest months (Forget 1997a) may also be avoided in this way. Repeated inspection and management of seed supplies by the rodents, however, severely limits the chance of seedlings emerging unnoticed, especially when shoots are used as cues. When digging up the seed, the rodents often damage or uproot the seedling.

It may be useful for seedlings to speed up establishment and the transfer of seed reserves into the seedling. The vulnerable phase of emergence can be rapid, and set on when rooting and transferring seed reserves to roots are already well on their way. Other than avoiding damage to the shoot, plants can also mitigate rodent-induced damage (e.g.

shoot cutting). Large seed reserves, transferred to the seedling, may function as a risk hedge (e.g. Kazuhiko *et al.* 1997). This advantage of large seeds combines well with having large seeds to stimulate scatterhoarding.

An alternative strategy, avoiding a race for seed reserves, may be to delay germination till the next period of food abundance, when no rodent will be managing old seed reserves. *L. alba* and *Astrocaryum* may do this. Delayed germination requires increasing seed storage life, which happens to be attractive for scatterhoarding rodents as well. Dispersability and seed survival may thus go hand in hand all the way to establishment.

Acknowledgements

The ideas presented in this paper have been formed by discussions with many different people, especially during various field trips. Although we are grateful to all of them, we would specifically like to thank Frans Bongers, Roel Brienen, Gérard Dubost, François Feer, Sander Koenraadt, Jan den Ouden, Anne Rutten, Almira Siepel, Leonie van der Voort, and Sip van Wieren for their contributions to discussions and field experiments. We thank Frans Bongers and Almira Siepel for their comments on the manuscript. The authors were supported by the Netherlands Foundation for the Advancement of Tropical Research with grant W84-407 (PAJ), by CNRS with research support to UMR CNRS-MNHN 8571 and by the French Ministère de l'Environnement with programs SOFT 91208 and SOFT 95042 (P.-M. Forget).

CHAPTER 27

EFFECTS OF TREE HEIGHT AND LIGHT AVAILABILITY ON PLANT TRAITS AT DIFFERENT ORGANISATION LEVELS

Frank J. Sterck, Toon Rijkers and Frans Bongers

1. Introduction

In the tropical rain forest understorey light is highly variable, in both time and space. While pioneer trees only survive and grow in open, high light spots, shade tolerant trees grow to maturity under ever changing light conditions. The shade tolerant species form the great majority of tree species in tropical rain forest, but they have been studied in less detail than pioneer species (Ackerly 1996). Trees respond to irradiance in various physiological and architectural traits, and depending on the scale of light variation such responses occur at various organisational levels, such as leaf, shoot, branch, and whole plant level (Tomlinson 1987, Gross 1986, Chazdon 1988). The type of responses to light availability depends on the environmental conditions, the species-specific properties, and the ontogenetic phase of the individual.

Little information is available in the literature on responses to light conditions for tree individuals after the seedling phase. The responses at the various organisation levels determine whole plant performance. For instance, at a high level, the display of leaves determines the interception of light, and thus the ability to acquire and use carbon for future growth and survival. Furthermore, whole plant responses affect the mechanical stability, i.e. resistance against wind (e.g. Niklas 1992). At a low organisation level responses include changes in leaf characteristics, morphological as well as physiological. For instance, specific leaf area and leaf nitrogen allocation patterns influence leaf photosynthetic performance and changes therein can affect whole plant photosynthesis (Hirose & Werger 1987, Pons *et al.* 1989). Responses at different levels of organisation can be linked through scaling-up-and-down, e.g. by integrating photosynthesis from leaf level via shoot to branch and crown level (see Ehleringer & Field 1993).

Trees undergo many changes in architecture as they grow. For a large part, these changes are typical to tree development irrespective of the environmental conditions. For another part, the responses are specific for the environmental conditions, e.g. the availability of light. It is hard to differentiate between ontogenetic and environmentally induced effects (Coleman *et al.* 1994), because light typically changes with increasing height in forest trees.

Bongers *et al.* (eds.), Nouragues. Dynamics and plant-animal interactions in a neotropical rainforest, 289–300.
2001 Kluwer Academic Publishers. Printed in the Netherlands.

In this chapter we focus on the architectural, morphological and physiological processes in two canopy tree species. We distinguish between light availability and tree height effects on these traits at different organisation level.

2. Study site, tree species and field methods

The two canopy tree species selected were *Dicorynia guianensis* Amshoff and Vouacapoua *americana* Aubl. (both Caesalpiniaceae). The species are further referred to by their generic names only. Both species are shade tolerant, and grow according to Troll's model (Hallé & Oldeman 1970, Oldeman 1974, Hallé *et al.* 1978). Following this model, stem and branches develop as sympodial plagiotropic axes, and stem and branches are hard to distinguish within the crown. Drénou (1988), however, showed that more models were needed to describe the growth patterns of *Dicorynia*.

The trees of both species with a stem diameter ≥ 10 cm at breast height were inventoried in a 12-ha plot (Petit Plateau). Trees with a diameter < 10 cm, but with a height above 50 cm, were inventoried in a central 1.5-ha plot. The crown position index (CPI) was used to estimate light availability for each tree. This index is based on visual estimates of the vegetation cover around and above the tree, and consist of five categories: 1, no direct light; 2, crown lit only from the side; 3, partial (10-90%) vertical illumination; 4, full vertical illumination; and 5, crown fully exposed to vertical and lateral light (Dawkins & Field 1978). The category '2' was subdivided into high (2.5), intermediate (2.0), and low (1.5) (Clark & Clark 1992).

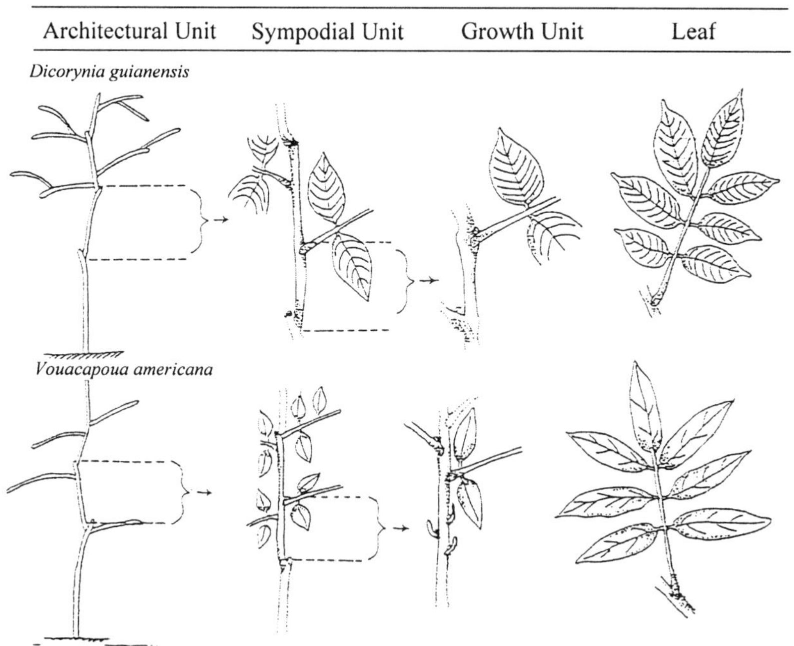

Figure 1. Illustration of plant components for *Dicorynia guianensis* and *Vouacapoua americana* (adapted from Sterck 1997).

From the inventory three sub-samples were selected for studying plant traits at different organisation levels. The first sub-sample (*Dicorynia* n = 33, *Vouacapoua* n = 40) was used for studying both the light environment at whole crown level and the architectural responses from leaf to whole crown level. Tree height ranged from 0.5-20 m. For this purpose,crowns of the tall trees were accessed by climbing neighbouring trees, using spikes, rope techniques, or canopy walkways. Data are presented of annual measurements between 1992 and 1994 (see Sterck 1997). In 1993 a hemispherical photograph was taken over each individual using a camera (Canon Ti-70) with a fish-eye lens (Canon, 7.5 mm f5.6). From these photographs canopy openness was calculated as the percentage of unobstructed sky, weighted for angle of incidence, according to Ter Steege (1993), and this was used as an estimation of light availability. The following architectural traits were measured: tree height, crown length, and crown width. From the last two measurements the relative crown length was calculated (i.e. ratio of crown length to crown width). Production rate, loss rate and size were measured for various plant components: leaves, growth units and sympodial units (for detailed methods see Sterck 1997). A sympodial unit is defined as the axis produced by one apical meristem; a growth unit is the smallest repetitive component produced by the meristems (Sterck 1997, Bongers & Sterck 1998). For *Dicorynia*, the growth unit is a metamer (White 1979, 1984, Bell 1991, Room *et al.* 1994), consisting of an internode, a node, one compound leaf, and an axillary bud (Fig. 1). For *Vouacapoua*, the growth unit is a sequence of metamers, also referred to as the 'unit of extension' (Bell 1991), with scale leaves at the base and compound leaves at the distal end (Fig. 1). In addition, for each tree, 20 leaves were harvested to measure leaf area and leaf mass, and to calculate total leaf area, leaf area index (total leaf area per crown area), and specific leaf area (leaf area per leaf mass, SLA).

A second sub-sample (N=4 per species) was used to assess the daily light variation at the crown level. For this purpose, the incident photon flux density (PFD) was measured in the upper crown with gallium-arsenide phosphide photodiodes (Model G-1118, Hamamatsu Photonics, Hamamatsu, Japan), which were calibrated against a Li-Cor quantum sensor under full sunlight. Trees were about 1 and 15m tall and were growing in either a large gap or under a closed canopy. At the top of the tree crown 5 light sensors were attached to leaf petioles. The sensors were oriented horizontally. PFD was recorded every 5 sec, and 1 min means were stored with a Li-Cor 1000 data logger from 7:30 until 17:00 h over 5 consecutive days during the dry season from September to November in 1996.

The third sub-sample was used to determine photosynthetic leaf characteristics. For each species individuals were selected for two tree heights (about 1 and 15m) and in two contrasting light environments (i.e. gap site and under a closed forest canopy) (n = 2-11, see Fig. 5). Light availability of trees was estimated using hemispherical photographs (canopy openness of gap and closed canopy trees was about 12% and 5%, respectively). Light saturated photosynthesis was measured *in situ* for the recently fully expanded leaves in the top of the tree crown. Gas-exchange rates were measured for 3-4 leaflets per plant with a portable open infra-red gas exchange system (CIRAS-1, PP-system, Hitchin, UK) between 9:00 and 12:00 h. After measuring gas exchange, leaflets were harvested, and leaf area was measured. Leaf dry weight was measured after drying the leaflets for 48 h at 70°C, and total nitrogen (N) content was determined with either an elemental analyser (Carbo Erba, model 1106, Milano, Italy) or, after extraction with hot water, spectrophotometrically, with a segmented flow analyser (Skalar San$^+$ System, Breda, the Netherlands).

3. Light availability

Light availability varied considerably within the populations of both study species (Fig. 2). Trees shorter than 20 m had CPI values between 1 and 3, 20-35 m tall individuals showed an increase in CPI values with height, and trees taller than 35 m had a CPI between 3 and 5. Figure 2 also illustrates a high variation in light conditions at any height level in the forest. Similar patterns are found for other shade tolerant tree species and for pioneer tree species in the Nouragues forest (Bongers & Sterck 1998) and in other tropical forests (Clark & Clark 1992, Lieberman *et al.* 1995). Thus, in general, trees occur at a broad range of light levels in the forest, but they may experience increasing levels of light as they grow from sapling to canopy tree.

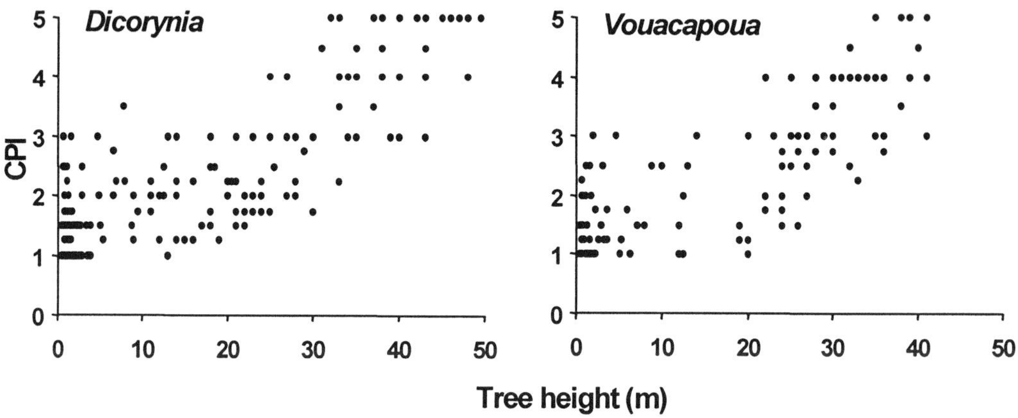

Figure 2. Crown Position Index (CPI) values in populations of *Dicorynia guianensis* and *Vouacapoua americana* (adapted from Bongers & Sterck 1998).

From the PFD readings it appears that individuals under a closed canopy received very low light during most of the day. For *Dicorynia*, short closed canopy individuals received less than 10 μmol m^{-2} s^{-1} nearly 80% of the time, and they were rarely exposed to more than 60 μmol m^{-2} s^{-1} (Fig. 3). In contrast, the tall closed canopy individuals received more light than the short closed canopy individuals since they were more frequently exposed to light levels above 60 μmol m^{-2} s^{-1}. Gap individuals were more frequently exposed to high PFD than closed canopy trees, which resulted in a greater total daily PFD. In both habitats the taller trees received higher PFD's than the saplings. Generally, height increment involves a decrease in canopy cover resulting in an increase in irradiance. Moreover, in gap sites, this involves a considerable increase in the period of direct exposure to the sun. The results for *Vouacapoua* were very similar (data not shown). Saplings of both species can survive many years under low light conditions in the understorey (Bongers *et al.* unpublished data), and they may do so by the photosynthetic utilisation of short periods of high light intensity (i.e. sunflecks) (Rijkers *et al.* 2000b). In general, our measured light levels are well comparable to those in other studies: overall low PFD (<10 μmol m^{-2} s^{-1}) dominate at the forest floor during most time of the day (Chazdon & Fetcher 1984a, 1984b, Oberbauer & Strain 1986, Chazdon 1988). However, data on PFD at greater height on trees in other forests are much rarer (but see Yoda 1974, Pearcy 1987, Torquebiau 1988, Koop & Sterck 1994).

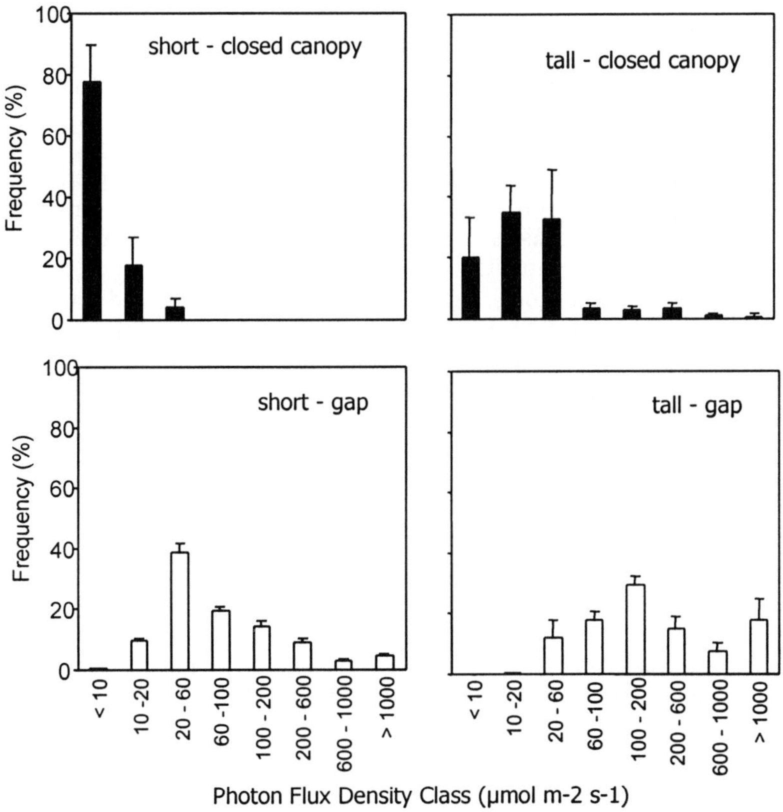

Figure 3. Frequency histogram of 1-minute means (±SD) of photon flux density (PFD) (μmol m^{-2} s^{-1}) of 4 *Dicorynia guianensis* trees: a short (about 1 m) closed canopy tree; a tall (about 15m) closed canopy tree; a short gap tree, and a tall gap tree. Percentage canopy openness of the short and tall closed canopy trees was 3.1 and 3.9 respectively; openness of the short and tall gap trees was 8.4 and 21.5 respectively.

4. Architectural development

Tree height and irradiance had an effect on plant traits at various organisation levels (Table 1). In the regression model we used standard partial regression coefficients (beta coefficients) showing the relative strength of the effect of tree height and canopy openness on the same plant trait. In only a few plant traits changes were dependent on both tree height and canopy openness. In those cases, the beta coefficient for height was always greater than the one for light, indicating that height had a greater effect on the trait than did light.

At the tree crown level, crowns became slender with increasing tree height in both species; for *Vouacapoua* the crown slenderness was also slightly affected by light. Total branch growth refers to the annual branch extension for the whole crown, and leader growth refers to the annual extension of the highest axis in the crown. Both species increased total branch growth with height and light, and they increased leader growth with light only. *Dicorynia* increased its total leaf area and leaf area index with height only. *Vouacapoua* increased its total leaf area with height and light; its LAI increased with a joint increase in tree height and canopy openness only, which is illustrated by the significant height x openness interaction term. At the sympodial unit level, the survival of the sympodial units did not change with either height or openness in both species. With increasing height *Dicorynia* decreased the number of lateral sympodial units produced. In other words, its amount of lateral meristems released per sympodial unit (i.e. the number of axillary meristems turning into apical meristems) decreased with increasing height. *Vouacapoua*, however, increased the number of lateral sympodial units with increasing openness. At the growth unit level, *Dicorynia* produced longer growth units with a joint increase in height and openness, while *Vouacapoua* did so with openness only. In *Vouacapoua* the production and actual number of present leaves increased with height. In *Dicorynia* there were no significant patterns with height or openness for these latter two traits. Furthermore, the number of fallen leaves was independent of height or openness in both species. Finally, at the leaf level, the leaf size was unaffected by height and light. Specific leaf area decreased with height and openness in *Dicorynia*, and with the interaction term in *Vouacapoua*.

The plant trait responses to light are to a large extent in agreement with responses found in other studies (e.g. King 1994). Trees grew more in height and produced smaller crowns, more leaf layers, more new sympodial units, longer woody units between leaves, higher leaf area display costs, and smaller specific leaf areas at higher light availability. The lack of some responses in *Dicorynia* resulted from the scarcity in high light levels, rather than from the inherent inability to respond (Sterck 1999).

The investigated plant trait changes with increasing tree height have rarely been reported in the literature, except for some leaf traits (Pearcy 1987, Thomas 1995, Rijkers *et al.* 2000a, Kull & Niinemets 1993, Schoetle 1994, Niinemets 1997). It can be argued that the general gradient of increasing light levels with increasing tree height selected for the plant trait changes with increasing tree height. Only at the leaf and whole crown level plant trait changes with increasing tree height did parallel plant trait responses to increasing light availability. Yet these changes could also be explained by other factors than light *sensu stricto*. Firstly, at the leaf level, the smaller specific leaf area at greater height can be to due higher wind speeds causing mechanical stress and physiologically dry conditions (e.g. Jaffe & Forbes 1993, Telewski 1995). Secondly, at the whole crown level, the increase in branch extension and total leaf area with height resulted from the greater number of apical meristems (Sterck & Bongers in press). This implies that there is little empirical support for the idea that light availability has acted as an imported selection force on changes in plant traits with tree height (see also Sterck & Bongers in press).

In this paper we treated plant traits, measured at different organisation levels, as independent from one another. However, traits at the lower organisation levels may, to some extent, determine traits at higher organisation levels, and vice versa (e.g. Sterck 1999, Sterck & Bongers in press). For instance, for *Vouacapoua* it appeared that increases in total leaf area, total branch growth, and leaf area index were due to higher sympodial unit production at the sympodial unit level; traits at lower organisation levels were less important. And, at high light, the meristems in the top of the crown produced more and longer growth units than the meristems at lower positions in the crown, resulting in narrower crowns. These observations are in agreement with the results in the regression analysis (Table 1), but this analysis does not explore the relationships between organisation levels explicitly.

TABLE 1. Multiple regression analysis of tree height (m) and canopy openness (%) on plant traits at different organisation levels for juvenile trees (height range 0.5-20m) of*Dicorynia guianensis* (n=33) and *Vouacapoua americana* (n=40). Coefficients of determination (R^2), intercepts and standardised partial regression coefficients (beta coefficients).

	R^2	intercept	height	Beta coefficients openness	height × openness
Dicorynia crown level					
relative crown length	0.62***	-	0.76***	-	-
branch growth (m yr^{-1})	0.84***	1.15***	0.73***	0.24*	-
leader growth (cm yr^{-1})	0.24**	-	-	0.41**	-
total leaf area (m^2) [a]	0.77***	-0.77***	0.84***	-	-
LAI	0.32***	0.53***	0.56**	-	-
sympodial unit level					
number of lateral SU's (yr^{-1})	0.24*	0.25***	-0.65**	-	-
number of surviving SU's (yr^{-1})	-	-	-	-	-
growth unit level					
length (cm)	0.41***	1.51***	-	-	0.66***
number of leaves produced (yr^{-1})					
number of leaves (yr^{-1})	-	-	-	-	-
number of fallen leaves (yr^{-1})	-	-	-	-	-
leaf level	-	-	-	-	-
size (cm^2)	-	-	-	-	-
SLA (cm^2 g^{-1})	0.63	270***	-0.64***	-0.18*	-
Vouacapoua crown level					
relative crown length	0.79***	-	0.75***	0.20**	-
branch growth (m yr^{-1})	0.84***	0.74***	0.68***	0.29*	-
leader growth (cm yr^{-1})	0.20**	-	-	0.45**	-
total leaf area (m^2) [a]	0.90***	-0.74***	0.79***	0.21*	-
LAI	0.52***	0.43***	-	-	0.33***
sympodial unit level					
number of lateral SU's (yr^{-1})	0.17*	-	-	0.40*	-
number of surviving SU's (yr^{-1})	-	-	-	-	-
growth unit level					
length (cm)	0.34***	-	-	0.58***	-
number of leaves produced (yr^{-1})	0.44***	2.55***	0.66***	-	-
number of leaves (yr^{-1})	0.50***	1.99***	0.57***	-	-
number of fallen leaves (yr^{-1})	-	-	-	-	-
leaf level	-	-	-	-	-
size (cm^2)	-	-	-	-	-
SLA (cm^2 g^{-1})	0.40***	199***	-	-	-0.15***

Relative crown length = crown length to crown width ratio; LAI = leaf area index; SU = sympodial unit; GU = growth unit; SLA = specific leaf area. Note that the number of leaves produced per growth unit is 1 by definition for *Dicorynia* and therefore excluded from analysis. Beta coefficients have no units.[a] = data are log transformed. Significance levels: * $P< 0.05$, ** $P< 0.01$, *** $P< 0.001$ and - = not significant. (Table adapted from Sterck & Bongers in press).

Safety margins for the support of the tree's own weight were calculated, using various mechanical tree models (Bongers & Sterck 1998, Sterck & Bongers 1998). Safety margins are quantified by safety factors on the basis of a specific model, e.g. the elastic stability model (McMahon 1973, Niklas 1992). A safety factor smaller than 1 means that a tree will buckle under its mass. Figure 4 illustrates that minimum safety margins were close to this theoretical minimum. Apparently, these trees are just thick enough to carry their own weight, and particularly those between 10-30 m tall. Others also found small safety margins in tropical forest trees (e.g. Claussen & Maycock 1995). The safety margins decreased with height early during ontogeny and increased later on (Fig. 4). Several explanations are possible for the decrease (Sterck & Bongers 1998): (1) Larger trees need more light to maintain their increasing tree body, and may thus suffer more from light suppressed conditions than smaller trees. Therefore they may reduce investments in stem thickness. (2) Shorter trees may be more susceptible to falling debris and climbing, or passing, animals, and may invest in larger safety margins to compensate for these stresses. (3) Trees may change their wood properties with increasing height, while such changes are not included in the elastic stability model. (4) Crown size and - mass are not explicitly specified in the model, but they do affect tree stability (Mattheck & Kubler 1995).

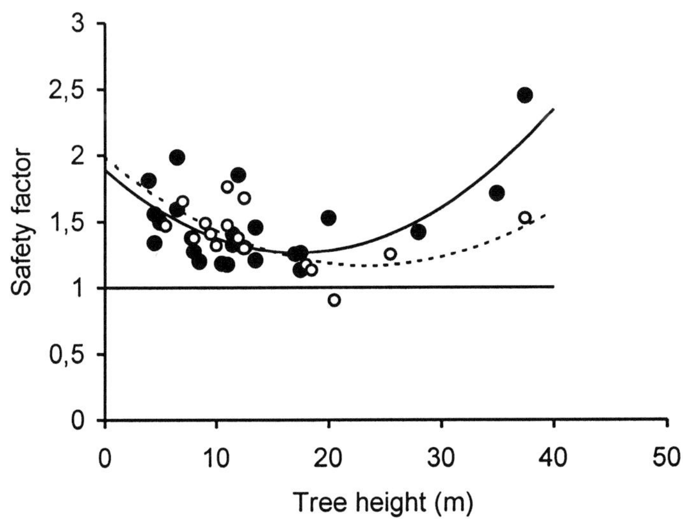

Figure 4. Safety factors for *Dicorynia guianensis* (dashed line) and *Vouacapoua americana* (solid line) in relation to tree height. Lines were fitted with a quadratic model: for *Dicorynia*, safety factor = 1.99 - 0.07 × height + 0.0015 × height^2 (R^2= 0.41, P = 0.031); for *Vouacapoua*, safety factor = 1.89 - 0.07 × height + 0.0021 × height^2 (R^2 = 0.53, P < 0.001) (adapted from Bongers & Sterck 1998).

Safety factors increased with light in *Vouacapoua* (Bongers & Sterck 1998). This is in agreement with other experiments on trees outside their natural habitat (Holbrook & Putz 1989). These results suggest that trees of *Vouacapoua* had relatively high investments in their mechanical safety when they occurred in more open conditions. Apparently, exposed trees invested more in stem thickness and less in stem extension. This may also reflect a response to the higher wind rates in the exposed sites (e.g. Telewski 1995, Ennos 1996), or to the bigger mass and size of the crown at more exposed conditions (Sterck & Bongers 1998).

5. Leaf characteristics

Photosynthesis is the driving process in carbon gain and hence influences plant development. This process depends on leaf properties, which may vary in relation to tree height and light availability (Chazdon *et al.* 1996, Rijkers *et al.* 2000a). Overall, leaf traits changed with increasing tree height in *Dicorynia* and with increasing light availability in *Vouacapoua* (Fig. 5). Short *Dicorynia* individuals had a lower A_{max} and N_{area}, and a higher SLA, than did tall individuals. In *Vouacapoua*, gap trees had a lower SLA and a higher A_{max} than did closed canopy trees, though this was not significant for tall trees. Short gap trees attained a nearly twice as high A_{max} as did short closed canopy trees, and their A_{max} was similar to A_{max} attained in tall gap trees. Between species, short gap saplings of *Vouacapoua* had considerably higher A_{max} than did short gap saplings of *Dicorynia*, whereas under a closed canopy the two species had nearly equal photosynthetic rates. Similar results were obtained for SLA and N_{area}.

Light response curves of photosynthesis at different tree heights and in contrasting light environments were similar for both species (Fig. 6), except for the short gap trees of *Dicorynia*. Light saturation of leaves of the tall gap individuals occurred at PFD greater than 500 μmol m^{-2} s^{-1}, whereas for the other trees light saturation occurred at or below 200 μmol m^{-2} s^{-1}. Similar results were found for Australian tropical trees (Pearcy 1987). The higher A_{max} values at greater PFD in tall gap trees (Fig. 5) corresponded with the more light exposed conditions in which the leaves developed (Fig. 3).

The higher A_{max} and N_{area}, and lower SLA for leaves developed under high light conditions versus those in shade conditions, found in *Vouacapoua*, are in agreement with other studies (Pearcy 1987, Popma & Bongers 1988, Ellsworth & Reich 1993, Raaimakers *et al.* 1995). The leaves of short *Vouacapoua* saplings showed plasticity in A_{max} and SLA. On the other hand, the leaf characteristics of short *Dicorynia* saplings were hardly affected by light. Lack of photosynthetic and morphological adjustments to contrasting light environments has also been found for other species (e.g. Chazdon 1992). For *Dicorynia*, short trees were unable to develop large and heavy leaves as in tall trees, suggesting that developmental phase constrained plasticity of the leaf properties. It is still not fully understood to what extent developmental phase or light conditions determine the vertical patterns in leaf properties (see Coleman *et al.* 1994). Leaf traits of trees of different heights may change in response to the environment, as air temperature, vapour pressure deficit and/or wind speed often increase along a vertical gradient in the forest (e.g. Kira & Yoda 1989). Wind, for instance, is causing mechanical stress for leaves, resulting in morphological and anatomical changes in leaves (Jaffe & Forbes 1993). On the other hand, several factors may explain changes in leaf characteristics as dependent on tree height. For example, because the ratio of total leaf area to total non-photosynthetic tissue decreases with increasing tree height less available foliage has to provide the tree with assimilates for growth and maintenance. Moreover, with increasing height extra carbon investments are needed to construct and maintain support tissues such as twigs and branches (Givnish 1986) or to lessen the risk of windthrow (Telewski 1995). Furthermore, leaves of tall trees must have a sufficiently negative water potential to take water from the soil to the leaf because, in general, these trees have a steeper water potential gradient from soil to leaf (e.g. Friend 1993, Niinemets 1997).

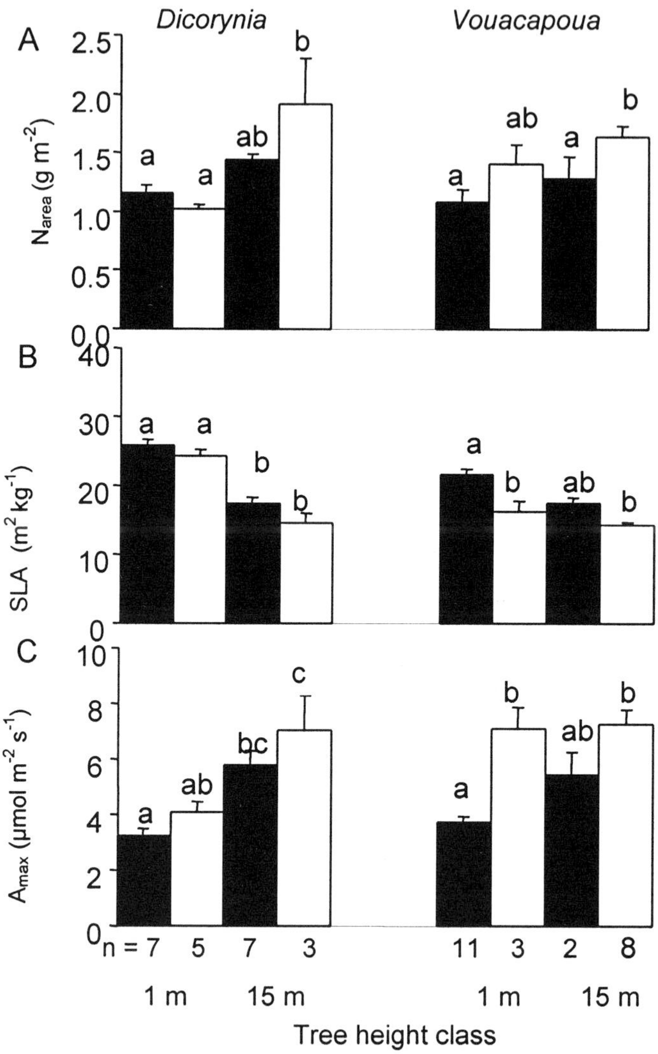

Figure 5. Effect of light environment and tree height on leaf traits in *Dicorynia guianensis* and *Vouacapoua americana*. Solid bars represent trees under a closed canopy and open bars trees in gap sites. Bars are mean values with SE. (A) Total nitrogen content per leaf area (N_{area}, g m^{-2}); (B) Specific leaf area (SLA, m^2 kg^{-1}) and (C) Light saturated photosynthetic rate (A_{max}, µmol m^{-2} s^{-1}). Values with different letters (within species) are significant different ($P<0.05$, one-way ANOVA and Tukey HSD). n = number of tree individuals (based on Rijkers 2000).

For *Dicorynia* the variation in SLA described 65% of the variation in A_{max} and 55% in N_{area}, and similarly for *Vouacapoua* the explained variation (R^2) was 70% and 60% respectively (data not shown). In both species A_{max} and leaf nitrogen content expressed per unit leaf mass were unaffected by tree height and light condition, except for the A_{max}-openness relationship as mass based A_{max} slightly increased with increasing light. Furthermore, in both species there was a positive relationship between A_{max} and N_{area} (for *Dicorynia*: $A_{max} = 0.08 + 3.56N_{area}$, $R^2=0.58$, $P <0.001$; for *Vouacapoua*: $A_{max} = -0.004 + 4.15N_{area}$, $R^2 = 0.64$, $P < 0.001$). In general, these results indicate that SLA is a key variable as the variation in photosynthetic capacity and leaf nitrogen content could be mainly traced back to variation in SLA (e.g. Ellsworth & Reich 1993, Raaimakers *et al.* 1995).

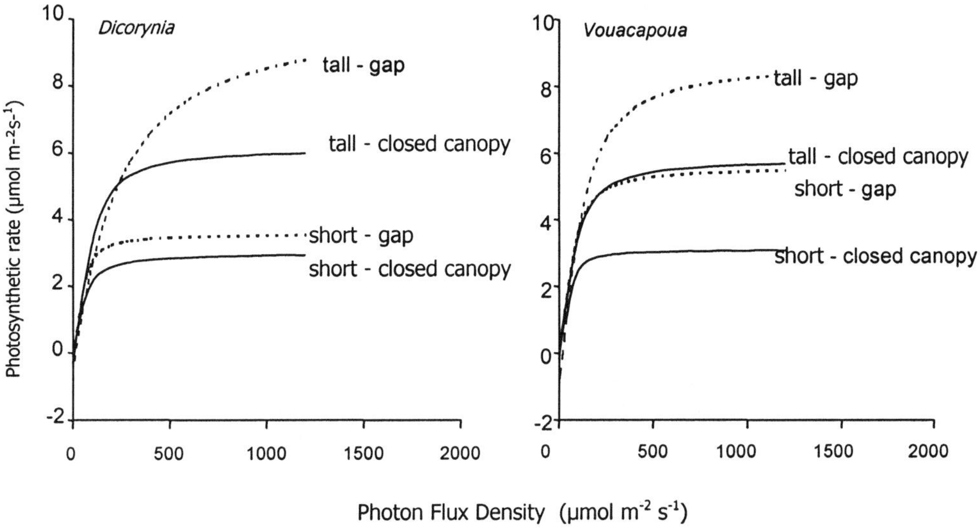

Figure 6. Photosynthetic light response curves for *Dicorynia guianensis* and *Vouacapoua americana*. A curve (dashed or solid) is the average non-rectangular hyperbola obtained from 3 leaves of one tree (based on Rijkers 2000).

6. Conclusions

The two shade tolerant tree species studied experienced a heterogeneous light environment, both along a horizontal and vertical gradient in the Nouragues forest. A distinction is made between tree height and light availability effects on plant traits from the leaf to whole-crown level. The studied trees of different height (0-20 m) exhibited plant trait responses to light at multiple levels of organisation, and these responses were generally similar to the responses observed in saplings in other studies. There is, however, very little information available on changes in plant traits with increasing tree height. At the leaf level, the traits investigated (A_{max}, SLA and N_{area}) changed with both tree height and light availability. In *Dicorynia* these traits were mainly affected by tree height being light availability of less importance, whereas in *Vouacapoua* traits changed with both increasing light and height. With regard to the plant traits at higher organisation level (growth unit, sympodial unit and crown level) the same pattern of response to light and height could be observed as found at the leaf level for the two species. Several factors can explain the change in plant traits with increasing tree height, for instance the decrease in ratio of total leaf weight to whole plant weight may result in a reduced availability of assimilates for the production of new tissue. Although tree height and light availability showed independent effects on the plant characteristics, we also point out that traits may also change due to strong changes in environmental conditions when trees grow to the forest canopy, especially the increase in wind speed may affect tree morphology and physiology.

Our approach was characterised by studying several plant traits in trees at multiple levels of organisation. Here we presented analyses for each of the organisational levels (and traits) individually. In sequel studies that are currently in process a more integrating approach is used. First, we address how plant traits at lower levels determine traits at higher levels (and vice versa). Secondly, we address the consequences of plant traits at multiple levels of organisation for the survival, growth, and reproduction of individual trees, and for the demographic patterns of whole populations. Doing so we wish to obtain more insight in the ecological consequences of the different plant traits studied here.

Acknowledgements

We would like to thank Kaoru Kitajima and Thijs Pons for their constructive criticism on an earlier version of the manuscript. This study was supported by grants W85--279 and W85-239 of the Netherlands Foundation for Advancement of Tropical Research (WOTRO).

APPENDICES

APPENDIX 1

FLORISTIC CHECKLIST OF THE NOURAGUES AREA

Pierre Belbenoit, Odile Poncy, Daniel Sabatier, Marie- Françoise Prévost, Bernard Riéra, Patrick Blanc, Denis Larpin and Corinne Sarthou

The list includes data gathered at the 'Inselberg site' on an area of about 10 km² (Charles-Dominique Chapter 1). It was compiled mainly from (a) the 'AUBLET' database of the 'Herbier de Guyane' (CAY) (Hoff *et al.* 1989); (b) the tree inventory in high forest (see Poncy *et al.* Chapter 4); (c) the botanical study of the inselberg (rock-savanna, Sarthou Chapter 6; low forests, Larpin Chapter 5). A few additional tree species recorded only from a 6-ha plot studied in the Saut-Pararé area are also included (see Poncy *et al.* Chapter 4). The botanical information was provided by several botanists and ecologists, including S. Barrier, C. Julliot, D. Larpin, O. Poncy, M.-F. Prévost, B. Riéra, D. Sabatier, B. Simmen, J.-F. Villiers (tree inventories), and also P. Blanc, M. Born, A. Cockle, G. Cremers, C. Feuillet, S. Gonzalez, J.-J. de Granville, M. Hoff, C. Sarthou, S.Y. Zhang.

Nomenclatural and systematic arrangements follow the *Checklist of the Plants of the Guianas (Guyana, Surinam, French Guiana)* (Boggan *et al.* 1997), except when taxonomic novelties have been published subsequently.

Incompletely identified taxa are mentioned only when they are sufficiently well documented to be sure that they are different from the other taxa. When collection references are missing or not reliable (e.g. records based on field observations without a voucher collection, or on a poor collection), names of taxa are placed between brackets. Differences may occur between this list and data presented in certain papers of this book. For example, as far as Myrtaceae are concerned, only 23 species out of the 33 mentioned by Larpin (Chapter 5) are listed. The 10 excluded taxa were considered insufficiently documented to be included in the checklist. The occurrence of the species in three major habitats is given : high dense forest (H), low forest on the slopes of the inselberg (L), and rock-savanna (RS). 'H' comprises various vegetation types, including mixed transitional forest at the foothill of the inselberg, swamp forest with *Euterpe* palms, liana forest, and bamboo forest, the two latter eventually being considered (*cf* Poncy *et al.* Chapter 4) as having originated from ancient anthropic influence. Information on habitat type may be missing, for instance when not indicated by the collector, or when the record of the species is based on seeds from the seed bank or from faeces, or on a specimen collected in a disturbed area (such as around the camp). The comparative importance of the species in these three major habitats was estimated using the information at our disposal. The symbol (X) indicates that the species can be expected in that habitat, based either on our knowledge

Bongers *et al.* (eds.), Nouragues. Dynamics and plant-animal interactions in a neotropical rainforest, 301–341.
2001 Kluwer Academic Publishers. Printed in the Netherlands.

from the Nouragues area (e.g., a species recorded in RS and H is predicted to be present also in L) or from other sites in French Guiana and adjacent areas.

Forest types

H = Taxon recorded from the high forest
L = Taxon recorded from the low forest
RS = Taxon recorded from the rock savanna
(X) = Taxon assumed to be present in high and low forest habitats

Lifeform

Emergent tree = tree > 35 m
Tall tree = tree 25-35 m
Medium tree = tree 15-25 m
Small tree = tree 10-15 m
Tree = single-stemmed woody plant ≥ 5 cm DBH and > 2 m tall
Treelet = single-stemmed woody plant < 5 cm DBH and > 2 m tall
Shrub = woody plant < 2 m tall
Liana = woody climber
Vine = non-woody climber
Epilithic = attached on rocks
Epiphyte = growing on other plants, not rooted in the soil
Hemi-epiphyte = growing for only a part of the life as an epiphyte
Saprophyte = plant feeding on dead organic matter

FLORISTIC CHECKLIST OF THE NOURAGUES AREA	H	L	RS	LIFE FORM
SPERMATOPHYTA Dicotyledones				
Acanthaceae				
Ruellia inflata L.C. Rich.	X			Terrestrial herb
Anacardiaceae				
Anacardium spruceanum Benth. ex Engl.	X			Tall tree
Astronium ulei Mattick				Emergent tree
Spondias mombin L.	X			Emergent tree
Tapirira bethanniana J.D. Mitchell	X			Tall tree
Tapirira guianensis Aubl.	X	X	X	Small tree
Tapirira obtusa (Benth.) J.D. Mitchell	X			Small tree
Thyrsodium guianense Sagot ex March.	X	X		Tall tree
Thyrsodium puberulum J.D. Mitchell & Daly	X			Tall tree
Thyrsodium spruceanum Benth.	X			Medium tree
Annonaceae				
Anaxagorea dolichocarpa Sprague & Sandw.	X			Treelet
Anaxagorea prinoides (Dunal) A. DC.	X			Treelet
Annona densicoma Mart.	X			Small tree
Annona haematantha Miq.	X			Tree
Cardiopetalum surinamense R.E. Fries	X	(X)		Treelet
Crematosperma brevipes (DC. ex Dunal) R.E.Fries	X			Treelet
Duguetia cadaverica Huber	X			Treelet
Duguetia calycina Benoist	X			Small tree
Duguetia eximia Diels	X			Treelet
Duguetia inconspicua Sagot	X			Treelet
Duguetia pycnastera Sandw.	X			Small tree
Duguetia surinamensis R.E. Fries	X			Medium tree
Ephedranthus guianensis R.E. Fries	X			Medium tree
Fusaea longifolia (Aubl.) Safford	X			Medium tree
Guatteria ouregou (Aubl.) Dunal	X			Treelet
Guatteria punctata (Aubl.) R.A. Howard	X			Tree
[*Guatteria* cf. *schomburgkiana* Mart.], after Larpin 1993	X			Medium tree
Guatteria sp. nov.	X			
Oxandra asbeckii (Pulle) R.E. Fries	X	X	X	Medium tree
Pseudoxandra cuspidata Maas	X			Medium tree
Rollinia exsucca (DC. ex Dunal) A. DC.	X			Treelet
Unonopsis perrottetii (A. DC.) R.E. Fries	X			Medium tree
Unonopsis rufescens (Baill.) R.E. Fries	X			Medium tree
Unonopsis stipitata Diels	X	X		Tree
Xylopia nitida Dunal	X			Tall tree

FLORISTIC CHECKLIST OF THE NOURAGUES AREA	H	L	RS	LIFE FORM
Apocynaceae				
Ambelania acida Aubl.	X			Treelet
Aspidosperma album (Vahl) Benoist ex Pichon	X	X		Tall tree
Aspidosperma carapanauba Pichon	X			Tree
Aspidosperma cruentum Woodson	X			Tree
Aspidosperma helstonei van Donselaar	X			Tall tree
Aspidosperma marcgravianum Woodson	X			Tall tree
Aspidosperma oblongum A. DC.	X			Emergent tree
Aspidosperma sandwithianum Markgr.	X			Medium tree
Couma guianensis Aubl.	X			Medium tree
Geissospermum laevis (Vell.) Miers	X			Tall tree
Himatanthus articulatus (Vahl) Woodson	X			Medium tree
Himatanthus bracteatus (A. DC.) Woodson	X			Medium tree
Lacmellea aculeata (Ducke) Monachino	X			Small tree
Lacmellea floribunda (Poepp.) Benth. & J.D. Hook.	X			Small tree
Macoubea guianensis Aubl.	X			Tall tree
[*Mandevilla hirsuta* (A. Rich.) K. Schum.], after Sarthou 1992	(X)	(X)	X	Liana
Mandevilla subspicata (Vahl) Markgr.		X	X	Liana
Odontadenia perrottetii (A. DC.) Woodson	X			Liana
Odontadenia puncticulosa (A. Rich.) Pulle	X			
Parahancornia fasciculata (Poir.) Benoist ex Pichon	X			Tall tree
Tabernaemontana attenuata (Miers) Urb.	X			Tree
Tabernaemontana disticha A. DC.	X	X		Tree
Tabernaemontana macrocalyx Muell. Arg.	(X)	X		Tree
Aquifoliaceae				
Ilex sp.	X			
Araliaceae				
Oreopanax capitatus (Jacq.) Decne. & Planch.	X			Liana
*Schefflera decaphylla (*Seem.) Harms	X	X		Tall tree
Schefflera morototoni (Aubl.) Maguire, Steyerm. & Frodin	X			Tall tree
Aristolochiaceae				
Aristolochia iquitensis O.C. Schmidt	X			Liana
Asclepiadaceae				
Ditassa cf. *consanguinea* Decne.			X	
Matelea sastrei Morillo			X	Vine
Asteraceae				
Fleischmannia microstemon (Cass.) R. King & H. Robinson				Terrestrial herb
Piptocoma schomburgkii (Schultz Bip.) Pruski	X	X	X	Treelet
Rolandra fruticosa (L.) Kuntze	(X)			Terrestrial herb
Synedrella nodiflora (L.) Gaertn.	(X)	X		Terrestrial herb
Unxia camphorata L.f.	(X)	X		Epilithic herb

FLORISTIC CHECKLIST OF THE NOURAGUES AREA	H	L	RS	LIFE FORM
Balanophoraceae				
Helosis cayennensis (Swartz) Spreng.	X			Parasite herb
Begoniaceae				
Begonia glabra Aubl.	X			Vine
Begonia hirtella Link	X			Epilithic herb
Bignoniaceae				
Arrabidaea mollis (Vahl.) Bureau ex K. Schum.	X			
Cydista aequinoctialis (L.) Miers	X	X		Liana
Distictella elongata (Vahl) Urb.	X	X		Liana
Distictella magnoliifolia (Kunth) Sandw.	X			Liana
Distictis pulverulenta (Sandw.) A. Gentry	X			Liana
Jacaranda copaia (Aubl.) D. Don	X			Tall tree
Mansoa alliacea (Lam.) A. Gentry	X			Liana
[*Memora alba* (Aubl.) Miers], after Larpin 1993	X			Liana
cf. *Memora flavida* (DC.) Bureau & K.Schum.	X			Liana
Memora flaviflora (Miq.) Pulle	X	X		Liana
Parabignonia steyermarkii Sandw.	X			Liana
Schlegelia paraensis Ducke	X			Liana
Tabebuia capitata (Bureau & K. Schum.) Sandw.	X	X	X	Emergent tree
Tabebuia serratifolia (Vahl) Nichols.	X	X	X	Medium tree
Bixaceae				
Cochlospermum orinocense (Kunth) Steudel	*X*	X		
Bombacaceae				
Catostemma sp.	X			Tall tree
Ceiba pentandra (L.) Gaertn.	X			Emergent tree
Eriotheca crassa (Uittien) A. Robyns	X			Tree
Eriotheca globosa (Aubl.) A. Robyns	(X)	X		Small tree
Eriotheca surinamensis (Uittien) A. Robyns	X	X	X	Tall tree
Pachira aquatica Aubl.	X			Tree
Pachira dolichocalyx A. Robyns	X			Tree
Quararibea duckei Huber	X			Tree
Quararibea guianensis Aubl.	X			Tree
Boraginaceae				
Cordia fulva I.M. Johnston	X			Tree
Cordia goeldiana Huber	X			Emergent tree
Cordia lomatoloba I.M. Johnston	X			Small tree
Cordia nodosa Lam.	X	(X)		Treelet
Cordia sagotii I.M. Johnston	X			Tree
Cordia tetrandra Aubl.	X			Tree
Burseraceae				
Crepidospermum rhoifolium (Benth.) Triana & Planch.	X			Medium tree

FLORISTIC CHECKLIST OF THE NOURAGUES AREA	H	L	RS	LIFE FORM
Dacryodes cuspidata (Cuatrec.) Daly	X			Tree
Dacryodes nitens Cuatrec.	X			Tall tree
Protium apiculatum Swart	X			Medium tree
Protium aracouchini (Aubl.) March.	X			Medium tree
Protium araguense Cuatrec.	X			Tree
Protium decandrum (Aubl.) March.	X			Tall tree
Protium cf. *demerarense* Swart	X			Tree
Protium giganteum Engl. var. *crassifolium* (Engl.) Daly	X			Tree
Protium guianense (Aubl.) March.	X			Medium tree
Protium heptaphyllum (Aubl.) March.	X	X		Medium tree
Protium morii Daly	X			Tree
Protium opacum Swart	X			Tree
Protium plagiocarpum Benoist	(X)			Small tree
Protium robustum (Swart) Porter	X			Tree
Protium sagotianum March.	X			Tall tree
Protium subserratum (Engl.) Engl.	X			Tree
Protium tenuifolium (Engl.) Engl.	X			Tree
Protium trifoliolatum Engl.	X			Tree
Tetragastris altissima (Aubl.) Swart	X			Tall tree
Tetragastris panamensis (Engl.) Kuntze	X			Tall tree
Trattinnickia burserifolia Mart.	X			Small tree
Trattinnickia demerarae Sandw.	X			Tall tree
Trattinnickia rhoifolia Willd.	X			Emergent tree
Cactaceae				
Epiphyllum phyllanthus (L.) Haw.	(X)	X		Epiphyte
Cactaceae sp.		X		
Caesalpiniaceae				
Batesia floribunda Spruce ex Benth.	X			Tree
Bauhinia cupreonitens Ducke	X			Liana
Bocoa prouacensis Aubl.	X			Medium tree
Cassia sp.	X			
Chamaecrista desvauxii (Collad.) Killip var. *saxatilis* (Amsh.) Irwin & Barneby			X	
Chamaecrista diphylla (L.) Greene	(X)	(X)	X	Epilithic herb
Crudia sp.	X			
Dialium guianense (Aubl.) Sandw.	X			Tall tree
Dicorynia guianensis Amsh.	X			Tall tree
Dimorphandra pullei Amsh.	X			Emergent tree
Eperua falcata Aubl.	X			Tall tree
Eperua grandiflora (Aubl.) Benth.	X			Tall tree
Eperua rubiginosa Miq.	X			Medium tree
Hymenaea courbaril L.	X			Medium tree

FLORISTIC CHECKLIST OF THE NOURAGUES AREA	H	L	RS	LIFE FORM
Macrolobium bifolium (Aubl.) Pers.	X			Tall tree
Peltogyne paniculata Benth. subsp. *pubescens* (Benth.) M.F. da Silva	X			Emergent tree
Peltogyne venosa (Vahl) Benth.	X			Tall tree
Sclerolobium albiflorum (Benoist) Zaruch. & Herend.	X	X		Tall tree
Swartzia arborescens (Aubl.) Pittier	X			Medium tree
Swartzia grandifolia Bong. ex Benth.	X			Medium tree
Swartzia guianensis (Aubl.) Urb.	X			Small tree
Swartzia leiocalycina Benth.	X			
Swartzia cf. *oblanceolata* Sandw.	X			Tree
Swartzia panacoco (Aubl.) Cowan var. *sagotii* (Sandw.) Cowan	X			Tall tree
Swartzia polyphylla DC.	X			Tall tree
Tachigali melinonii (Benth.) Zaruch. & Herend.	X			Tree
Tachigali myrmecophila (Ducke) Ducke	X			Tree
Tachigali cf. *paniculata* Aubl.	X			Tree
Tachigali paraensis (Benth.) Barneby	X			Tree
Vouacapoua americana Aubl.	X			Tall tree
Campanulaceae				
Centropogon cornutus (L.) Druce	X			Terrestrial herb
Canellaceae				
Cinnamodendron tenuifolium Uittien	X			Small tree
Caricaceae				
Jacaratia spinosa (Aubl.) A. DC.	X			Small tree
Caryocaraceae				
Caryocar glabrum (Aubl.) Pers.	X			Tall tree
Cecropiaceae				
Cecropia distachya Huber	X			Tree
Cecropia granvilleana C.C. Berg	X			Small tree
Cecropia obtusa Trécul	X			Small tree
Cecropia sciadophylla Mart.	X			Tall tree
Coussapoa angustifolia Aubl.	X	X		Epiphyte & tree
Coussapoa asperifolia Trécul	X			Epiphyte & tree
Coussapoa latifolia Aubl.	X	X	X	Epiphyte & tree
Pourouma bicolor Mart.	X			Small tree
Pourouma cf. *cecropiaefolia* Mart.	X			Tree
Pourouma guianensis Aubl.	X			Tall tree
Pourouma melinonii Benoist	X			Medium tree
Pourouma minor Benoist	X			Medium tree
Pourouma mollis Trécul	X			Medium tree
Pourouma tomentosa Miq.	X			Medium tree
Pourouma velutina Miq.	X			Small tree
Pourouma villosa Trécul	X			Tall tree

FLORISTIC CHECKLIST OF THE NOURAGUES AREA	H	L	RS	LIFE FORM
Pourouma sp.	X			
Celastraceae				
Goupia glabra Aubl.	X			Tall tree
Maytenus myrsinoides Reissek	X	X		Tall tree
Maytenus oblongata Reissek	X			Treelet
[*Maytenus pruinosa* Reissek], after Larpin 1993	X			Tree
Chrysobalanaceae				
Couepia caryophylloides Benoist	X			Medium tree
Couepia guianensis Aubl. subsp. *guianensis*	X			Tall tree
Couepia obovata Ducke	X			Medium tree
Couepia parillo A.P. DC.	X			Medium tree
Hirtella araguariensis Prance	X			Treelet
Hirtella bicornis Mart. & Zucc.	X			Medium tree
Hirtella davisii Sandw.	X			tree
Hirtella glandulosa Spreng.	X			Medium tree
Hirtella hispidula Miq.	X			Tree
Hirtella physophora Mart. & Zucc.	X			Tree
Hirtella racemosa Lam. var. *hexandra* (Willd. ex Roem. & Schult.)	X			Treelet
Hirtella racemosa Lam. var. *racemosa*	X			Treelet
Hirtella silicea Griseb.	X	X		Small tree
Hirtella suffulta Prance	X			Tall tree
Hirtella tenuifolia Prance	X			Small tree
Licania alba (Bernoulli) Cuatrec.	X			Tall tree
Licania apetala (E. Mey.) Fritsch	X			Tree
Licania canescens Benoist	X			Medium tree
Licania caudata Prance	X			Small tree
Licania cyathodes Benoist	X			Small tree
Licania densiflora Kleinh.	X			Tall tree
Licania granvillei Prance	X			Tall tree
Licania guianensis (Aubl.) Griseb.	X			Tree
Licania heteromorpha Benth.	X			Tall tree
Licania hypoleuca Benth.	X	X		Tall tree
Licania irwinii Prance	X	X		Medium tree
Licania kunthiana J.D. Hook.	X			Tall tree
Licania latifolia Benth. ex J.D. Hook.	X			Medium tree
Licania latistipula Prance	X			Tall tree
Licania laxiflora Fritsch	X			Tall tree
Licania longistyla (J.D. Hook.) Fritsch	X			Medium tree
Licania majuscula Sagot	X			Medium tree
Licania membranacea Sagot ex Laness.	X			Tall tree
Licania micrantha Miq.	X			Tall tree

FLORISTIC CHECKLIST OF THE NOURAGUES AREA	H	L	RS	LIFE FORM
Licania minutiflora (Sagot) Fritsch	X			Emergent tree
Parinari excelsa Sabine	X			Emergent tree
Parinari parvifolia Sandw.	X			Emergent tree
Clusiaceae				
Clusia grandiflora Splitg.	X	X	X	Epiphyte & treelet
Clusia minor L.		X	X	Epiphyte & tree
Clusia nemorosa G. Mey.	X	X	X	Epiphyte & tree
Clusia panapanari (Aubl.) Choisy	X			Epiphyte & treelet
Clusia platystigma Eyma	X			Epiphyte & treelet
Clusia purpurea (Splitg.) Engl.				Epiphyte & tree
Moronobea coccinea Aubl.	X	(X)		Emergent tree
Platonia insignis Mart.	X			Tall tree
Rheedia acuminata (Ruiz & Pav.) Planch. & Triana	X	X		Small tree
cf. *Rheedia benthamiana* Planch. & Triana	X			Tall tree
Rheedia macrophylla (Mart.) Planch. & Triana	X			Tree
Symphonia globulifera L.f.	X			Tall tree
Symphonia sp.	X			
Tovomita cephalostigma Vesque	X			Tree
Tovomita choisyana Planch. & Triana	X			Medium tree
Vismia guianensis (Aubl.) Choisy	X			Small tree
Vismia latifolia (Aubl.) Choisy	X			Small tree
Vismia macrophylla Kunth	X			
Vismia ramuliflora Miq.	X			Tree
Vismia sandwithii Ewan	X			Treelet
Combretaceae				
Buchenavia amazonia Alwan & Stace	X			Tree
Buchenavia grandis Ducke	X			Tree
Buchenavia nitidissima (L.C. Rich.) Alwan & Stace	X			Tree
Buchenavia cf. *tetraphylla* (Aubl.) R. Howard	X			Emergent tree
Terminalia amazonia (J.F. Gmelin) Exell	X	X		Medium tree
Terminalia dichotoma G. Mey.	X	X		Medium tree
Terminalia guyanensis Eichl.	X			Tree
Connaraceae				
Connarus fasciculatus (DC.) Planch.	X			
Connarus perrottetii (DC.) Planch.	X			Liana
Rourea sp.	X			
Convolvulaceae				
Bonamia maripoides H. Hallier		X		
Dicranostyles cf. *guianensis* Mennega	X			
Dicranostyles cf. *villosus* Mennega	X			
Ipomoea angulata Mart. ex Choisy			X	Epilithic herb

FLORISTIC CHECKLIST OF THE NOURAGUES AREA	H	L	RS	LIFE FORM
Maripa cf. *glabra* Choisy	(X)			Liana
Maripa scandens Aubl.	X			Liana
Cucurbitaceae				
Cayaponia cf. *ophthalmica* R.E. Schultes	X			
Cayaponia racemosa (Mill.) Cogn.	(X)			Vine
Cayaponia rigida (Cogn.) Cogn.	(X)			
Gurania bignoniacea (Poepp. & Endl.) C. Jeffrey	X			Vine
Psiguria triphylla (Miq.) C. Jeffrey	X			Liana
Selysia prunifera (Poepp. & Endl.) Cogn.	X			Liana
Dichapetalaceae				
[*Dichapetalum pedunculatum* (DC.) Baill.], after Larpin 1993	X			Liana
Tapura amazonica Poepp. & Endl.	X			Small tree
Tapura capitulifera Baill.	X			Medium tree
Tapura singularis Ducke	X			Tree
Dilleniaceae				
Davilla sp.	X			
[*Doliocarpus* cf. *dentatus* (Aubl.) Standl.], after Larpin 1993	(X)	X		Liana
[*Doliocarpus guianensis* (Aubl.) Gilg.], after Larpin 1993	X			Liana
Doliocarpus major J.F. Gmelin	X			
Doliocarpus sp.	X	X	X	
Ebenaceae				
Diospyros carbonaria Benoist	X			tree
Diospyros dichroa Sandw.	X			Tall tree
Diospyros duckei Sandw.	X			Medium tree
Diospyros guianensis (Aubl.) Guerke	X			Tree
Diospyros cf. *matheriana* A.C. Smith	X			Tree
Elaeocarpaceae				
Sloanea garckeana K. Schum.	X			Small tree
Sloanea gracilis Uittien	X			Tree
Sloanea grandiflora J.E. Smith	X			Tree
Sloanea sinemariensis Aubl.	X			Tree
Sloanea terniflora (Moçino & Sessé ex DC) Standl.	X			Tree
Ericaceae				
Cavendishia callista J.D. Smith	X	X	X	Liana
Satyria cerander (Dunal) A.C. Smith	X	X	X	Liana
Sphyrospermum cordifolium Benth.	(X)	X	X	Woody epiphyte
Erythroxylaceae				
Erythroxylum citrifolium St. Hil.	(X)	X	X	Tree & treelet
Erythroxylum ligustrinum DC.	X	X	X	Tree & shrub
Erythroxylum macrophyllum Cav.	X			Small tree
Erythroxylum mucronatum Benth.	(X)	X		Tree & treelet

FLORISTIC CHECKLIST OF THE NOURAGUES AREA	H	L	RS	LIFE FORM
Erythroxylum squamatum Swartz	X	X	X	Treelet
Erythroxylum vernicosum O.E. Schulz	X		X	Tree
Euphorbiaceae				
Acalypha sp.	X			
Alchornea triplinervia (Spreng.) Muell. Arg.	X			Tree
Alchorneopsis floribunda (Benth.) Muell. Arg.	X			Medium tree
cf. *Amanoa guianensis* Aubl.	X			
Conceveiba guianensis Aubl.	X			Small tree
Croton hostmannii Miq.	X	X	X	Treelet
Croton matourensis Aubl.	X			
Croton schiedeanus Schlechtend.	X			
Croton tafelbergicus Croizat	X	X	X	Treelet
Dalechampia parvibracteata Lanj.	X			Liana
Drypetes cf. *fanshawei* Sandw.	X			Tree
Drypetes variabilis Uittien	X			Tall tree
Glycydendron amazonicum Ducke	X			Tall tree
Hevea guianensis Aubl.	X			Medium tree
Hyeronima alchorneoides Allemao var. *alchorneoides*	X			Tree
Mabea piriri Aubl.	X			Tree
Mabea speciosa Muell. Arg. subsp. *speciosa*	X			Small tree
Mabea subsessilis Pax & K. Hoffm.	X			Tree
Maprounea guianensis Aubl.	X			Medium tree
Micrandra brownsbergensis Lanj.	X	X		Medium tree
Omphalea diandra L.	X			Liana
Phyllanthus cf. *attenuatus* Miq.	X	X		Small tree
Plukenetia polyadenia Muell. Arg.	X			Liana
Pogonophora schomburgkiana Miers ex Benth.	X	X		Medium tree
[cf. *Sagotia racemosa* Baill.], after Larpin 1993	X			Tree
Sapium montanum Lanj.	X	X	X	Treelet
Sapium paucinervium Hemsley	X			Medium tree
Fabaceae				
Andira coriacea Pulle	X			Medium tree
Andira surinamensis (Bondt) Splitg. ex Amshoff	X			Tall tree
Clitoria sagotii Fantz	X			Liana
Dalbergia monetaria L.f.	X			Small tree
Dioclea elliptica Maxwell	X			Liana
Dioclea sp.	X			
Diplotropis purpurea (Rich.) Amsh.	X			Emergent tree
Dipteryx odorata (Aubl.) Willd.	X			Tall tree
Dipteryx punctata (Blake) Amsh.	X	X		Emergent tree
Dussia discolor (Benth.) Amsh.	X			Emergent tree

FLORISTIC CHECKLIST OF THE NOURAGUES AREA	H	L	RS	LIFE FORM
Hymenolobium flavum Kleinh.	X			Tall tree
Hymenolobium petraeum Ducke	X			Emergent tree
Lonchocarpus aff. heptaphyllus (Poir.) DC.	X			Medium tree
cf. *Monopteryx* sp.	X			
Ormosia coccinea (Aubl.) Jackson	X			Tree
Ormosia paraensis Ducke	X			Tree
Platymiscium ulei Harms	X			
Poecilanthe hostmannii (Benth.) Amsh.	X			Medium tree
Pterocarpus officinalis Jacq.	X			Tall tree
Stylosanthes guianensis (Aubl.) Swartz	(X)	(X)	X	Epilithic herb
Vatairea erythrocarpa Ducke	X			
Vatairea paraensis Ducke	X			Emergent tree
Vataireopsis surinamensis Lima	X			Emergent tree
Flacourtiaceae				
Banara guianensis Aubl.	X			Small tree
Carpotroche crispidentata Ducke	X			Tree & treelet
Casearia acuminata DC.	X			Tree & treelet
Casearia commersoniana Camb.	X			Treelet
Casearia javitensis Kunth	X			Small tree
Casearia pitumba Sleumer	X			Tree & treelet
[*Homalium guianense* (Aubl.) Oken], after Larpin 1993	X			Medium tree
Laetia procera (Poepp.) Eichl.	X			Emergent tree
Ryania speciosa Vahl	X	X		Treelet
Gentianaceae				
Irlbachia alata (Aubl.) Maas subsp. *alata*			X	Epilithic herb
Irlbachia alata (Aubl.) Maas subsp. *longistyla* Persoon & Maas			X	Epilithic herb
Irlbachia purpurascens (Aubl.) Maas	(X)	X	X	Terrestrial herb
Voyria aphylla (Jacq.) Pers.		X	X	Terrestrial herb
Voyria aurantiaca Splitg.	X			Saprophyte
Voyria caerulea Aubl.	X			Saprophyte
Voyria clavata Splitg.	X	X		Saprophyte
Voyria corymbosa Splitg.	X	X		Saprophyte
Voyria rosea Aubl.	X	X		Saprophyte
Voyria tenella Hook.	X			Saprophyte
Voyria tenuiflora Griseb.	X			Saprophyte
Voyriella parviflora (Miq.) Miq.	X	X	X	Terrestrial herb
Gesneriaceae				
Besleria cf. *laxiflora* Benth.	X			Epilithic treelet
Codonanthe calcarata (Miq.) Hanst.	X			Epiphyte herb
Codonanthe crassifolia (Focke) C. Morton	X	X	X	Epiphyte herb
Codonanthopsis dissimulata (H.E. Moore) Wiehler	X			Epiphyte herb

FLORISTIC CHECKLIST OF THE NOURAGUES AREA	H	L	RS	LIFE FORM
Columnea calotricha J.D. Smith	X			Epiphyte herb
Columnea sanguinea (Pers.) Hanst.	X			Epiphyte herb
Drymonia coccinea (Aubl.) Wiehler	X			Liana & epiphyte
Drymonia serrulata (Jacq.) Mart.	X			Vine
Episcia sphalera Leeuwenberg	X	X		Epilithic herb
Episcia xantha Leeuwenberg	X			Epilithic herb
Lembocarpus amoenus Leeuwenberg	X	X		Epilithic herb
Napeanthus jelskii Fritsch	X			Epilithic herb
Nautilocalyx mimuloides (Benth.) C. Morton	X	X		Terrestrial herb
Nautilocalyx pictus (Hook.) Sprague	X			Terrestrial herb
Paradrymonia campostyla (Leeuwenberg) Wiehler	X	X		Vine
Paradrymonia densa (C.H. Wright) Wiehler	X	X		Epilithic herb
Paradrymonia maculata (J.D. Hook.) Wiehler	X			Vine
Hernandiaceae				
Hernandia guianensis Aubl.	X			
Hippocrateaceae				
Cheiloclinium cognatum (Miers) A.C. Smith	X	X		Liana
Elachyptera floribunda (Benth.) A.C. Smith	X			
Hippocratea volubilis L.	X			Liana
Salacia cf. *cordata* (Miers) A.M. Mennega	X			Liana
[*Salacia impressifolia* (Miers) A.C. Smith], after Larpin 1993	X	X		Liana
Salacia insignis A.C. Smith	X			Liana
Salacia multiflora (Lam.) DC.	X			Lianalet
Salacia sp.	X			
Tontelea cylindrocarpa (A.C. Smith) A.C. Smith	X			Liana
Tontelea laxiflora (Benth.) A.C. Smith	X	X	X	Liana
Hugoniaceae				
Hebepetalum humiriifolium (Planch.) Benth.	X			Medium tree
Humiriaceae				
Humiria balsamifera (Aubl.) St. Hil. var. *balsamifera*	X	X	X	Treelet & tree
Sacoglottis cydonioides Cuatrec.	X			Tall tree
Sacoglottis guianensis Benth.	X	X		Tall tree
[*Schistostemon dichotomum* (Urb.) Cuatrec.], after Larpin, 1993	X			Medium tree
Vantanea parviflora Lam.	X			Emergent tree
Icacinaceae				
Dendrobangia boliviana Rusby	X			Tall tree
Poraqueiba guianensis Aubl.	X			Medium tree
Lacistemataceae				
Lacistema aggregatum (Bergius) Rusby	X			Treelet
Lacistema grandifolium Schnizlein	X			Medium tree

FLORISTIC CHECKLIST OF THE NOURAGUES AREA	H	L	RS	LIFE FORM
Lauraceae				
Aiouea guianensis Aubl.	X			Tall tree
Aiouea sp.	X			
Aniba guianensis Aubl.	X			Tree
Aniba panurensis (Meissn.) Mez	X			Tree
Aniba rosaeodora Ducke	X			Tall tree
Aniba terminalis Ducke	X			Tree
Cassytha filiformis L.	X	(X)		Parasite herb
Endlicheria punctulata (Mez) C.K. Allen	X			Treelet
Endlicheria pyriformis (Nees) Mez	X			Tree
Endlicheria sericea Nees	X			Tree
Kubitzkia mezii (Kosterm.) Van der Werff	X			Tree
Licaria cannella (Meissn.) Kosterm.	X	X		Tall tree
Licaria chrysophylla (Meissn.) Kosterm.	X			Tall tree
Licaria guianensis Aubl.	X	X		Tall tree
Licaria vernicosa. (Mez) Kosterm	X			Treelet
cf. *Mezilaurus* sp.	X			
Nectandra globosa (Aubl.) Mez	X			Tree
Nectandra maguireana C.K. Allen	X			Medium tree
Ocotea aciphylla (Nees) Mez	X			Medium tree
Ocotea amazonica (Meissn.) Mez	X			Tree
Ocotea argyrophylla Ducke	X			Tree
Ocotea ceanothifolia (Nees) Mez	X			Tree
Ocotea cymbarum Kunth	X			Tree
Ocotea discrepens C.K. Allen	X			Medium tree
Ocotea fendleri (Meissn.) Rohwer	X			Tree
Ocotea floribunda (Swartz) Mez	X			Tall tree
Ocotea guianensis Aubl.	X			Tree
Ocotea indirectinervia C.K. Allen	X			Tree
Ocotea leucoxylon (Swartz) de Lanessan	X			Tree
Ocotea nigra Benoist	X			Tree
Ocotea oblonga (Meissn.) Mez	X			Tree
Ocotea rubra Mez	X			Tall tree
Ocotea schomburgkiana (Nees) Mez	X			Tree
Ocotea splendens (Meissn.) Mez	X			Tree
Ocotea tomentella Sandw.	X			Small tree
Rhodostemonodaphne grandis (Mez) Rohwer	X			Tree
Rhodostemonodaphne kunthiana (Nees) Rohwer	X			
Lecythidaceae				
Corythophora rimosa W. Rodrigues	X			Tall tree
Couratari guianensis Aubl. emend. Prance	X			Emergent tree

FLORISTIC CHECKLIST OF THE NOURAGUES AREA	H	L	RS	LIFE FORM
Couratari multiflora (J.E. Smith) Eyma	X			Tall tree
Couratari oblongifolia Ducke & Knuth	X			Tree
Eschweilera apiculata (Miers) A.C. Smith	X			Emergent tree
Eschweilera cf. *chartaceifolia* Mori	X			Tree
Eschweilera collina Eyma	X			Tall tree
Eschweilera congestiflora (Benoist) Eyma	X			Medium tree
Eschweilera coriacea (A.P. DC.) Mori	X			Tall tree
Eschweilera decolorans Sandw.	X			Tall tree
Eschweilera grandiflora (Aubl.) Sandw.	X			Medium tree
Eschweilera laevicarpa Mori	X			Tree
Eschweilera micrantha (Berg) Miers	X			Tall tree
Eschweilera parviflora (Aubl.) Miers	X			Tall tree
Eschweilera pedicellata (Rich.) Mori	X			Medium tree
Eschweilera sagotiana Miers	X			Tall tree
Eschweilera simiorum (Benoist) Eyma	X			Medium tree
Eschweilera squamata Mori	X			Tree
Gustavia augusta L.	X			Medium tree
Gustavia hexapetala (Aubl.) J.E. Smith	X			Medium tree
Lecythis chartacea Berg	X			Tree
Lecythis confertiflora (A.C. Smith) Mori	X			Tall tree
Lecythis corrugata Poit.	X			Tall tree
Lecythis holcogyne (Sandw.) Mori	X			Tall tree
Lecythis idatimon Aubl.	X	X		Medium tree
Lecythis persistens Sagot subsp. *aurantiaca* Mori	X			Tall tree
Lecythis persistens Sagot subsp. *persistens*	X			Tall tree
Lecythis poiteaui Berg	X			Tall tree
Lecythis zabucajo Aubl.	X			Emergent tree
Lentibulariaceae				
Utricularia amethystina Salzm. ex St. Hil. & Girard			X	Epilithic herb
Utricularia choristotheca P. Taylor			X	Epilithic herb
Utricularia hispida Lam.			X	Epilithic herb
Utricularia jamesoniana Oliver			X	Epiphyte herb
Utricularia subulata L.			X	Epilithic herb
Utricularia tenuissima Tutin			X	Terrestrial herb
Loganiaceae				
Antonia ovata Pohl	X			Medium tree
Potalia amara Aubl.	X			Shrub or treelet
Strychnos erichsonii Rich. Schomb. ex Progel	X			Liana
Strychnos glabra Sagot ex Progel	X			Liana
Strychnos guianensis (Aubl.) Mart.	X			Liana
Strychnos oiapocensis Froes	X	X		Liana

FLORISTIC CHECKLIST OF THE NOURAGUES AREA	H	L	RS	LIFE FORM
Strychnos toxifera Rob. Schomb. ex Benth.	X			Liana
Loranthaceae				
Phthirusa pyrifolia (Kunth) Eichl.	X	X	X	Woody parasite
Phthirusa stelis (L.) Kuijt	X	X	X	Woody parasite
Psittacanthus redactus Rizzini			X	Woody parasite
Lythraceae				
Cuphea blackii Lourteig			X	Epilithic herb
Malpighiaceae				
Byrsonima aerugo Sagot	X			Tall tree
Byrsonima densa (Poir.) DC.	X			Tall tree
Byrsonima laevigata (Poir.) DC.	X			Tree
Byrsonima stipulacea Adr. Juss.	X			Tall tree
Heteropterys sp.	X	X	X	
Malvaceae				
Pavonia castanaefolia St. Hil. & Naud.	X			Terrestrial herb
Pavonia schiedeana Steud.	X			Terrestrial herb
Marcgraviaceae				
Marcgravia coriacea Vahl	X			Liana
Norantea guianensis Aubl.	X	X	X	Liana
Souroubea guianensis Aubl.	X	X	X	Liana
Melastomataceae				
Aciotis laxa (DC.) Cogn. var. *laxa*	X			Treelet
Aciotis purpurascens (Aubl.) Triana	X			Shrub
Appendicularia thymifolia (Bonpl.) DC.			X	Epilithic herb
Bellucia grossularioides (L.) Triana	X			Tree
Clidemia conglomerata DC.	X			Terrestrial herb
Clidemia dentata D. Don	X			Treelet
Clidemia hirta (L.) D. Don var. *elegans* (Aubl.) Griseb.	X			
Clidemia involucrata DC.	X			
Clidemia septulinervia Cogn.	X			
Clidemia venosa (Gleason) Wurdack	X			Treelet
Ernestia confertiflora Wurdack			X	Epilithic subshrub
Ernestia glandulosa Gleason	X			Treelet
Ernestia granvillei Wurdack	(X)	X	X	Treelet
Ernestia pullei Gleason			X	Shrub & treelet
Henriettea maroniensis Sagot	(X)	X		Tree
Henriettea succosa (Aubl.) DC.	X	X		Tree
Henriettella caudata Gleason	X			treelet
Henriettella flavescens (Aubl.) Triana	X			Treelet
Leandra divaricata (Naud.) Cogn.	X	X		
Leandra paleacea Wurdack	X			Treelet

FLORISTIC CHECKLIST OF THE NOURAGUES AREA	H	L	RS	LIFE FORM
Leandra purpurea Gleason	X			
Leandra cf. rufescens (DC.) Cogn.	X			Treelet
Leandra cf. *verticillata* (DC.) Cogn.	X	X		
Loreya mespiloides Miq.	X			Small tree
Macrocentrum cristatum (DC.) Triana var. *cristatum*	X	X	X	Subshrub
Maieta guianensis Aubl.	X			Shrub & treelet
Miconia acuminata (Steud.) Naud.	X			Medium tree
Miconia affinis DC.	X			Small tree
Miconia argyrophylla DC.	X			Treelet
Miconia bracteata (DC.) Triana	X			Treelet
Miconia cacatin (Aubl.) Renner	X			Treelet
Miconia ceramicarpa (DC.) Cogn. var. *candolleana* Cogn.	X			Shrub
Miconia ceramicarpa (DC.) Cogn. var. *ceramicarpa*	X	X		Shrub
Miconia ciliata (Rich.) DC.	X	X	X	Shrub & treelet
Miconia diaphanea Gleason	X			Treelet
Miconia dispar Benth.	(X)	X		Treelet
Miconia holosericea (L.) DC.	X	X	X	Treelet
Miconia cf. *lepidota* DC.	X			
Miconia mirabilis (Aubl.) L.O. Williams	X			Small tree
Miconia plukenetii Naud.	X			Tree
Miconia prasina (Swartz) DC.	X	X		Tree
Miconia trimera Wurdack	X			Shrub
Mouriri crassifolia Sagot	X			Tall tree
Mouriri huberi Cogn.	X			Tall tree
Mouriri sagotiana Triana	X			Epiphyte & terrestrial
Topobea parasitica Aubl.	(X)	(X)	X	Epiphyte & tree
Meliaceae				
Carapa guianensis Aubl.	X			Tall tree
Carapa procera A. DC.	X			Tall tree
Cedrela odorata L.	X			Emergent tree
Guarea gomma Pulle	X			Tree
Guarea grandifolia A. DC.	X			Tree
Guarea guidonia (L.) Sleumer	X			Medium tree
Guarea kunthiana Adr. Juss.	X			Medium tree
[*Guarea* cf. *macrophylla* Vahl], after Larpin 1993	X			Medium tree
Guarea michel-moddei Pennington & Mori	X			Tree
Guarea pubescens (L.C. Rich.) Adr. Juss.	X	X		Small tree
Guarea silvatica C. DC.	X			Medium tree
Trichilia cipo (Adr. Juss.) C. DC.	X			
Trichilia lepidota Mart.	X			Medium tree
Trichilia micrantha Benth.	X			Medium tree

FLORISTIC CHECKLIST OF THE NOURAGUES AREA	H	L	RS	LIFE FORM
Trichilia pallida Swartz	X			Tree
Trichilia quadrijuga Kunth	X			Medium tree
Trichilia schomburgkii C. DC.	X			Medium tree
Trichilia septentrionalis C. DC.	X			Tall tree
Trichilia surinamensis (Miq.) C. DC.	X	X		Tall tree
Mendonciaceae				
Mendoncia hoffmannseggiana Nees	X	(X)		Vine
Menispermaceae				
Abuta bullata Moldenke	X	X		Liana
Abuta grandifolia (Mart.) Sandw.	X			Tree
Abuta cf. *obovata* Diels	(X)			Liana
Abuta cf. *rufescens* Aubl.	X			Liana
Anomospermum cf. *chloranthum* Diels	X			Liana
Elephantomene cf. *eburnea Barneby* & Krukoff	X			Liana
Mimosaceae				
Abarema barbouriana (Standl.) Barneby & Grimes	X	X		Tree
Abarema curvicarpa (Irwin) Barneby & Grimes	X			Tree
Abarema gallorum Barneby & Grimes	X	X		Tree
Abarema jupunba (Willd.) Britton & Killip var. *jupunba*	X	X		
Abarema mataybifolia (Sandw.) Barneby & Grimes	X			Tree
Acacia tenuifolia (L.) Willd.	X			
Balizia pedicellaris (DC.) Barneby & Grimes	X	X	X	Medium tree
Calliandra aff. *surinamensis* Benth.	(X)	X	X	Treelet & tree
Enterolobium oldemanii Barneby & Grimes	X			Tree
Enterolobium schomburgkii (Benth.) Benth.	X			Tall tree
Inga acreana Harms	X			Tree
Inga acrocephala Steud.	X			Medium tree
Inga alata Benoist	X			Tree
Inga alba (Swartz) Willd.	X			Tree
Inga albicoria O. Poncy	X			Tree
Inga auristellae Harms	X			Small tree
Inga bourgoni (Aubl.) DC.	X			Tall tree
Inga brachystachys Ducke	X			Treelet
Inga bracteosa Benth.	X			Small tree
Inga capitata Desv.	X			Medium tree
Inga cayennensis Sagot ex Benth.	X	X	X	Tree
Inga edulis (Vell.) Mart.	X			Tall tree
Inga fanchoniana O. Poncy	X			Tree
Inga fastuosa (N.J. Jacq.) Willd.	X			Treelet
Inga flagelliformis Mart.	X			Tree
Inga graciliflora Benth.	X			Tree

FLORISTIC CHECKLIST OF THE NOURAGUES AREA	H	L	RS	LIFE FORM
Inga gracilifolia Ducke	X			Small tree
Inga grandiflora Ducke	X			Tree
Inga huberi Ducke	X			Medium tree
Inga jenmanii Sandw.	X			Medium tree
Inga lateriflora Miq.	X	X		Medium tree
Inga leiocalycina Benth.	X			Medium tree
Inga lomatophylla (Benth.) Pitt.	X			Tree
Inga longipedunculata Ducke	X			Tree
Inga marginata Willd.	X			Tree
Inga melinonis Sagot	X			Tree
Inga nouragensis O. Poncy	X			Tree
Inga paraensis Ducke	X			Tree
Inga pezizifera Benth.	X			Tall tree
Inga poeppigiana Benth.	X			Tree
Inga retinocarpa. O. Poncy	X			Tree
Inga rubiginosa (L.C. Rich.) DC.	X			Tall tree
Inga stipularis DC.	X			Medium tree
Inga thibaudiana DC.	X	X		Medium tree
Inga tubaeformis Benoist	X	X		Tree
Inga umbellifera (Vahl) Steud. ex DC.	X	X	X	Tree
Inga virgultosa (Vahl) Desv.	X	X		Small tree
Mimosa annularis Benth. var. *xinguensis* (Ducke) Barneby	(X)			Liana
[*Mimosa guilandinae* (DC.) Barneby var. *guilandinae*], after Julliot 1992	(X)			Liana
Mimosa myriadena Benth.	X			
Mimosa polydactyla Humb. & Bonpl. ex Willd.	X			
Mimosa pudica L.	X			
Parkia nitida Miq.	X			Emergent tree
Parkia pendula (Willd.) Benth. ex Walpers	X			Emergent tree
Parkia reticulata Ducke	X			Tree
Parkia ulei (Harms) Kuhlmann	X			Emergent tree
Parkia velutina Benoist	X			Emergent tree
Pseudopiptadenia psilostachya (DC.) Lewis & M.P. Lima	X			Tall tree
Pseudopiptadenia suaveolens (Miq.) Grimes	X			Tall tree
Stryphnodendron guianense (Aubl.) Benth.	X			Tree
Stryphnodendron moricolor Barneby & Grimes	X			Tree
Stryphnodendron polystachyum (Miq.) Kleinh.	X			Emergent tree
Zygia racemosa (Ducke) Barneby & Grimes	X			Tall tree
Zygia sabatieri Barneby & Grimes	X			Tree
Zygia tetragona Barneby & Grimes	X	X		Tree
Monimiaceae				
Siparuna cristata (Poepp. & Endl.) A. DC.	X			Tree

FLORISTIC CHECKLIST OF THE NOURAGUES AREA	H	L	RS	LIFE FORM
Siparuna cuspidata (Tul.) A. DC.	X			Small tree
Siparuna decipiens (Tul.) A. DC.	X			Small tree
Siparuna guianensis Aubl.	X			Small tree
Siparuna pachyantha A.C. Smith	X			Small tree
Siparuna sprucei A. DC.	X	X		Small tree
Moraceae				
Bagassa guianensis Aubl.	X			Emergent tree
Batocarpus amazonicus (Ducke) Fosb.	X			Medium tree
Brosimum acutifolium Huber subsp. *acutifolium*	X			Tall tree
Brosimum guianense (Aubl.) Huber	X	X		Tall tree
Brosimum parinarioides Ducke	X			Emergent tree
Brosimum rubescens Taub.	X			Emergent tree
Brosimum utile (Kunth) Pitt	X			Emergent tree
Clarisia ilicifolia (Spreng.) Lanj. & Rossberg	X			Tree
Ficus amazonica (Miq.) Miq.	X	X		Hemi-epiphytic tree
Ficus broadwayi Urb.	X			Hemi-epiphytic tree
Ficus gomelleira Kunth & Bouché	X			Medium tree
Ficus guianensis Desv. ex Hamilton	X	X	X	Tall tree
Ficus insipida Willd.	X			Tall tree
Ficus cf. *leiophylla* C.C. Berg	X			Medium tree
Ficus mathewsii (Miq.) Miq.	X			Tall tree
Ficus nymphaeifolia Mill.	X			Tall tree
Ficus pertusa L.f.	X			Medium tree
Ficus trigona L.f.	X			Hemi-epiphytic tree
Helicostylis pedunculata Benoist	X			Medium tree
Helicostylis tomentosa (Poepp. & Endl.) Rusby	X			Tall tree
Maquira calophylla (Poepp. & Endl.) C.C. Berg	X			Tree
Maquira guianensis Aubl.	X			Tall tree
Naucleopsis glabra Spruce ex Pitt.	X			Medium tree
Naucleopsis guianensis (Mildbr.) C.C. Berg	X			Medium tree
Perebea guianensis Aubl.	X			Medium tree
Perebea mollis (Poepp. & Endl.) Huber	X			Tall tree
Pseudolmedia laevis (Ruiz & Pav.) Macbr.	X			Emergent tree
Trymatococcus oligandrus (Benoist) Lanj.	X			Medium tree
Trymatococcus paraensis Ducke	X			Tall tree
Myristicaceae				
Iryanthera hostmannii (Benth.) Warb.	X			Medium tree
Iryanthera sagotiana (Benth.) Warb.	X			Tall tree
Osteophloeum platyspermum (Spruce ex A. DC.) Warb.	X			Tall tree
Virola kwatae Sabatier	X			Tall tree
Virola michelii Heckel	X			Tall tree

FLORISTIC CHECKLIST OF THE NOURAGUES AREA	H	L	RS	LIFE FORM
Virola multicostata Ducke	X			Tree
Virola sebifera Aubl.	X			Tree
Virola surinamensis (Rolander) Warb.	X			Emergent tree
Myrsinaceae				
Cybianthus guianensis (A.DC.) Miq. subsp. *pseudoicacoreus* (A. DC.) Agostini	X	X	X	
Cybianthus venezuelanus Mez	X			Treelet
Myrtaceae				
[*Calycorectes bergii* Sandw.], after Larpin 1993	X			Treelet
Calycorectes grandifolius O. Berg	X			Shrub
Calyptranthes amshoffae McVaugh	X	X		Medium tree
Calyptranthes lepida McVaugh	X	X	X	Tree
Eugenia albicans (O.C. Berg) Urban	X	X		Treelet
Eugenia biflora (L.) DC.			X	Treelet
Eugenia coffeifolia DC.	X			Tree
Eugenia cucullata Amsh.	X			Tall tree
Eugenia cupulata Amsh.	X			Tree
Eugenia florida DC.	X	X	X	Tree
Eugenia latifolia Aubl.	X			Treelet
Eugenia macrocalyx (Rusby) McVaugh	X			Small tree
Eugenia marowynensis Miq.	X	X	X	Small tree
Eugenia monticola (Swartz) DC.	X	X		Tree
Eugenia patrisii Vahl	X			Small tree
Eugenia pseudopsidium Jacq.	X			Treelet
Eugenia ramiflora Desv.	X	X	X	Treelet
Eugenia tetramera (McVaugh) M.L. Kawasaki & B. Holst	X			Tree
Marlierea ferruginea (Poir.) McVaugh	X			
Myrcia amazonica DC.	X			Treelet
Myrcia citrifolia (Aubl.) Urb.	X	X	X	Treelet
Myrcia fallax (L.C. Rich.) DC.	X	X	X	Treelet
[*Myrcia* cf. *graciliflora* Sagot], after Larpin 1993	X			Treelet
Myrcia guianensis (Aubl.) DC.	X	X	X	Treelet
Myrcia cf. *paivae* O. Berg	X			Treelet
Myrcia cf. *platyclada* DC.	X			Treelet
Myrcia quitarensis (Bentham) Sagot	X	X	X	Treelet
Myrcia saxatilis (Amsh.) McVaugh		X	X	Treelet
Myrcia sylvatica (G. Mey.) DC.	X	X	X	Treelet
Myrciaria floribunda (West ex Willd.) O. Berg	X	X	X	Small tree
cf. *Psidium acutangulum* Mart. ex DC.	X			
Nyctaginaceae				
Guapira eggersiana (Heimerl) Lundell	X	X	X	
Neea floribunda Poepp. & Endl.	X			Small tree

FLORISTIC CHECKLIST OF THE NOURAGUES AREA	H	L	RS	LIFE FORM
Neea ovalifolia Spruce ex Schmidt	X	X	X	Small tree
Nyctaginaceae sp.	X	X		
Ochnaceae				
Elvasia macrostipularis Sastre & Lescure	X			Tree
Ouratea candollei (Planch.) Tiegh.	X	X	X	Tree
Ouratea decagyna Maguire	X			Medium tree
Ouratea guianensis Aubl.	X			Treelet
Ouratea impressa (Tiegh.) Lemée	X			Shrub & tree
Ouratea leblondii (Tiegh.) Lemée	X	X	X	Tree
Ouratea macrocarpa Sastre	X			
Ouratea melinonii (Tiegh.) Lemée	X			Treelet
Sauvagesia aliciae Sastre	X	X	X	Shrub
Sauvagesia erecta L.		X		Epilithic herb
Sauvagesia tafelbergensis Sastre		X		Terrestrial herb
Olacaceae				
Cathedra acuminata (Benth.) Miers	X	X		Tree
Chaunochiton kappleri (Sagot ex Engl.) Ducke	X			
Dulacia guianensis (Engl.) Kuntze	X	X		Tall tree
Heisteria cauliflora J.E. Smith	X			Shrub & treelet
Heisteria densifrons Engl.	X			Treelet
Minquartia guianensis Aubl.	X			Tall tree
Ptychopetalum olacoides Benth.	X			Medium tree
Opiliaceae				
[*Agonandra silvatica* Ducke], after Larpin 1993	X			Tall tree
Passifloraceae				
Dilkea retusa Mast.	X			
Passiflora amoena Escobar	X			Liana
Passiflora auriculata Kunth	X			Liana
Passiflora candida (Poepp. & Endl.) Masters	(X)			Liana
Passiflora coccinea Aubl.	X			Liana
Passiflora crenata Feuillet & Cremers	X			Liana
Passiflora exura Feuillet	(X)			Liana
Passiflora fanchonae Feuillet	X	X		Liana
Passiflora fuchsiiflora Hemsley	X	X		Liana
Passiflora garckei Masters	X	X		Liana
Passiflora glandulosa Cav.	X			Liana
Passiflora aff. *variolata* Poepp. & Endl.	X			Liana
Passiflora vespertilio L.	(X)			Liana
Passiflora spp.				
Phytolaccaceae				
Petiveria cf. *alliacea* L.	X			

FLORISTIC CHECKLIST OF THE NOURAGUES AREA	H	L	RS	LIFE FORM
Piperaceae				
Peperomia glabella (Swartz) A. Dietr.	X			Epiphyte herb
Peperomia macrostachya (Vahl) A. Dietr.	X	X		Epiphyte herb
Peperomia obtusifolia (L.) A. Dietr.	X			Epiphyte herb
Peperomia ouabianae C. DC	X			
Peperomia pernambucensis Miq.	X			Epiphyte herb
Peperomia rotundifolia (L.) Kunth	X	X		Epiphyte herb
Peperomia serpens (Swartz) J.C. Loudon	X			Epiphyte herb
Peperomia sp.	X			
Piper adenandrum (Miq.) C. DC.	X			Treelet
Piper aequale Vahl	X			Treelet
Piper angustifolium Lam.	X			Subshrub
Piper anonifolium (Kunth) C. DC.	X			Treelet
Piper arboreum Aubl.	X			Tree
Piper augustum Rudge	X			Treelet
Piper avellanum (Miq.) C. DC.	X			Treelet
Piper brownsbergense Yunck.	X			Treelet
Piper consanguineum Kunth	X			Subshrub
Piper demeraranum (Miq.) C. DC.	X			Treelet
Piper divaricatum G. Mey.	X			
Piper hispidum Swartz	X			Treelet
Piper hostmannianum (Miq.) C. DC.	X			Liana
Piper humistratum Görts & K.U. Kramer	X			Terrestrial herb
Piper nematanthera (Miq.) C. DC.	X			Liana
Piper obliquum Ruiz & Pav.	X			
Piper cf. *oblongifolium* (Klotzsch) C. DC.	X			Treelet
Piper paramaribense C. DC.	X			Subshrub
Piper cf. *pulleanum* Yunck.	X			Treelet
Piper tectoniifolium Kunth	X			Treelet
Piper trichoneuron (Miq.) C. DC.	X			
Piper wachenheimii Trel.	X			Treelet
Polygalaceae				
Moutabea guianensis Aubl.	X			Liana
Moutabea cf. *longifolia* Poepp. & Endl.	X			Liana
Polygala adenophora DC.	(X)	(X)	X	Epilithic herb
Polygala spectabilis DC.	X	X	X	Shrub & treelet
Securidaca uniflora Oort	X	X	X	Liana
Securidaca sp.	X			
Polygonaceae				
Coccoloba marginata Benth.	X			Liana
Coccoloba lucidula Benth.	X			Liana

FLORISTIC CHECKLIST OF THE NOURAGUES AREA	H	L	RS	LIFE FORM
Coccoloba ovata Benth.	X			Liana
Coccoloba parimensis Benth.	X			Liana
Proteaceae				
Euplassa pinnata (Lam.) Johnston		X		Medium tree
Roupala montana Aubl.	X	X	X	Medium tree
Quiinaceae				
Lacunaria crenata (Tul.) A.C. Smith	X			Small tree
Lacunaria jenmanii (Oliver) Ducke	X			Small tree
Quiina cf. guianensis Aubl.	X			
Quiina obovata Tul.	X			Small tree
Quiina oiapocensis Pires	X			Tall tree
Touroulia guianensis Aubl.	X			Small tree
Rhabdodendraceae				
Rhabdodendron amazonicum (Spruce ex Benth.) Huber	X			Small tree
Rhamnaceae				
Ampelozizyphus amazonicus Ducke	X			Tree
Gouania blanchetiana Miq.				Liana
Ziziphus cinnamomum Triana & Planch.	X			Tree
Rhizophoraceae				
Cassipourea guianensis Aubl.	X	X		Tree
Rubiaceae				
Alibertia surinamensis (Bremek.) Steyerm.	X			Treelet
Alibertia myrciifolia K. Schumann	X	X	X	
Amaioua corymbosa Kunth	X			
Amaioua guianensis Aubl.	X			
Borreria alata (Aubl.) DC.	(X)	X		Epilithic herb
Borreria capitata (Ruiz & Pav.) DC. var. *tenella* (Kunth) Steyerm.			X	Epilithic herb
Capirona decorticans Spruce	X			Tall tree
Chimarrhis turbinata DC.	X			Tall tree
Chomelia sp.		X		
Coccocypselum guianense (Aubl.) K. Schum.	X	X		Terrestrial herb
Coussarea racemosa A. Rich.	X			Small tree
Duroia aquatica (Aubl.) Bremek.	X			Small tree
Duroia eriopila L.f.	X			Small tree
Duroia genipoides J.D. Hook. ex K. Schum.	X			
Duroia longiflora Ducke	X			Medium tree
Duroia micrantha Ladbr. Zaracchi & Kirkbride	X			
Faramea corymbosa Aubl.	X			Treelet &r tree
Faramea guianensis (Aubl.) Bremek.	X			Shrub & treelet
Ferdinandusa paraensis Ducke	X			Medium tree

FLORISTIC CHECKLIST OF THE NOURAGUES AREA	H	L	RS	LIFE FORM
Geophila cordifolia Miq.	X	X		Epilithic herb
Geophila sp.	X			
Guettarda acreana Krause	X			Small tree
Guettarda aromatica Poell. & Endl.	X			
Hillia illustris (Vell.) K. Schum.	X			Epiphyte & treelet
Isertia spiciformis DC.	X	X		Treelet
Ixora graciliflora Benth.	X	X		Tree
Ixora piresii Steyerm.				Treelet
Ixora ulei Krause		X		
Malanea macrophylla Bartling ex Griseb.	X	X		Liana
Manettia alba (Aubl.) Wernham	(X)	X		Liana
Morinda brachycalyx (Bremek.) Steyerm.	X	X		Treelet
Morinda sp.	X	X		
Oldenlandia cf. *corymbosa* L.	X			
Palicourea calophylla DC.	X			Shrub
Palicourea expetens Standl.	X			Tree
Palicourea guianensis Aubl.	X			
[*Palicourea quadrifolia* (Rudge) DC.], after Larpin 1993	X			Tree
Palicourea triphylla DC.	X			Treelet
Perama hirsuta Aubl.			X	Epilithic herb
Posoqueria latifolia (Rudge) Roem. & Schult.	X			Small tree
Psychotria anceps Kunth	X			Tree
Psychotria apoda Steyerm.	X			Treelet
Psychotria cf. *bahiensis* DC.	X			Treelet
Psychotria blepharophylla (Standl.) Steyerm.	X			Shrub
Psychotria bracteocardia (DC.) Muell. Arg.	(X)	X		Treelet
Psychotria capitata Ruiz & Pav.	X	X		Tree
Psychotria ctenophora Steyerm.	X	X	X	Shrub & treelet
Psychotria cupularis (Muell. Arg.) Standl.	X			Tree
Psychotria deflexa DC.	X			Treelet
Psychotria erecta (Aubl.) Standl. & Steyerm.	X	X		Shrub & treelet
Psychotria galbaoensis Steyerm.	X			Tree
Psychotria hoffmannseggiana (Willd. ex Roem. & Schult.) Muell. Arg.	X	X	X	Shrub & treelet
Psychotria cf. *kappleri* (Miq.) Muell. Arg. ex Benoist	(X)			Shrub & treelet
Psychotria ligularis (Rudge) Steyerm. var. *carapichea* (Poir.) Steyerm.	X			Treelet
Psychotria mapourioides DC.	X			Treelet
Psychotria microbotrys Ruiz ex Standley	X			Shrub & treelet
Psychotria moroidea Steyerm.	X	X		Shrub & treelet
Psychotria poeppigiana Muell. Arg.	X			Shrub
Psychotria racemosa (Aubl.) Raeusch.	X	X		Shrub & treelet
Psychotria surinamensis Bremek.	X			

FLORISTIC CHECKLIST OF THE NOURAGUES AREA	H	L	RS	LIFE FORM
Psychotria ulviformis Steyerm.	X			
Psychotria variegata Steyerm.	X			Terrestrial herb
Randia armata (Swartz) DC.	X			Tree
Rudgea crassiloba (Benth.) Robinson	X	X	X	
Rudgea tanaosepala Sandw.	X			Treelet
Sipanea pratensis Aubl.	(X)	(X)	X	Epilithic herb
Sipanea wilson-brownei Cowan		X		Terrestrial herb
cf. *Tocoyena guianensis* K. Schum.	X			
Uncaria guianensis (Aubl.) J.F. Gmelin	X			Liana
Rutaceae				
Ertela trifolia (L.) Kuntze	(X)	X		
Esenbeckia cowanii Kaastra	(X)	X		Small tree
Esenbeckia grandiflora Mart.	(X)	X		Small tree
Galipea trifoliata Aubl.	(X)	X		Tree
Zanthoxylum pentandrum (Aubl.) R. Howard	X			Tree
Sapindaceae				
Cupania diphylla Vahl	X	X		Small tree
Cupania hirsuta Radlk.	X			Medium tree
Cupania scrobiculata L.C. Rich.	X			Small tree
Matayba arborescens (Aubl.) Radlk.	X			Liana
Matayba guianensis Aubl.	X			Small tree
Matayba inelegans Spruce ex Radlk.	X			Tree
Paullinia anodonta Radlk.	X			Liana
Paullinia capreolata (Aubl.) Radlk.	X			Liana
Paullinia cf. *plagioptera* Radlk.	X			Liana
Paullinia rufescens Rich. ex Juss.	X			
Paullinia cf. *tricornis* Radlk.	X			Liana
Paullinia cf. *venosa* Radlk.	X			Tree
Paullinia sp.	X			
Porocystis toulicioides Radlk.	X			
Pseudima frutescens (Aubl.) Radlk.	X			Medium tree
Talisia carinata Radlk.	X			Small tree
Talisia clathrata Radlk.	X			Tree
Talisia hexaphylla Vahl	X			Tree
[*Talisia longifolia* (Benth.) Radlk.], after Larpin 1993	X			Small tree
Talisia micrantha Radlk.	X			Treelet
Talisia mollis Kunth ex Camb.	X			
Talisia pedicellaris Sagot ex Radlk.	X			Tree
Talisia praealta (Sagot) Radlk.	X			Medium tree
Talisia simaboides K.U. Kramer	X			Tall tree
Vouarana guianensis Aubl.	X			Small tree

FLORISTIC CHECKLIST OF THE NOURAGUES AREA	H	L	RS	LIFE FORM
Sapotaceae				
Chrysophyllum argenteum Jacq. subsp. *auratum* (Miq.) Pennington	X			Tree
Chrysophyllum argenteum Jacq. subsp. *nitidum* (G. Mey.) Pennington	X			Medium tree
Chrysophyllum cuneifolium (Rudge) A. DC.	X			Medium tree
Chrysophyllum eximium Ducke	X			Medium tree
Chrysophyllum lucentifolium Cronq.	X			Emergent tree
Chrysophyllum pomiferum (Eyma) Pennington	X			Tall tree
Chrysophyllum prieurii A. DC.	X			Tall tree
Chrysophyllum sanguinolentum (Pierre) Baehni	X			Medium tree
Chrysophyllum venezuelanense (Pierre) Pennington	X			Treelet
Diploön cuspidatum (Hoehne) Cronquist	X			
Ecclinusa guianensis Eyma	X			Tall tree
Ecclinusa lanceolata (Mart. & Eichl.) Pierre	X			Small tree
Ecclinusa ramiflora Mart.	X			Tree
Elaeoluma nuda (Baehni) Aubr.	X			
Manilkara bidentata (A. DC.) Chevalier subsp. *bidentata*	X	X		Tall tree
Manilkara huberi (Ducke) Chevalier	X	X		Tall tree
Manilkara zapota (L.) van Royen [not a native species]	X			
Micropholis cayennensis Pennington	X			Tall tree
Micropholis egensis (A. DC.) Pierre	X			Medium tree
Micropholis guyanensis (A. DC.) Pierre subsp. *guyanensis*	X			Tall tree
Micropholis longipedicellata Aubr.	X			Tree
Micropholis melinoniana Pierre	X			Tall tree
Micropholis mensalis (Baehni) Aubr.	X	X		Tall tree
Micropholis obscura Pennington	X			Tall tree
Micropholis porphyrocarpa (Baehni) Monachino	X			Tree
Micropholis sanctae-rosae (Baehni) Pennington	X			Tree
Micropholis venulosa (Mart. & Eichl.) Pierre	X			Tall tree
Pouteria ambelaniifolia (Sandw.) Pennington	X			Tall tree
Pouteria benai (Aubr. & Pellegr.) Pennington	X			Tall tree
Pouteria aff. *bilocularis* (Winkler) Baehni	X			Tall tree
Pouteria caimito (Ruiz & Pav.) Radlk.	X			Tree
[*Pouteria* aff. *cicatricata* Pennington], after Larpin 1993	X			Tree
Pouteria cladantha Sandw.	X			Tree
Pouteria aff. *congestifolia* Pilz	X			Tree
Pouteria coriacea (Pierre) Pierre	X			Tree
Pouteria cuspidata (A. DC.) Baehni subsp. *dura* (Eyma) Pennington	X			Tree
Pouteria decorticans Pennington	X			Tree
Pouteria deliciosa Pennington	X			Tree
Pouteria egregia Sandw.	X			Tall tree
Pouteria eugeniifolia (Pierre) Baehni	X			Tall tree

FLORISTIC CHECKLIST OF THE NOURAGUES AREA	H	L	RS	LIFE FORM
Pouteria engleri Eyma	X			Tree
Pouteria filipes Eyma	X			Tall tree
Pouteria glomerata (Miq.) Radlk. subsp. *stylosa* (Pierre) Pennington	X			Tree
Pouteria gonggrijpii Eyma	X			Medium tree
Pouteria grandis Eyma	X			Tree
Pouteria guianensis Aubl.	X			Tall tree
Pouteria hispida Eyma	X			Small tree
Pouteria jariensis Pires & Pennington	X			Tree
Pouteria laevigata (Mart.) Radlk.	X			Tall tree
Pouteria macrophylla (Lam.) Eyma	X			Tree
Pouteria melanopoda Eyma	X			Medium tree
Pouteria oblanceolata Pires	X			Tree
Pouteria reticulata (Engl.) Eyma	X			Tree
Pouteria retinervis Pennington	X			Small tree
Pouteria sagotiana (Baill.) Eyma	X			Small tree
Pouteria singularis Pennington	X			Tree
Pouteria torta (Mart.) Radlk. subsp. *glabra* Pennington	X			Small tree
Pouteria torta (Mart.) Radlk. subsp. *tuberculata* (Sleumer) Pennington	X			Tree
Pouteria venosa (Mart.) Baehni	X			Tall tree
Pouteria virescens Baehni	X			Tree
Pouteria williamii (Aubr. & Pellegr.) Pennington	X			tree
Pradosia huberi (Ducke) Ducke	X			Tree
Pradosia ptychandra (Eyma) Pennington	X			Medium tree
[*Pradosia verticillata* Ducke], after Larpin 1993	X			Tree
Sarcaulus brasiliensis (A. DC.) Eyma	X			
Scrophulariaceae				
Scrophulariaceae sp.	X			
Simaroubaceae				
Picramnia guianensis (Aubl.) Jansen-Jacobs	X	X		Treelet
Simaba guianensis Aubl.	X	X		Small tree
Simaba orinocensis Kunth	X			Treelet
Simaba polyphylla (Cavalcante) W. Thomas	X			Treelet
Simarouba amara Aubl.	X			Tall tree
Solanaceae				
Brunfelsia guianensis Benth.	X	(X)	X	Treelet
Cyphomandra endopogon Bitter	X			Liana
Cyphomandra hartwegii (Miers) Walp.	X			Treelet
Lycianthes pauciflora (Vahl) Bitter	X			
Markea camponoti Ducke	X			
Markea coccinea L.C. Rich.	X			Liana
Markea formicarum Dammer	X			Liana

FLORISTIC CHECKLIST OF THE NOURAGUES AREA	H	L	RS	LIFE FORM
Markea sessiliflora Ducke	(X)	X		
Solanum anceps Ruiz & Pav.	X			Terrestrial herb
Solanum asperum L.C. Rich.	X			Treelet
Solanum coriaceum Dunal	X			Liana
Solanum lepidotum Dunal	X			Woody
Solanum schlechtendalianum M. Walpers	X			Treelet
Solanum velutinum Dunal	X			Treelet
Sterculiaceae				
Byttneria morii L. Barnett & Dorr	X			Tree
Sterculia excelsa Mart.	X			Tall tree
Sterculia frondosa A. Rich.	X			Tall tree
Sterculia pruriens (Aubl.) K. Schum.	X			Tall tree
Sterculia sp.	X			
Theobroma subincanum Mart.	X			Medium tree
Theobroma velutinum Benoist	X			Small tree
Styracaceae				
Styrax sp.	X			
Symplocaceae				
Symplocos martinicensis Jacq.	X			Medium tree
Theaceae				
Ternstroemia dentata (Aubl.) Swartz	X	X	X	Tree
Theophrastaceae				
Clavija lancifolia Desf.	X			
Thymeleaceae				
Thymeleaceae sp.	X	X	X	
Tiliaceae				
Apeiba glabra Aubl.	X			Tall tree
Apeiba petoumo Aubl.	X			Tall tree
Luehea cf. speciosa Willd.	X			
Lueheopsis rugosa (Pulle) Burret	X			Tall tree
Turneraceae				
Turnera rupestris Aubl.	X			Treelet
Urticaceae				
Pilea sp.	X			
Verbenaceae				
Aegiphila villosa (Aubl.) J.F. Gmelin	X			Tree
Petrea bracteata Steud.	(X)	(X)	X	Liana
Vitex guianensis Moldenke	X			Medium tree
Vitex triflora Vahl	X	X		Tree
Violaceae				
Amphirrhox longifolia (St. Hil.) Spreng.	X			

FLORISTIC CHECKLIST OF THE NOURAGUES AREA	H	L	RS	LIFE FORM
Leonia glycycarpa Ruiz & Pav.	X			Medium tree
Noisettia orchidiflora (Rudge) Gingins	X			Terrestrial herb
Paypayrola guianensis Aubl.	X			Small tree
Rinorea amapensis Hekking	X			Tree
Rinorea flavescens (Aubl.) Kuntze	X			
Rinorea neglecta Sandw.	X			
Rinorea riana Kuntze	X			Tree
Viscaceae				
Phoradendron piperoides (Kunth) Trel.	X	X		Woody parasite
Vochysiaceae				
Erisma floribundum Rudge	X			Tall tree
Erisma uncinatum Warm.	X			Tall tree
Qualea polychroma Stafl.	X			
Qualea rosea Aubl.	X			Tall tree
Ruizterania albiflora (Warm.) Marcano-Berti	X			Emergent tree
[*Vochysia guianensis* Aubl.], after Larpin 1993	X	X		Tall tree
Vochysia neyratii Normand	X			Tall tree
Vochysia surinamensis Stafl.	X			Tall tree
SPERMATOPHYTA Monocotyledones				
Araceae				
Anaphyllopsis americanum (Engl.) Hay	X			Terrestrial herb
Anthurium clavigerum Poepp. & Endl.	X			Vine
Anthutium aff. *crassinervium* (Jacq.) Schott	X	(X)		
Anthurium gracile (Rudge) Lindl.	X	X		Epiphyte herb
Anthurium jenmanii Engl.	X	X	X	Epilithic herb
Anthurium cf. *lanjouwii* Jonker & Jonker	X			Epiphyte herb
Anthurium rubrinervium (Link) G. Don	X			Epilithic herb
Anthurium scandens (Aubl.) Engl.	X			Epiphyte herb
Anthurium sinuatum Benth. ex Schott	X			Vine
Anthurium thrinax Madison	X			Epilithic herb
Anthurium trinerve Miq.	X			
Anthurium sp.				
Caladium bicolor (Aiton) Ventenat	X	X		Epilithic herb
Dieffenbachia duidae (Steyerm.) Bunting	X			Terrestrial herb
Dieffenbachia paludicola N.E. Brown ex Gleason	X			Aquatic herb
Dieffenbachia seguine (Jacq.) Schott	X			Terrestrial herb
Dracontium polyphyllum L.	X	X		Epilithic herb
Heteropsis flexuosa (Kunth) Bunting	X			Vine
Homalomena picturata (Linden & André) Regel	X			Terrestrial herb
Monstera adansonii Schott	X	X		Vine
Monstera dubia (Kunth) Engl. & Krause	X			Vine

FLORISTIC CHECKLIST OF THE NOURAGUES AREA	H	L	RS	LIFE FORM
Monstera obliqua Miq.	X			Vine
Monstera spruceana (Schott) Engl.	X			Vine
Philodendron billietiae Croat	X	X	X	Epiphyte herb
Philodendron callosum Krause	X			Epiphyte herb
Philodendron deflexum Poepp. ex Schott	X	X		Epiphyte herb
Philodendron duckei Croat & Grayum	X			
Philodendron fragrantissimum (Hook.) Kunth	X	X	X	Epiphyte herb
Philodendron goeldii Barroso	X			Epiphyte herb
Philodendron grandifolium (Jacq.) Schott	X	X		Vine
Philodendron guianense Croat & Grayum	X			
Philodendron insigne Schott	X	X	X	Epiphyte herb
Philodendron linnaei Kunth	X	X	X	Epiphyte herb
Philodendron melinonii Brongn. ex Regel	X	X		Epiphyte herb
Philodendron ornatum Schott	X	X		Epilithic herb
Philodendron pedatum (Hook.) Kunth	X	X		Vine
Philodendron placidum Schott	X			
Philodendron rudgeanum Schott	X			Vine
Philodendron scandens K. Koch & Sello	X			Vine
Philodendron solimoesence A.C. Smith	X			Epilithic herb
Philodendron sphalerum Schott	X			Vine
Philodendron squamiferum Poepp.	X	X		Epiphyte herb
Philodendron cf. surinamense (Schott) Engler	X			
Philodendron wittianum Engl.	X			
Rhodospatha cf. *oblongata* Poepp.	X			
Rhodospatha venosa Gleason	X			
Spathiphyllum humboldtii Schott	X			Aquatic herb
Syngonium podophyllum Schott	X			
Xanthosoma conspurcatum Schott	X	X		Epilithic herb
Arecaceae				
Astrocaryum gynacanthum Mart.	X			Palm
Astrocaryum paramaca Mart.	X			Palm
Astrocaryum sciophilum (Miq.) Pulle	X			Palm
Bactris acanthocarpoides Barb. Rodr.	X			Palm
Bactris aubletiana Trail	X			Palm
Bactris constanciae Barb. Rodr.	X			Palm
Bactris floccosa Spruce	X			Palm
Bactris gastoniana Barb. Rodr.	X			Palm
Bactris hirta Mart.	(X)	X		Palm
Bactris oligocarpa Barb. Rodr. & Trail ex Barb. Rodr.	X	(X)		Palm
Bactris rhaphidacantha Wessels Boer	X			Palm
Bactris simplicifrons Mart.	X			Palm

FLORISTIC CHECKLIST OF THE NOURAGUES AREA	H	L	RS	LIFE FORM
Desmoncus sp.	X			Liana palm
Euterpe oleracea Mart.	X			Palm
Geonoma deversa (Poit.) Kunth	X			Palm
Geonoma stricta (Poit.) Kunth	X			Palm
Hyospathe elegans Mart.	X	(X)		Palm
Jessenia bataua (Mart.) Burret	X	X		Palm
Maximiliana maripa (Corrêa) Drude	X			Palm
Oenocarpus bacaba Mart.	X			Palm
Orbignya sp.	X			Palm
Socratea exorrhiza (Mart.) H. Wendl.	X			Palm
Syagrus stratincola Wessels Boer		X		Palm
Bromeliaceae				
Aechmea aquilega (Salisb.) Griseb.	X	X	X	Epiphyte & epilithic
Aechmea bromeliifolia (Rudge) Baker	X	X	X	Epiphyte & epilithic
Aechmea melinonii Hook.	X	X	X	Epiphyte & epilithic
Ananas ananassoides (Baker) L.B. Smith	(X)	X	X	Epilithic
Araeococcus micranthus Brongn.	(X)	X		Epiphyte
Billbergia violacea Beer	X	X		Epiphyte & epilithic
Bromelia agavifolia Brongn. ex Houllett	X	X		Terrestrial herb
Catopsis berteroniana (Schult.f.) Mez	(X)	(X)	X	Epiphyte
Disteganthus basilateralis Lem.	X	X		Epilithic
Disteganthus lateralis (L.B. Smith) Gouda	X	X		
Guzmania lingulata (L.) Mez	X	X	X	Epiphyte & epilithic
Guzmania melinonis Regel	X	X		Epiphyte
Pitcairnia geyskesii (L.B. Smith) Varadarajan & Gilmartin			X	Epilithic
Pitcairnia sastrei L.B. Smith & R.W. Read	X	X		Epilithic
Racinae spiculosa (Griseb.(Spencer &L.B. Smith	(X)	X	X	Epiphyte
Streptocalyx longifolius (Rudge) Baker	(X)	X		Epiphyte
Tillandsia anceps Lodd.	X	X		Epiphyte
Tillandsia bulbosa Hook.	(X)	X	X	
Tillandsia flexuosa Swartz			X	Epiphyte
Tillandsia monadelpha (E. Morren) Baker	X	X		Epiphyte
Tillandsia tenuifolia L.	(X)	X		Epiphyte
Vriesea pleiosticha (Griseb.) Gouda	X	X	X	Epiphyte & epilithic
Vriesea splendens (Brongn.) Lem. var. *splendens*	X	X	X	Epiphyte & terrestrial
Werauthia gladioliflora (Wendl.) J.R. Grant	X	X		Epilithic
Burmanniaceae				
Apteria aphylla (Nutt.) Barnhart ex Small	X	X	X	Saprophyte
Burmannia capitata (Walter ex J.F. Gmel.) Mart.	X	(X)	X	Epilithic
Campylosiphon purpurascens Benth.	(X)			Saprophyte
Dictyostega orobanchoides (Hook.) Miers	X	X		Terrestrial herb

FLORISTIC CHECKLIST OF THE NOURAGUES AREA	H	L	RS	LIFE FORM
Gymnosiphon breviflorus Gleason	X	X		Terrestrial herb
Gymnosiphon minutus Snelders & Maas	(X)			Saprophyte
Hexapterella gentianoides Urb.	(X)			Saprophyte
Commelinaceae				
Buforrestia candolleana C.B. Clarke	X			Terrestrial herb
Commelina rufipes Seub. var. *glabrata* (D.R. Hunt) Faden & D.R. Hunt	X	X		Epilithic
Costaceae				
Costus claviger Benoist	X			Terrestrial herb
Costus congestiflorus L.C. Rich. ex Gagnep.	X			Terrestrial herb
Costus curcumoides Maas	X			Terrestrial herb
Costus erythrothyrsus Loes.	X	X		Epilithic herb
Cyclanthaceae				
Asplundia brachyphylla Harling	X			Terrestrial herb
Asplundia glandulosa (Gleason) Harling	X			Terrestrial herb
Asplundia heteranthera Harling	X			Vine
Asplundia sp.	X			
Cyclanthus bipartitus Poit.	X			Terrestrial herb
Evodianthus funifer (Poit.) Lindm. subsp. *funifer*	X	X		Vine
Ludovia lancifolia Brongn.	X	X	X	
Stelestylis surinamensis Harling		X	X	Terrestrial herb
Thoracocarpus bissectus (Vell.) Harling	X			Liana
Cyperaceae				
Becquerelia cymosa Brongn.	X	(X)		Terrestrial herb
Bisboeckelera longifolia (Rudge) Kuntze	X			
Bisboeckelera microcephala (Boeck.) T. Koyama	X			Terrestrial herb
Calyptrocarya bicolor (H. Pfeiff.) T. Koyama	X			
Calyptrocarya glomerulata (Brongn.) Urb.	X			Terrestrial herb
Cyperus luzulae (L.) Rottb. ex Retz.	X			Terrestrial herb
Cyperus odoratus L.	X			Terrestrial herb
Cyperus polystachyos Rottb.	X			
Fimbristylis dichotoma (L.) Vahl	X			Terrestrial herb
Fimbristylis miliacea (L.) Vahl [not a native species]	X			Terrestrial herb
Hypolytrum jenmanii C.B. Clarke subsp. *jenmanii*	X			Terrestrial herb
Hypolytrum longifolium (L.C. Rich.) Nees	X			Terrestrial herb
Kyllinga brevifolia Rottb.	X			Terrestrial herb
Kyllinga odorata Vahl	X			Terrestrial herb
Mapania sylvatica Aubl.	X			Terrestrial herb
Rhynchospora barbata (Vahl) Kunth	(X)	(X)	X	Terrestrial herb
Rhynchospora capitata (Homboldt & Kunth) Roem. & Schult.			X	Terrestrial herb
Rhynchospora cephalotes (L.) Vahl	(X)	X	X	Terrestrial herb
Rhynchospora filiformis Vahl			X	

FLORISTIC CHECKLIST OF THE NOURAGUES AREA	H	L	RS	LIFE FORM
Rhynchospora globosa (Kunth) Roem. & Schult.			X	Terrestrial herb
Rhynchospora junciformis (Kunth) Boeck.			X	Terrestrial herb
Rhynchospora tenuis Willd. ex Link	X	(X)	X	Terrestrial herb
Scleria cyperina Willd. ex Kunth	(X)	(X)	X	Terrestrial herb
Scleria distans Poir.			X	Epilithic
Scleria secans (L.) Urb.	X	X	X	Vine
Dioscoreaceae				
Dioscorea sp.	X			
Eriocaulaceae				
Paepalanthus lamarckii Kunth			X	Epilithic
Syngonanthus caulescens (Poir.) Ruhland	X			Terrestrial herb
Heliconiaceae				
Heliconia acuminata L.C. Rich.	X			Terrestrial herb
Heliconia bihai (L.) L.	X			
Heliconia hirsuta L.f.	X			Terrestrial herb
Heliconia pendula Wawra	X			Terrestrial herb
Heliconia richardiana Miq.	X	(X)		Terrestrial herb
Marantaceae				
Calathea cyclophora Baker	X			Terrestrial herb
Calathea cylindrica (Roscoe) K. Schum.	X			Terrestrial herb
Calathea grandis Petersen	X			Terrestrial herb
Calathea propinqua (Poepp. & Endl.) Körn.	X	X		Terrestrial herb
Calathea squarrosa L. Andersson & H. Kennedy	X	X	X	Epilithic herb
Ischnosiphon arouma (Aubl.) Körn.	X			Terrestrial herb
Ischnosiphon gracilis (Rudge) Körn.	X			Terrestrial herb
Ischnosiphon martianus Eichl. ex Petersen	X	(X)		Terrestrial herb
Ischnosiphon petiolatus (Rudge) L. Andersson	X			Terrestrial herb
Monophyllanthe oligophylla K. Schum.	X			Terrestrial herb
Monotagma spicatum (Aubl.) Macbr.	X			Terrestrial herb
Orchidaceae				
Aganisia pulchella Lindl.	(X)	X	X	Epiphytic herb
Batemannia colleyi Lindl.	(X)	X		Epiphytic herb
Brassia lanceana Lindl.	(X)	X		Epiphytic herb
Bulbophyllum roraimense Rolfe		X		Epiphytic herb
Campylocentrum micranthum (Lindl.) Rolfe	(X)	X	X	Epiphytic herb
Cheiradenia cuspidata Lindl.	X	X		Epiphytic & epilithic
Cleistes rosea Lindl.	(X)	(X)	X	Terrestrial herb
Cleistes tenuis (Rchb.f. ex Griseb.) Schlechter			X	Terrestrial herb
Cranichis sp.		X		
Cyrtopodium andersonii (Lamb.) R. Brown	(X)	X	X	
Dichaea pendula (Aubl.) Cogn.	X		X	Epiphytic herb

FLORISTIC CHECKLIST OF THE NOURAGUES AREA	H	L	RS	LIFE FORM
Dichaea splitgerberi Rchb.f.	(X)	X		Epiphytic herb
Dichaea trulla Rchb.f.	(X)		X	Epiphytic herb
Elleanthus brasiliensis (Lindl.) Rchb.f.		X	X	Epiphytic herb
Elleanthus caravata (Aubl.) Rchb.f.	(X)	X	X	Epiphytic herb
Elleanthus graminifolius (Barb. Rodr.) Loijtnant	(X)	X	X	Epiphytic herb
*Encyclia ionosma (*Lindl.) Schlechter	(X)		X	Epilithic herb
Epidendrum anceps Jacq.		X		Epiphytic herb
Epidendrum nocturnum Jacq.	(X)		X	Epiphytic herb
Epidendrum purpurascens Focke	(X)	X	X	Epiphytic herb
Erythrodes sp.	X			
Habenaria rodriquezii Cogn.			X	Terrestrial herb
*Jacquiniella globosa (*Jacq.) Schlechter	(X)	X	X	Epiphytic herb
Koellensteinia carraoënsis Garay & Dunsterv.		X		Terrestrial herb
Koellensteinia graminea (Lindl.) Rchb.f.	(X)	X		Epiphytic herb
Koellensteinia sp.		X		Epiphytic herb
Masdevallia minuta Lindl.	(X)	X	X	Epiphytic herb
Maxillaria brunnea Linden & Rchb.f.	(X)	X		Epiphytic herb
Maxillaria camaridii Rchb.f.	(X)	X		Epiphytic herb
Maxillaria crassifolia (Lindl.) Rchb.f.	X	X		Epiphytic herb
Maxillaria desvauxiana Rchb.f.		X		Epiphytic herb
Maxillaria uncata Lindl.	X	X	X	Epiphytic herb
Maxillaria villosa (Barb. Rodr.) Cogn.	(X)	X	X	Epiphytic herb
Maxillaria violaceopunctata Rchb.f.	(X)	X	X	Epiphytic herb
Maxillaria sp.	X	X		Epiphytic herb
Notylia sagittifera (Kunth) Link, Klotzsch, & Otto		X		Epiphytic herb
Notylia yauaperiensis Barb. Rodr.			X	Epiphytic herb
Octomeria sarthouae Luer			X	Epiphytic herb
Octomeria sp.			X	
Ornithocephalus kruegeri Rchb.f.		X		Epiphytic
Palmorchis pabstii Veyret	X			Terrestrial
Palmorchis prospectorum Veyret	X			Terrestrial
Paphinia sp.		X		
Pelexia callifera (C. Schweinf.) Garay	X			Terrestrial
Phragmipedium lindleyanum (Schomb.) Rolfe		X	X	Epilithic herb
Pleurothallis aristata W.J. Hook.	(X)	X		Epiphytic herb
Pleurothallis picta Lindl.	(X)	X		Epiphytic herb
Pleurothallis spiculifera Lindl.	(X)	X	X	Epiphytic herb
Polystachya flavescens (Blume) J.J. Smith			X	Epiphytic herb
Polystachya foliosa (Hook.) Rchb.f.	(X)	X	X	Epiphytic herb
Prescottia sp.	X			
Rodriguezia candida (Lindl.) Rchb.f.			X	Epiphytic herb

FLORISTIC CHECKLIST OF THE NOURAGUES AREA	H	L	RS	LIFE FORM
Rodriguezia lanceolata Ruiz & Pav.	(X)	X	X	Epiphytic herb
Scaphyglottis bifida (Rchb.f.) Schweinf.			X	Epiphytic herb
Scaphyglottis fusiformis (Griseb.) Schultes			X	Epiphytic herb
Scaphyglottis prolifera Cogn.	(X)	X	X	Epiphytic herb
Scaphyglottis sickii Pabst			X	Epiphytic herb
Scaphyglottis stellata Lodd. ex Lindl.	X	X		
Scaphyglottis violacea Lindl.			X	Epiphytic herb
Scuticaria steelei (Hook.) Lindl.		X		Epiphytic herb
Sigmatostalix huebneri Mansf.		X		Epiphytic herb
Stelis argentata Lindl.	(X)	X		Epiphytic herb
Trichosalpinx egleri (Pabst) Luer		X		Epiphytic herb
Trichosalpinx orbicularis (Lindl.) Luer	(X)	X	X	Epiphytic herb
Vanilla ovata Rolfe	(X)	X		Vine
Wullschlaegelia calcarata Benth.	X			Saprophyte
Zygosepalum labiosum (L.C. Rich.) Schweinf.	(X)	X		Epiphytic herb
Zygosepalum lindeniae (Rolfe) Garay & Dunsterv.	(X)	X		Epiphytic
Poaceae				
Axonopus compressus (Swartz) P. Beauv.	X			Terrestrial herb
Axonopus ramosus Swallen			X	Terrestrial herb
Cenchrus echinatus L.	X			Terrestrial herb
Digitaria ciliaris (Retz.) Koeler	X			Terrestrial herb
Guadua latifolia (Humb. & Bonpl.) Kunth	X			Terrestrial herb
Ichnanthus nemoralis (Schrad.) Hitchc. & Chase	(X)	X	X	Terrestrial herb
Ichnanthus panicoides P. Beauv.	X			
Ichnanthus tenuis (J. Presl) Hitchc. & Chase	(X)	X		Terrestrial herb
Olyra obliquifolia Steud.	X	X		Terrestrial herb
Panicum miliaceum L. [not a native species]	X			Terrestrial herb
Panicum rivale Swallen			X	Terrestrial herb
Pariana campestris Aubl.	X	X	X	Terrestrial herb
Pariana radiciflora Sagot ex Doell	X			
Paspalum conjugatum Berg.	X			Terrestrial herb
Paspalum decumbens Swartz	X			Terrestrial herb
Paspalum maritimum Trin.	X			Terrestrial herb
Paspalum multinervium A.G. Burman	(X)	(X)	X	Terrestrial herb
Pharus latifolius L.	X			Terrestrial herb
Pharus parvifolius Nash	X			Terrestrial herb
Pharus virescens Doell	X			Terrestrial herb
Setaria italica (L.) P. Beauv. [not a native species]	X			Terrestrial herb
Sorghum halepense (L.) Pers. [not a native species]	X			Terrestrial herb
Thrasya scandens (Tutin) Soderstrom ex A.G. Burm.			X	Terrestrial herb

FLORISTIC CHECKLIST OF THE NOURAGUES AREA	H	L	RS	LIFE FORM
Rapateaceae				
Rapatea paludosa Aubl.	x			Terrestrial herb
Spathanthus unilateralis (Rudge) Desv.	x			Aquatic herb
Smilacaceae				
Smilax spp. [5 taxa]				
Thurniaceae				
Thurnia sphaerocephala (Rudge) J.D. Hook.	x			Terrestrial
Triuridaceae				
Sciaphila albescens Benth.	x			Saprophyte
Triuris hyalina Miers				Saprophyte
Xyridaceae				
Abolboda americana (Aubl.) Lanj.			x	Epilithic herb
Xyris savanensis Miq.			x	Epilithic herb
Zingiberaceae				
Renealmia guianensis Maas	x			Terrestrial herb
Renealmia monosperma Miq.	x			Terrestrial hern
Renealmia orinocensis Rusby	x			Terrestrial herb
SPERMATOPHYTA Gnetales				
Gnetaceae				
Gnetum nodiflorum Brongn.	x			Woody epiphyte
Gnetum paniculatum Spruce ex Benth.	x			Liana
HEPATOPHYTA				
Herbertaceae				
Herbertus divergens (Steph.) Herz.			x	
Lejeuneaceae				
Ceratolejeunea sp.			x	
Cystolejeunea lineata (Lehm. & Lindenb.) Evans			x	
BRYOPHYTA				
Calymperaceae				
Calymperes erosum C. Müll.		x		
Calymperes lonchophyllum Schwaegr.		x		
Calymperes nicaraguense Ren. & Card.	x			
Calymperes venezuelanum (Mitt.) Broth. ex Pitt.		x		
Syrrhopodon hornschuchii Mart.		x		
Syrrhopodon leprieurii Mont.	x			
Syrrhopodon prolifer Schwaegr.	x	(x)	x	
Dicranaceae				
Campylopus richardii Brid.			x	
Campylopus savannarum (C. Müll.) Mitt.			x	
Leucoloma cruegerianum (C. Müll.) Jaeg. & Sauerb.	x			
Leucoloma serrulatum Brid.	x			

FLORISTIC CHECKLIST OF THE NOURAGUES AREA	H	L	RS	LIFE FORM
Hookeriaceae				
Callicostella pallida (Hornsch.) Angstr.		X		
Hypnella pallescens (Hook.) Jaeg.	X			
Hypnaceae				
Chrysohypnum diminutivum (Hampe) Buck	X			
Isopterygium tenerum (Swartz) Mitt.		X		
Rhacopilopsis trinitensis (C. Müll.) Britt. & Dix.	X			
Leucobryaceae				
Leucobryum martianum (Hornsch.) C. Müll.	X	X	X	
Octoblepharum cocuiense Mitt.	X	X		
Octoblepharum erectifolium Mitt. ex Williams		X		
Octoblepharum stramineum Mitt.			X	
Meteoriaceae				
Squamidium leucotrichum (Tayl.) Broth.	X			
Orthotrichaceae				
Macromitrium cirrosum (Hedw.) Brid.		X		
Macromitrium pellucidum Mitt.		X		
Phyllodrepaniaceae				
Phyllodrepanium falcifolium (Schwaegr.) Crosby		X		
Pterobryaceae				
Renauldia paradoxica Allen		X		
Sematophyllaceae				
Acroporium pungens (Hedw.) Broth.		X		
Sematophyllum subsimplex (Hedw.) Mitt.	X	X	X	
Taxithelium planum (Brid.) Mitt.	X	X		
Trichosteleum papillosum (Hornsch.) Jaeg.	X	X		
Stereophyllaceae				
Pilosium chlorophyllum (Hornsch.) C. Müll.	X			
Thuidiaceae				
Cyrtohypnum involvens (Hedw.) Buck & Crum		X		
LYCOPHYTA				
Lycopodiaceae				
Huperzia linifolia (L.) Trevis.	(X)	(X)	X	Epiphyte
Selaginellaceae				
Selaginella epirrhizos Spring	X			Terrestrial
Selaginella flagellata Spring	X			Terrestrial
Selaginella producta Baker	X	X	X	
Selaginella radiata (Aubl.) Spring	X	X		Terrestrial
Selaginella revoluta Baker	X	X		Terrestrial

FLORISTIC CHECKLIST OF THE NOURAGUES AREA	H	L	RS	LIFE FORM
PTEROPHYTA				
Adiantaceae				
Adiantopsis radiata (L.) Fée	X			Terrestrial fern
Adiantum adiantoides (J.E. Smith) C. Chr.	X			Terrestrial fern
Adiantum cajennense Willd. ex Klotzsch	X			Terrestrial fern
Adiantum sp. nov. aff. cordatum Maxon	X			Terrestrial fern
Adiantum fructuosum Poepp. ex Spreng.	X			Terrestrial fern
Adiantum fuliginosum Fée	X			Terrestrial fern
Adiantum glaucescens Klotzsch	X			Terrestrial fern
Adiantum latifolium Lam.	X			Terrestrial fern
Adiantum leprieurii Hook.	X			Terrestrial fern
Adiantum oyapokense Jenm.	X			Terrestrial fern
Adiantum tomentosum Klotzsch	X			Terrestrial fern
Pityrogramma calomelanos (L.) Link				Terrestrial fern
Aspleniaceae				
Asplenium angustum Swartz	(X)	X		Epiphytic fern
Asplenium auritum Swartz	X			Epiphytic fern
Asplenium juglandifolium Lam.	X	X		Epiphytic & terrestrial fern
Asplenium laetum Swartz	X			Epiphytic & epilithic
Asplenium serratum L.	X	X		Epiphytic fern
Blechnaceae				
Salpichlaena volubilis (Kaulf.) J. Smith	X			Terrestrial fern
Cyatheaceae				
Cyathea andina (H. Karst.) Domin	X			Treefern
Cyathea cyatheoides (Desv.) K.U. Kramer	X			Treefern
Cyathea pungens (Willd.) Domin	X			Treefern
Cyathea surinamensis (Miq.) Domin	X			Treefern
Dennstaedtiaceae				
Dennstaedtia obtusifolia (Willd.) T. Moore	X			Terrestrial herb
Hypolepis repens (L.) K. Presl	X			Terrestrial herb
Lindsaea dubia Spreng.	X	X		Terrestrial herb
Lindsaea guianensis (Aubl.) Dryander	X			Terrestrial herb
Lindsaea lancea (L.) Bedd. var. *falcata* (Dryander) Rosenstock	X			Terrestrial herb
Lindsaea lancea (L.) Bedd. var. *lancea*	X	X		Terrestrial herb
Lindsaea pallida Klotzsch	X	X		Terrestrial herb
Lindsaea reniformis Dryander	X			Terrestrial herb
Saccoloma inaequale (Kunze) Mett.	X			Terrestrial herb
Dryopteridaceae				
Cyclodium guianense (Klotzsch) L.D. Gomez	X	X		Terrestrial fern
Cyclodium inerme (Fée) A.R. Smith	X			Terrestrial fern
Cyclodium meniscioides (Willd.) K. Presl	X	X		Terrestrial & epiphytic

FLORISTIC CHECKLIST OF THE NOURAGUES AREA	H	L	RS	LIFE FORM
Didymochlaena truncatula (Swartz) J. Smith	X			Terrestrial fern
Olfersia cervina (L.) Kunze	X			Hemi-epiphytic fern
Polybotrya caudata Kunze	X			Hemi-epiphytic fern
Grammitidaceae				
Cochlidium furcatum (Hook. & Grev.) C. Chr.		X	X	Epiphyte
Cochlidium linearifolium (Desv.) Maxon ex C. Chr.	X	X	X	Epiphyte
Cochlidium serrulatum (Swartz) L.E. Bishop	(X)	X	X	Epiphyte
Grammitis blanchetii (C. Chr.) A.R. Smith		X		Epiphyte
Grammitis suspensa (L.) Proctor	X	X	X	Epiphyte
Hymenophyllaceae				
Hymenophyllum brevifrons Kunze	X			Epiphytic fern
Hymenophyllum decurrens (Jacq.) Swartz	(X)	X		Epiphytic fern
Hymenophyllum hirsutum (L.) Swartz	X			Epiphytic fern
Hymenophyllum polyanthos (Swartz) Swartz	X	X	X	Epiphytic fern
Trichomanes ankersii Parker ex Hook. & Grev.	X			Epiphytic fern
Trichomanes arbuscula Desv.	X	X		Epiphytic fern
Trichomanes crispum L.	(X)	X	X	Terrestrial fern
Trichomanes cristatum Kaulf.	(X)	X		Terrestrial & epiphytic fern
Trichomanes diversifrons (Bory) Mett. ex Sadebeck	X	X		Terrestrial fern
Trichomanes elegans L.C. Rich.	X			Terrestrial fern
Trichomanes kapplerianum J.W. Sturm	X			Epiphytic fern
Trichomanes macilentum Bosch	X			Fern
Trichomanes pedicellatum Desv.	X			Epiphytic fern
Trichomanes pinnatum Hedw.	X			Terrestrial fern
Trichomanes radicans Swartz	X			Epiphytic & epilithic
Trichomanes trollii Bergd.	X			Terrestrial & epiphytic
Lomariopsidaceae				
Bolbitis nicotianifolia (Swartz) Alston	X			Terrestrial fern
Bolbitis semipinnatifida (Fée) Alston	X			Terrestrial fern
Elaphoglossum flaccidum (Fée) T. Moore	X			Epiphytic fern
Elaphoglossum glabellum J. Smith	X	X	X	Epiphytic fern
Elaphoglossum herminieri (Bory & Fée ex Fée) T. Moore	X	X	X	Epiphytic fern
Elaphoglossum laminarioides (Bory ex Fée) T. Moore	(X)	X		Epiphytic fern
Elaphoglossum luridum (Fée) Christ	X	X		Epiphytic fern
Elaphoglossum pteropus C. Chr.	(X)	X		Epiphytic fern
Elaphoglossum raywaense (Jenm.) Alston	X			Epiphytic fern
Lomagramma guianensis (Aubl.) Ching	X			Epiphytic fern
Lomariopsis japurensis (Mart.) J. Smith	X			Terrestrial & epilithic fern
Marattiaceae				
Danaea nodosa (L.) J.E. Smith	X			Terrestrial fern
Danaea simplicifolia Rudge	X			Terrestrial

FLORISTIC CHECKLIST OF THE NOURAGUES AREA	H	L	RS	LIFE FORM
Metaxyaceae				
Metaxya rostrata (Kunth) K. Presl	X			
Oleandraceae				
Nephrolepis biserrata (Swartz) Schott	X			Terrestrial & epiphytic fern
Nephrolepis rivularis (Vahl) Mett. ex Krug	(X)	X		Epiphytic fern
Oleandra articulata (Swartz) K. Presl	(X)	X		Epiphytic fern
Polypodiaceae				
Campyloneurum phyllitidis (L.) K. Presl	X	X		Epiphytic fern
Dicranoglossum desvauxii (Klotzsch) Proctor	X			Epiphytic fern
Microgramma fuscopunctata (Hook.) Vareschi	X			Epiphytic & epilithic fern
Microgramma lycopodioides (L.) Copeland	X	X	X	Epiphytic & epilithic fern
Microgramma reptans (Cav.) A.R. Smith	(X)		X	Epiphytic & epilithic fern
Niphidium crassifolium (L.) Lellinger	X	X		Terrestrial & epiphytic fern
Pecluma pectinata (L.) Price	X			Epiphytic fern
Polypodium dulce Poir.	X		X	Fern
Polypodium triseriale Swartz	(X)	X	X	Epiphytic fern
Pteridaceae				
Pteris pungens Willd.	X			Terrestrial fern
Schizaeaceae				
Actinostachys pennula (Swartz) Hook.			X	Terrestrial fern
Actinostachys subtrijuga (Mart.) K. Presl			X	Terrestrial fern
Schizaea incurvata Schkuhr			X	Terrestrial fern
Tectariaceae				
Tectaria incisa Cav.	X			Terrestrial fern
Tectaria plantaginea (Jacq.) Maxon	X	X		Terrestrial fern
Tectaria trifoliata (L.) Cav.	X			Terrestrial fern
Triplophyllum funestum (Kunze) Holttum	X			Terrestrial fern
Thelypteridaceae				
Thelypteris leprieurii (Hook.) R. Tryon	X			Terrestrial fern
Thelypteris pennata (Poir.) C. Morton	X			Terrestrial fern
Vittariaceae				
Anetium citrifolium (L.) Splitg.	X			Epiphytic fern
Antrophyum cajenense (Desv.) Spreng.	X			Epiphytic fern
Antrophyum guayanense Hieron.	X	X		Epiphytic fern

Pierre Charles-Dominique captures bats. He identifies, measures and weighs them. Just before release he gives them a high-energy drink. *Photo Frans Bongers.*

APPENDIX 2

PRACTICAL GUIDE TO THE PALMS

Jean-Jacques de Granville

Photo 1. Bactris circonstanciae infrutescence. *Photo Ph. Birnbaum.*

Bongers *et al.* (eds.), Nouragues. Dynamics and plant-animal interactions in a neotropical rainforest, 343–350.
2001 Kluwer Academic Publishers. Printed in the Netherlands.

1. List of palms

Genus *Astrocaryum*
Astrocaryum gynacanthum Martius
Astrocaryum paramaca Martius ['Counana']
Astrocaryum sciophilum (Miquel) Pulle ['Mourou-mourou']

Genus *Bactris*
Bactris acanthocarpoides Barbosa Rodrigues
Bactris aubletiana Trail
Bactris constanciae Barbosa Rodrigues
Bactris floccosa Spruce
Bactris gastoniana Barbosa Rodrigues
Bactris hirta Martius
Bactris oligocarpa Barbosa Rodrigues
Bactris rhaphidacantha Wessels Boer
Bactris simplicifrons Martius

Genus *Desmoncus*
Desmoncus sp.

Genus *Euterpe*
Euterpe oleracea Martius ['Pinot', 'Wassaïe']

Genus *Geonoma* ['Waï']
Geonoma deversa (Poiteau) Kunth
Geonoma stricta (Poiteau) Kunth

Genus *Jessenia*
Jessenia bataua (Martius) Burret subsp. *oligocarpa* (Grisebach & H. Wendland) Balick ['Patawa']

Genus *Maximiliana*
Maximiliana maripa (Correa de Serra) Drude ['Maripa']

Genus *Oenocarpus*
Oenocarpus bacaba Martius ['Comou']

Genus *Orbignya*
Orbignya sp. ['Macoupi']

Genus *Scheelea*
Scheelea sp. ['Macoupi'

Genus *Socratea*
Socratea exorrhiza (Martius) H. Wendland ['Awara-monpè']

Genus *Syagrus*
Syagrus stratincola Wessels Boer

2. Relative abundance of palm species on the eight main trails

CAMP: surroundings of the camp
RD/RG: trails along the creek ('Rive droite' and 'Rive gauche')
C.T.1: 1st trail from camp to the southern slope of the outcrop ('Chemin des terrasses n°1 ')
C.T.2: 2nd trail from camp to the southern slope of the outcrop ('Chemin des terrasses n°2')
SOMM.: trail from camp to the top of the outcrop ('Chemin du sommet')
CASC.: trail from camp to the water falls ('Chemin des cascades')
BOUCLE: loop from camp to camp
S.C.: trail from camp to the eastern slope of the outcrop ('Chemin de la Savane Coco')

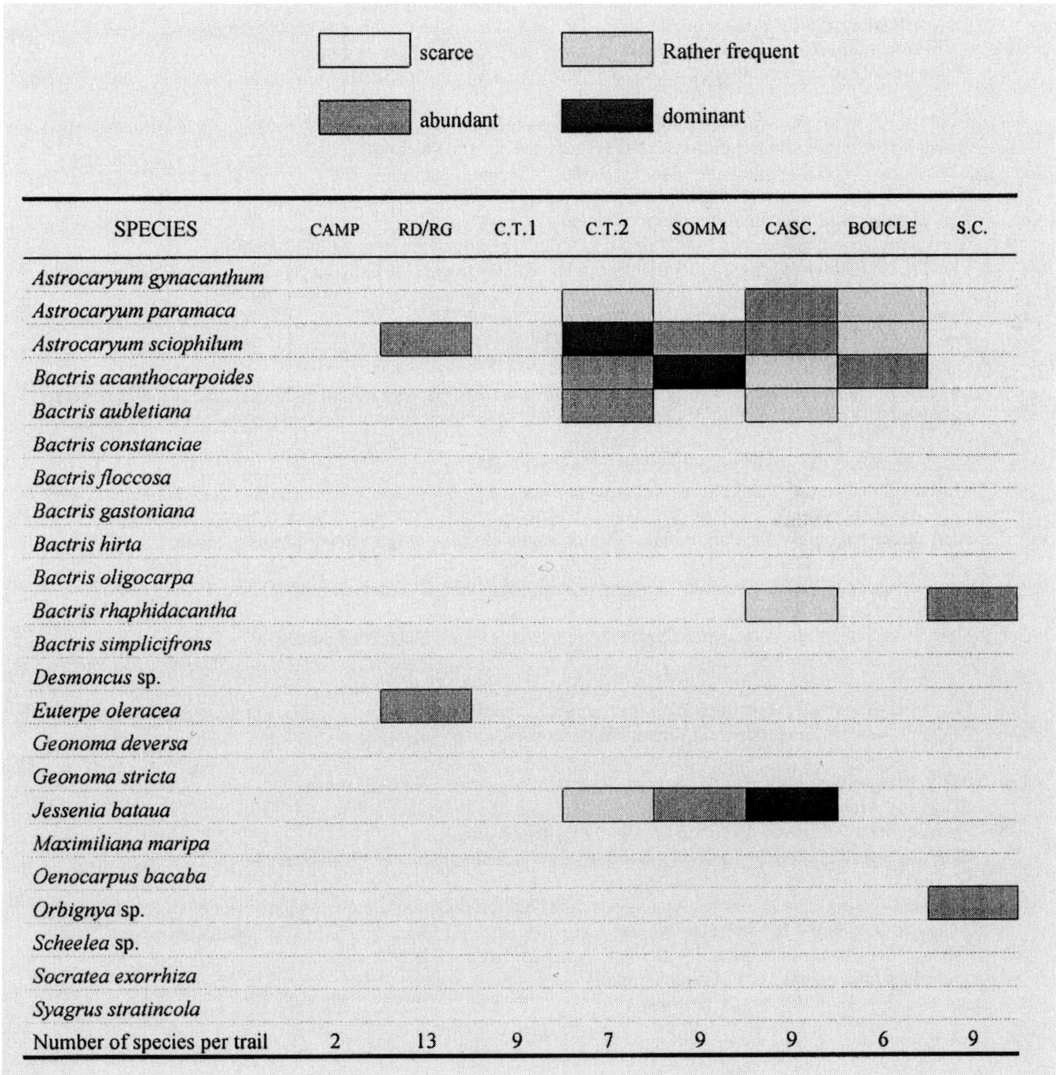

3. Key to the species

1a. Trunk supported by cone of prickly stilt roots. Pinnae preamorse at tip.......... *Socratea exorrhiza*
1b. Trunk without stilt roots. Pinnae not praemorse at tip.......... 2

2a. Palms unarmed. Margins of blades without bristles.......... 3
2b. Palms usually more or less spiny. Margins of blade with minute, hardly visible, bristles at apex (presence detected by moving finger from tip to base of pinnae).......... 12

3a. Small understory palms. Trunk < 2 cm in diameter. Leaves < 1 m long. Blade simple and bifid at apex or irregularly divided into 2-6 pairs of broad segments.......... 4
3b. Palms medium-sized to tall, either « stemless » or with trunk > 4 cm in diameter. Blade pinnate with more than 10 pinnae per side.......... 6

4a. Blade < 10 cm in width, always simple and bifid at apex.......... *Geonoma stricta*
4b. Blade > 10 cm in width, pinnatisect, usually with 3-4 pairs of segments.......... 5

5a. Stem 4-10 mm in diameter. Blade < 30 cm in width. Inflorescence bearing 3-6 rachillae. Leaf sheath split opposite to petiole in old leaves.......... *Geonoma deversa*
5b. Stem 8-20 mm in diameter. Blade > 30 cm in width. Inflorescence bearing 8-17 rachillae. Leaf sheath tubular in young as well as in old leaves, never split opposite to petiole.......... *Hyospathe elegans*

6a. Palms always « stemless ».(even when « adult »).......... 7
6b. Palms with well-developed stems (at least when adult).......... 8

7a. Leaf blade divided into narrow pinnae to tip. Staminate flowers with coiled anthers.......... *Orbignya* sp
7b. Leaf blade with apical pairs of pinnae usually united. Staminate flowers with straight anthers.......... *Scheelea* sp.

8a. Single-stemmed (« solitary ») palms, never suckering at base.......... 9
8b. Muli-stemmed ('clustered') palms suckering at base.......... 11

9. Blade green and shiny beneath. Pinnae more or less in clusters, at least toward base of rachis.......... 10

10a. Stem usually > 25 cm in diameter. Leaf blades with more than 120 pairs of pinnae, in clusters along entire rachis.......... *Maximiliana maripa*
10b. Stem usually < 25 cm in diameter. Leaf blades with less than 120 pairs of pinnae, in clusters toward base of rachis.......... *Oenocarpus bacaba*

11a. Palms growing in swamp forest. Pinnae in..serted at regular intervals.......... *Euterpe oleracea*
11b. Palms growing in low, transition forest. Pinnae conspicuously arranged in clusters.......... *Syagrus stratincola*

12a. Palms always armed with black, strongly flattened, sometimes winged spines. Blade dull white or whitish pruinose beneath.......... 13
12b. Palms generally armed with round or slightly flattened spines, sometimes with recurved hooks (in climbing palms). Blade green and shiny beneath.......... 15

13a. 'Stemless' palms, armed with winged spines. Leaf blade divided into narrow pinnae to tip, the apical segments narrower than others.......... *Astrocaryum paramaca*
13b. Palms with obvious stem, armed with long, flattened (but not winged) spines. Leaf blade with apical pairs of pinnae usually united, the apical segments wider than others.......... 14

14a. Multi-stemmed palms, with umbrella-shaped crown. Stems < 10 cm in diameter, armed with rings of black spines. Leaf sheaths not persistent................................*Astrocaryum gynacanthum*
14b. Single-stemmed palms, with funnel-shaped crown. Stem $\geq$ 10 cm in diameter, unarmed. Leaf sheaths persistent................................*Astrocaryum sciophilum*

15a. Climbing palms. Leaf rachis extended into a cirrus bearing retrorse hooks................................*Desmoncus* sp
15b. Stemless or erect palms, never climbing. Leaf rachis not extended into a cirrus................................ 16

16a. Leaf blade pubescent on both sides................................*Bactris floccosa*
16b. Leaf blade glabrous on both sides................................ 17

17a. Very small palms. Leaves < 1 m long, the blade either entire and bifid or with 2-6 segments per side....... 18
17b. Small to medium-sized palms. Leaves > 1 m long, the blade pinnate, with more than 6 pinnae per side.... 21

18a. Leaf blade simple, bifid at apex, the rachis 25-40 cm long. Sheaths and petiole densely armed with spines. Fruit setose................................ *Bactris hirta*
18b. Leaf blade simple or with up to 6 pinnae per side, when simple, the rachis 2-15 cm long. Sheath and petiole unarmed or with few spines................................19

19a. Blade always simple and deeply bifid, the segments more or less elliptic and conve above. Peduncular bract setose................................*Bactris aubletiana*
19b. Blade either simple and bifid or with 2-6 pinnae per side. When simple, the segments slightly sigmoid and flat. Peduncular bract glabrous................................ 20

20a. Stem slender, > 50 cm tall, at least in fertile individuals. Leaves more or less spirally distributed along upper part of stem. Inflorescence recurved, infrafoliar. Fruit globose, orange to red at maturity................. *Bactris simplicifrons*
20b. Stem inconspicuous or very short, usually < 30 cm tall. Leaves forming a crown. Inflorescence erect, interfoliar. Fruit ovoid, purplish black at maturity................................*Bactris oligocarpa*

21a. Stem < 3 cm in diameter or inconspicuous. Pinnae sigmoid, long acuminate at apex, less than 20 per side 22
21b. Stem > 3 cm in diameter. Pinnae straight, more than 20 per side................................ 23

22a. Stem inconspicuous or very short. Leaf rachis generally < 60 cm long, with 5-12 pinnae per side. Fruits ovoid, glabrous................................ *Bactris gastoniana*
22b. Stem up to 2 m tall. Leaf rachis > 60 cm long, with 13-18 pinnae per side. Fruits globose, covered with fleshy bristles................................*Bactris constanciae*

23a. Stem short, with leaf scars very closely spaced. Pinnae disposed in a single plane, at regular intervals along rachis................................*Bactris rhaphidacantha*
23b. Stem well developed, with leaf scars widely spaced. Pinna clustered and oriented in several planes................................*Bactris acanthocarpoides*

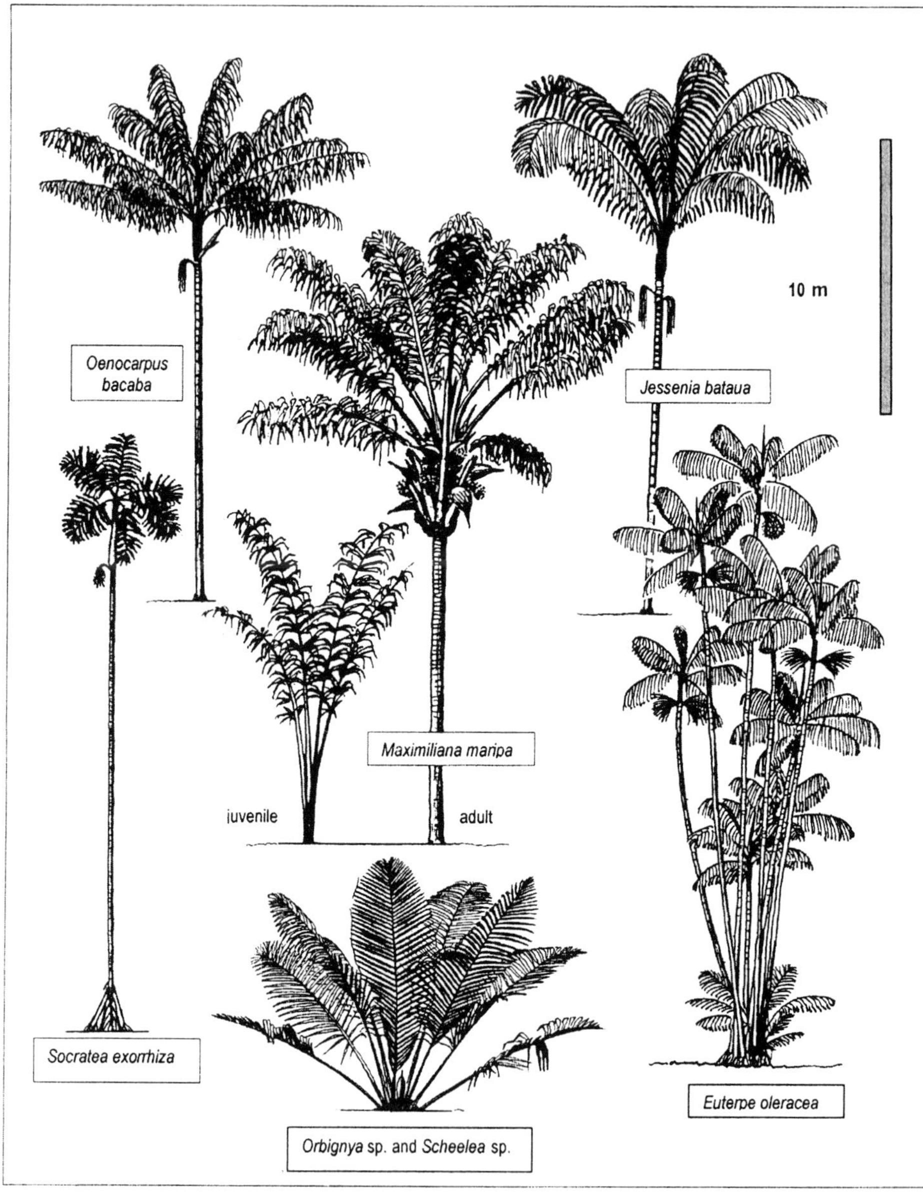
Oenocarpus bacaba
Jessenia bataua
10 m
Maximiliana maripa
juvenile
adult
Socratea exorrhiza
Euterpe oleracea
Orbignya sp. and Scheelea sp.

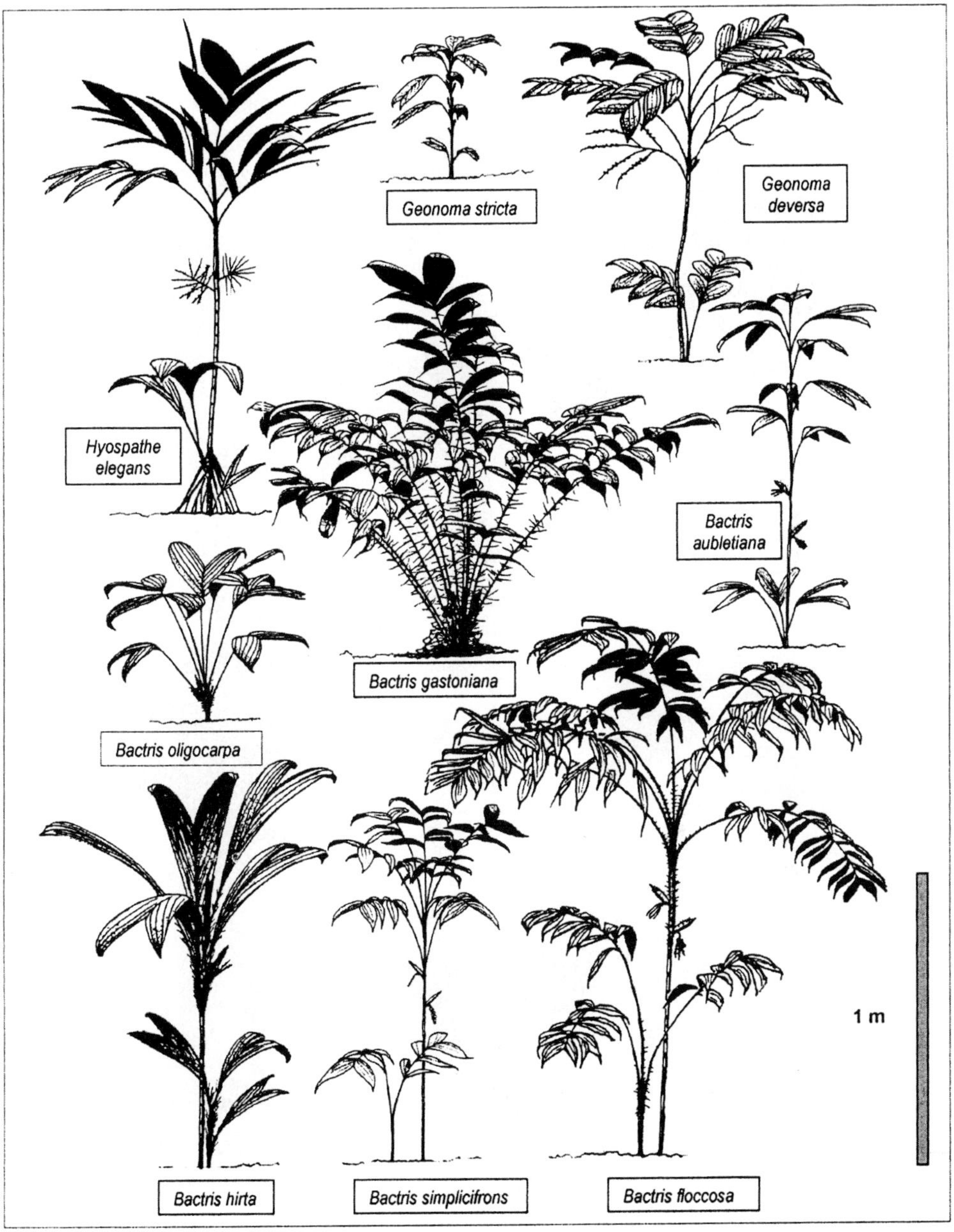
Geonoma stricta
Geonoma deversa
Hyospathe elegans
Bactris aubletiana
Bactris gastoniana
Bactris oligocarpa
1 m
Bactris hirta
Bactris simplicifrons
Bactris floccosa

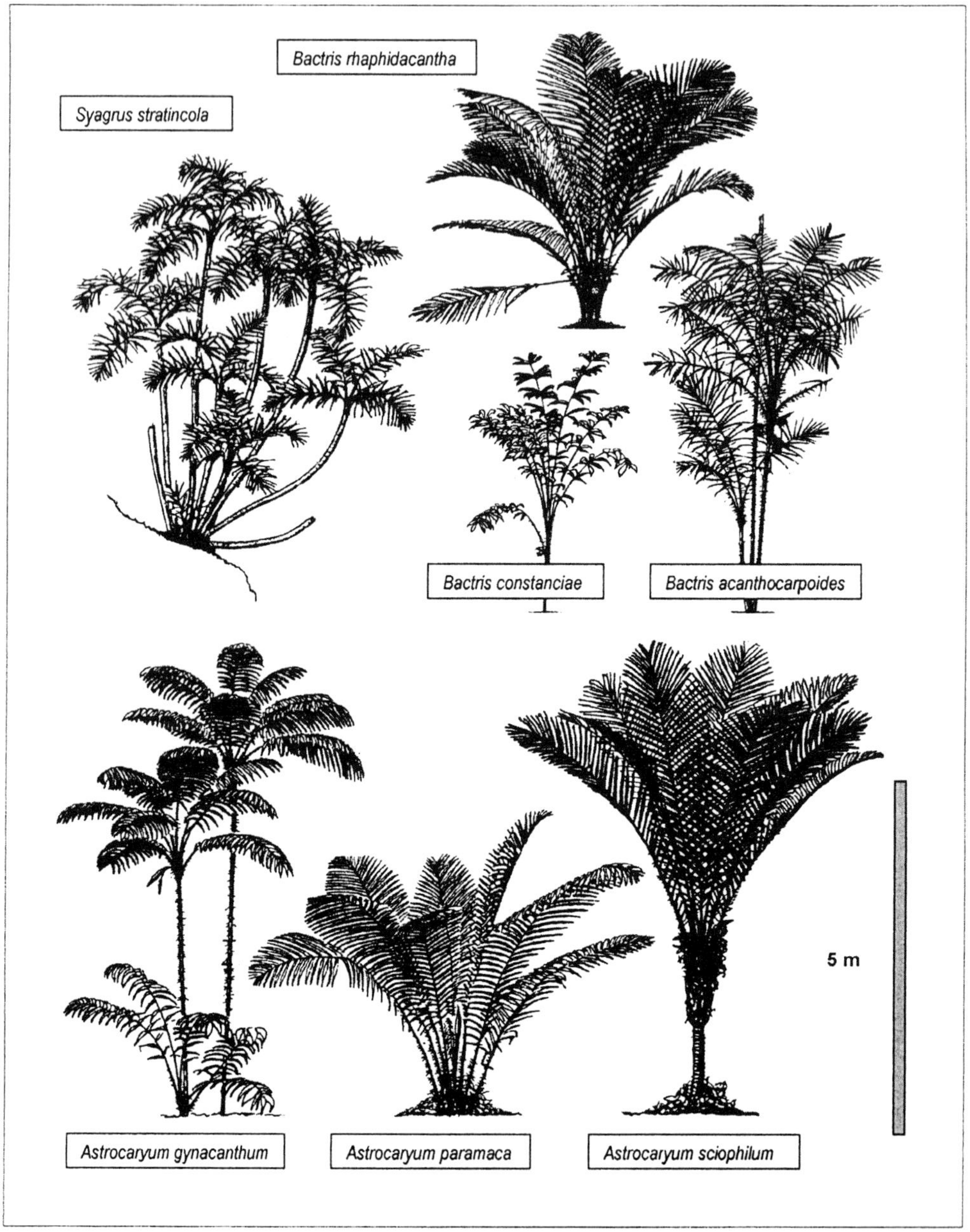
Bactris rhaphidacantha
Syagrus stratincola
Bactris constanciae
Bactris acanthocarpoides
Astrocaryum gynacanthum
Astrocaryum paramaca
Astrocaryum sciophilum
5 m

APPENDIX 3

MAMMALS OF THE NOURAGUES AND LOWER ARATAYE AREAS

François Feer and Pierre Charles-Dominique

The list includes published records from inventories conducted at the Saut Pararé waterfall area of the Arataye river since 1977 and at Nouragues area since 1986 (Guillotin 1982, Guillotin & Petter 1984, Dubost 1988, Brosset & Charles-Dominique 1990, Julien-Laferrière 1991, Guillotin *et al.* 1994, Brosset *et al.* 1996, Voss & Emmons 1996). Our list below differs from Voss and Emmons's (1996) by including new records. The nomenclature is based on Wilson and Reeder (1993).

Research on nonvolant mammals began at Saut Pararé with trapping at ground level (Guillotin 1982, Julien-Laferrière 1991, Voss & Emmons 1996) and continued at Nouragues with several trap sizes (Tomahawks, Shermans and BTS) catching species not exceeding much 1kg (D. Julien-Laferrière, F. Feer and M. L. Guillemin). Trapping between 3-35 m height was conducted at Nouragues in 1998 and 1999 (F. Catzeflis and J.-F. Mauffrey).

Strip censuses along the trail system surrounding the Nouragues station were conducted during night and day in order to estimate the abundance of large rodents, kinkajous, marsupials (D. Julien-Laferrière and P.M. Forget) and primates (C. Julliot, B. Simmen and E. Pagès). Occurrences of other large mammals at Nouragues were recorded in a logbook by all residents since 1992.

Bats were studied by P. Charles-Dominique, A. Brosset, A. Cockle and J.F. Cosson between 1987 and 1996 (Brosset & Charles-Dominique 1990, Brosset *et al.* 1996, Charles-Dominique 1993, Cockle 1997). Mist-nets were used on the ground as well in the canopy and different roosts were also explored during this period (59 species, 2109 captures).

According to the current state of the mammal inventory, the number of nonvolant species in Arataye and Nouragues areas reaches 67. A small number of species are yet expected to occur in the area. They are species known from neighbouring areas of French Guiana, especially Saint Eugène at Petit Saut dam (Granjon *et al.* 1997, Vié *et al.* 1996, J. C. Vié personal communication) and Paracou (Voss *et al.* in press, O. Henry, R. Day personal communication). They include three small marsupials (*Gracilinanus kolinowski*, *Gracilinanus emiliae*, *Marmosops pinheiroi*), and three rodents (*Coendou melanurus*, *Neusticomys oyapocki*, *Isothrix sinnamariensis*). Other species known from Guyana and likely to occur in French Guiana, are small murids (*Oecomys roberti*, *O. trinitatis*) (Voss & Emmons 1996).

Bongers *et al.* (eds.), Nouragues. Dynamics and plant-animal interactions in a neotropical rainforest, 351–355.
2001 Kluwer Academic Publishers. Printed in the Netherlands.

Species	Life mode[1]	Trophic status[2]	Activity cycle[3]	Abundance[4]	Body mass (kg)[5]
MARSUPIALA					
Didelphidae					
Caluromys philander	A	F, I, N	N	C	0.33
Chironectes minimus	Aq	f	N	R	0.5
Didelphis marsupialis	T, A	F, I, V	N	C	1.1
Didelphis albiventris	T	F, I, V	N	U	0.8
Marmosa lepida	A		N	R	0.02
Marmosa murina	A	I, F	N	U	0.04
Marmosops parvidens	A	I, F	N	R	0.02
Metachirus nudicaudatus	T	I, F	N	C	0.46
Micoureus demerae	A	F, I	N	U	0.15
Monodelphis brevicaudata	T	I, F	D	R	0.08
Philander opossum	T, A	I, F	N	C	0.52
XENARTHRA					
Bradipodidae					
Bradypus tridactylus	A	L	D, N	U	4.6
Megalonychidae					
Choloepus didactylus	A	L	N		7.9
Dasypodidae					
Cabassous unicinctus	T	I	N	R	1.6-4.8
Dasypus kappleri	T	I	N	C	9.9
Dasypus novemcinctus	T	I	N, D	C	5.0
Priodontes maximus	T	I	N	R	45.0-60.0
Myrmecophagidae					
Cyclopes didactylus	A	I	N	R	0.3-0.5
Myrmecophaga tridactyla	T	I	D, N	U	20.0-40.0
Tamandua tetradactyla	T, A	I	D, N	C	4.8
CHIROPTERA					
Emballonuridae					
Cormura brevirostris	V	I	N	R	0.009
Diclidurus scutatus	V	I	N	R	0.013
Peropteryx macrotis	V	I	N	U	0.005
Rhynchonycteris naso	V	I	N	R	0.004
Saccopteryx bilineata	V	I	N	U	0.009
Saccopteryx leptura	V	I	N	R	0.005
Mormoopidae					
Pteronotus gymnonotus	V	I	N	R	0.013
Pteronotus parnellii	V	I	N	C	0.022
Phyllostomidae					
Phyllostominae					
Chrotopterus auritus	V	I, V	N	R	0.068
Glyphonycteris (Micr.) daviesi	V	I	N	R	0.019
Glyphonycteris (Micr.) sylvestris	V	I	N	R	0.008
Lonchorina inusita	V	I	N	U	0.019
Macrophyllum macrophyllum	V	I	N	R	0.008
Micronycteris brosseti	V	I	N	R	0.005
Micronycteris hirsuta	V	I	N	R	0.012
Micronycteris megalotis	V	I	N	R	0.006

Species	Life mode[1]	Trophic status[2]	Activity cycle[3]	Abundance[4]	Body mass (kg)[5]
Micronycteris microtis	V	I	N	R	0.006
Micronycteris schmidtorum	V	I	N	R	0.006
Mimon bennetti	V	I	N	R	0.019
Mimon crenulatum	V	I	N	U	0.012
Phylloderma stenops	V	I	N	R	0.045
Phyllostomus discolor	V		N	U	0.034
Phyllostomus elongatus	V	I	N	C	0.038
Phyllostomus hastatus	V	I, V	N	U	0.083
Phyllostomus latifolius	V	I	N	U	0.031
Tonatia saurophila	V	I	N		0.027
Tonatia schulzi	V	I	N	R	0.019
Tonatia silvicola	V	I	N	C	0.036
Trachops cirrhosus	V	I, V	N	U	0.037
Trinycteris (Micr.) nicefori	V	I	N	R	0.008
Vampyrum spectrum	V	I, V	N	R	0.153
Glossophaginae					
Anoura caudifera	V	N	N	R	0.010
Anoura geoffroyi	V	N	N	C	0.015
Choeroniscus minor	V	N	N	R	0.009
Lionycteris spurrelli	V	N	N	C	0.009
Lonchophylla thomasi	V	N	N	C	0.007
Carolliinae					
Carollia brevicauda	V	F	N	U	0.012
Carollia perspicillata	V	F	N	C	0.017
Rhinophylla pumilio	V	F	N	C	0.009
Stenodermatinae					
Ametrida centurio	V	F	N	U	0.010
Artibeus concolor	V	F	N	U	0.020
Artibeus gnomus	V	F	N	C	0.010
Artibeus jamaicensis	V	F	N	C	0.056
Artibeus lituratus	V	F	N	C	0.066
Artibeus obscurus	V	F	N	C	0.036
Chiroderma trinitatum	V	F	N	U	0.013
Chiroderma villosum	V	F	N	U	0.022
Mesophylla macconnelli	V	F	N	U	0.008
Platyrrhinus helleri	V	F	N	C	0.014
Sturnira lilium	V	F	N	R	0.019
Sturnira tildae	V	F	N	U	0.023
Uroderma bilobatum	V	F	N	U	0.017
Vampyressa brocki	V	F	N	R	0.009
Vampyrodes caraccioli	V	F	N	U	0.032
Desmodontinae					
Desmodus rotundus	V	H	N	U	0.028
Thyropteridae					
Thyroptera tricolor	V	I	N	U	0.004
Vespertilionidae					
Eptesicus chiriquinus	V	I	N	R	0.11
Eptesicus furinalis	V	I	N	R	0.09
Myotis nigricans	V	I	N	U	0.005
Myotis riparius	V	I	N	R	0.006

Species	Life mode[1]	Trophic status[2]	Activity cycle[3]	Abundance[4]	Body mass (kg)[5]
Molossidae					
Eumops auripendulus					0.029
Eumops hansae	V	I	N	R	0.013
Molossops abrasus	V	I	N	R	0.032
Molossops greenhalli	V	I	N	U	0.018
Molossops paranus	V	I	N	R	0.012
Molossops planirostris	V	I	N	R	0.013
Molossus rufus (ater)	V	I	N	R	0.037
Molossus molossus	V	I	N	R	0.015
Nyctinomops laticaudatus	V	I	N	U	0.015
PRIMATES					
Callitrichidae					
Saguinus midas	A	F, I	D	C	0.6
Cebidae					
Alouatta seniculus	A	L, F	D	C	7.7
Ateles paniscus	A	F, L	D	C	9.5
Cebus apella	A	F, I	D	C	3.0
Cebus olivaceus	A	F, I	D	U	3.2
Pithecia pithecia	A	F, S	D	U	1.7
Saimiri sciureus	A	F, I	D	R	0.8
CARNIVORA					
Canidae					
Speothos venaticus	T	V	D	R	5.0-7.0
Felidae					
Herpailurus yaguaroundi	T	V	D, N	R	7.0
Leopardus tigrinus	T	V	N, D	R	1.3-3.5
Leopardus pardalis	T	V	N, D	R	9.3
Leopardus wiedii	T,A	V	N ?	R	3.0-9.0
Panthera onca	T	V	N, D	R	31.0-158.0
Puma concolor	T	V	N, D	R	29.0-120.0
Mustelidae					
Eira barbara	T,A	F,V	D, N ?	C	3.8
Galictis vittata	T	V, I		R	1.4-3.3
Lontra longicaudis	Aq	f		R	5.0-14.5
Pteronura brasiliensis	Aq	f		R	22-34
Procyonidae					
Nasua nasua	T,A	F, I	D	U	4.5
Potos flavus	A	F, N	N	C	1.8
PERISSODACTYLA					
Tapiridae					
Tapirus terrestris	T	L, F	D, N	U	160.0
ARTIODACTYLA					
Tayassuidae					
Pecari tajacu	T	S, L	D	C	20.8
Tayassu pecari	T	S, L	D, N	U	36.4
Cervidae					
Mazama americana	T	L, F	N, D	U	40.0-50.0
Mazama gouazoubira	T	L, F	D	U	15.0-18.0

Species	Life mode[1]	Trophic status[2]	Activity cycle[3]	Abundance[4]	Body mass (kg)[5]
RODENTIA					
Sciuridae					
Sciurillus pusillus	A	?	D	C	0.04
Sciurus aestuans	A,T	S, F	D	C	0.17
Muridae					
Neacomys guianae	T	S, F, I	N	C	0.015 -0.020
Nectomys squamipes	T	F, I	N	U	0.2
Oecomys bicolor	A	S	N	U	0.02
Oecomys paricola	A	S	N	U	0.04
Oecomys rutilus	A	S	N	U	0.02
Oecomys rex	A	S	N	U	
Oligoryzomys fulvescens	A	S	N	U	
Oryzomys megacephalus	T	S, F, I	N	C	0.07
Oryzomys macconnelli	T	S, F, I	N	U	0.07
Oryzomys yunganus	T	S, F, I	N	U	
Rhipidomys nitela	T		N	U	0.04
Rhipidomys leucodactylus	T		N	U	0.08
Erethizontidae					
Coendou prehensilis	A	L, F, S	N	U	3.6
Hydrochaeridae					
Hydrochaeris hydrochaeris	Aq	L	D, N	R	27.0-79.0
Dasyproctidae					
Dasyprocta leporina	T	S, F	D	C	4.4
Myoprocta acouchy	T	S, F	D	C	0.96
Agoutidae					
Agouti paca	T	F	N	C	7.5
Echimyidae					
Echimys chrysurus	A	S, F	N	R	0.6
Makalata didelphoides	A	S, F	N	R	0.28
Mesomys hispidus	A	F, L, I	N	R	0.2
Proechimys cayennensis	T	S, F	N	U	0.16-0.5
Proechimys cuvieri	T	S, F	N	C	0.35-0.6

[1] Life mode: A = arboreal, T = terrestrial, Aq = aquatic, V = volant.
[2] Trophic status: F = fruit pulp, S = seeds, L = leaves, N = nectar, I = invertebrates, V = vertebrates, H = hematophagous, f = fish.
[3] Activity cycle: D = diurnal, N = nocturnal.
[4] Abundance: P = probable, R = rare, U = uncommon, C = common.
[5] Body mass: data from F. Catzeflis (personal communication), A. Cockle (personal communication), Brosset *et al.* 1974, Brosset & Charles-Dominique 1990, Dubost 1997, Emmons & Feer 1997, Guillemin *et al.* (this volume), Guillotin & Petter 1984, Guillotin *et al.* 1994, Julien-Laferrière & Atramentowicz (this volume), Richard-Hansen *et al.* 1999.

This *Pourouma* pioneer tree fills a hole in the Nouragues forest canopy. *Photo Frans Bongers*.

APPENDIX 4

BIRD SPECIES (FROM NOURAGUES INSELBERG TO ARATAYE RIVER)

Jean-Marc Thiollay, Mathilde Jullien, Marc Théry and Christian Erard

The present list is based on published data by Tostain *et al.* (1992), Goguillon (1994), Thiollay (1994) and Jullien (1997). Some unpublished information (e.g. by C. Erard, M. Jullien and M. Théry) was also included. This list does not include an unidentified *Sittasoma* species, caught several times on the quadrats (Thiollay 1994), nor a black white-winged *Xipholena* (possibly *X. nigripennis*) seen in March-April 1987 (Tostain *et al.* 1992). Occasional species or vagrant individuals are indicated by [o] after the species name. The actual identity of several species, which are well out of their known geographical distribution still needs to be confirmed and is not supported by a specimen. Two species recently observed along the Arataye river should be added to this list: *Nyctibius bracteatus* (Nyctibiidae) and *Synallaxis albescens* (Furnariidae). They have not been considered in the analysis of the bird community, but a total of 435 species have been observed in the Nouragues natural reserve. Nomenclature and classification follow Tostain *et al.* (1992).

Mass refers to mean body weight categories: 1 (< 8 g), 2 (8-16 g), 3 (16-32 g), 4 (32-64 g), 5 (64-128 g), 6 (128-256 g), 7 (256-512 g), 8 (512-1024 g), 9 (1024-2048 g) and 10 (> 2048 g).

Abundance (Abund): 1 (< 1 pair/km²), 2 (1-9 pairs/km²), 3 (> 10 pairs/km²).

Location (Loc): indicates where the species forages. A: aerial, B: on bark, trunks and epiphytes, C: in canopy, F: gleaning, probing or snatching prey on foliage and branches, G: on the ground, L: in lower strata of the understorey, M: in middle and upper strata of understorey, P: flycatching, pouncing and other perch hunting types, W: along watercourses, banks and swamps.

Diet: based on the main type of food. C: carnivore, F: frugivore (Fd = seed-disperser; Fs = seed-predator), I: insectivore (Id = occasional frugivore and seed-disperser), N: nectarivore, O: omnivore (Od = seed-disperser; Op = seed-predator), P: piscivore or feeding on aquatic organisms.

Social system (Soc Syst): A: ant-follower, F: permanent or regular attendant of mixed species flocks, either insectivores or frugivores, L: lekking species, M: monogamous, P: polygamous.

Habitat (Hab): D: disturbed or secondary forest, F: mature primary forest, G: treefall gaps, I: bare rocks with or without clumps of shrubs and small trees, on granitic inselbergs, O: second growth in large openings and forest edges, R: large rivers and their forested banks, S: forest streams and swamps, seasonally flooded forest.

Bongers *et al.* (eds.), Nouragues. Dynamics and plant-animal interactions in a neotropical rainforest, 357–369.
2001 Kluwer Academic Publishers. Printed in the Netherlands.

Family and Species	Mass	Abund	Loc	Diet	Soc Syst	Hab
TINAMIDAE (2 gen., 5 sp.)						
Tinamus major	8	2	G	Fs	P	F
Crypturellus cinereus	7	1	G	Fs	P	D
Crypturellus soui	6	2	G	Fs	P	F
Crypturellus variegatus	7	3	G	Fs	P	F
Crypturellus erythropus	7	1	G	Fs	P	D
PHALACROCORACIDAE (1 gen., 1 sp.)						
Phalacrocorax olivaceus [o]	9	1	W	P	M	R
ANHINGIDAE (1 gen., 1 sp.)						
Anhinga anhinga [o]	9		W	P	M	R
ARDEIDAE (6 gen., 6 sp.)						
Egretta alba [o]	8	1	W	P	M	R
Butorides striatus	6	1	W	P	M	R
Agamia agami	8	1	W	P	M	S
Bubulcus ibis [o]	7	1	G	I	M	I
Tigrisoma lineatum	8	2	W	P	M	S
Zebrilus undulatus	6	1	W	P	M	R
THRESKIORNITHIDAE (1 gen., 1 sp.)						
Mesembrinibis cayennensis [o]	8	1	W	P	M	R
CATHARTIDAE (3 gen., 4 sp.)						
Coragyps atratus [o]	9	1	A	C	M	O
Cathartes aura [o]	9	1	A	C	M	O
Cathartes melambrotus	9	1	A	C	M	F
Sarcoramphus papa	10	1	A	C	M	O
ACCIPITRIDAE (14 gen., 23 sp.)						
Elanoides forficatus	7	1	A	I	M	O
Leptodon cayanensis	7	1	C,F	C	M	F
Chondrohierax uncinatus	7	1	C,F	C	M	S
Harpagus bidentatus	6	1	C,F	C	M	F
Harpagus diodon	6	1	C,F	C	M	F
Ictinia plumbea	6	1	A	I	M	F
Accipiter bicolor	5	1	M,F	C	M	F
Accipiter superciliosus	6	1	C,F	C	M	G
Accipiter poliogaster	8	1	M,F	C	M	F
Accipiter striatus [o]	5	1	C,F	C	M	O
Buteo albicaudatus [o]	7	1	C,P	C	M	I
Asturina nitida	7	1	C,P	C	M	O
Leucopternis albicollis	8	1	C,P	C	M	O
Leucopternis melanops	7	1	M,P	C	M	F
Leucopternis schistacea [o]	8	1	M,P	C	M	S
Buteogallus meridionalis [o]	8	1	C,P	C	M	O
Buteogallus urubutinga	9	1	C,P	C	M	S

Family and Species	Mass	Abund	Loc	Diet	Soc Syst	Hab
Harpyhaliaetus solitarius [o]	9	1	C,F	C	M	O
Morphnus guianensis	9	1	C,F	C	M	F
Harpia harpyja	9	1	C,F	C	M	F
Spizastur melanoleucus	8	1	C,F	C	M	F
Spizaetus ornatus	9	1	C,F	C	M	F
Spizaetus tyrannus	9	1	C,F	C	M	D
PANDIONIDAE (1 gen., 1 sp.)						
Pandion haliaetus [o]	9	1	W	P	M	R
FALCONIDAE (3 gen., 7 sp.)						
Micrastur semitorquatus	8	1	M,F	C	M	F
Micrastur mirandollei	8	1	C,F	C	M	F
Micrastur ruficollis	6	1	M,P	C	M	F
Micrastur gilvicollis	6	2	M,P	C	M,A	F
Daptrius americanus	8	1	C,F	Od	P	F
Falco deiroleucus	7	1	A	C	M	I
Falco rufigularis	6	1	A	C	M	R
CRACIDAE (4 gen., 4 sp.)						
Ortalis motmot	7	1	C,F	Fd	P	D
Penelope marail	9	2	C,F	Fd	M	F
Aburria pipile	9	1	C,F	Fd	M	S
Crax alector	10	2	G	Fs	M	F
PHASIANIDAE (1 gen., 1 sp.)						
Odontophorus gujanensis	7	2	G	Os	M	F
OPISTHOCOMIDAE (1 gen., 1 sp.)						
Opisthocomus hoazin	8	1	L,F	Fs*	P	R
PSOPHIIDAE (1 gen., 1 sp.)						
Psophia crepitans	9	2	G	Fd**	P	F
RALLIDAE (1 gen., 1 sp.)						
Aramides cajanea	7	1	G	I	M	S
HELIORNITHIDAE (1 gen., 1 sp.)						
Heliornis fulica	6	1	W	P	M	R,S
EURYPYGIDAE (1 gen., 1 sp.)						
Eurypyga helias	6	1	G	P	M	S
SCOLOPACIDAE (2 gen., 3 sp.)						
Tringa melanoleuca [o]	6	1	W	I	-	R
Tringa solitaria [o]	4	1	W	I	-	R
Actitis macularia [o]	4	1	W	I	-	R
COLUMBIDAE (3 gen., 7 sp.)						
Columba speciosa	7	1	C,F	Fs	M	D
Columba cayennensis [o]	7	1	C,F	Fs	M	D
Columba subvinacea	6	2	C,F	Fs	M	F
Columba plumbea	6	2	C,F	Fs	M	F
Leptotila rufaxilla	6	2	G	Fs	M	F
Leptotila verreauxi	6	1	G	Fs	M	I
Geotrygon montana	5	2	G	Fs	M	F

Family and Species	Mass	Abund	Loc	Diet	Soc Syst	Hab
PSITTACIDAE (10 gen., 18 sp.)						
Ara ararauna	9	1	C,F	Fs	M	F,S
Ara macao	9	1	C,F	Fs	M	F
Ara chloroptera	9	1	C,F	Fs	M	F
Ara severa	7	1	C,F	Fs	M	F
Pyrrhura picta	5	1	C,F	Fs	M	F
Forpus sclateri [o]	3	1	C,F	Fs	M	O
Brotogeris chrysopterus	4	2	C,F	Fs	M	O
Touit batavica	4	1	C,F	Fs	M	F
Touit purpurata	4	1	C,F	Fs	M,F	F
Pionites melanocephala	6	2	C,F	Fs	M	F
Pionopsitta caica	6	1	C,F	Fs	M	F
Pionus menstruus	7	2	C,F	Fs	M,F	F
Pionus fuscus	7	2	C,F	Fs	M,F	F
Amazona dufresniana	7	1	C,F	Fs	M	F
Amazona ochrocephala	7	1	C,F	Fs	M	F
Amazona amazonica	7	2	C,F	Fs	M,F	F
Amazona farinosa	8	2	C,F	Fs	M,F	F
Deroptyus accipitrinus	6	2	C,F	Fs	M,F	F
CUCULIDAE (4 gen., 6 sp.)						
Piaya cayana	5	2	C,F	I	M	D
Piaya melanogaster	5	2	C,F	I	M	F
Piaya minuta	4	1	C,F	I	M	R,S
Dromococcyx pavoninus [o]	5	1	C,F	I	-	F
Tapera naevia	4	1	C,F	I	M	O
Crotophaga ani	5	1	L,F	I	P	O
STRIGIDAE (5 gen., 6 sp.)						
Otus watsoni	6	2	C,F	I	M	D
Lophostrix cristata	8	1	M,F	C	M	F
Pulsatrix perspicillata	8	1	C,F	C	M	F
Strix virgata	7	1	M,F	C	M	F
Strix nigrolineata	7	1	M,F	C	M	F
Glaucidium hardyi	5	2	M,F	C	M	F
NYCTIBIIDAE (1 gen., 2 sp.)						
Nyctibius grandis	8	1	C,P	I	M	F
Nyctibius griseus	6	1	C,P	I	M	D,O
CAPRIMULGIDAE (5 gen., 5 sp.)						
Lurocalis semitorquatus	5	1	A	I	M	O
Chordeiles acutipennis	4	1	A	I	M	O
Nyctidromus albicollis	5	2	A	I	M	I,O
Caprimulgus nigrescens	4	3	A	I	M	I
Hydropsalis climacocerca	4	2	A	I	M	R

Family and Species	Mass	Abund	Loc	Diet	Soc Syst	Hab
APODIDAE (3 gen., 5 sp.)						
Streptoprocne zonaris	5	1	A	I	M	F
Chaetura chapmani [o]	3	1	A	I	M	F
Chaetura andrei [o]	3	1	A	I	M	F
Chaetura spinicauda	3	2	A	I	M	F
Panyptila cayennensis	3	1	A	I	M	F
TROCHILIDAE (18 gen., 23 sp.)						
Glaucis hirsuta	1	2	M,F	N	L	O
Threnetes niger	1	1	M,F	N	L	S
Phaethornis superciliosus	1	3	L,F	N	L	F
Phaethornis malaris	2	2	C,F	N	L	F
Phaethornis bourcieri	1	2	L,F	N	L	F
Phaethornis ruber	1	2	L,F	N	L	G
Phaethornis longuemareus	1	2	M,F	N	L	G
Campylopterus largipennis	2	3	M,F	N	M	F
Eupetomena macroura	1	1	C,F	N	?	S
Florisuga mellivora	1	2	C,F	N	M	F
Anthracothorax nigricollis	2	1	M,F	N	M	R
Avocettula recurvirostris	1	2	C,F	N	M	F
Lophornis ornatus	1	2	C,F	N	M	R
Discosura longicauda	1	2	C,F	N	M	F
Chlorestes notatus	1	2	C,F	N	M	D,O
Chlorostilbon mellisugus	1	1	C,F	N	M	O
Thalurania furcata	1	3	M,F	N	M	F
Hylocharis sapphirina	1	1	L,F	N	L	O
Hylocharis cyanus	1	1	C,F	N	L	F
Topaza pella	2	2	C,F	N	L	S,R
Heliothryx aurita	1	2	C,F	N	M	F
Heliomaster longirostris	1	1	C,F	N	?	F
Calliphlox amethystina [o]	1	1	C,F	N	?	O
TROGONIDAE (1 gen., 5 sp.)						
Trogon melanurus	5	2	C,F	Od	M	F
Trogon viridis	5	3	C,F	Od	M	F
Trogon collaris	4	2	C,F	Od	M	F
Trogon rufus	4	2	M,F	Od	M	F
Trogon violaceus	4	1	C,F	Od	M	D
ALCEDINIDAE (2 gen., 5 sp.)						
Megaceryle torquata	7	1	W	P	M	R
Chloroceryle amazona	5	1	W	P	M	R,S
Chloroceryle americana	3	1	W	P	M	R
Chloroceryle inda	5	1	W	P	M	R,S
Chloroceryle aenea	2	1	W	P	M	S

Family and Species	Mass	Abund	Loc	Diet	Soc Syst	Hab
MOMOTIDAE (1 gen., 1 sp.)						
Momotus momota	6	2	M,P	Od	M	F
GALBULIDAE (3 gen., 6 sp.)						
Brachygalba lugubris	3	1	C,P	I	M	O
Galbula albirostris	3	2	M,P	I	M	G
Galbula ruficauda [o]	3	1	M,P	I	M	O
Galbula leucogastra	3	1	M,P	I	M	R
Galbula dea	3	2	C,P	I	M	F
Jacamerops aurea	5	1	C,P	I	M	F
BUCCONIDAE (5 gen., 7 sp.)						
Notharchus macrorhynchus	5	2	C,P	I	M	F
Notharchus tectus	4	1	C,P	I	M	D,O
Bucco tamatia	4	1	C,P	I	M	D
Bucco capensis	4	1	M,P	I	M	F
Malacoptila fusca	4	2	M,P	I	M	F
Monasa atra	4	2	C,F	I	M	F
Chelidoptera tenebrosa	4	1	C,P	I	M	O,R
CAPITONIDAE (1 gen., 1 sp.)						
Capito niger	4	2	C,F	Fd	M	F,D
RAMPHASTIDAE (3 gen., 5 sp.)						
Pteroglossus aracari	6	2	C,F	Fd	M	F
Pteroglossus viridis	6	1	C,F	Fd	M	F
Selenidera culik	6	2	C,F	Fd	M	F
Ramphastos vitellinus	7	2	C,F	Fd	M	F
Ramphastos tucanus	8	2	C,F	Fd	M	F
PICIDAE (7 gen., 16 sp.)						
Picumnus exilis	2	2	M,B	I	M,F	D
Picumnus spilogaster	3	1	M,B	I	M	F
Picumnus minutissimus [o]	2	1	C,B	I	M	D
Melanerpes cruentatus	4	1	C,B	I	M	D
Veniliornis passerinus	4	1	C,B	I	M	D
Veniliornis cassini	4	2	C,B	I	M	F
Piculus flavigula	4	2	C,B	I	M,F	F,D
Piculus chrysochloros	5	1	C,B	I	M	F
Piculus rubiginosus	5	2	M,B	I	M	F
Celeus undatus	5	2	C,B	I	M	F
Celeus grammicus [o]	5	1	M,B	I	M	F
Celeus elegans	6	2	M,B	I	M	F
Celeus torquatus	5	1	C,B	I	M	F
Dryocopus lineatus	6	1	C,B	I	M	G,D
Campephilus rubricollis	6	2	C,B	I	M	F
Campephilus melanoleucos	6	1	C,B	I	M	R

Family and Species	Mass	Abund	Loc	Diet	Soc Syst	Hab
FURNARIIDAE (6 gen., 14 sp.)						
Synallaxis macconnelli	3	1	L,F	I	M	D
Certhiaxis gutturata	3	1	L,F	I	M	D
Philydor ruficaudatus	3	1	M,B	I	M	F
Philydor erythrocercus	3	2	M,B	I	M,F	F
Philydor pyrrhodes	3	2	M,B	I	M,F	F
Automolus ochrolaemus	4	2	M,B	I	M,F	F
Automolus infuscatus	4	2	M,B	I	M,F	F
Automolus rubiginosus	4	2	L,F	I	M,F	F
Automolus rufipileatus	4	1	M,F	I	M,F	R,S
Sclerurus mexicanus	4	2	G	I	M	F
Sclerurus rufigularis	4	2	G,L,F	I	M	F
Sclerurus caudacutus	4	2	G	I	M	F
Xenops milleri	2	1	C,F	I	M	F
Xenops minutus	2	2	M,F	I	M,F	F
DENDROCOLAPTIDAE (12 gen., 19 sp.)						
Dendrocincla fuliginosa	4	2	M,B	I	M,A	F
Dendrocincla merula	4	2	M,B	I	M,A	F
Deconychura longicauda	3	2	M,B	I	M,F	F
Deconychura stictolaema	3	2	M,B	I	M,F	S
Sittasomus griseicapillus	3	1	C,B	I	M,F	F
Glyphorynchus spirurus	2	3	L,B	I	M,F	F
Nasica longirostris	5	1	C,B	I	M	S
Dendrexetastes rufigula	5	1	C,B	I	M,F	F
Hylexetastes perrotii	5	1	L,B	I	M	F
Xiphocolaptes promeropirhynchus	6	1	C,B	I	M	F
Dendrocolaptes certhia	5	2	M,B	I	M,F	F
Dendrocolaptes picumnus	5	2	M,B	I	M	F
Xiphorhynchus picus [o]	4	1	M,B	I	M	O
Xiphorhynchus obsoletus	4	1	M,B	I	M	F
Xiphorhynchus pardalotus	4	2	M,B	I	M,F	F
Xiphorhynchus guttatus	5	1	M,B	I	M	S
Lepidocolaptes albolineatus	4	1	M,B	I	M	G
Campylorhamphus trochilirostris	4	1	C,B	I	M,F	F
Campylorhamphus procurvoides	4	2	M,B	I	M,F	F
THAMNOPHILIDAE (21 gen., 40 sp.)						
Cymbilaimus lineatus	4	2	M,F	I	M,F	D,G
Frederickena viridis	5	1	L,F	I	M	G
Taraba major	4	1	L,F	I	M	O
Sakesphorus melanothorax	3	2	L,F	I	M	D
Thamnophilus doliatus	3	1	L,F	I	M	O,I
Thamnophilus nigrocinereus [o]	3	1	M,F	I	M	S,R
Thamnophilus murinus	3	2,3	M,F	I	M	F

Family and Species	Mass	Abund	Loc	Diet	Soc Syst	Hab
Thamnophilus punctatus	3	1	L,F	I	M	I,R
Thamnophilus amazonicus	3	2	C,F	I	M,F	F,D
Pygiptila stellaris	3	1	M,F	I	M	G
Thamnomanes ardesiacus	3	2,3	M,F	I	M,F	F
Thamnomanes caesius	3	2,3	M,F	I	M,F	F
Myrmotherula brachyura	1	3	C,F	I	M	F
Myrmotherula surinamensis	2	1	M,F	Id	M	R,S
Myrmotherula guttata	2	2,3	L,F	I	M,F	F
Myrmotherula gutturalis	2	2,3	M,F	I	M,F	F
Myrmotherula axillaris	2	2,3	M,F	I	M,F	F
Myrmotherula longipennis	2	2,3	M,F	I	M,F	F
Myrmotherula menetriesii	2	2	M,F	I	M,F	F
Herpsilochmus sticturus	2	2	C,F	I	M	F
Herpsilochmus stictocephalus	2	2	C,F	I	M,F	F
Microrhopias quixensis	2	2	M,F	I	M,F	D
Terenura spodioptila	1	2	C,F	I	M,F	F
Cercomacra cinerascens	3	3	C,F	I	M	F,D
Cercomacra tyrannina	2	2	M,F	I	M	S,D
Cercomacra nigrescens	3	1	L,F	I	M	D,S
Hypocnemis cantator	2	3	L,F	I	M	G
Hypocnemoides melanopogon	2	1	L,F	I	M	R,S
Percnostola rufifrons	3	2	L,F	Id	M,A	D
Percnostola leucostigma	3	2	L,F	I	M	S,G
Sclatera naevia	3	1	L,F	I	M	R,S
Myrmeciza longipes [o]	3	1	G,L,F	I	M	G
Myrmeciza ferruginea	3	3	G	I	M	G
Myrmeciza atrothorax	3	1	G,L,F	I	M	O
Pithys albifrons	3	3	G,L,F	I	M,A	F
Gymnopithys rufigula	3	2	G,L,F	I	M,A	F
Hylophylax naevia	2	3	G,L,F	I	M,A	G
Hylophylax punctulata [o]	2	1	L,F	I	M	S
Hylophylax poecilonota	3	3	G,L,F	I	M,A	F
Myrmornis torquata	4	2	G	I	M	F
FORMICARIIDAE (4 gen., 5 sp.)						
Formicarius colma	4	2	G	I	M	F
Formicarius analis	4	2,3	G	I	M	F
Grallaria varia	5	2	G	I	M	F
Hylopezus macularius	4	2	G	I	M	F
Myrmothera campanisona	4	2	G	I	M	S,G
CONOPOPHAGIDAE (1 gen., 1 sp.)						
Conopophaga aurita	3	2	G	I	M	F

Family and Species	Mass	Abund	Loc	Diet	Soc Syst	Hab
COTINGIDAE (11 gen., 12 sp.)						
Phoenicircus carnifex	5	2	M,F	Fd	L,F	F
Iodopleura fusca	2	2	C,F	Fd	P,F	F
Cotinga cotinga	4	2	C,F	Fd	L,F	F
Cotinga cayana	5	2	C,F	Fd	L,F	F
Xipholena punicea	5	2	C,F	Fd	L,F	F
Lipaugus vociferans	5	3	M,F	Fd	L	F
Gymnoderus foetidus	7	1	C,F	Fd	L	F
Haematoderus militaris [o]	7	1	C,F	Fd	L	F
Querula purpurata	5	2	C,F	Fd	P,F	F
Perissocephalus tricolor	5	2	M,F	Od	L,F	F
Procnias alba	5	2	C,F	Fd	L	F
Rupicola rupicola	7	2	M,F	Fd	L,F	F
OXYRUNCIDAE (1 gen., 1 sp.)						
Oxyruncus cristatus	3	1	C,F	Od	M	F
PIPRIDAE (9 gen., 11 sp.)						
Pipra erythrocephala	2	2,3	C,F	Fd	L,F	F
Pipra pipra	2	2,3	M,F	Fd	L	F
Pipra serena	2	2,3	L,F	Fd	L	F
Chiroxiphia pareola [o]	2	1	M,F	Fd	L	R
Corapipo gutturalis	2	2	L,F	Fd	L	F
Manacus manacus	3	1	L,F	Fd	L	O,R
Xenopipo atronitens [o]	2	1	C,F	Fd	L	O,D
Neopipo cinnamomea [o]	2	1	M,F	I	M	D
Tyranneutes virescens	2	3	C,F	Fd	L	F,S
Piprites chloris	3	2	C,F	Fd	L	D
Schiffornis turdinus	3	2	L,F	Od	M	F
TYRANNIDAE (41 gen., 66 sp.)						
Zimmerius gracilipes	2	2	C,F	Od	M	F
Ornithion inerme [o]	1	1	M,F	I	M	S
Phaeomyias murina [o]	2	1	M,F	I	M	O
Sublegatus modestus [o]	2	1	M,F	I	M	I
Tyrannulus elatus	2	1	M,F	Fd	M	D
Myiopagis gaimardii	2	2	C,F	Id	M,F	F
Myiopagis caniceps	2	1	C,F	I	M	F
Myiopagis flavivertex [o]	2	1	M,F	I	M	S
Elaenia flavogaster	3	1	C,P	Id	M	S
Mionectes macconnelli	2	2	L,F	Od	L,F	F
Leptopogon amaurocephalus	2	2	M,F	I	M	F
Phylloscartes virescens	2	1	M,F	I	M	R
Corythopis torquata	3	2	G,L,F	I	M	F
Myornis ecaudatus	1	2	C,F	I	M,F	F
Lophotriccus vitiosus	2	1	M,F	I	M	D,R

Family and Species	Mass	Abund	Loc	Diet	Soc Syst	Hab
Lophotriccus galeatus	2	1	M,F	I	M	D,G
Hemitriccus minor	2	2	M,F	I	M	G,D
Hemitriccus josephinae	2	2	M,F	I	M	D
Hemitriccus zosterops	2	3	M,F	I	M	F
Todirostrum sylvia	2	1	L,F	I	M	O
Todirostrum cinereum	1	1	M,F	I	M	O
Todirostrum pictum	1	2	C,F	I	M,F	F
Ramphotrigon ruficauda	3	2	C,F	Id	M	F
Rhynchocyclus olivaceus	3	2	C,F	Id	M,F	F
Tolmomyias assimilis	2	2	C,F	I	M,F	F
Tolmomyias poliocephalus	2	1	C,F	I	M	F
Tolmomyias flaviventris	2	1	C,F	I	M,F	F
Platyrhinchus saturatus	2	2	M,F	I	M	S
Platyrhinchus mystaceus	2	2	M,F	I	M	F
Platyrhinchus coronatus	2	3	M,F	I	M	F
Platyrhinchus platyrhynchos	2	1	L,F	I	M	F
Onychorhynchus coronatus	2	1	M,F	I	M	S
Terenotriccus erythurus	1	2,3	M,F	I	M	D
Myiobius barbatus	2	2,3	M,F	I	M,F	F
Myiophobus fasciatus	2	1	M,F	I	M	I
Contopus cinereus	3	1	C,F	I	M	F
Contopus albogularis	3	2	C,P	I	M	G
Colonia colonus	3	1	C,P	I	M	F
Hirundinea ferruginea	3	2	L,P	I	M	I
Attila cinnamomeus	3	2	C,F	Id	M	S,F
Attila spadiceus	4	2	C,F	I	M	F
Rhytipterna simplex	4	2	C,F	Id	M	D
Laniocera hypopyrrha	4	2	M,F	Od	M	F
Sirystes sibilator	4	2	C,P	I	M,F	F
Myiarchus swainsoni [o]	3	1	C,P	I	M	O
Myiarchus ferox	3	1	C,P	Id	M	I
Pitangus lictor	4	2	M,P	I	M	R
Pitangus sulphuratus	4	1	M,P	Id	M	O
Megarhynchus pitangua	4	1	M,P	Id	M	O
Myiozetetes cayanensis	3	1	M,P	Od	M	O
Myiozetetes similis	3	1	M,P	Od	M	O
Myiozetetes luteiventris [o]	3	1	M,P	Od	M	O
Conopias albovittata	3	2	C,P	Od	M,F	F
Myiodynastes maculatus [o]	4	1	C,P	Id	M	R
Legatus leucophaius	3	1	M,P	Id	M	O
Tyrannopsis sulphurea	4	1	C,P	Od	M	R,O
Tyrannus albogularis	4	2	C,P	Id	M	F
Tyrannus melancholicus	4	1	C,P	Id	M	O

Family and Species	Mass	Abund	Loc	Diet	Soc Syst	Hab
Tyrannus dominicensis	4	1	C,P	I	M	R,O
Pachyramphus rufus	3	1	C,F	Od	M	O
Pachyramphus marginatus	3	1	C,F	I	M,F	D
Pachyramphus surinamus	3	1	C,F	I	M,F	D
Pachyramphus minor	4	2	C,F	Id	M,F	F
Tityra cayana	5	2	C,F	Od	M	F
Tityra semifasciata [o]	5	1	C,F	Od	M	G
Tityra inquisitor	4	1	C,F	Od	M	F
HIRUNDINIDAE (8 gen., 9 sp.)						
Tachycineta albiventer	2	1	A	I	M	R
Phaeoprogne tapera	3	1	A	I	M	R
Progne chalybea [o]	4	1	A	I	M	O
Notiochelidon cyanoleuca [o]	3	1	A	I	M	I
Atticora fasciata	2	1	A	I	M	R
Atticora melanoleuca	2	1	A	I	M	R
Neochelidon tibialis	2	1	A	I	M	O
Stelgidopteryx ruficollis	2	1	A	I	M	O
Hirundo rustica [o]	2	1	A	I	M	O
CORVIDAE (1 gen., 1 sp.)						
Cyanocorax cayanus	6	1	C,F	Od	P	F
TROGLODYTIDAE (4 gen., 5 sp.)						
Thryothorus coraya	3	2,3	L,F	I	M	D,G
Thryothorus leucotis	3	1	L,F	I	M	R,S
Henicorhina leucosticta	3	1	G	I	M	F,G
Microcerculus bambla	2	3	G,L,F	I	M	F,G
Cyphorinus arada	3	2	G,L,F	I	P	F
TURDIDAE (1 gen., 2 sp.)						
Turdus fumigatus	4	1	M,F	Od	M	D
Turdus albicollis	4	3	G,L,F	Od	M	F
SYLVIIDAE (3 gen., 4 sp.)						
Microbates collaris	2	2	G,L,F	I	M,F	D
Ramphocaenus melanurus	2	2	L,F	I	M,F	G
Polioptila plumbea	1	2	C,F	I	M,F	F
Polioptila guianensis	1	1	C,F	I	M,F	F
VIREONIDAE (4 gen., 8 sp.)						
Cyclarhis gujanensis	3	2	C,F	I	M	D
Vireolanius leucotis	3	2	C,F	Od	M,F	F
Vireo olivaceus	2	3	C,F	Od	M	F
Hylophilus thoracicus	2	1	C,F	Id	M	D
Hylophilus semicinereus	2	2	C,F	I	M,F	F
Hylophilus pectoralis	2	1	C,F	Id	M	F
Hylophilus muscicapinus	2	2	C,F	I	M,F	F
Hylophilus ochraceiceps	2	2	M,F	I	M,F	F

Family and Species	Mass	Abund	Loc	Diet	Soc Syst	Hab
PARULIDAE (2 gen., 2 sp.)						
Parula pitiayumi	2	1	C,F	I	M,F	R
Basileuterus rivularis	2	2	G	Id	M	S,R
ICTERIDAE (4 gen., 8 sp.)						
Scaphidura oryzyvora	6	1	C,F	Od	P	O
Psarocolius decumanus	7	1	C,F	Od	P	O
Psarocolius viridis	7	2	C,F	Od	P,F	F
Cacicus cela	5	1	C,F	Od	P,F	R
Cacicus haemorrhous	5	2	C,F	Od	P,F	G
Icterus chrysocephalus [o]	4	1	C,F	Od	P	O
Icterus cayanensis	4	1	C,F	Od	P	G
Icterus nigrogularis [o]	4	1	C,F	Od	P	O
EMBERIZIDAE (26 gen., 45 sp.)						
Coereba flaveola	2	3	C,F	Od	M,F	F,G
Dacnis cayana	2	2	C,F	Od	M,F	F
Dacnis lineata	2	2	C,F	Od	M,F	F
Cyanerpes caeruleus	2	2	C,F	Od	M,F	F
Cyanerpes cyaneus	2	2	C,F	Od	M,F	F
Chlorophanes spiza	3	2	C,F	Od	M,F	F
Conirostrum speciosum	2	1	C,F	Od	M,F	F
Tersina viridis	3	1	C,F	Id	M	G
Tangara velia	3	2	C,F	Od	M,F	F
Tangara chilensis	3	3	C,F	Od	M,F	F
Tangara punctata	3	2	C,F	Od	M,F	F
Tangara varia	3	1	M,F	Od	M	R
Tangara mexicana	3	1	C,F	Od	M,F	O
Tangara gyrola	3	2	C,F	Od	M,F	F
Tangara cayana [o]	3	1	C,F	Od	M	O
Euphonia minuta	2	1	C,F	Fd	M,F	D
Euphonia chlorotica	2	1	C,F	Fd	M	F
Euphonia violacea	2	1	C,F	Fd	M,F	O
Euphonia cayennensis	2	2	C,F	Fd	M,F	F
Euphonia chrysopasta	2	2	C,F	Fd	M,F	F
Thraupis episcopus	4	1	C,F	Od	M	O
Thraupis palmarum	4	1	M,F	Od	M	O
Ramphocelus carbo	4	1	L,F	Od	M,F	O
Cyanicterus cyanicterus	4	1	C,F	Od	M	F
Lanio fulvus	3	2	C,F	I	M,F	F
Tachyphonus cristatus	3	2	C,F	Id	M,F	F
Tachyphonus surinamus	3	2	M,F	Id	M,F	F
Tachyphonus phoenicius	3	1	M,F	Id	M	I,O
Tachyphonus luctuosus	3	1	M,F	Id	M,F	D
Eucometis penicillata	3	1	C,F	Id	M	R,S

Family and Species	Mass	Abund	Loc	Diet	Soc Syst	Hab
Nemosia pileata	3	1	C,F	Od	M,F	S
Hemithraupis guira	2	2	C,F	Od	M,F	F
Hemithraupis flavicollis	3	2	C,F	I	M,F	F
Lamprospiza melanoleuca	4	1	C,F	Od	M	F
Cyanocompsa cyanoides	3	2	G,L,F	Fs	M	G
Saltator maximus	4	1	M,F	Od	M	G,D
Saltator coerulescens [o]	4	1	M,F	Od	M	O
Periporphyrus erythromelas [o]	4	1	C,F	Fd	M	F
Pitylus grossus	4	2	M,F	Od	M,F	F
Caryothraustes canadensis	4	2	C,F	Od	M,F	F
Arremon taciturnus	3	2	G	Od	M	S,G
Oryzoborus angolensis	2	1	G	Fs	M	O
Oryzoborus crassirostris [o]	3	1	G	Fs	M	I
Sporophila castaneiventris	2	1	G	Fs	M	O
Zonotrichia capensis [o]	3	1	G	Os	M	O

* Though mostly folivorous, *Opisthocomus hoazin* is here included with the frugivores.

** *Psophia crepitans* is here considered as a frugivore though its diet includes much animal matter

Caterpilars attacking a leaf in the forest understorey. *Photo Toon Rijkers.*

APPENDIX 5

AMPHIBIAN AND REPTILE SPECIES AT THE NOURAGUES NATURE RESERVE

Marga Born and Philippe Gaucher

A. List of families and species

ANURA
Allophrynidae
Allophryne ruthveni
Bufonidae
Atelopus flavescens (N)
Atelopus franciscus
Bufo guttatus
Bufo margaritifera
Bufo marinus
Bufo sp.
Centrolenidae
Cochranella oyampiensis
Hyalinobatrachium taylori (A)
Hyalinobatrachium nouragensis
Dendrobatidae
Colostethus baeobatrachus (A)
Colostethus beebei
Colostethus degranvillei
Dendrobates tinctorius
Dendrobates ventrimaculatus
Epipedobates femoralis
Epipedobates hahneli
Hylidae
Hyla boans
Hyla brevifrons (A)
Hyla calcarata (A)
Hyla fasciata
Hyla geographica
Hyla granosa
Hyla hadroceps
Hyla leucophyllata
Hyla melanargyrea (A)
Hyla sp.
Hyla multifasciata (A)
*Hyla ornatissima (*A)
Osteocephalus cabrerai
Osteocephalus leprieurii (A)
Osteocephalus oophagus
Osteocephalus taurinus
Phyllomedusa bicolor
Phyllomedusa tomopterna
Phyllomedusa vaillanti
Phrynohyas coriacea
Phrynohyas resinifictrix
Phrynohyas venulosa
Scinax boesemani (A)
Scinax cruentommus (A)
Scinax proboscideus
Scinax cf. x-*signatus* (A)
Scinax ruber (N)
Scinax sp. (A)
Leptodactylidae
Adenomera andreae
Adenomera hylaedactyla
Ceratophrys cornuta
Eleutherodactylus chiastonotus
Eleutherodactylus gutturalis (N)
Eleutherodactylus inguinalis
Eleutherodactylus marmoratus
Eleutherodactylus zeuctotylus
Eleutherodactylus sp.
Leptodactylus knudseni
Leptodactylus longirostris (N)
Leptodactylus myersi (N)
Leptodactylus mystaceus
Leptodactylus pentadactylus
Leptodactylus petersii (A)
Leptodactylus rhodomystax
Leptodactylus stenodema
Leptodactylus wagneri
Lithodytes lineatus (A)
Physalaemus petersi
Microhylidae
Chiasmocleis shudikarensis
Hamptophryne boliviana (A)
Pipidae
Pipa pipa (A)
Ranidae
Rana palmipes
SAURIA
Gekkonidae
Coleodactylus amazonicus
Gonatodes annularis
Gonatodes humeralis
Pseudogonatodes guianensis (A)
Thecadactylus rapicauda
Gymnophthalmidae
Alopoglossus angulatus (A)
Arthrosaura kockii
Bachia flavescens
Iphisa elegans
Leposoma guianense
Neusticurus bicarinatus (N)
Neusticurus rudis (N)
Prionodactylus argulus
Tretioscincus agilis
Iguanidae
*Iguana iguana (*A*)*
Polychrotidae
Anolis fuscoauratus
Anolis nitens
Anolis ortonii
Anolis punctatus
Scincidae
Mabuya nigropunctata
Teiidae
Ameiva ameiva
Kentropyx calcarata
Tupinambis teguixin
Tropiduridae
Tropidurus plica
Tropidurus umbra
Uracentron azureum
Uranoscodon superciliosus (*A*)

Bongers *et al. (*eds.), Nouragues. Dynamics and plant-animal interactions in a neotropical rainforest, 371–379.
2001 Kluwer Academic Publishers. Printed in the Netherlands.

AMPHISBAENIA
Amphisbaenidae
Amphisbaena fuliginosa (A)
Amphisbaena alba (A)
SERPENTES
Boidae
Boa constrictor
Corallus caninus (A)
Corallus hortulanus
Epicrates cenchria (A)
Eunectes murinus (A)
Colubridae
Atractus badius
Atractus flammigerus (N)
Atractus schach (A)
Chironius carinatus (N)
Chironius fuscus
Chironius multiventris (A)
Chironius scurrulus (A)
Clelia clelia
Dendrophidion dendrophis (N)
Drymarchon corais
Drymobius rhombifer (A)
Drymoluber dichrous (A)
Erythrolamprus aesculapii
Helicops angulatus
Imantodes lentiferus (N)
Leptodeira annulata
Leptophis ahaetulla (A)
Liophis breviceps (N)
Liophis reginae (A)
Liophis typhlus (N)
Oxybelis aeneus
Oxybelis fulgidus (N)
Philodryas viridissimus (N)
Pseustes sulphureus (A)
Rhinobotryum lentiginosum (N)
Siphlophis cervinus (N)
Taeniophallus brevirostris
Tantilla melanocephala (A)
Tripanurgos compressus
Xenodon rhabdocephalus (N)
Xenopholis scalaris (A)
Xenoxybelis argenteus
Elapidae
Micrurus collaris (N)
Micrurus lemniscatus (A)
Micrurus psyches
Micrurus surinamensis (A)
Viperidae
Bothrops atrox
Bothrops bilineatus (N)
Bothrops brazili (N)
Lachesis muta
TESTUDINES
Chelidae
Platemys platycephala
Phrynops gibbus (A)
Phrynops nasutus
Emydidae
Rhinoclemmys punctularia
Kinosternidae
Kinosternon scorpioides (A)
Pelomedusidae
Podocnemis cayennensis (A)
Testudinidae
Geochelone denticulata
CROCODYLIA
Alligatoridae

Note: (N) = species only present at the Nouragues Station; (A) = species only present at the Arataye Station.

B. Ecological information and occurrence of 231 amphibian and reptile species at five sites in French Guiana

(NOU = Nouragues Field Station; A = Arataye Field Station; RES = Nouragues Nature Reserve (total of Nou + A); PS = Petit Saut; TS = Trois Sauts; PSE = Piste St. Elie) and at one site in Brazil (BRA = INPA-WWF Reserves, near Manaus). Data taken from Born 1996a and 1996b, unpublished data fieldwork 1991, 1994, 1998 (N) and 1991 (PSE), Hoogmoed & Avila Pires 1991 (PS), de Massary 1999 (PS), Vidal unpublished data fieldwork 1994-1995 (PS), Starace 1998 (PS), Lescure 1982 (TS), Chippaux 1986 (TS&PS), Gaucher unpublished data 1995-1998 (A), Zimmerman & Rodrigues 1990 (BRA). Ecological information completed with data from Chippaux 1986, Duellman 1990 and Starace 1998. ND = No data available. Total sites = Total number of sites where a species is present. Abbreviations in table are explained under NOTES.

	ECOLOGICAL SUMMARY				*ABSENCE / PRESENCE PER SITE*							
FAMILY & SPECIES	Habitat	Diel Activity	Breeding Habitat	Reproductive System	NOU	A	(RES)	PS	PSE	TS	BRA	Total sites
ANURA												
Allophrynidae												
Allophryne ruthveni	B,PF,S	N	SP,P	AQ E & T	+	+	(+)	-	-	+	-	3
Bufonidae												
Atelopus flavescens	G,PF	D	S	AQ E & T	+	-	(+)	-	-	-	-	1
Atelopus franciscus	G,PF	D	S	AQ E & T	+	+	(+)	+	+	-	-	4
Atelopus spumarius	G,PF,S	D	S	AQ E & T	-	-	(-)	-	-	+	+	2
Bufo dapsilis	G,PF	N,D	SP	AQ E & T	-	-	(-)	-	-	-	+	1
Bufo guttatus	G,PF	N,D	S	AQ E & T	+	+	(+)	+	-	+	-	4
Bufo margaritifera	G,PF	D	SP,P,RB	AQ E & T	+	+	(+)	+	+	+	-	5
Bufo marinus	G,RB	N	SP,S,RB	AQ E & T	+	+	(+)	+	+	-	-	4
Bufo sp.	G,PF	D	SP,P	AQ E & T	+	+	(+)	+	+	+	-	5
Dendrophryniscus minutus	G,S	D	SP,S	AQ E & T	-	-	(-)	-	-	+	+	2
Centrolenidae												
Cochranella oyampiensis	B,Tr,S	N	S,R	ARB E & AQ T	+	+	(+)	+	-	+	+	5
Hyalinobatrachium taylori	B,Tr,S	N	S,R	ARB E & AQ T	-	+	(+)	+	-	+	-	3
Hyalinobatrachium nouragensis	B,Tr,S	N	S,R	ARB E & AQ T	+	+	(+)	-	-	-	-	2
Dendrobatidae												
Colostethus baeobatrachus	G,PF	D	G,L	TER E, T carried to water	-	+	(+)	-	-	-	-	1
Colostethus beebei	G,PF	D	G,L	TER E, T carried to water	+	+	(+)	+	+	+	-	5
Colostethus degranvillei	G,S	D	G,S	TER E, T carried to water	+	+	(+)	+	+	-	-	4
Colostethus marchesianus	G,PF	D	P,SP	TER E, T carried to water	-	-	(-)	-	-	-	+	1
Colostethus sp. nov.	G,PF	D	G,L	TER E, T carried to water	-	-	(-)	-	-	-	+	1
Colostethus sp. 1	G,S	D	S	TER E, T carried to water	-	-	(-)	-	-	+	-	1
Dendrobates tinctorius	G,C,PF	D	G,L,C, TH,Brom	TER E, T carried to water	+	+	(+)	-	-	+	-	3
Dendrobates ventrimaculatus	B,Tr,C,PF	D	Brom	ARB E, T carried to water	+	+	(+)	-	-	+	-	3
Epipedobates femoralis	G,PF,C	D	G,L,C, TH,Brom	TER/ARB E, T carried to water	+	+	(+)	+	+	+	+	6
Epipedobates hahneli	G,PF,C,FE	D	TH, Brom	TER E, T carried to water	+	+	(+)	-	+	+	-	4
Hylidae												
Hyla boans	Tr,B,S	N	S,R	AQ E & T in constructed basin	+	+	(+)	+	+	+	+	6
Hyla brevifrons	Tr,PF	N	P	ARB E & AQ T	-	+	(+)	-	-	+	+	3
Hyla calcarata	B,Tr	N	SP,P	AQ E & T	-	+	(+)	-	-	+	-	2
Hyla dentei	Tr	N	?	?	-	-	(-)	-	-	+	-	1

	ECOLOGICAL SUMMARY				***ABSENCE / PRESENCE PER SITE***							
FAMILY & SPECIES	Habitat	Diel Activity	Breeding Habitat	Reproductive System	NOU	A	(RES)	PS	PSE	TS	BRA	Total sites
Hyla fasciata	B,Tr	N	SP,SW	AQ E & T	+	+	(+)	+	-	+	-	4
Hyla geographica	B,Tr,S	N	S,R	AQ E & T	+	+	(+)	-	+	+	+	5
Hyla granosa	B,Tr,PF,S	N	SP,S,SW	AQ E & T	+	+	(+)	-	+	+	+	5
Hyla hadroceps	Tr,PF	N,D	TH	AQ E & T in water filled treehole	+	+	(+)	-	+	-	-	3
Hyla leucophyllata	B,Tr,PF,C	N	P in C	ARB E & AQ T	+	+	(+)	+	+	+	-	5
Hyla melanargyrea	B,Tr,PF	N	P	AQ E & T	-	+	(+)	+	-	-	-	2
Hyla minuta	B,Tr,PF	N	P	AQ E & T	-	-	(-)	+	-	-	+	2
Hyla multifasciata	B,Tr,PF	N	SP.P,SW	AQ E & T	-	+	(+)	-	+	+	-	3
Hyla ornatissima	B,Tr,PF,S	N	SP,P,SW	?	-	+	(+)	-	-	+	-	2
Hyla sp. 1	Tr,B,PF	N	P,SW	ARB E & AQ T	+	+	(+)	-	+	-	-	3
Hyla sp. 2	Tr	N	TH	?	-	-	(-)	-	-	+	-	1
Hyla sp. 3	Tr	N	?	?	-	-	(-)	-	-	+	-	1
Hyla sp. 4	Tr	N	P	AQ E & T	-	-	(-)	-	-	+	-	1
Hyla sp. 5	B,Tr,S	N	P	AQ E & T	-	-	(-)	-	-	-	+	1
Osteocephalus cabrerai	Tr,B,PF,S	N	TH?,P?	AQ E & T in water-filled treehole ?	+	+	(+)	-	+	+	+	5
Osteocephalus leprieurii	Tr,B,PF,SF	N	P	AQ E & T	-	+	(+)	-	-	+	-	2
Osteocephalus oophagus	Tr,B,PF	N	Brom	AQ E & T in water-filled bromeliad	+	+	(+)	+	-	-	-	3
Osteocephalus taurinus	Tr,B,PF	N	P	AQ E & T	+	+	(+)	+	+	+	+	6
Osteocephalus sp. 1	Tr,B,PF	N	T	AQ E & T ?	-	-	(-)	-	-	+	-	1
Osteocephalus sp. 2	Tr,S	N	SP,S ?	AQ E & T	-	-	(-)	-	-	-	+	1
Phyllomedusa bicolor	Tr,B,PF	N	P	ARB E & AQ T	+	+	(+)	+	+	+	+	6
Phyllomedusa tarsius	Tr,PF	N	P	ARB E & AQ T	-	-	(-)	-	-	-	+	1
Phyllomedusa tomopterna	B,PF	N	P	ARB E & AQ T	+	+	(+)	+	+	+	+	6
Phyllomedusa vaillanti	Tr,B,PF	N	P	ARB E & AQ T	+	+	(+)	+	+	+	+	6
Phrynohyas coriacea	Tr,PF	N	P	AQ E & T	+	+	(+)	-	-	-	+	3
Phrynohyas resinifictrix	Tr,PF	N	TH	AQ E & T in water-filled treehole	+	+	(+)	-	+	+	+	5
Phrynohyas venulosa	Tr,PF,SF	N	SP,P	AQ E & T	+	+	(+)	-	-	+	-	3
Scinax boesemani	Tr,PF	N	P	AQ E & T	-	+	(+)	+	+	+	-	4
Scinax cruentommus	Tr,B,PF	N	P	AQ E & T	-	+	(+)	-	-	-	+	2
Scinax proboscideus	B,Tr,PF	N	SP,P	AQ E & T	+	+	(+)	+	-	+	-	4
Scinax cf. x-signatus	B,Tr,PF	N	P	AQ E & T	-	+	(+)	+	-	-	-	2
Scinax ruber	G,B,T	N	P	AQ E & T	+	-	(+)	+	+	+	-	4
Scinax sp. nov.	PF	N	?	?	-	-	(-)	+	-	-	-	1
Scinax sp. 1	Tr,B,PF	N	SP,P	AQ E & T	-	+	(+)	-	-	-	-	1
Leptodactylidae												
Adelophryne gutturosa	G,PF	N,D	G,L	?	-	-	(-)	-	-	+	-	1
Adenomera andreae	G,PF	D,N	G,L	E & T in terrestrial claybuilt FM	+	+	(+)	+	-	+	+	5
Adenomera hylaedactyla	G,PF	N,D	G,L	E & T in terrestrial claybuilt FM	+	+	(+)	-	+	+	-	4
Ceratophrys cornuta	G,PF	N,D	P	AQ E & T	+	+	(+)	-	-	+	+	4
Eleutherodactylus chiastonotus	G,PF,C	N,D	G,L	TER E, DD into frog	+	+	(+)	+	+	+	-	5
Eleutherodactylus fenestratus	G,B,PF	N,D	G,L	TER E, DD into frog	-	-	(-)	-	-	-	+	1
Eleutherodactylus gutturalis	G,PF	N,D	G,L	TER E, DD into frog	+	-	(+)	-	-	+	-	2
Eleutherodactylus inguinalis	Tr,PF	N	G,L ?	TER E, DD into frog (probably)	+	+	(+)	+	+	+	-	5
Eleutherodactylus marmoratus	B,PF	N	G,L ?	TER E, DD into frog	+	+	(+)	+	+	+	-	5
Eleutherodactylus zeuctotylus	G,RO,PF	N	G,L ?	TER E, DD into frog	+	+	(+)	+	+	+	-	5
Eleutherodactylus sp. 1	G,PF	N	G,L	TER E, DD into frog	-	-	(-)	-	-	+	-	1
Eleutherodactylus sp. 2	G,B,PF	N	G,L	TER E, DD into frog	-	-	(-)	-	-	-	+	1

	ECOLOGICAL SUMMARY				***ABSENCE / PRESENCE PER SITE***							
FAMILY & SPECIES	Habitat	Diel Activity	Breeding Habitat	Reproductive System	NOU	A	(RES)	PS	PSE	TS	BRA	Total sites
Eleutherodactylus sp. 3	B,PF	N	G,L	TER E, DD into frog (probably)	+	+	(+)	-	-	-	-	2
Eleutherodactylus sp. 4	G,PF	N	G,L	TER E, DD into frog	-	-	(-)	-	-	-	+	1
Leptodactylus fuscus	G,FE	N	P	E in FM near water, AQ T	-	-	(-)	+	+	-	-	2
Leptodactylus knudseni	G,PF,SUB	N	P	E in FM near water, AQ T	+	+	(+)	+	+	+	+	6
Leptodactylus longirostris	G,PF,FE	D,N	P	E in FM, near water, AQ T	+	-	(+)	-	+	-	-	2
Leptodactylus myersi	G,PF,FE	N	P	E in FM near/on water, AQ T	+	-	(+)	-	-	+	-	2
Leptodactylus mystaceus	G,PF,FE	N,D	P,SW	E in FM near water, AQ T	+	+	(+)	+	+	+	+	6
Leptodactylus pentadactylus	G,PF,SUB	N	P	E & T in FM near water	+	+	(+)	-	+	+	+	5
Leptodactylus petersii	G,PF,SUB	N	SP,P,SW	E in FM near/on water, AQ T	-	+	(+)	-	-	-	-	1
Leptodactylus rhodomystax	G,PF	N	SP,P,SW	E in FM near/on water, AQ T	+	+	(+)	+	+	+	+	6
Leptodactylus riveroi	G,S,SUB	N	SP	E in FM near/on water, AQ T	-	-	(-)	-	-	-	+	1
Leptodactylus stenodema	G,PF,FE	N	P	E & T in FM near/on water	+	+	(+)	-	-	+	+	4
Leptodactylus cf. wagneri	G,PF,S	N	SP,S,SW	E in FM near/on water, AQ T	+	+	(+)	+	-	+	+	5
Lithodytes lineatus	G,PF,SUB	N	?	?	-	+	(+)	-	-	+	+	3
Physalaemus petersi	G,PF	N,D	P	E in FM near/on water, AQ T	+	+	(+)	-	-	+	-	3
Microhylidae												
Chiasmocleis hudsoni	G,PF,SUB	N	P	AQ E & T	-	-	(-)	-	-	+	-	1
Chiasmocleis shudikarensis	G,PF,SUB	N,D	SP,P	AQ E & T	+	+	(+)	+	-	+	+	5
Chiasmoleis sp. 1	G,PF	N	P	AQ E & T	-	-	(-)	-	-	-	+	1
Ctenophryne geayi	G, PF	N	P	AQ E & T	-	-	(-)	+	-	-	+	2
Hamptophryne boliviana	G,PF,SUB	N	P	AQ E & T	-	+	(+)	+	-	+	-	3
Otophryne pyburni	G,PF,SUB	D	S	AQ E & T	+	+	(+)	-	-	+	-	3
Synapturanus mirandaribeiroi	G,PF,SUB	N,D	G,L	E & T in the ground	-	-	(-)	-	-	+	+	2
Synapturanus salseri	G,PF,SUB	N,D		E & T in the ground	-	-	(-)	-	-	-	+	1
Pipidae												
Pipa pipa	AQ,PF,P,SP	N,D	SP,P	E embedded on dorsum	-	+	(+)	-	-	+	-	2
Pipa arrabali	AQ,PF,P,SP	N,D	SP,P	E embedded on dorsum	-	-	(-)	-	-	-	+	1
Ranidae												
Rana palmipes	R,RB,S	N,D	R,RB,S	AQ E & T	+	+	(+)	-	-	+	-	3
SAURIA												
Gekkonidae												
Coleodactylus amazonicus	G,PF,SH	D	L,BDW	O	+	+	(+)	+	+	ND	+	5
Gonatodes annularis	BA,PF,SH	D	L,BDW	O	+	+	(+)	+	+	ND	-	4
Gonatodes humeralis	BA,PF,SH	D	L,BDW	O	+	+	(+)	+	+	ND	+	5
Lepidoblepharis heyerorum	G,PF,S,SH	D,N	L,BDW	O	-	-	(-)	+	-	ND	-	1
Pseudogonatodes guianensis	G,PF,S,SH	D	L,BDW	O	-	+	(+)	-	-	ND	-	1
Thecadactylus rapicauda	BA,PF,SH	N	L,BDW	O	+	+	(+)	+	+	ND	+	5
Gymnophthalmidae												
Alopoglossus angulatus	G,PF,FE,S	D	L,BDW	O	-	+	(+)	+	-	ND	+	3
Arthrosaura kockii	G,PF,S,SH	D	L,BDW	O	+	+	(+)	+	+	ND	-	4
Arthrosaura reticulata	G,PF,S,SH	D	L,BDW	O	-	-	(-)	-	-	ND	+	1
Bachia flavescens	G,PF,S,FOS	D,N	L,BDW	O	+	+	(+)	-	+	ND	+	4
Bachia panoplia	G,PF,S,FOS	D	L,BDW	O	-	-	(-)	-	-	ND	+	1
Iphisa elegans	G,PF,S,FOS	D	L,BDW	O	+	+	(+)	+	-	ND	-	3

FAMILY & SPECIES	Habitat	Diel Activity	Breeding Habitat	Reproductive System	NOU	A	(RES)	PS	PSE	TS	BRA	Total sites
	ECOLOGICAL SUMMARY				*ABSENCE / PRESENCE PER SITE*							
Leposoma guianense	G,PF,SH	D	L,BDW	O	+	+	(+)	+	+	ND	+	5
Leposoma percarinatum	G,PF,S,SH	D	L,BDW	O	-	-	(-)	-	-	ND	+	1
Neusticurus bicarinatus	G,PF,S,SAQ	D	L,BDW	O	+	-	(+)	+	+	ND	+	4
Neusticurus rudis	G,PF,S,SAQ	D	L,BDW	O	+	-	(+)	+	-	ND	-	2
Prionodactylus argulus	BA,PF,C	D	L,BDW	O	+	+	(+)	+	-	ND	-	3
Tretioscincus agilis	BA,FE,C	D	L,BDW	O	+	+	(+)	+	-	ND	+	4
Iguanidae												
Iguana iguana	Tr,FE,S	D	SA	O	-	+	(+)	-	-	ND	+	2
Polychrotidae												
Anolis fuscoauratus	Tr,PF,FE,SH	D	L,BDW	O	+	+	(+)	+	-	ND	+	4
Anolis nitens	G,BA,PF,SH	D	L,BDW	O	+	+	(+)	+	+	ND	+	5
Anolis ortonii	Tr,FE,C	D	L,BDW	O	+	+	(+)	-	-	ND	+	3
Anolis philopunctatus	Tr,PF,SH	D	L,BDW	O	-	-	(-)	-	-	ND	+	1
Anolis punctatus	Tr,PF,SH	D	L,BDW	O	+	+	(+)	+	+	ND	-	4
Polychrus marmoratus	Tr,B,FE,S	D	L,BDW?	O	-	-	(-)	-	+	ND	-	1
Scincidae												
Mabuya nigropunctata	G,FE,C	D	L	V	+	+	(+)	+	+	ND	+	5
Teiidae												
Ameiva ameiva	G,FE,C	D	L,BDW	O	+	+	(+)	+	+	ND	+	5
Cnemidophorus lemniscatus	G,FE,C,BE	D	L,BDW	O	-	-	(-)	-	+	ND	-	1
Kentropyx calcarata	G,PF,FE,C	D	L,BDW	O	+	+	(+)	+	+	ND	+	5
Tupinambis teguixin	G,FE,C,S	D	TN	O	+	+	(+)	+	+	ND	+	5
Tropiduridae												
Tropidurus plica	Tr,PF,FE,SH	D	BDW	O	+	+	(+)	+	+	ND	+	5
Tropidurus umbra	Tr,PF,SH	D	BDW	O	+	+	(+)	+	-	ND	+	4
Uracentron azureum	Tr,FE,CA	D	BDW	O	+	+	(+)	+	-	ND	+	4
Uranoscodon superciliosus	Tr,PF,FE,S	D	BDW	O	-	+	(+)	+	+	ND	+	4
AMPHISBAENIA												
Amphisbaenidae												
Amphisbaena fuliginosa	G,PF,SF,FOS	D	L	O	-	+	(+)	-	+	ND	-	2
Amphisbaena alba	G,PF,SF,FOS	D	L	O	-	+	(+)	-	-	ND	-	1
SERPENTES												
Aniliidae												
Anilius scytale	G,PF,SF,FOS	N,D	L	V	-	-	(-)	+	+	+	+	4
Boidae												
Boa constrictor	G,B,PF,SF, SAV	N,D	G	V	+	+	(+)	+	+	+	+	6
Corallus caninus	Tr,B,PF	N	?	V	-	+	(+)	+	+	+	+	5
Corallus hortulanus	Tr,B,G,PF,SF	N	?	V	+	-	(+)	+	+	+	+	5
Epicrates cenchria	G,SF,SAQ	N,D	?	V	-	+	(+)	+	-	-	+	3
Eunectes murinus	G,B,PF,S, SAQ,S,P	N	?	V	-	+	(+)	+	-	+	-	3
Colubridae												
Apostolepis pymi	G,PF,FOS	D,N	L	O	-	-	(-)	-	-	-	+	1
Apostolepis sp. 1	G,PF,FOS	D,N?	L	O	-	-	(-)	-	-	-	+	1
Atractus badius	G,PF,SF,FOS	N,D	L?	O	+	+	(+)	+	+	+	-	5
Atractus flammigerus	G,PF,FOS	N,D	L?	O	+	-	(+)	+	-	+	-	3
Atractus latifrons	G,PF	N,D	L?	O	-	-	(-)	-	-	-	+	1
Atractus major	G,PF,FOS	N,D	L?	O	-	-	(-)	-	-	-	+	1
Atractus schach	G,PF,SF,SAV	?	L?	O	-	+	(+)	+	-	-	-	2
Atractus snethlagae	G,FOS	?	L?	O	-	-	(-)	-	-	-	+	1
Atractus torquatus	G,PF,SF,FOS	N,D	L?	O	-	-	(-)	+	-	-	+	2
Atractus zidoki	G,PF	?	?	O	-	-	(-)	+	-	+	-	2
Atractus sp. 1	G,FOS	?	L?	O	-	-	(-)	-	-	-	+	1
Chironius carinatus	G,B,PF,SF	D	L	O	+	-	(+)	+	+	+	+	5

FAMILY & SPECIES	ECOLOGICAL SUMMARY				ABSENCE / PRESENCE PER SITE							
	Habitat	Diel Activity	Breeding Habitat	Reproductive System	NOU	A	(RES)	PS	PSE	TS	BRA	Total sites
Chironius cinnamomeus	G,T	D	?	O	-	-	(-)	-	-	-	+	1
Chironius fuscus	G,B,PF	D	L	O	+	+	(+)	+	+	+	+	6
Chironius multiventris	G,B,PF	D	L	O	-	+	(+)	+	+	-	-	3
Chironius scurrulus	G,B,PF,SF	D	L	O	-	+	(+)	+	-	+	+	4
Clelia Clelia	G,PF,SF,SAV	N,D	L?	O	+	+	(+)	+	-	+	+	5
Dendrophidion dendrophis	Tr,B,PF	D	?	O	+	-	(+)	+	-	-	+	3
Dipsas catesbyi	G,B,PF	N	L?	O	-	-	(-)	+	-	+	+	3
Dipsas indica	G,B,PF	N	L?	O	-	-	(-)	+	-	+	+	3
Dipsas pavonina	G,B,PF	N	?	O	-	-	(-)	+	-	+	-	2
Dipsas variegata	G,B,PF,SF	N	?	O	-	-	(-)	+	-	+	-	2
Drepanoides anomalus	?	?	?	?	-	-	(-)	+	-	-	-	1
Drymarchon corais	G,B,SF,SAV	D	L?	O	+	+	(+)	+	-	-	+	4
Drymobius rhombifer	G,B,PF,SF	D	L	O	-	+	(+)	+	-	-	-	2
Drymoluber dichrous	G,B,PF	D	L?	O	-	+	(+)	+	+	+	+	5
Erythrolamprus aesculapii	G,PF	D	L?	O	+	+	(+)	+	-	+	+	5
Helicops angulatus	PF,SF,SAQ,S,P	N,D	L,S	O/V	+	+	(+)	+	+	+	+	6
Helicops hagmanni	PF,S,P	N,D	L,S	O/V	-	-	(-)	-	-	-	+	1
Hydrodynastes bicinctus	PF,SAQ,S,P	D	L,S	O	-	-	(-)	+	-	-	-	1
Hydrops triangularis	G,PF,S,P	?	L?	O	-	-	(-)	-	-	-	+	1
Imantodes cenchoa	B,PF	N	?	O	-	-	(-)	+	+	+	+	4
Imantodes lentiferus	Tr,B,PF	N	?	O	+	-	(+)	-	-	+	-	2
Leptodeira annulata	G,PF,SF, SAV, SAQ	N	L?	O	+	+	(+)	+	+	+	+	6
Leptophis ahaetulla	G,B,PF,SF, SAV	D	B	O	-	+	(+)	+	+	+	+	5
Liophis breviceps	G,PF,S,P	D	L?	O	+	-	(+)	+	-	+	+	4
Liophis cobella	G,PF,SF, SAV,S,P	N,D	L?	O	-	-	(-)	+	+	-	+	3
Liophis lineatus	G,SAV	D	L?	O	-	-	(-)	-	+	-	-	1
Liophis miliaris	G,PF,SF,S,P	N,D	L?	O	-	-	(-)	+	+	+	+	4
Liophis poecilogyrus	G	D	L?	O	-	-	(-)	-	-	-	+	1
Liophis reginae	G,PF,SF	D	L?	O	-	+	(+)	+	+	+	-	4
Liophis typhlus	G,PF,SF	D	L?	O	+	-	(+)	+	-	+	+	4
Mastigodryas boddaerti	G,PF,SAV	D	L?	O	-	-	(-)	-	+	-	+	2
Oxybelis aeneus	G,B,PF,SF, SAV	D	?	O	+	+	(+)	+	+	-	+	5
Oxybelis fulgidus	G,B,PF,SF, SAV	D	?	O	+	-	(+)	+	-	+	+	4
Oxyrhopus formosus	G,PF	N,D	?	O	-	-	(-)	+	-	-	+	2
Oxyrhopus petola	G,B,PF,SF, SAV	N,D	?	O	-	-	(-)	+	-	-	+	2
Oxyrhopus trigeminus	G	N	?	O	-	-	(-)	-	-	-	+	1
Philodryas viridissimus	G,B,Tr,PF,SF	D	?	O	+	-	(+)	+	-	-	+	3
Phimophis guianensis	G,SAV,FOS	D	L ?	O	-	-	(-)	+	-	-	-	1
Pseudoboa coronata	G,B,PF,SF, SAV	N	?	O	-	-	(-)	+	-	+	-	2
Pseudoboa neuwiedii	G,SF,SAV,	N	L ?	O	-	-	(-)	+	-	-	-	1
Pseudoeryx plicatilis	G,PF,SF, SAV,P,S	N,D	L	O	-	-	(-)	+	-	+	-	2
Pseustes poecilonotus	G,B,PF,SF	D	?	O	-	-	(-)	+	-	-	+	2
Pseustes sulphureus	G,B,PF,SF	D	?	O	-	+	(+)	+	+	+	+	5
Rhinobotryum lentiginosum	G,PF,Tr,B	N	L?	O	+	-	(+)	+	-	-	+	3
Sibon nebulata	G,B,PF,SF, SAV	N	L ?	O	-	-	(-)	+	-	-	-	1
Siphlophis cervinus	G,B,PF,SF	N	?	O	+	-	(+)	+	-	-	+	3
Spilotes pullatus	G,SF,SAV	D	L?	O	-	-	(-)	+	+	-	+	3
Taeniophallus brevirostris	G,PF	D	L?	O	+	+	(+)	+	+	+	+	6
Tantilla melanocephala	G,PF,SF,FOS	D	L?	O	-	+	(+)	+	+	+	+	5
Tripanurgos compressus	G,B,PF	N	?	O	+	+	(+)	+	-	+	+	5
Waglerophis merremii	G,SF,SAV, SAQ	D	L?	O	-	-	(-)	+	-	-	-	1

FAMILY & SPECIES	Habitat	Diel Activity	Breeding Habitat	Reproductive System	NOU	A	(RES)	PS	PSE	TS	BRA	Total sites
	ECOLOGICAL SUMMARY				ABSENCE / PRESENCE PER SITE							
Xenodon rhabdocephalus	G,PF,SF	N,D	?	O	+	-	(+)	+	-	+	+	4
Xenodon severus	G,PF,SF	N,D	?	O	-	-	(-)	+	+	-	+	3
Xenodon werneri	G,PF,SF,P,S	N,D	?	O	-	-	(-)	-	-	+	-	1
Xenopholis scalaris	G,L,PF	D	?	O	-	+	(+)	+	-	-	-	2
Xenoxybelis argenteus	G,B,PF,	D	?	O	+	+	(+)	+	+	+	+	6
Elapidae												
Micrurus averyi	G,FOS	N,D	L?	O	-	-	(-)	-	-	-	+	1
Micrurus collaris	G,PF,FOS	N	L?	O	+	-	(+)	-	-	-	-	1
Micrurus hemprichii	G,PF,SF,FOS	N	L?	O	-	-	(-)	+	-	-	+	2
Micrurus lemniscatus	G,PF,SF	N,D	L?	O	-	+	(+)	+	+	+	+	5
Micrurus psyches	G,PF,SF,FOS	N,D	L?	O	+	+	(+)	+	+	-	-	4
Micrurus spixii	G,FOS	?	L?	O	-	-	(-)	-	-	-	+	1
Micrurus surinamensis	G,PF,SF, SAV,S,P	N,D	L?	O	-	+	(+)	-	+	-	+	3
Leptotyphlopidae												
Leptotyphlops septemstriatus	G,PF,FOS	?	L	O	-	-	(-)	-	-	+	+	2
Leptotyphlops tenellus	G,B,PF,FOS	?	L	O	-	-	(-)	-	-	+	+	2
Typhlopidae												
Typhlops reticulates	PF,SF,SAV, G,FOS	N	L	O	-	-	(-)	+	-	ND	-	1
Viperidae												
Bothrops atrox	G,PF,SF,SAV	N,D	L	V	+	+	(+)	+	+	+	+	6
Bothrops bilineatus	Tr,B,G,PF,SF	N	L?	V	+	-	(+)	+	+	+	+	4
Bothrops brazili	G,PF,SF,S*	N,D	L	V	+	-	(+)	+	+	+	-	4
Lachesis muta	G,PF,SF, S*,FOS	N	L?	O	+	+	(+)	+	-	+	+	5
TESTUDINES												
Chelidae												
Platemys platycephala	PF,SF,FE,P,S	D,N	L	O	+	+	(+)	+	+	ND	ND	4
Phrynops gibbus	PF,SF,S,P,M	N	L	O	-	+	(+)	-	-	ND	ND	1
Phrynops nasutus	PF,SF,S,P	N,D	L?	O	+	+	(+)	-	-	ND	ND	2
Emydidae												
Rhinoclemmys punctularia	PF,SF,S,P,IA	D,N	L	O	+	+	(+)	-	-	ND	ND	2
Kinosternidae												
Kinosternon scorpioides	G,PF,SF,FE,S, P	D,N	L?	O	-	+	(+)	-	+	ND	ND	2
Pelomedusidae												
Podocnemis cayennensis	R,IA	D	RB	O	-	+	(+)	-	-	ND	ND	1
Testudinidae												
Geochelone denticulata	G.,F,SF,SH	D	L	O	+	+	(+)	-	+	ND	ND	3
CROCODYLIA												
Alligatoridae												
Paleosuchus trigonatus	PF,S	N,D	C,TN	O	+	+	(+)	+	+	ND	ND	4

NOTES:
HABITAT
AQ = Aquatic; B = Bushes and shrubs, low vegetation; BA = On the bark of trees, between buttresses at lower heights; C= Canopy gaps; CA = Canopy; FE = Forest edge; FOS = Fossorial; G = Ground; IA = inundated area; P = Pond, stagnant (lentic) water; PF = Primary forest; R = River; RB = Riverbank; RO = Rock boulders and caves; S = Streams and criques (running water); SAV = Savannas; SAQ = Semi aquatic; SF = Secondary forest; SH = Shade; SUB = Subterranean, in holes between roots and buttresses; Tr = Trees; * = occasionally.

DIEL ACTIVITY
D = Diurnal; D,N = mainly diurnal, occasionally nocturnal; N = Nocturnal; N,D = Mainly nocturnal, occasionally diurnal.

BREEDING HABITAT
BDW = Bark of a tree or decaying wood; Brom = Bromeliads; G = Ground; L = Litter layer; P = Pond, stagnant (lentic) water; R = River; RB = Riverbank; S = Streams and criques (running water); SA = Sand; SP = Small pond with stagnant (lentic) water in the inundated edge of a stream or river; SW = Swamp forest; TH = Water filled tree hole in tree or log; TN = Termite nest.

REPRODUCTIVE SYSTEM
AQ = Aquatic; ARB E = Arboreal eggs laid on leaves overhanging water, hatching tadpoles drop into water; ARB E, T carried to water = Eggs laid on bromeliads and tadpoles subsequently transported on the back of a parent to a water filled bromeliad axil; DD = Direct development, no free-living stage of a tadpole; E = Eggs; FM = Foam mass: eggs are deposited in a 'nest' consisted of foam, produced by the male by kicking air, mucus, sperm and eggs as they are emitted during the amplexus; O = Oviparous or egg laying; T = Tadpoles; TER = Terrestrial; TER E, T carried to water = Eggs laid on the litter surface, tadpoles subsequently transported on the back of a parent to a water body; V = Viviparous, gives birth to live young.

The red howling monkey *Alouatta seniculus*. *Photo Pierre Charles-Dominique.*

APPENDIX 6

FISHES OF THE ARATAYE RIVER

François J. Meunier and Thierry Boujard

Data are available on the distribution of fishes for eight river basins in French Guiana, the five longer rivers: Maroni (520 km, 65830 km^2), Mana (430 km, 12090 km^2), Oyapock (370 km, 26820 km^2), Approuague (270 km, 10250 km^2) and Sinnamary (262 km, 6565 km^2) and the three shorter rivers: Iracoubo (140 km, 1470 km^2), Kourou (112 km, 2000 km^2) and Comté (100km, 2215 km^2). The Kourou river limits two ichtyogeographical regions, one in the west with Maroni, Mana, Iracoubo and Sinnamary, the second in the east with Comté, Approuague and Oyapock.

In the list we used the classification of Nelson (1994). We have characterized the distribution of each species according to its presence in the various rivers of French Guiana (see Planquette *et al.* 1996, Keith *et al.* 2000, Le Bail *et al.* 2000):

++ = species that live at least in all the main rivers in the west ans in the east of the country;

\+ = species that live in the two geographical regions but that are unknown in one or two main rivers;

e = species that live only in the eastern region;

ee = species that are known to occur in the Approuague basin only (endemic species). Six species are endemic species of the Arataye river (and of Approuague), two Characiformes and four Siluriformes.

List of families and species of Arataye river

CHARACIFORMES

Characidae

Acestrorhynchus falcatus (Bloch, 1794) ++
Acestrorhynchus guianensis Menezes, 1969 ++
Astyanax abramoides Eigenmann, 1909 +
Astyanax bimaculatus (Linné, 1758) ++
Astyanax keithi Géry, Planquette & Le Bail, 1996 +
Astyanax leopoldi Géry, Planquette & Le Bail, 1988 e
Astyanax meunieri Géry, Planquette & Le Bail, 1996 +
Bryconamericus aff. stramineus Eigenmann, 1909 ++
Bryconops affinis (Günther, 1864) ++
Bryconops aff. caudomaculatus (Günther, 1869) e
Bryconops melanurus (Bloch, 1795) +

Bongers *et al. (*eds.), Nouragues. Dynamics and plant-animal interactions in a neotropical rainforest, 381–384.
2001 Kluwer Academic Publishers. Printed in the Netherlands.

Charax pauciradiatus Gunther, 1864 [++]
Creagrutus planquettei Géry et Renno, 1989 [ee]
Hemibrycon surinamensis Géry, 1962 [+]
Hemigrammus guyanensis Géry, 1959 [+]
Hemigrammus rodwayi Durbin, 1909 [++]
Hemigrammus unilineatus cayennensis Géry, 1959 [++]
Hyphessobrycon copelandi Durbin *in* Eigenmann, 1908 [+]
Moenkhausia chrysargyrea (Günther, 1864) [++]
Moenkhausia colettii (Steindachner, 1882) [++]
Moenkhausia georgiae Géry, 1966 [++]
Moenkhausia aff. grandisquamis (Müller & Troschel, 1845) [ee]
Moenkhausia inrai Géry, 1992 [+]
Moenkhausia aff. intermedia Eigenmann, 1908 [+]
Moenkhausia oligolepis (Günther, 1864) [++]
Moenkhausia surinamensis Géry, 1965 [+]
Phenacogaster aff. megalostictus Eigenmann, 1909 [++]
Poptella brevispina Reis, 1989 [++]
Thayeria ifati Géry, 1959

Characidiidae
Characidium fasciadorsale Fowler, 1914 [++]
Melanocharacidium cf. blennioides (Eigenmann, 1909) [+]

Curimatidae
Chilodus zunevei Puyo, 1945 [+]
Curimata cyprinoides (Linné, 1758) [++]
Cyphocharax helleri (Steindachner, 1910) [++]
Cyphocharax sp. (gr. spilurus) after Géry *et al.*, 1991 [+]

Erythrinidae
Erythrinus erythrinus (Schneider, 1801) [++]
Hoplerythrinus unitaeniatus (Spix, 1829) [++]
Hoplias aimara (Valenciennes, 1846) [++]
Hoplias malabaricus (Bloch, 1794) [++]

Gasteropelecidae
Gasteropelecus sternicla (Linné, 1758) [++]

Hemiodidae
Bivibranchia bimaculata Vari, 1985 [+]
Hemiodopsis quadrimaculatus (Pellegrin, 1908) [+]
Hemiodus aff. unimaculatus (Bloch, 1794) [+]
Parodon guyanensis Géry, 1959 [+]

Lebiasinidae
Copella carsevennensis (Regan, 1912) [++]
Nannostomus bifasciatus Hoedeman, 1954 [+]
Pyrrhulina filamentosa Val. *in* Cuv. & val., 1846 [++]

Serrasalmidae
Myleus rhomboidalis (Cuvier, 1818) [+]
Myleus rubripinnis (Val. *in* Cuv. & Val., 1849) [+]
Myleus ternetzi (Norman, 1929) [++]

Serrasalmus humeralis Val. *in* Cuv. & Val., 1849 [+]

SILURIFORMES

Ageneiosidae

Ageneiosus brevifilis Cuvier & Valenciennes, 1857 [+]

Doradidae

Doras carinatuus (Linné, 1766) [+]

Auchenipteridae

Auchenipterus nuchalis (Spix, 1829) [++]

Glanidium leopardum (Hoedeman, 1961) [+]

Parauchenipterus galeatus (Linné, 1766) [++]

Tatia cf intermedia (Steindachner,1876) [+]

Aspredinidae

Dysichthys coracoideus amaurus Eigenmann, 1912 [+]

Callichthyidae

Corydoras aeneus (Gill, 1858) [+]

Corydoras approuaguensis Nijssen & Isbrucker, 1983 [ee]

Corydoras amapaensis Nijssen, 1971 [+]

Corydoras geoffroy Lacépède, 1803 [+]

Corydoras guianensis Nijssen, 1970 [+]

Corydoras spilurus Norman, 1926[++]

Cetopsidae

Hemicetopsis sp. after Boujard et al., 1990b [+]

Helogeneidae

Helogenes marmoratus Günther, 1863 [++]

Loricariidae

Ancistrus cf. temminckii (Valenciennes, 1840) [e]

Ancistrus cf. leucostictus (Günther, 1864) [+]

Farlowella reticulata Boeseman, 1971 [+]

Hartia surinamensis Boeseman, 1971 [+]

Hypostomus gymnorhynchus (Norman, 1926) [++]

Lasiancistrus brevispinis Heitmans, Nijssen & Isbrücker, 1983[+]

Lithoxus boujardi (Muller & Isbrücker, 1993) [e]

Coricaria parnahybae 1907 [+]

Loricaria parnahybae Steindachner, 1907 [+]

Rineloricaria stewarti Eigenmann, 1910 [++]

Pimelodidae

Heptapterus longior (Eigenmann, 1912) [+]

Hepapterus tenuis Mees, 1986 [+]

Imparfinis minutus (Lütken, 1874) [+]

Microglanis poecilius Eigenmann, 1912

Pimelodella cristata (Müller et Troschel, 1848) [++]

Pimelodella geryi Hodeman*, 1961* [++]

Pimelodella magalops Eigenmann, 1912[ee]

Pimelodus ornatus Kner, 1858 [++]

Pseudopimelodus raninus raninus (Valenciennes, 1840) [+]

Pseudoplatystoma fasciatum (Linné, 1766) [+]

Rhamdella leptosoma Fowler, 1914 [+]
Rhamdia quelen (Quoy & Gaimard, 1824) [++]
Trichomycteridae
Ituglanis sp [ee]

GYMNOTIFORMES
Apteronotidae
Apteronotus aff. albifrons (Linné, 1766) [+]
Gymnotidae
Gymnotus anguillaris Hoedeman, 1962 [++]
Gymnotus carapo Linné, 1758 [++]
Hypopomidae
Brachyhypopomus beebei (Schultz, 1944) [+]
Hypopomus artedi (Kaup, 1856) [+]
Hypopygus lepturus Hoedeman, 1962 [+]
Sternopygidae
Eigenmannia virescens (Valenciennes, 1847) [++]
Eigenmannia n. sp. after Le Bail *et al.*, 1984 [+]
Sternopygus macrurus (Bloch & Schneider, 1801) [++]

CYPRINODONTIFORMES
Cyprinodontidae
Rivulus agilae Hoedeman, 1954 [++]
Rivulus igneus Huber, 1991 [++]
Rivulus lungi Berkenkamp, 1984 [++]
Rivulus xiphidius Huber, 1979 [+]

SYNBRANCHIFORMES
Synbranchidae
Synbranchus marmoratus [++]

PERCIFORMES
Cichlidae
Aequidens tetramerus (Heckel, 1840) [+]
Apistogramma gossei Kullander, 1982 [e]
Crenicichla albopunctata Pellegrin, 1904 [+]
Crenicichla johanna Heckel, 1840 [e]
Crenicichla saxatilis (Linné, 1758) [++]
Geophagus camopiensis Pellegrin, 1903 [e]
Guyanacara geayi (Pellegrin, 1902) [e]
Krobia aff. guianensis Kullander & Nijssen, 1989 [+]
Satanoperca aff. jurupari Kullander & Nijssen, 1989 [+]
Sciaenidae
Pachypops fourcroi (Lacepède, 1802) [+]
Plagioscion squamosissimus (Heckel, 1840)[++]

REFERENCES

REFERENCES

Ackerly, D.D. (1996) Canopy structure and dynamics: integration of growth processes in tropical pioneer trees, in S.S. Mulkey, R.L. Chazdon, and A.P. Smith (eds.), *Tropical forest plant ecophysiology*, Chapman and Hall, New York, pp. 619–658.

Ackerman, J.D., Sabat, A. and Zimmerman, J.K. (1996) Seedling establishment in an epiphytic orchid: an experimental study of seed limitation, *Oecologia* 106, 192–198.

Adler, G.H. (1995) Fruit and seed exploitation by central american spiny rats, *Proechimys semispinosus,Studies on Neotropical Fauna and Environment* 30, 237–244.

Adler, G.H. and Kestell, D.W. (1998) Fates of neotropical tree seeds influenced by spiny rats (*Proechimys semispinosus*), *Biotropica* 30, 677–681.

Alexandre, D.Y. (1989) *Dynamique de la régénération naturelle en forêt dense de Côte d'Ivoire*, ORSTOM, Etudes et Thèses.

Altmann, J. (1974) Observational study of behavior: sampling methods, *Behaviour* 49, 227–267.

Andersson, C. and Frost, I. (1996) Growth of *Quercus robur* seedlings after experimental grazing and cotyledon removal, *Acta Botanica Neerlandica* 45, 85–94.

Andresen, H. (1999) Seed dispersal by monkeys and the fate of dispersed seeds in a Peruvian rainforest, *Biotropica* 31, 145–158.

Asquith, N.M., Wright, S.J. and Clauss, M.J. (1997) Does mammal community composition control recruitment in neotropical forests? Evidence from Panama, *Ecology* 78, 941–946.

Atramentowicz, M. (1986) Dynamique de population chez trois Marsupiaux didelphidés de Guyane,*Biotropica* 18, 136–149.

Atramentowicz, M. (1988) La frugivorie opportuniste de trois marsupiaux didelphidés de Guyane,*Rev. Ecol. - Terre Vie* 43, 47–57.

Augspurger, C.K. (1984) Seedling survival of tropical tree species: interactions of dispersal distance, light-gaps and pathogens, *Ecology* 65, 1705–1712.

August, P.V. (1984) Population ecology of small mammals in the Llanos of Venezuela, *Special Publ. Mus. Texas Technical Univ.* 22, 71–1.

Avila-Pires, T.C.S. (1995) Lizards of Brazilian Amazonia (Reptilia: Squamata). *Zool. Verhandelingen Leiden* 299.

Baer, B. (1995) *Die Nahrungsökologie des Rothandtamarins* (*Saguinus midas midas*) *im primären Regenwald Guyanas*, Diplomarbeit. Zoologisches Institut der Universität Zürich.

Balda, R.P. and Kamil, A.C. (1992) Long-term spatial memory in Clarck's nutcracker, *Nucifraga columbiana*, *Anim. Behav.* 44, 761-769.

Balée, W. and Campbell, D.G. (1990) Evidence for the successional status of liana forest (Xingu River Basin, Amazonian Brazil), *Biotropica* 22, 36–47.

Balslev, H. and Renner, S.S. (1989) Diversity of east Ecuadorean lowland forests, in L.B. Holm-Nielsen, I.C. Nielsen, and H. Balslev (eds.), *Tropical forests, botanical dynamics, speciation and diversity,* Academic Press, London, San Diego, New York, pp 287–295.

Bariteau, M. (1993) *La régénération naturelle avant et après exploitation sur le dispositif d'expérimentation sylvicole de Paracou en Guyane française.* PhD. thesis, Université Pierre et Marie Curie, Paris, France.

Barkley, C.L. and Jacobs, L.F. (1998) Visual environment and delay affect cache retrieval accuracy in a food-storing rodent, *Anim. Learn. Behav.* 26, 439–447.

Barnard, C.J. (1983) *Animal Behaviour*, Wiley Inter-Science, New York.

Barnett, R.J. (1977) The effect of burial by squirrels on germination and survival of oak and hickory nuts,*Am. Mid. Nat.* 98, 319-330.

Barthlott, W., Gröger, A. and Porembski, S. (1993) Some remarks on the vegetation of tropical inselbergs: diversity and ecological differenciation, *Biogeographica* 69, 105–124.

Barton, R.A. (1992) Allometry of food intake in free-ranging anthropoid primates, *Folia Primatol.* 58, 56–59.

Bates, H. W. (1863) *The Naturalist on the River Amazon.* Dent, London.

Bayart, F. and Pagès, E. (1998) Ecological and social correlates of intergroup encounters in *Saguinus midas* (Callitrichidae) in French Guiana, *Folia Primatol.* 69, 42.

Beadle, N.C.W. (1968) Some aspects of the ecology and physiology of australian xeromorphic Plants, *Aust. J. Sci.* 30, 348–355.

Bongers *et al.* (eds.), Nouragues. Dynamics and plant-animal interactions in a neotropical rainforest, 385-416.
2001 Kluwer Academic Publishers. Printed in the Netherlands.

Beck-King, H., von Helversen, O. and Beck-King, R. (1999) Population density and food resource utilization of *Agouti paca* (rodentia, Agoutidae) using alternative methods, in a pacific lowland forest in Costa Rica, *Biotropica* 31, 675–685.

Behling, H. (1996) First report on new evidence for the occurrence of *Podocarpus* and possible human presence at the mouth of the Amazon during the Late-glacial, *Veg. Hist. Archaeobot.* 5, 241–246.

Bell, A.D. (1991) *Plant Form, an illustrated guide to flowering plant morphology*, Oxford University Press, Oxford.

Bergeson, D.J. (1996) *The positional behavior and prehensile tail use of Alouatta palliata, Ateles geoffroyi, and Cebus capucinus.* Ph.D. Dissertation, Washington University, St Louis.

Bertram, B.C.R. (1978) Living in groups: predators and prey, in J.R. Krebs and N.B. Davis (eds.),*Behavioural ecology: an evolutionary approach*, Oxford, Blackwell Science Publications.

Besag, J. and Diggle, P.J. (1977) Simple Monte Carlo tests for spatial pattern, *Applied Statistics* 26, 327–333.

Bierregaard, R.O. Jr. (1990) Avian communities in the understory of Amazonian forest fragments, in A. Keast (ed.) *Biogeography and ecology of forest bird communities,* SPB Academic Publishing bv. The Hague, The Netherlands, pp. 333–343.

Bigarella, J.J. and Ferreira, A.M.M. (1985) Amazonian geology and the pleistocene and the cenozoic environments and paleoclimates, in G.T. Prance and T.E. Lovejoy (eds.), *Amazonia*, Pergamon Press, Oxford, pp 49–71.

Bisbal, F.J. (1986) Food habits of some neotropical carnivores in Venezuela (Mammalia, Carnivora),*Mammalia*, 50, 329–339.

Blake, J.G., Stiles, F.G. and Loiselle, B.A. (1990) Birds of La Selva biological station: Habitat use, trophic composition, and migrants, in A.H. Gentry (ed.) *Four neotropical rainforests*, Yale University Press, New Haven, Conn, pp. 161–182.

Blanc, P. (1989) *Biologie des plantes de sous-bois tropicaux*, Thèse de Doctorat d'Etat, Université Paris VI, 395 p.

Blanc, P. (1997) La biodiversité tropicale: mémoire des changements passés,*Lettre PIGB-PMRC-France* 6, 30–39.

Bodmer, R.E. (1989) *Frugivory in Amazon ungulates*, Ph.D. Thesis, University of Cambridge.

Bodmer, R.E. (1990a) Fruit patch size and frugivory in the lowland tapir (*Tapirus terrestris*),*J. Zool. London* 219, 457–467.

Bodmer, R.E. (1990b) Responses of ungulates to seasonal inundations in the amazon floodplain,*J. Trop. Ecol.* 6, 191–201.

Bodmer, R.E. (1991) Strategies of seed dispersal and seed predation in amazonian ungulates,*Biotropica* 23, 255–261.

Boggan, J., Funk, V., Kelloff, C., Hoff., M., Cremers, G. and Feuillet, C. (1997)*Checklist of the Plants of the Guianas (Guyana, Surinam, French Guiana),* 2nd edition, Biological Diversity of the Guianas Program, Smithsonian Institution, Washington, D.C.

Bonaccorso, F.J. (1979) Foraging and reproductive ecology in a Panamanian bat community.*Bull. Fla. State Mus. Biol. Sci.* 24, 359–408.

Bonaccorso, F.J., Glanz, W.E. and Sandford, C.M. (1980) Feeding assemblages of mammals at fruiting*Dipteryx panamensis* (Papilionaceae) trees in Panama: seed predation, dispersal, and parasitism, *Rev. Biol. Trop.* 28, 61–72.

Bongers, F. (1998) Manipulation of light in tropical rain forest, in Tropenbos. *Research in tropical rain forests: its challenges for the future.* Seminar Proceedings, 25-26 November (1997), The Tropenbos Foundation, Wageningen, The Netherlands,Pp 169-184.

Bongers, F. and Popma, J. (1988) *Trees and gaps in a Mexican tropical rain forest. Species differentiation in relation to gap-associated environmental heterogeneity.* Ph.D. thesis, University of Utrecht.

Bongers, F. and Sterck, F.J. (1998) Architecture and development of rainforest trees: responses to light variation, in D.M. Newbery, H.H.T. Prins, and N. Brown (eds.), *Dynamics of Tropical Communities*, Blackwell Scientific Publications, Oxford, pp. 126–162.

Bongers, F., Charles-Dominique, P., Forget, P.M., and Théry, M. (2001)*Nouragues: Dynamics and Plant-animal Interactions in a Neotropical Rainforest,* Kluwer Academic Publisher, Biological Monographs Series, Dordrecht, The Netherlands.

Bongers, F., Engelen, D., and Klinge, H. (1985) Phytomass structure of natural plant communities on Spodosols in Southern Venezuela: the Bana woodland, *Vegetatio* 63, 13–34.

Born, M. (1996a) *Reptiles and amphibians of Nouragues, French Guiana*, Born, Amsterdam.

Born, M. (1996b) Species diversity in a Neotropical rainforest: amphibians of Nouragues, a nature reserve in French Guiana, *Bos Nieuwsletter* 15, 25–33.

Bornhardt, W. (1900) *Zur Oberflächengestaltung und Geologie Deutsch-Ostafrikas*, Reimer, Berlin.

Boucher, D.H. (1981) Seed predation by mammals and forest dominance by *Quercus oleoides*, a tropical lowland oak, *Oecologia* 49, 409–414.

Boujard, T. (1992) Space-time organization of riverine fish communities in french Guiana, *Environ. Biol. Fish.* 34, 235–246.

Boujard, T. and Meunier, F.J. (1991) Croissance de l'épine pectorale, histologie osseuse et dimorphisme sexuel chez l'atipa, *Hoplosternum littorale* Hancock, 1828 (Callichthyidae, Siluriforme), *Cybium* 15, 55–68.

Boujard, T. and Rojas-Beltran, R. (1988a) Zonation longitudinale du peuplement ichtyque du fleuve Sinnamary (Guyane Française), *Rev. Hydrobiol. Trop.* 21, 47–61.

Boujard, T. and Rojas-Beltran, R. (1988b) Description des captures au filet tramail sur le cours supérieur du Sinnamary (Guyane Française), *Rev. Hydrobiol. Trop.* 21, 349–356.

Boujard, T., Le Bail, P.-Y. and Planquette, P. (1988) Données biologiques sur quelques espèces continentales de Guyane française d'intérêt piscicole, *Aquat. Living Resour.* 1, 107–113.

Boujard, T., Lecomte, F., Renno, J.-F., Meunier, F.J. and Neveu, P. (1991) Growth in four populations of *Leporinus friderici* (Bloch, 1794) (Anostomidae, Teleostei) in French Guiana, *J. Fish Biol.* 38, 387–397.

Boujard, T., Meunier, F.J., Pascal, M. and Cosson, J.F. (1990a) Les Téléostéens d'un haut bassin fluvial guyanais, l'Arataye. I- Inventaire des Characoïdes, *Cybium* 14, 175–182.

Boujard, T., Meunier, F.J., Pascal, M. and Cosson, J.F. (1990b) Les Téléostéens d'un haut bassin fluvial guyanais, l'Arataye. II- Inventaire des non-Characoïdes, *Cybium* 14, 345–351.

Boujard, T., Pascal, M. and Meunier, F.J. (1990c) Microrépartition spatio-temporelle du peuplement ichtyologique d'un haut bassin fluvial de Guyane: l'Arataye. *Rev. Ecol.- Terre Vie* 45, 357–373.

Boujard, T., Pascal, M., Meunier, F.J. and Le Bail, P.-Y. (1997) *Poissons de Guyane. Guide écologique de l'Approuague et de la réserve des Nouragues.* Ed. INRA, Paris.

Boujard, T., Sabatier, D., Rojas-Beltran, R., Prévost, M.F. and Renno, J.F. (1990d) The food habits of three allochthonous feeding characoids in French Guiana, *Rev. Ecol.- Terre Vie* 45, 247–258.

Bourgeois, J. and Jeanjou, D. (1981) Subduction océanique, subduction continentale et surrection andine: l'exemple du Pérou septentrional, *C.R. Acad. Sci., Paris* 243, 859–864.

Boyé, M., Cabaussel, G. and Perrot, Y. (1979) *Atlas des départements Français d'Outre-Mer. La Guyane.* Ed. CNRS-ORSTOM, Paris, 76 p.

Boyé, M., Cabaussel, G. and Perrot, Y. (1979) Climatologie, in *Atlas de la Guyane*, Collection des Atlas des Départements d'Outre Mer, II. La Guyane. CNRS and ORSTOM, Paris.

Barnea, A., Yom-Tov, Y. and Friedman, J. (1990) Differential germination of two related species of *Solanum* in response to bird ingestion, *Oikos* 57, 222–228.

Bradbury, J.W. (1981) The evolution of leks, in R.D. Alexander and D.W. Tinkle (eds.), *Natural selection and social behavior: recent research and new theory*, Chiron, New York, pp. 138–169.

Branan, W.V., Werkhoven, M.C.M. and Marchinton, R.L. (1985) Food habits of brocket and white-tailed deer in Suriname, *J. Wildlife Manage.* 49, 972–976.

Braun-Blanquet, J. (1964) *Pflanzensoziologie*, 3 Aufl., Springer Verlag, Wien.

Brawn, J.D., Karr, J.K. and Nichols, J.D. (1995) Demography of birds in a neotropical forest: effects of allometry, taxonomy, and ecology, *Ecology* 76, 41–51.

Bremer, H. and Jennings, J. (1978) Inselbergs/Inselberge, *Z. Geomorphol. N. F.* Suppl. 31.

Brewer, S.W. and Rejmánek, M. (1999) Small rodents are significant dispersers of tree seeds in a Neotropical forest, *J. Veg. Sci.* 10, 165–174.

Brokaw, N.V.L. (1982) The definition of treefall gap and its effect on measures of forest dynamics, *Biotropica* 14, 158–160.

Brokaw, N.V.L. (1985) Treefalls, regrowth and community structure in tropical forests, in S. Pickett and P.S. White (eds.) *Ecology of natural disturbance and patch dynamics*, Academic Press, San Diego, London, Boston, New York, Sydney, Tokyo & Toronto, pp. 53–71.

Brosset, A. (1969) La vie sociale des oiseaux dans une forêt équatoriale du Gabon, *Biol. Gabonica* 5, 29–69.

Brosset, A. (1986) Chiroptères Rhinolophoidae de l'ancien monde et Phyllostomatoidae d'Amérique du Sud: un exemple d'évolutions divergentes. In Vertébrés et forêts tropicales humides d'Afrique et d'Amérique, *Mem. Mus. Natl. Hist. Nat.* Ser. A Zool. 132, 146–150.

Brosset, A. (1996) Role of the sibling species in the dynamics of the forest-bird communities in M'Passa (northeastern Gabon), in M.L. Cody (ed.) *Long-term studies of vertebrate communities*, Academic Press, San Diego, London, Boston, New York, Sydney, Tokyo & Toronto, pp. 251–289.

Brosset, A. and Charles-Dominique, P. (1990) The bats from French Guiana: a taxonomic, faunistic and ecological approach, *Mammalia* 54, 509–560.

Brosset, A. and Gaucher, Ph. (1997) Colonisation des bâtiments du Centre Spatial de Kourou (Guyane) par les Chiroptères, *Mammalia* 61, 274–276.

Brosset, A., Charles-Dominique, P., Cockle A. and J.F. Cosson (1996). Bat communities and deforestation in French Guiana, *Can. J. Zool.* 74, 1874–1982.

Brosset, A., Cosson, J.F., Gaucher, P. and Masson, D. (1995) Les chiroptères d'un marécage côtier de Guyane : composition du peuplement, *Mammalia* 60, 527–535.

Brunig, E.F. (1974) *Ecological studies in the kerangas forests of Sarawak and Brunei,* Borneo Literature Bureau, Jalan Tun Haji Openg, Kuching, Sarawak, Malaysia.

Buchanan, D.B., Mittermeier, R.A. and van Roosmalen, M.G.M. (1981) The saki monkeys, genus *Pithecia*, in A.F. Coimbra-Filho and R.A. Mittermeier (eds.), *Ecology and Behavior of Neotropical Primates*, vol.1, Academia Brasileira de Cîencias, Rio de Janeiro, pp. 391–417.

Büdel, B., Becker, U., Porembski, S. and Barthlott, W. (1997) Cyanobacteria and cyanobacterial lichens from inselbergs of the Ivory Coast, Africa, *Bot. Acta* 110, 458–465.

Büdel, B., Lüttge, U., Stelzer, R., Huber, O. and Medina, E. (1994) Cyanobacteria of rocks and soils in the Orinoco region and in the Guyana highlands, Venezuela, *Bot. Acta* 107, 422–431.

Burnham, K.P., Anderson, D.R. and Laake, J.L. (1980) Estimation of density from line transect sampling of biological populations, *Wildlife Monogr.* 72, 1–202.

Buskirk, W.H. (1976) Social systems in a tropical forest avifauna, *Am. Nat.* 110, 293–310.

Cant, J.G.H. (1986) Locomotion and feeding postures of spider and howling monkeys: Field study and evolutionary interpretation, *Folia primatol.* 46,1–14.

Cant, J.G.H. (1987) Positional behavior of female Bornean orangutans (*Pongo pygmaeus*), *Am. J. Primatol.* 12, 71–90.

Cant, J.G.H. (1992) Positional behavior and body size of arboreal primates: A theoretical framework and an illustration of its application, *Am. J. Phys. Anthropol.* 88, 273–283.

Carey, E.V., Brown, S., Gillespie, A.J.R. and Lugo, A.E. (1994) Tree mortality in mature lowland tropical moist and lower montane forests of Venezuela, *Biotropica* 26, 255–265.

Carvalho, F.M. (1980) Alimentaçao do mapara (*Hypophthalmus edentatus* Spix, 1829) do lago Castanho, Amazonas (Siluriformes, Hypohthalmidae), *Acta Amazonica* 10, 545–555.

Casal, J.J. and Smith, H. (1989) The function, action and adaptive significance of phytochrome in light-grown plants, *Plant Cell Environment* 12, 855–862.

Catzeflis, F., Richard-Hansen, C., Fournier-Chambrillon, C., Lavergne, A. and Vié, J.-C. (1997) Biométrie, reproduction et sympatrie chez *Didelphis marsupialis* et *D. albiventris* en Guyane française (Didelphidae: Marsupialia) *Mammalia* 61, 231–243.

Chambers, J.C. (1995) Disturbance, life history strategies, and seed fates in alpine herbfield communities, *Am. J. Botany* 82, 421–433.

Chapman, C.A. (1989) Primate seed dispersal: The fate of dispersed seeds, *Biotropica* 21, 148–154.

Charles-Dominique, P. (1986) Inter-relation between frugivorous vertebrates and pioneer plants: *Cecropia* birds and bats in French Guiana, in A. Estrada and T.H. Fleming (eds.), *Frugivores and seed dispersal.* Dr W. Junk Publ., Dordrecht, pp. 119–135.

Charles-Dominique, P. (1991) Feeding strategies and activity budget of the frugivorous bat *Carollia perspicillata* (Phyllostomidae) in French Guiana, *J. Trop. Ecol.* 7, 243–256.

Charles-Dominique, P. (1993) Speciation and coevolution: an interpretation of frugivory phenomena, in T.H. Fleming and A. Estrada (eds.), *Frugivores and seed dispersal : ecological and evolutionary aspects*, *Vegetatio* 107/108, 75–84.

Charles-Dominique, P. (1995) Interactions plantes-animaux frugivores, conséquences sur la dissémination des graines et la régénération forestière, *Rev. Ecol.- Terre Vie* 50, 223–235.

Charles-Dominique, P. and Servant, M. (2000) Reconstitution des perturbations anciennes d'une forêt humide de Guyane par l'étude des dépôts sédimentaires et l'analyse des graines de plantes pionnières d'une carotte, in M. Servant and S. Servant-Vildary (eds.), *Dynamique à long terme des écosystèmes forestiers tropicaux* Symp. Internat., Paris 20-22 mars 1996, IRD, MAB-UNESCO, CNRS, 63–69.

Charles-Dominique, P., Absy, M.L., Larpin, D., Ledru, M.P., Riéra, B., Sarthou, C., Servant, M., Tardy, C. and Turcq, B. (2000) Incidents climatiques et modifications de l'écosystème forestier depuis 3000 ans en Guyane française, in M. Servant and S. Servant-Vildary (eds.), *Dynamique à long terme des écosystèmes forestiers tropicaux* Symp. Internat., Paris 20-22 mars 1996, IRD, MAB-UNESCO, CNRS, 71–76.

Charles-Dominique, P., Atramentowicz, M., Charles-Dominique, M., Gérard, H., Hladik, A., Hladik, C. M. and Prévost, M. F. (1981) Les mammifères frugivores arboricoles nocturnes d'une forêt guyanaise: inter-relations plantes-animaux, *Rev. Ecol. - Terre Vie* 35, 341–435.

Charles-Dominique, P., Blanc, P., Larpin, D., Ledru, M.P., Riéra, B., Sarthou, C., Servant, M. and Tardy, C. (1998) Forest perturbations and biodiversity during the last ten thousand years in French Guiana, *Acta Oecologica* 19, 295–302.

Chattaway, M.M. (1968) Bud development and lignotuber formation in Eucalypts, *Aust. J. Sci.*30, 103–115.

Chauvel, A., Grimaldi, M. and Tessier, D. (1991) Changes in soil pore-space distribution following deforestation and revegetation: An example from the Central Amazon Basin, Brazil, *Forest Ecol. Manag.* 38, 259–271.

Chauvet, S. (1997) *Facteurs écologiques de la régénération des arbres en forêt tropicale. Effets majeurs des rongeurs sur la survie des graines de deux espèces de Sapotacées en Guyane française,* DEA de lUniversité Paris VI.

Chazdon, R.L. (1988) Sunflecks and their importance to forest understorey plants, *Adv. Ecol. Res.* 18: 1–63.

Chazdon, R.L. (1992) Photosynthetic plasticity of two rain forest shrubs across natural gap transects,*Oecologia* 92, 586–595.

Chazdon, R.L. and Fetcher, N. (1984a) Light environments of tropical forests, in: E. Medina, H.A. Mooney and C. Vasquez-Yanes (eds.), *Physiological Ecology of Plants of the Wet Tropics*, Junk, The Hague, pp 27-35.

Chazdon, R.L. and Fetcher, N. (1984b) Photosynthetic light environments in a low land tropical rain forest in Costa Rica, *J. Ecol.*72, 553–564.

Chazdon, R.L., Pearcy, R. W., Lee, D.W. and Fetcher, N. (1996) Photosynthetic responses of tropical plants to contrasting light environments, in S.S. Mulkey, R.L. Chazdon and A.P. Smith (eds.),*Tropical Forest Plant Ecophysiology*, Chapman and Hall, New York, pp 5–55.

Chippaux, J.P. (1986) *Les serpents de la Guyane française*. Faune Tropicale 27, ORSTOM, Paris.

Chivers, D.J. and Hladik, C.M. (1980) Morphology of the gastrointestinal tract in Primates: comparisons with other mammals in relation to diet, *J. Morphol.* 166, 337–386.

Choubert, B. (1974) *Les Précambriens des Guyanes*, Mémoires, B.R.G.M., 81.

Cimprich, D.A. and Grubb, T.C.R. (1994) Consequences for Carolina chikadees of foraging with Tufted Titmice in winter, *Ecology* 75, 1615–1625.

Clark, D.A. and Clark, D.B. (1992) Life history diversity of canopy and emergent trees in a Neotropical rain forest, *Ecological Monographs* 62, 315–44.

Clark, D.B. (1990) The role of disturbance in the regeneration ofneotropical moist forests. Pp. 291–315 in Bawa, K.S. and Hadley, M. (eds.), *Reproductive ecology of tropical forest plants.* MAB–UNESCO Series Vol 7. UNESCO, Paris. 421 pp.

Clark, D.B., Clark, D.A., Rich, P.M. Weiss, S. and Oberbauer, F. (1996) Landscape-evaluation of understorey light and canopy structure: methods and application to a neotropical lowland rain forest,*Can. J. Forest Res.* 26, 747–757.

Claussen, J.W. and Maycock, C.R. (1995) Stem allometry in a North Queensland tropical rain forest,*Biotropica* 4, 421–426.

Clutton-Brock, T.H. and Harvey, P.H. (1977) Species differences in feeding and ranging behaviour in Primates, in T.H. Clutton-Brock (ed.), *Primate Ecology: Studies of Feeding and Ranging Behaviour in Lemurs, Monkeys and Apes*, Academic Press, London, pp. 557–584.

Cockle, A. (1997) *Diversité des modalités d'installation des plantes le long de gradients de perturbation en forêt tropicale de Guyane. Dissémination des graines, caractérisation des supports de germination, installation des plantules.* Thèse de l'Université de Paris VI.

Cody, M.L. (1971) Finch flocks in the Mohave Desert, *Theoretical Population Biology* 2, 142–158.

Coleman, J.S., McConnaughay, K.D.M. and Ackerly, D. (1994) Interpreting phenotypic variation in plants,*Trends Ecol. Evol.* 9, 187–191.

Comiskey, J.A., Aymard, G. and Dallmeier, F. (1994) Structure and composition of lowland mixed forest in the Kwakwani region of Guyana, *Biollania* 10,13–28.

Condit, R., Hubbell, S.P., Lafrankie, J.V., Sukumar, R., Manokaran, N., Foster, R.B. and Ashton, P.S. (1996) Species-area and species-individual relationships for tropical trees: a comparison of three 50-ha plots,*J. Ecol.* 84, 549–562.

Connell, J.H. (1971) On the role of natural enemies in preventing competitive exclusion in some marine animals and in rain forest trees, in P.J. den Boer and P.R. Gradwell (eds.), *Dynamics of populations*, PUDOC, Wageningen, pp. 298–312.

Connell, J.H. (1978) Diversity in tropical rainforest and coral reefs, *Science* 199, 1302–1310.

Cooper, H.M., Charles-Dominique, P. and Viénot, F (1986) Signification de la coloration des fruits en fonction de la vision des vertébrés consommateurs, in Vertébrés et forêts tropicales humides d'Afrique et d'Amérique, *Mem. Mus. Natl. Hist. Nat.* Ser. A Zool. 132, 131–143.

Cosson, J.F. (1994) *Dynamique de population et dispersion de la chauve-souris frugivore Carollia perspicillata en Guyane française*. Thèse de Doctorat, Université Paris-Sud XI.

Cosson, J.F., Charles-Dominique, P. and Larpin, D. (1992) Dynamics of a bat community in French Guiana, *Mesogée, Bulletin Mus. Hist. Nat. Marseille* 52, 101.

Cracraft, J. and Prum, R. (1988) Patterns and processes of diversification: Speciation and historical congruence in some tropical birds, *Evolution* 42, 603–620.

Craig, C. L., Weber, R.S. and G. D. Bernard (1996) Evolution of predator-prey systems: spider foraging plasticity in response to the visual ecology of prey, *Am. Nat.* 147, 205–229.

Crockett, C.M. and Eisenberg, J.E. (1987) Howlers: variations in group size and demography, in B.B. Smuts, D.L. Cheney, R.W. Seyfarth, R.W. Wrangham, and T.T. Struhsaker (eds.),*Primate Societies*, University Chicago Press, Chicago, pp. 54–68.

Crump, M.L. (1974) Reproductive strategies in a tropical anuran community,*Univ. Kansas Publ. Mus. Nat. Hist.* 61, 1–68.

Dagosto, M. (1995) Seasonal variation in positional behavior of malagasy lemurs, *Int. J. Primatol.* 16, 807–833.

Damuth, J.E. (1977) Late Quaternary Sedimentation in the Western equatorial Atlantic,*Geol. Soc. Am. Bull.* 99, 695–710.

Damuth, J.E. and Fairbridge, R.W. (1970) Equatorial atlantic deep sea arkosik sands in ice-age aridity in tropical South America, *Geol. Soc. Am. Bull.* 81, 189–206.

Dawkins, H.C. and Field, D.R.B. (1978) *A long-term surveillance system for british woodland vegetation*, Department of Forestry, Oxford University, UK.

Davies, S.J., Palmiotto, P.A., Ashton, P.S., Lee, H.S. and Lafrankie, J.V. (1998) Comparative ecology of 11 sympatric species of Macaranga in Borneo: tree distribution in relation to horizontal and vertical resource heterogeneity, *J. Ecol.* 86, 662–673.

De Foresta, H., Charles-Dominique, P., Erard, C. and Prévost, M.F. (1984) Zoochorie et premiers stades de la régénération naturelle après coupe en forêt guyanaise, *Rev. Ecol.-Terre Vie* 39, 369–400.

De Mercey, P. (1994) *Utilisation de l'espace et activités de l'Agami Psophia crepitans en période de raréfaction alimentaire,* DEA d'Ecologie générale et Production végétale, Université Paris VI.

Dearing, M.D. (1997) The manipulation of plant toxins by a food-hoarding herbivore, *Ochotona princeps,Ecology* 78, 774–781.

Denslow, J.S. (1987) Tropical rainforest gaps and tree species diversity, *Annu. Rev. Ecol. Syst.* 18,431–51.

Descamps, M., Gasc, J.P., Lescure, J. and Sastre, C. (1978) Etude des écosystèmes guyanais. II. Données biogéographiques sur la partie orientale des Guyanes, *C.R. Soc. Biogéogr.* 467, 55–82.

Dickinson, K.J.M., Mark, A.F. and Dawkins, B. (1993) Ecology of lianoid/epiphytic communities in coastal podocarp rain forest, Haast Ecological District, New Zealand, *J. Biogeography* 20, 687–705.

Dinerstein, E. (1991) Seed dispersal by greater one-horned rhinoceros (*Rhinoceros unicornis*) and the flora of Rhinoceros latrines, *Mammalia* 55, 355–362.

Dixon, M.D., Johnson W.C. and Adkisson, C.S. (1997) Effects of weevil larvae on acorn use by blue jays, *Oecologia* 111, 201–208.

Dörrstock S., Porembski S. and Barthlott W. (1996) Ephemeral flush vegetation on inselbergs in the Ivory Coast (West Africa), *Candollea* 51, 407–419.

Dos Reis, N.R. and Guillaumet, J.L. (1983) Les chauves-souris frugivores de la région de Manaus et leur rôle dans la dissémination des espèces végétales, *Rev. Ecol. - Terre Vie* 36, 147–169.

Dowsett-Lemaire, F. (1988) Fruit choice and seed dispersal by birds and mammals in the evergreen forests of upland malawi, *Rev. Ecol.- Terre Vie* 43, 251–281.

Drénou, C. (1988) *Etude de l'architecture d'un arbre guyanais: l'angelique, Dicorynia guianensis Amshoff (Caesalpiniaceae),* DEA de l'Université Montpellier II.

Dubost, G. (1987) Une analyse écologique de deux faunes de mammifères forestiers tropicaux,*Mammalia* 51, 415–436.

Dubost, G. (1988) Ecology and social life of the red acouchy,*Myoprocta exilis*; comparisons with the orange-rumped agouti, *Dasyprocta leporina*, *J. Zool. London* 214, 107–113.

Dubost, G. (1997) Comportements comparés du pécari à lèvres blanches,*Tayassu pecari*, et du pécari à collier, *T. tajacu* (Artiodactyles, Tayassuidés), *Mammalia* 61, 313–343.

Duellman, W.E. (1965) A biogeographic account of the herpetofauna of Michoacàn, Mexico,*Univ. Kansas Publ. Mus. Nat. Hist.* 15, 627–709.

Duellman, W.E. (1990) Herpetofaunas in neotropical rainforests: comparative composition, history and resource use, in A. Gentry (ed.), *Four neotropical rainforests*, Yale Univ. Press, London, pp. 455–505.

Duellman, W.E. and Hoogmoed, M.S. (1992) Some Hylid frogs from the Guiana Highlands, northeastern South America: new species, distributional records, and a generic reallocation,*Occ. Pap. Mus. of Nat. Hist. Univ. of Kansas*, Lawrence 147, 1–21.

Duellman, W.E. and Trueb, L. (1986) *Biology of amphibians*. McGraw-Hill, Inc. New York, pp. 13–47.

Duivenvoorden, J.F. (1994) Vascular plant species counts in the rain forests of the middle Caqueta area, Colombian Amazonia, *Biodivers. Conserv.* 3, 685–715.

Duivenvoorden, J.F. (1996) Patterns of tree species richness in rain forests of the Middle Caqueta area, Colombia, NW Amazonia, *Biotropica* 28, 142–158.

Duivenvoorden, J.F. and Cleef, A.M. (1994) Amazonian savanna vegetation on the sandstone plateau near Araracuara, Colombia, *Phytocoenologia* 24, 197–232.

Duivenvoorden, J.F. and Lips, J.M. (1993) *Ecología del paisaje del Medio Caquetá, Memoria explicativa de los mapas*, Tropenbos-Colombia, Santa-Fé de Bogotà.

Duivenvoorden, J.F. and Lips, J.M. (1995) *A land-ecological study of soils, vegetation, and plant diversity in Colombian Amazonia*, Tropenbos series 12, Tropenbos Foundation, Wageningen, The Netherlands.

Ehleringer, J.R. and Field, C.B. (1993) *Scaling physiological processes: leaf to globe*, Academic Press, Inc. London.

Eigenmann, C.H. (1912) The fresh water fish of British Guiana, including a study of the ecological grouping of species and relation of the fauna of the plateau to that of the lowlands, *Mem. Carnegie Mus.* 5, 1–578.

Eisenberg, J.F. and Thorington, R.W. (1973) A preliminary analysis of a neotropical fauna, *Biotropica* 5, 150–161.

Ek, R.C. (1997) Patterns of species diversity in the neotropical lowland rain forest of Guyana, in R.C. Ek (ed.), *Botanical Diversity in the Tropical Rain Forest of Guyana,* Tropenbos-Guyana series 4, Wageningen, chap. 6, pp. 171–187.

Ek, R.C. and Ter Steege, H. (1997) The Flora of Mabura Hill area, Guyana, in R.C. Ek (ed.),*Botanical Diversity in the Tropical Rain Forest of Guyana*, Tropenbos-Guyana series 4, Wageningen, chap. 2, pp. 27–64.

Ek, R.C. and Van der Hout, P. (1997) Botanical diversity of Greenhart dominated mixed rainforest near Mabura Hill, Guyana, in R.C. Ek (ed.), *Botanical Diversity in the Tropical Rain Forest of Guyana*, Tropenbos-Guyana series 4, Wageningen, chap. 4, pp. 91–142.

Ekman, J. (1986) Tree use and predator vulnerability of wintering passerines, *Ornis Scandinavia* 17, 261–267.

Ellsworth, D.S. and Reich, P.B. (1993) Canopy structure and vertical patterns of photosynthesis and related leaf traits in a deciduous forest, *Oecologia* 96, 169–178.

Emlen, S.T. and Vehrencamp, S.L. (1985) Cooperative breeding strategies among birds, in B. Hölldobler and M. Lindauer (eds.) *Experimental behavioural Ecology and Sociobiology*, G. Fisher Verlag, Stuttgart, pp. 359–379.

Emmons, L.H. (1982) Ecology of *Proechimys* (Rodentia, Echimyidae) in south-eastern Peru, *J. Trop. Ecol.* 23, 280–290.

Emmons, L.H. (1984) Geographic variation in densities and diversities of non-flying mammals in Amazonia, *Biotropica* 16, 210–222.

Emmons, L.H. and Feer, F. (1997) Neotropical rainforest mammals, a field guide. Second edition. Chicago University Press.

Emperaire, L. (1987) *Végétation et gestion des ressources naturelles dans la Caatinga du sud-est du Piaui (Brésil),* Thèse d'Etat, Université P. et M. Curie (Paris 6), Paris.

Endler, J.A. (1978) A predator's view of animal color patterns, *Evol. Biol.* 11, 319–364.

Endler, J.A. (1984) Progressive background matching in moths, and a quantitative measure of crypsis,*Biological J. Linn. Soc. London* 22, 187–231.

Endler, J.A. (1986). Defense against predation, in M.E. Feder and G.V. Lauder (eds.), *Predator-prey relationships, perspectives and approaches from the study of lower vertebrates,* University of Chicago Press, Chicago, pp. 109–134.

Endler, J.A. (1990). On the measurement and classification of color in studies of animal color patterns,*Biol. J. Linn. Soc. London* 41, 315–352.

Endler, J.A. (1991) Variation in the appearance of guppy color patterns to guppies and their predators under different visual conditions, *Vision Res.* 31, 587–608.

Endler, J.A. (1992) Signals, signal conditions, and the direction of evolution, *Am. Nat.* 139, S125–S153.

Endler, J.A. (1993) The color of light in forests and its implications, *Ecol. Monogr.* 63, 1–27.

Endler, J.A. (1995) Multiple-trait coevolution and environmental gradients in guppies,*Trends Ecol. Evol.* 10, 22–29.

Endler, J.A. (1997) Light, behavior and conservation of forest-dwelling organisms, in J.R. Clemmons and R. Buchholz (eds.), *Behavioral Approaches to Conservation in the Wild,* Cambridge University Press, Cambridge, pp. 330–356.

Endler, J.A. and Théry, M. (1996) Interacting effects of lek placement, display behavior, ambient light and color patterns in three Neotropical forest-dwelling birds, *Am. Nat.* 148, 421–452.

Ennos, A.R. (1996) Wind as an ecological factor, *Trends Ecol. Evol.* 12, 108–111.

Erard, C. (1986) Richesse spécifique de deux peuplements d'oiseaux forestiers équatoriaux: une comparaison Gabon-Guyane, *Mém. Mus. Natl. Hist. Nat. (A)* 132, 53–66.

Erard, C. (1989) Bird community structure in two rainforests: Africa (Gabon) and South America (French Guiana): A comparison, in M.L. Harmelin-Vivien and F. Bourlière (eds.), *Vertebrates in complex tropical systems.* Ecological studies, vol. 69, Springer-Verlag, New York, Berlin, Heidelberg, London, Paris, Tokyo, pp. 89–122.

Erard, C. and Brosset, A. (in press) Oiseaux: structure et fonctionnement des peuplements, in M. Lamotte and J. Roy (eds.), *La faune du mont Nimba (Libéria, Guinée et Côte d'Ivoire). Mém. Mus. Natl. Hist. Nat.*, Paris.

Erard, C. and Sabatier, D. (1989) Rôle des oiseaux frugivores terrestres dans la dynamique forestière en Guyane française, *Proceed. Intern. Ornithol. Congr.* 19, 803–815.

Erard, C. and Théry, M. (1994) Frugivorie et ornithochorie en forêt guyanaise: l'exemple des grands oiseaux terrestres et de la Pénélope marail, *Alauda* 62, 27–31.

Erard, C., Théry, M. and Sabatier, D. (1989) Régime alimentaire de *Rupicola rupicola* (Cotingidae) en Guyane française. Relations avec la frugivorie et la zoochorie, *Rev. Ecol.-Terre Vie* 44, 47–74.

Erard, C., Théry, M. and Sabatier, D. (1991) Régime alimentaire de *Tinamus major* (Tinamiidae), *Crax alector* (Cracidae) et *Psophia crepitans* (Psophiidae), grands oiseaux terrestres de la forêt guyanaise,*G. F. S.* 8, 183–210.

Erikson, G.E. (1963) Brachiation in the New World monkeys and in anthropoid apes, *Symp. Zool. Soc. Lond.* 10, 135–164.

Eriksson, O. and Fröborg, H.(1996) 'Windows of opportunity' for recruitment in long-lived clonal plants: experimental studies of seedling establishment in *Vaccinium* shrubs, *Can. J. Bot.* 74, 1369–1374.

Estrada, A. and Coates-Estrada, R. (1984) Fruit eating and seed dispersal by howling monkeys (*Alouatta palliata*) in the tropical rain forest of Los Tuxtlas, Mexico, *Am. J. Primatol.* 6, 77–91.

Estrada, A. and Coates-Estrada, R. (1985) A preliminary study of resource overlap between howling monkeys (*Alouatta palliata*) and the other arboreal mammals in the tropical rain forest of Los Tuxtlas, Mexico,*Am. J. Primatol.* 9, 27–37.

Estrada, A. and Coates-Estrada, R. (1986) Frugivory by howling monkeys (*Alouatta palliata*) at Lost Tuxtlas, Mexico: dispersal and fate of seed, in A. Estrada and T.H. Fleming (eds.),*Frugivores and seed dispersal,* Dordrecht, The Netherlands, Dr W. Junk, pp. 93–104.

Estrada, A. and R. Coates-Estrada (1991) Howler monkeys (*Alouatta palliata*), dung beetles (Scarabaeidae) and seed dispersal: ecological interactions in the tropical rain forest of Los Tuxtlas, Mexico,*J. Trop. Ecol.* 7, 459–574.

Faaborg, J. (1988) *Ornithology. An ecological approach*, Prentice Hall, Englewood Cliffs, N.J.

Faaborg, J. and Arendt, W.J. (1995) Survival rates of Puerto-Rican birds: are islands that different?*Auk* 112, 503–50.

Fankhauser, C. and Chory, J. (1997) Light control of plant development, *An. Rev. Cell Develop. Biol.* 13, 203–229.

Fanshawe, J.H. (1995) Effects of selective logging on the bird community of Arabuko-Sokoke forest, Kenya, Thesis dissertation, University of Oxford.

Feeny, P. (1970) Seasonal changes in oak leaf tannins and nutrients as a cause of spring feeding by winter moth caterpillars, *Ecology* 51, 565–581.

Feer, F. (1999) Effects of dung beetles (Scarabaeidae) on seeds dispersed by howler monkeys (*Alouatta seniculus*) in the French Guianan rain forest, *J. Trop. Ecol.* 15, 129–142.

Fenton, M.B.(1972) The structure of aerial-feeding bat faunas as indicated by ears and wing elements, *Can. J. Zool.* 50, 287–296.

Fenton, M.B. and Bell, P.G. (1981) Recognition of species of insectivorous bats by their echolocation calls,*J. Mammal.* 62, 233–243.

Fernandes, C.C. (1997) Lateral migration of fishes in Amazon floodplains, *Ecol. Freshw. Fish* 6, 36–44.

Findley, J.S. (1976) The structure of bat communities, *Am. Nat.* 110, 129–139.

Findley, J.S. (1993) *Bats: A community perspective*, Cambridge University Press, Cambridge.

Findley, J.S., and Wilson D.E. (1982) Ecological significance of chiropteran morphology, in T.H. Kunz (ed.) *Ecology of Bats*, Plenum, New York, pp. 243–260.

Fischer, E.A., and Araujo, A.C. (1995) Spatial organization of a bromeliad community in the Atlantic rainforest south-eastern Brazil, *J. Trop. Ecol.* 11, 559–567.

Fischer, K.E. and Chapman, C.A. (1993) Frugivores and fruit syndromes: differences in patterns at the genus and species level, *Oikos* 66, 472–482.

Fleagle, J.G. and Mittermeier, R.A. (1980) Locomotor behavior, body size, and comparative ecology of seven Surinam monkeys, *Am. J. Phys. Anthrop.* 52, 301–314.

Fleck, D.C. and Woolfenden, G.E. (1997) Can acorn tannin predict scrub-jay caching behavior?*J. Chem. Ecol.* 23, 793–806.

Fleck, D.W. and Harder, J.D. (1995) Ecology of marsupials in two amazonian rain forests in Northern Peru, *J. Mammal.* 76, 809–815.

Fleming, T.H. (1986) The structure of neotropical bat communities: A preliminary analysis, *Rev. Chil. Hist. Nat.* 59, 135–150.

Fleming, T.H. and Estrada, A. (eds.). (1993) *Frugivory and seed dispersal: Ecological and evolutionary aspects.* Advances in Vegetation Science 15. Reprinted from *Vegetatio* 107/108. Kluwer Academic Publishers, Dordrecht, Boston, London.

Fleming, T.H. and Heithaus, E. R. (1981) Frugivorous bats, seed shadows, and the structure of tropical forests, *Biotropica* (suppl.) 13, 45–53.

Fontaine, R. (1990) Positional behavior in *Saimiri boliviensis* and *Ateles geoffroyi*, *Am. J. Phys. Anthrop.* 82, 485–508.

Fooden, J. (1964) Stomach contents and gastro-intestinal proportions in wild-shot Guianan monkeys, *Am. J. Phys. Anthrop.* 22, 227–231.

Ford, S. (1988) Postcranial adaptations of the earliest platyrrhine, *J. Hum. Evol.* 17, 155–192.

Forget, P.M. (1990) Seed-dispersal of *Vouacapoua americana* (Caesalpiniaceae) by caviomorph rodents in French Guiana, *J. Trop. Ecol.* 6, 459–468.

Forget P.M. (1991) Comparative recruitment pattern of two non-pioneer tree species in French Guiana, *Oecologia* 85, 434–439.

Forget P.M. (1991) Scatterhoarding of *Astrocaryum paramaca* by *Proechimys* in French Guiana: comparison with *Myoprocta exilis, Trop. Ecol.* 6, 459–468.

Forget P.M. (1992) Seed removal and seed fate in *Gustavia superba* (Lecythidaceae), *Biotropica* 24, 408–414.

Forget P-M. (1993) Post-dispersal predation and scatterhoarding of *Dipteryx panamensis* (Papilionaceae) seeds by rodents in Panama, *Oecologia* 94, 255–261.

Forget P.M. (1994) Regeneration pattern of *Vouacapoua americana* (Caesalpiniaceae), a rodent-dispersed tree species in French Guiana, *Biotropica* 26, 420–426.

Forget P-M. (1996) Removal of seeds of *Carapa procera* (Meliaceae) by rodents and their fate in rainforest in French guiana, *J. Trop. Ecol.* 12, 751–761.

Forget P.M. (1997a) Effect of microhabitat on seed fate and seedling performance in two rodent-dispersed tree species in rainforest in French Guiana, *J. Ecol.* 85, 693–703.

Forget P.M. (1997b) Ten-year seedling dynamics in *Vouacapoua americana* in French Guiana: a hypothesis, *Biotropica* 29, 124–126.

Forget P.M. and Milleron, T. (1991) Evidence for secondary seed dispersal by rodents in Panama, *Oecologia* 87, 596–599.

Forget P.M. and Sabatier, D. (1997) Dynamics of seedling shadow of a frugivore-dispersed tree species in French Guiana, *J. Trop. Ecol.* 13, 767–773.

Forget P.M., Milleron, T. and Feer, F. (1998) Patterns in post-dispersal seed removal by neotropical rodents and seed fate in relation to seed size, in D. M. Newbery, H. T. T. Prins and N. D. Brown (eds.), *Dynamics of Tropical communities*, Blackwell Science, Cambridge, pp. 25–49.

Forget P.M., Milleron, T., Feer, F., Henry, O. and Dubost, G. (2000) Effects of dispersal pattern and mammalian herbivores on seedling recruitment in *Virola michelii* (Myristicaceae) in French Guiana, *Biotropica* 32, 452–462.

Foster, J.R. and Reiners, W.A. (1986) Size distribution and extension of canopy gaps in a northern Appalachian spruce-fir forest, *Vegetatio* 68, 109–114.

Foster, R.B. (1990) The floristic composition of the Rio Manu floodplain forest, in A.H. Gentry (ed.), *Four Neotropical Rainforests*, Yale Univ. Press, New-Haven, p. 99–111.

Fragoso, J.M.V. (1997) Tapir-generated seed shadows: scale-dependent patchiness in the Amazon rain forest, *J. Ecol.* 85, 519–529.

Frangi, J.L. and Lugo, A.E. (1991) Hurricane damage to a flood plain forest in the Luquillo mountains of Puerto Rico. *Biotropica* 23, 324–355.

Freese, C.H. and Oppenheimer, J.R. (1981) The capuchin monkeys, genus *Cebus*, in A.F. Coimbra-Filho and R.A. Mittermeier (eds.), *Ecology and Behavior of Neotropical Primates*, Academia Brasileira de Ciências, Rio de Janeiro, pp. 331–390.

Fretey, J. (1987) *Les Tortues de Guyane française*, Sepanguy, Cayenne.

Friend, A.D. (1993) The prediction and physiological significance of tree height, in A.M. Solomon and H.H. Shugart (eds.), *Vegetation Dynamics and Global Change*, Chapman and Hall, N.Y., London, pp. 101–115.

Frost, D.R. (1985) *Amphibian species of the world*, Allen Press. Inc. & Ass. of Syst. Coll., Lawrence, Kansas.

Frost, D.R. (2000) *Amphibian species of the world*: An online reference, Am. Mus. Nat. Hist., V2.20.

Garber, P.A. (1980) Locomotor behavior and feeding ecology of the Panamanian tamarin (*Saguinus oedipus geoffroyi*, Callitrichidae, Primates), *Int. J. Primatol.* 1, 185–201.
Garber, P.A. (1986) The ecology of seed dispersal in two species of callitrichid primates (*Saguinus mystax* and *Saguinus fuscicollis*), *Am. J. Primatol.* 10, 155–170.
Garber, P.A. (1991a) A comparative study of positional behavior in three species of tamarin monkeys, *Primates* 32, 219–230.
Garber, P.A. (1991b) Primate behavioral ecology, *Encycl. Hum. Biol.* 6, 127–133.
Garber, P.A. (1992) Vertical clinging, small body size, and the evolution of feeding adaptations in the Callithrichinae, *Am. J. Phys. Anthropol.* 88, 469–482.
Garber, P.A. (1993) Feeding ecology and behaviour of the genus *Saguinus*, in A.B. Rylands (ed.), *Marmosets and Tamarins, Systematics, Behaviour, and Ecology*, Oxford Science Publications, Oxford University Press, Oxford, pp. 273–295.
Garber, P.A. and J.D. Preutz (1995) Positional behavior of moustached tamarin monkeys: effects of habitat on locomotor variability and stability, *J. Hum. Evol.* 28, 411–426.
Gardner, A.L. (1977) Feeding habits, in R.J. Parker, J.K. Jones Fr, and D.C. Carter (eds.), *Biology of Bats of the New World family Phyllostomatidae*, Part II, Spec. Publ. Mus. Texas. Tech. Univ., pp. 293–350.
Garwood, N.C. (1989) Tropical soil seed banks: a review, in M.A. Leck, V.T. Parker and R.L. Simpson (eds.), *Ecology of soil seed banks*, Academic Press, Orlando, Florida USA, pp. 149–444.
Gasc, J.P. (1976) Contribution à la connaissance des Squamates (Reptilia) de la Guyane française, nouvelles localités pour les sauriens, *C.R. Soc. Biogéogr.* 454, 17–36.
Gasc, J.P. (1981) Quelque nouvelles données sur la répartition et l'écologie des sauriens en Guyane française, *Rev. Ecol.- Terre Vie* 35, 273–325.
Gasc, J.P. (1986) Le peuplement herpétologique d'*Astrocaryum paramaca* (Arecacées), un palmier important dans la structure de la forêt en Guyane française, *Mém. Mus. Nat. Hist. Nat.*, Paris, A Zool. 132, 97–107.
Gasc, J.P. (1990) *Les lézards de Guyane*. Chabaud, Paris.
Gasc, J.P. and Rodrigues, M.T. (1980) Liste préliminaire des serpents de la Guyane française, *Bull. Mus. Nat. Hist. Nat*, Paris, 4e série, 2A, 559–598.
Gasc, J.P. and Sastre, C. (1978) Les formations ouvertes à graminées bambusiformes (cambrouzes) sont-elles en Guyane Française le signe d'un ancien peuplement humain? *10è Congrès Natl Soc. Sav. Sci.* 1, 98–103.
Gautier-Hion, A., Duplantier, J.M., Quris, R., Feer, F., Sourd, C., Decoux, JP., Dubost, G., Emmons, L., Erard, C., Heckestweiller, P., Moungazi, A., Roussilhon, C. and Thiollay, J.M. (1985) Fruit characters as a basis of fruit choice and seed dispersal in a tropical forest vertebrate community, *Oecologia*, 65, 324–337.
Gautier-Hion, A., Gautier, J.-P. and Maisels, F. (1993) Seed dispersal versus seed predation: an inter-site comparison of two related African monkeys, in T.H. Fleming and A. Estrada (eds.), *Frugivory and Seed Dispersal: Ecological and Evolutionary Aspects*, Kluwer Academic Publishers, Dordrecht, pp. 237–244.
Gebo, D.L. (1992) Locomotor and postural behavior in *Alouatta palliata* and *Cebus capucinus*, *Am. J. Primatol.* 26, 277–290.
Gebo, D.L. and C.A. Chapman (1995) Positional behavior in five sympatric Old World monkeys, *Am. J. Phys. Anthropol.* 97, 49–76.
Genoways, H.H. and Williams, S.L. (1979) Records of bats (Mammalia : chiroptera) from Suriname, *Ann. Carnegie Mus.* 48, 323–335.
Gentry, A.H. (1988) Changes in plant community diversity and floristic composition on geographical and environmental gradients, *Ann. Mo. Bot. Gard.* 75, 1–34.
Gentry, A.H. (1988) Tree species richness of upper Amazonian forests, *Proc. Nat. Acad. Sci.* 85, 156–159.
Gentry, A.H. (1990) *Four neotropical rainforests*, Yale University Press, New Haven, USA.
Gentry, A.H. and Dodson, J.A. (1988) Diversity and biogeography of neotropical vascular epiphytes, *Ann. Mo. Botanical Garden* 74, 205–233.
Georgakopoulos, J. and Etienne, A.S. (1994) Identifying location by dead reckoning and external cues, *Behav. Process.* 31, 57–74.
Géry, J. (1992) Description de deux nouvelles espèces proches de *Moenkhausia lepidura* (Kner) (Poissons, Characiformes, Tetragonopterinae), avec une revue du groupe, *Rev. fr. Aquariol.* 19, 69–78.
Géry, J., Planquette, P. and Le Bail, P.Y. (1991) Faune characoide (poissons ostariophysaires) de l'Oyapock, l'Approuague et la rivière de kaw (Guyane Française), *Cybium*, 15 (1, Suppl.) 1–69.
Géry, J., Planquette, P. and Le Bail, P.Y. (1996) Nouvelles espèces et sous-espèces guyanaises d'*Astyanax* S.L. (Teleostei, Characiformes, Characidae) à épines pelviennes, avec une introduction concernant le groupe, *Cybium* 20, 3–36.
Gibbs, A.K. and Barron, C. (1983) The Guiana shield reviewed, *Episodes* 2, 7–14.

Giraldeau, L.A. (1984) Group foraging, the skill pool effect and frequency-dependent learning,*Am. Nat.* 124, 72–79.

Giraldeau, L.A. and Lefebvre, L. (1986)Exchangeable producers and scroungers roles in a captive flock of feral pigeons. A case for the skill pool effect, *Anim. Behav.* 34, 797–803.

Giraldeau, L.A., Soos, C. and Beauchamp, G. (1994) A test of the producer-scrunger foraging game in captive flocks of Spice finches, *Lonchura punctulata*, *Behav. Ecol. Sociobiol.* 34, 251–256.

Givnish, T.J. (1986) Biochemical constrains on crown geometry in forest herbs, in T.J. Givnish (ed.),*On the Economy of plant form and function*, Cambridge University Press, Cambridge, pp. 525–583.

Glander, K.E. (1978) Howling monkey feeding behavior and plant secondary compounds: A study of strategies, in G.G. Montgomery (ed.) *The ecology of arboreal folivores*, Washington, D.C.: Smithsonian Inst. Press, pp. 561–574.

Goguillon, B. (1994) A travers la Guyane des Oiseaux. Les Nouragues, *Harpia* 6, 9–18.

Goulding, M. (1980) *The fishes and the forest*, University of California Press, Berkeley.

Goulding, M., Carvalho, M. and Ferreira, E.G. (1988) Rich life in poor waters. SPB Acad. Publishers, The Hague.

Gradwohl, J. and Greenberg, R. (1980) The formation of antwren flocks on Barro Colorado Island, Panama,*Auk* 97, 385–395.

Grand Atlas de l'Archéologie (1990) Larousse, The time.

Grand, T.I. (1983) Body weight: its relationship to tissue composition, segmental distribution of mass and motor function. III. The Didelphidae of French Guiana, *Aust. J. Zool.* 31, 299–312.

Grand, T.I. (1984) Motion economy within the canopy, in P.S. Rodman and J.G.H. Cant (eds.),*Adaptations for foraging in non-human primates*, New York, Columbia University Press, pp. 54–72.

Granjon, L., Cosson, J.-F., Judas, J. and Ringuet, S. (1996) Influence of tropical rainforest fragmentation on mammal community in French Guiana: short term effects, *Acta Oecol.* 17, 673–684.

Granville, J.J. (de) (1978) *Recherches sur la flore et la végétation guyanaise*, Thèse de Doctorat d'Etat, USTL, Montpellier.

Granville, J.J. (de) (1982) Rain forest and xeric flora refuges in French Guiana, in G.T. Prance (ed.),*Biological diversification in the tropics,* Columbia University Press, pp. 159–181.

Granville, J.J. (de) (1988) Phytogeographical characteristics of the Guianan forests, *Taxon* 37, 578–594.

Granville, J.J. (de) and Sastre, C. (1974) Aperçu sur la végétation des inselbergs du sud-ouest de la Guyane française, *C. R. Soc. Biogéogr.* 439, 54–58.

Granville, J.J. de (1992) Un cas de distribution particulier: les espèces forestières peri-amazoniennes,*C.R. Soc. Biogéogr.* 68, 1–33.

Greenberg, R. and Gradwohl, J. (1985) A comparative study of the social organization of antwrens on Barro Colorado Island, Panama, in P.A. Buckley, E.S. Morton, R.S. Ridgely, and F.G. Buckley (eds.),*Neotropical ornithology*, AOU Ornithol. Monog., 36, pp. 845–855.

Greenberg, R. and Gradwohl, J. (1986) Constant density and stable territoriality in some tropical insectivorous birds, *Oecologia (Berlin)* 69, 618–625.

Greenberg, R. and Gradwohl, J. (1997) Territoriality, adult survival, and dispersal in the checker-throated Antwren in Panama. *J. Avian Biol.* 28, 103–110.

Greene, H.W. (1988) Species richness in tropical predators, in F. Almeda and C.M. Pringle (eds.)*Tropical rain forests. Diversity and conservation*, California Academy of Science, San Francisco, pp. 259–280.

Greig-Smith, P.W. (1978) Imitative foraging in mixed species flocks of Seychelles birds, *Ibis* 120, 233–235.

Grenand, P. (1979) Histoire des Amerindiens, *Atlas des D.O.M, la Guyane,* pl. 17, 3–4.

Grimaldi, M. and Boulet, R. (1989) Relation entre l'espace poral et le fonctionnement hydrodynamique d'une couverture pédologique sur socle de Guyane française, *Cah. ORSTOM, Sér. Pédol.* XXV (3), 263–275.

Grimaldi, M. and Riéra, B. (2001) Geography and climate, in F. Bongers, P. Charles-Dominique, P.M. Forget, and M. Théry (eds.), *Nouragues: Dynamics and plant-animal interactions in a neotropical rainforest,* Kluwer Academic Publishers, Dordrecht.

Gross, L.J. (1986) Photosynthetic dynamics and plant adaptation to environmental variability,*Lect. Math. Life Sci.* 19, 135–170.

Gruau, G., Martin H., Leveque B., Capdevila, R., and Marot, A. (1985) Rb-Sr and Sm-Nd geochronology of lower proterozoic granite-greenstone terrains in French Guiana, South America, *Precambrian Res.* 30, 63–80.

Grubb, P.J. (1977) The maintenance of species-richness in plant communities: the importance of the regeneration niche, *Biol. Rev.* 52, 107–145.

Grubb, P.J. and Metcalfe, D.J. (1996) Adaptation and inertia in the Australian lowland rain-forest flora: contradictory trends in intergeneric and intrageneric comparisons of seed size in relation to light demand, *Funct. Ecol.* 10, 512–520.

Guehl, J.M. (1984) Dynamique de l'eau dans le sol en forêt tropicale humide guyanaise. Influence de la couverture pédologique, *Ann. Sci. Forest.* 41, 195–236.

Guillotin, M. (1981) Données écologiques sur les petits rongeurs forestiers terrestres de guyane française, Thèse de Doctorat USTL, Montpellier.

Guillotin, M. (1982) Activités et régimes alimentaires: leurs interactions chez *Proechimys cuvieri* et *Oryzomys capito velutinus* (Rodentia) en forêt guyanaise, *Rev. Ecol.- Terre Vie* 36, 337–371.

Guillotin, M. (1982) Place de *Proechimys cuvieri* (Rodentia, Echimyidae) dans les peuplements micromammaliens terrestres de la forêt guyanaise, *Mammalia* 46, 299–318.

Guillotin, M. and Petter, F. (1984) Un *Rhipidomys* nouveau de Guyane française, *R. leucodactylus aratayae* ssp. nov. (Rongeurs, Cricétidés), *Mammalia* 48, 541–544.

Guillotin, M., Dubost, G. and Sabatier, D. (1994) Food choice and food competition among the three major primate species of French Guiana, *J. Zool. (Lond.)* 233, 551–579.

Hadj, C.L.Z., Steele, M.A. and Smallwood, P.D. (1996) Caching decisions by grey squirrels: a test of the handling time and perishability hypothesis, *Anim. Behav.* 52, 941–948.

Haffer, J. (1969) Speciation in Amazonian forest birds, *Science Wash.* 165, 131–137.

Haffer, J. (1987) Quaternary history of tropical America, in T.C. Whitmore and G.T. Prance, (eds.), *Biogeography and quaternary history in tropical America*, Clarendon Press, Oxford, pp 1–18.

Hallé, F. and Oldeman, R.A.A. (1970) *Essai sur l'architecture et dynamique de croissance des arbres tropicaux*, Masson, Paris.

Hallé, F., Oldeman, R.A.A. and Tomlinson, P.B. (1978) *Tropical trees and forests: an architectural analysis*, Springer-Verlag, Berlin.

Hallwachs, W. (1994) *The clumpsy dance between agoutis and plants: scatterhoarding by Costa Rican dry forest agoutis (Dasyprocta punctata: Dasyproctidae: Rodentia)*, Ph.D. thesis, Cornell University, Ithaca, NY.

Hammond, D.S. and Brown, V.K. (1998) Disturbance, phenology and life-history characteristics: factors influencing distance/density-dependent attack on tropical seeds and seedlings, in D.M. Newberry, H.H.T. Prins and N.D. Brown (eds.), *Dynamics of Tropical Communities*, Blackwell Science, Cambridge, pp. 51–78.

Hammond, D.S. and Brown, V.K. (1995) Seed-size of woody plants in relation to disturbance, dispersal, soil type in wet Neotropical forests, *Ecology* 76, 2544–2561.

Harper, J.L. (1977) *Population biology of plants*, Academic Press, London.

Hartzhorn, G.S. (1980) Neotropical forest dynamics, *Biotropica* 12, 23–30.

Hartzhorn, G.S. (1990) An overview of neotropical forest dynamics, in Gentry, A.H. (ed.), *Four neotropical rainforests*, Yale University Press, pp. 585–599.

Heithaus, E.R., Fleming, T.H., and P.A. Opler (1975) Foraging patterns and resource utilization in seven species of bats in a seasonal tropical forest, *Ecology* 56, 841–854.

Henry, O (1994) *Caractéristiques et variations saisonnières de la reproduction de quatre mammifères forestiers terrestres de Guyane française: Oryzomys capito velutinus (Rodentia, Cricetidae), Proechimys cuvieri (Rodentia, Echimyidae), Dasyprocta leporina (Rodentia, Dasyproctidae), Tayassu tajacu (Artiodactyla, Tayassuidae), influence de l'âge, des facteurs environnementaux et de l'alimentation*, Thèse de Doctorat, Université Paris VII.

Henry, O. (1997) The influence of sex and reproductive state on diet preference in four terrestrial mammals of the French Guianan rainforest, *Can. J. Zool.* 75, 929–935

Henry, O. (1999) Frugivory and the importance of seeds in the diet of the orange-rumped agouti (*Dasyprocta leporina*) in French Guiana, *J. Trop. Ecol.* 15, 291–300.

Henry, O., Feer, F. and Sabatier, D. (2000) Diet of the lowland tapir (*Tapirus terrestris* L.) in French Guiana, *Biotropica* 32, 364–368.

Herrera, C.M. (1979) Ecological aspects of heterospecific flock formation in a Mediterranean passerine bird community, *Oikos* 33, 85–96.

Herrera, C.M., Jordano, P., Lopez-Soria, L. and Amat, J.A. (1994) Recruitment of a mast-fruiting, bird-dispersed tree: bridging frugivore activity and seedling establishment, *Ecol. Monogr.* 64, 315–344.

Hirose, T. and Werger, M.J.A. (1987) Maximizing daily canopy photosynthesis with respect to the leaf nitrogen allocation pattern in the canopy, *Oecologia* 72, 520–526.

Hladik, C.M. (1988) Seasonal variations in food supply for wild Primates, in I. de Garine and G.A. Harrison (eds.), *Coping with uncertainty in food supply*, Clarendon Press, Oxford, pp. 1–25.

Hladik, C.M. and Hladik, A. (1969) Rapports trophiques entre végétation et primates dans la forêt de Barro Colorado (Panama), *Rev. Ecol.- Terre Vie* 1, 25–117.

Hladik, C.M. and Simmen, B. (1996) Taste perception and feeding behavior in nonhuman primates and human populations, *Evol. Anthropol.* 5, 161–174.

Hladik, C.M., Hladik, A., Bousset, J., Valdebouze, P., Viroben, G. and Delort-Laval, J. (1971) Le régime alimentaire des Primates de l'île de Barro Colorado (Panama), *Folia Primatol.* 16, 85–122.

Hoch, G.A. and Adler, G.H. (1997) Removal of black palm (*Astrocaryum standleyanum*) seeds by spiny rats (*Proechimys semispinosus*), *J. Trop. Ecol.* 13, 51–58.

Hödl, W. (1990) Reproductive diversity in Amazonian lowland frogs. *Fortschr. D. Zool.* 38, 45–51.

Hoff, M., Cremers G., Feuillet, C. and Granville, J.J. de (1989) La banque de données 'AUBLET' de l'Herbier du Centre ORSTOM de Cayenne (CAY), *Bull. Jard. Bot. Nat. Belgique* 59, 171–178.

Holbrook, N.M. and Putz, F.E. (1989) Influence of neighbours on tree form: effects of lateral shading and prevention of sway on the allometry of *Liquidambar styraciflua* (sweet gum), *Am. J. Bot.* 76, 1740–1749.

Holmes, M.G. and Smith, H. (1975). The function of phytochrome in plants growing in the natural environment, *Nature* 254, 512–514.

Hooghiemstra, H. and van der Hammen, T. (1998) Neogene and quaternary development of the neotropical rain forest: the forest refugia hypothesis, and a literature overview, *Earth-Sci. Rev.* 44, 147–183.

Hoogland, J.L. (1979) The effect of colony size on individual alertness of prairie dogs (Sciuridea:*Cynomys* spp.), *Anim. Behav.* 27, 394–407.

Hoogmoed, M.S. (1973) Notes on the herpetofauna of Surinam IV.The lizards and Amphisbaenians of Surinam, *Biogeographica* 4.

Hoogmoed, M.S. and Avila-Pires, T.C.S. (1991) Annotated checklist of the herpetofauna of Petit Saut, Sinnamary River, French Guiana, *Zool. Mededelingen Leiden,* 65, 53–88.

Hoogmoed, M.S., and Lescure, J. (1975) An annotated checklist of the lizards of French Guiana, mainly based on two recent collections, *Zool. Mededelingen Leiden*, 49, 141–171.

Horner, J.D., Gosz, J.R. and Cates, R.G. (1988) The role of carbon-based secondary metabolites in decomposition in terrestrial ecosystems, *Am. Nat.* 132, 869–883.

Horvitz, C.C. and Le Corff, J. (1993) Spatial scale and dispersion pattern of ant- and bird-dispersed herbs in two tropical lowland rain forests, *Vegetatio* 107/108, 351–362.

Horvitz, C.C. and Schemske, D.W. (1986a) Seed dispersal and environmental heterogeneity in a neotropical herb: a model of population and patch dynamics, in A. Estrada and T.H. Fleming (eds.), *Frugivores and seed dispersal*, Junk, Dordrecht, pp 169–186.

Horvitz, C.C. and Schemske, D.W. (1986b) Seed dispersal of a neotropical myrmecochore: variation in removal rates and dispersal distance, *Biotropica* 18, 319–323.

Horvitz, C.C. and Schemske, D.W. (1994) Effects of dispersers, gaps, and predators on dormancy and seedling emergence in a tropical herb, *Ecology* 75, 1949–1958.

Hostache, G., Pascal, M. and Planquette, P. (1993) Saisonalité de la reproduction chez l'Atipa (*Hoplosternum littorale*. Callichthyidae, Siluriforme, Teleostei) par l'analyse de l'évolution du rapport gonado-somatique, *Aquat. Liv. Resour.* 6, 155–162.

Hostache, G., Pascal, M. and Tessier, C. (1995) Influence de la température d'incubation, sur le rapport mâle-femelle chez l'Atipa, *Hoplosternum littorale* Hancock (1828), *Can. J. Zool.* 73, 1239–1246.

Houle, G. (1995) Seed dispersal and seedling recruitment: the missing link, *Ecoscience* 2, 238–244.

Houle, G. (1996) Environmental filters and seedling recruitment on a coastal dune in subarctic Quebec (Canada), *Can. J. Bot.* 74, 1507–1513.

Howe, H.F. (1980) Monkey dispersal and waste of a Neotropical fruit, *Ecology* 61, 944–959.

Howe, H.F. (1986) Consequences of seed dispersal: a case study from Central America,*J. Bombay Nat. Hist. Soc.* 83 (Suppl.) 19–42.

Howe, H.F. (1986) Seed dispersal by fruit-eating birds and mammals, in D.R. Murray (eds.), *Seed Dispersal*, Academic Press Australia, Sydney, pp. 123–189.

Howe, H.F. (1989) Scatter- and clump-dispersal and seedling demography: hypothesis and implications, *Oecologia* 79, 417–426.

Howe, H.F. (1993) Aspects of variation in a neotropical seed dispersal system, *Vegetatio* 107/108, 149–162.

Howe, H.F. and Smallwood, J. (1982) Ecology of seed dispersal, *An. Rev. Ecol. Syst.* 13, 201–228.

Howe, H.F. and Van de Kerckhove, G.A. (1979) Fecundity and seed dispersal by birds of a tropical tree,*Ecology* 60, 180–189.

Howe, H.F. and Van de Kerckhove, G.A. (1981) Removal of wild nutmeg *Virola surinamensis* crops by birds, *Ecology* 62, 1093–1106.

Howe, H.F. and Westley, L.C. (1988) *Ecological relationships of plants and animals.* Oxford University Press, New York and Oxford.

Howe, H.F., Schupp, E.W. and Westley, L.C. (1985) Early consequences of seed dispersal for a neotropical tree (*Virola surinamensis*), *Ecology* 66, 781–791.

Hubbell, S.P. (1979) Tree dispersion, abundance, and diversity in a tropical dry forest, *Science* 203, 1299–1309.

Hubbell, S.P. and Foster, R.B. (1986) Canopy gaps and the dynamics of a neotropical forest, in Crawley, M.J. (ed.), *Plant ecology*, Blackwell Scientific Publications, pp. 77–98.

Huber, J.H. (1991) Revue des espèces de *Rivulus* de Guyane française, avec descriptions de *Rivulus cladophorus* n. np. et de *R. igneus* n. sp., et nouvelles descriptions de *R. urophthalmus* Günther et de *R. micropus* Steindachner, *Rev. fr. Aquariol.* 18, 65–74.

Huber, O. (1980) Die Feldsvegetation am oberen Orinoko in Süd-Venezuela, in H. Reisigl (ed.),*Blumenparadiese und botanische Gärten der Erde*, Pinguin Verlag, Innsbruck und Umschau-Verlag, Frankfurt/Main, pp. 200–203.

Hunt, K.D. (1992) Positional behavior of *Pan troglodytes* in the Mahale Mountains and Gombe Stream National Parks, Tanzania, *Am. J. Phys. Anthrop.* 87, 83–105.

Hurault, J. (1963) Recherches sur les inselbergs granitiques nus en Guyane française, *Rev. Géomorph. Dynam.* 4, 49–61.

Hurault, J. (1967) *L'érosion régressive dans les régions tropicales humides et la genèse des inselbergs granitiques*, I.G.N., Paris.

Hurault, J. (1974) Les inselbergs rocheux des régions tropicales humides, témoins de paléoclimats,*C.R. Soc. Biogéogr.* 439, 49–54.

Ibisch, P.L., Rauer, G. Rudolph, D. and Barthlott, W. (1995) Floristic, biogeographical, and vegetational aspects of Pre-Cambrian rock outcrops (inselbergs) in eastern Bolivia, *Flora* 190, 299–314.

Jackson, J.F. (1981) Seed size as a correlate of temporal and spatial patterns of seed fall in a Neotropical forest, *Biotropica* 13, 121–30.

Jacobs, L.F. (1992) Memory for cache locations in Merriam's kangaroo rats, *Anim. Behav.* 43, 585–593.

Jacobs, L.F. and Liman, E.R. (1991) Gray squirrels remember the locations of buried nuts,*Am. J. Primatol.* 41, 103–110.

Jaffe, M.J. and Forbes, S. (1993) Thigmomorphogenesis: the effect of mechanical perturbation on plants,*Plant Growth Regul.* 12, 313–324.

Jaffre, T. and Veillon, J.M. (1991) Etude floristique et structurale de deux forêts denses humides sur roches ultrabasiques en Nouvelle Calédonie, *Adansonia* 3–4, 243–273.

Jans, L., Poorter, L., Van Rompaey, R.S.A.R. and Bongers, F. (1993) Gaps and forest zones in tropical moist forest in Ivory-Coast, *Biotropica* 25, 258–269.

Jansen, J. and Konings, K. (1993) *Forest dynamics, light and population structure in a rain forest in French Guiana.* MSc. Thesis. Wageningen University.

Jansen, P.A. and Forget, P.M. (2001) Scatterhoarding and tree regeneration, in F. Bongers, P. Charles-Dominique, P.M. Forget, and M. Théry (eds.), *Nouragues: Dynamics and plant-animal interactions in a neotropical rainforest* Kluwer Academic Publisher. Dordrecht.

Jansen, P.A. and Zuidema, P.A. (2001) Logging, seed dispersal by vertebrates, and natural regeneration of tropical timber trees, in R. Fimbel, A. Grajal and J. Robinson (eds.), *The cutting edge: conserving wildlife in logged tropical forest.* Columbian University Press, NewYork, in press.

Janson, C.H. (1983) Adaptation of fruit morphology to dispersal agents in a Neotropical forest,*Science* 219, 187–189.

Janson, C.H. and Emmons, L.H. (1990) Ecological structure of the nonflying mammal community at Cocha Cashu biological station, Manu National Park, Peru, in A. Gentry (ed.), *Four neotropical rainforests*, Yale University Press, New Haven, pp. 314–338.

Janson, C.H. and Boinski, S. (1992) Morphological and behavioral adaptations for foraging in generalist primates: The case of the Cebines, *Am. J. Phys. Anthropol.* 88, 483–498.

Janson, C.H., Stiles, E.W. and White, D.W. (1986) Selection on plant fruiting traits by brown capuchin monkeys: a multivariate approach, in A. Estrada and T.H. Fleming (eds.) *Frugivores and seed dispersal*, W. Junk Publishers, Dordrecht, pp. 83–92.

Janzen, D.H. (1970) Herbivores and the number of tree species in tropical forests, *Am. Nat.* 104, 501–528.

Janzen, D.H. (1971) Seed predation by animals, *Annu. Rev. Ecol. Syst.* 2, 465–492.

Janzen, D.H. (1981) Digestive seed predation by a Costa Rican Baird's tapir, *Biotropica* 13, 59–63.

Janzen, D.H. (1982) Removal of seeds from horse dung by tropical rodents: influence of habitat and amount of dung. *Ecology* 63, 1887–1900.

Janzen, D.H. (1983) Dispersal of seeds by vertebrates guts, in D.J. Futuyma and M. Slatkin (eds.),*Coevolution*, Sinauer Associates, Sunderland, pp. 232–262.

Janzen, D.H. (1985) *Spondias mombin* is culturally deprived in megafauna-free forest, *J. Trop. Ecol.* 1, 131–155.

Janzen, D.H. (1986) Mice, big mammals, and seeds: it matters who defecates what where. in A. Estrada and T.H. Fleming (eds.), *Frugivores and seed dispersal,* Dordrecht, The Netherlands, Dr W. Junk, pp. 251–271.

Janzen, D.H. and Martin, P. S. (1982) Neotropical anachronisms: the fruits the gomphotheres ate, *Science* 215, 19–27.

Janzen, D.H., Miller G.A., Hackforth-Jones, J., Pond, C.M., Hooper; K. and Janos, D.P. (1976) Two Costa Rican bat generated seed shadows of *Andira inermis* (Leguminosae), *Ecology* 57, 1068–1075.

Jenkins, S.H. and Peters, R.A. (1992) Spatial patterns of food storage by Merriam's kangaroo rats, *Behav. Ecol.* 3, 60–65.

Johns, A.D. and Skorupa, J.P. (1987) Responses of rainforest primates to habitat disturbance: a review, *Int. J. Primatol.* 8, 157–191.

Johnston, M. and Gillman, M. (1995) Tree population studies in low-diversity forests, Guyana. I. Floristic composition and structure, *Biodiv. Conservation* 4, 339–632.

Jokinen, S. and Suhonen, J. (1995) Food caching by willow and crested tits: a test of scatterhoarding models, *Ecology* 76, 892–898.

Jordano, P. (1987) Frugivory, external morphology and digestive system in mediterranean sylviid warblers *Ibis* 129, 175–189.

Jordano, P. (1992) Fruits and frugivory, in M. Fenner (ed.), *Seeds: The ecology of regeneration in natural plant communities*, CAB International, London, pp 105–151.

Jordano, P. and Herrera, C.M. (1995) Shuffling the offspring: Uncoupling and spatial discordance of multiple stages in vertebrate seed dispersal, *Ecoscience* 2, 230–237.

Julien-Laferrière, D. (1989) *Utilisation de l'espace et des ressources alimentaires chez Caluromys philander (Marsupialia, Didelphidae) en Guyane Française. Comparaison avec Potos flavus (Eutheria, Procyonidae)*, Thèse de Doctorat, Université Paris XIII.

Julien-Laferrière, D. (1991) Organisation du peuplement de marsupiaux en Guyane française, *Rev. Ecol - Terre Vie* 46, 125–144.

Julien-Laferrière, D. (1993) Ecology of the Kinkajou (*Potos flavus*) in French Guiana: first study by radio-tracking in a tropical rain forest, *J. Trop. Ecol.* 9, 19–32.

Julien-Laferrière, D. (1999) Foraging strategies and food partitioning in the neotropical frugivorous mammals *Caluromys philander* and *Potos flavus*, *J. Zool.* 247, 71–80.

Julien-Laferrière, D. (2001) Frugivory by Kinkajous (*Potos flavus*), in F. Bongers, P. Charles-Dominique, P.M. Forget, and M. Théry (eds.), *Nouragues: Dynamics and plant-animal interactions in a neotropical rainforest*, Kluwer Academic Publisher, Dordrecht.

Julien-Laferrière, D. and Atramentowicz, M. (1990) Feeding and reproduction of three didelphid marsupials in two neotropical forests (French Guiana), *Biotropica* 22, 404–415.

Jullien, M. (1997) *Signification adaptative de la vie en groupe chez les oiseaux: le cas des passereaux forestiers néotropicaux*, Thèse de Doctorat, Université Paris 6.

Jullien, M. and Clobert, J. (2000) The survival value of flocking in neotropical birds: reality or fiction? *Ecology*, 81, 3416-3430.

Jullien, M. and Thiollay, J.M. (1998) Multi-species territoriality and dynamic of neotropical forest understorey bird flocks, *J. Anim. Ecol.* 67, 227–252.

Julliot, C. (1992) *Utilisation des ressources alimentaires par le singe hurleur roux, Alouatta seniculus (Atelidae, Primates), en Guyane: impact de la dissémination des graines sur la régénération forestière*, Thèse de Doctotat, Université de Tours.

Julliot, C. (1994a) Diet diversity and habitat of howler monkeys, in B. Thierry, J.R. Anderson, J.J. Roeder, and N. Herrenschmidt (eds.), *Current Primatology, Vol I: Ecology and Evolution*, Selected Proceedings of the XIVth Congress of the International Primatological Society, ULP, Strasbourg, France, pp. 67–71.

Julliot, C. (1994b) Predation of a young spider monkey (*Ateles paniscus*) by a crested eagle (*Morphnus guianensis*), *Folia Primatol.* 63, 75–77.

Julliot, C. (1994c) Frugivory and seed dispersal by red howler monkeys: evolutionary aspect *Rev. Ecol. - Terre Vie* 49, 331–341.

Julliot, C. (1996a) Fruit choice by red howler monkeys (*Alouatta seniculus*) in a tropical rain forest, *Am. J. Primatol.* 40, 261–282.

Julliot, C. (1996b) Seed dispersal by red howler monkeys (*Alouatta seniculus*) in a tropical rain forest, *Int. J. Primatol.* 17, 239–258.

Julliot, C. (1997) Impact of seed dispersal by red howler monkeys (*Alouatta seniculus*) on the seedling population in the understorey of tropical rainforest, *J. Ecol.* 85, 431–440.

Julliot, C. and Sabatier, D. (1993) Diet of the red howler monkey (*Alouatta seniculus*) in French Guiana, *Int. J. Primatol.* 14, 527–550.

Julliot, C. and Simmen, B. (1998) Food partitioning among a community of neotropical primates, *Folia Primatol.* 69, 43–44.

Julliot, C., Simmen, B. and Zhang, S.-Y. (2001) Frugivory and seed dispersal by three neotropical primates, in F. Bongers, P. Charles-Dominique, P.M. Forget, and M. Théry (eds.), *Nouragues: Dynamics and Plant-Animal interactions in a neotropical rainforest,* Kluwer Academic Publisher, Dordrecht.

Junk, W.J. (1985) Temporary fat storage, an adaptation of some fish species to the water level fluctuations and related environmental changes of the Amazon river, *Amazoniana* 9, 315–351.

Kalko, E.K.V. and Condon, M.A. (1998) Echolocation, olfaction and fruit display: how bats find fruit of flagellichorous cucurbits, *Funct. Ecol.* 12, 364–372.

Kalko, E.K.V. and Handley, C.O. (2001) Neotropical bats in the canopy: diversity, community structure, and implications for conservation, *Plant Ecol.* (in press).

Kalko, E.K.V., Handley, C.O. and Handley, D. (1996) Organization, diversity and long-term dynamics of a neotropical bat community, in M. Cody and J. Smallwood (eds.), *Long-term studies in vertebrate communities*, Academic press; Los Angeles, C.A., pp. 503–553.

Karasov, W.H. and Levey, D.J. (1990) Digestive system trade-offs and adaptations of frugivorous passerine birds, *Physiol. Zool.* 63, 1248–1270.

Karr, J.R. (1977) Ecological correlates of rarity in a tropical forest bird community, *Auk* 90, 240–248.

Karr, J.R. and Brawn, J.D. (1990) Food resources of understory birds in Central Panama: quantification and effects on avian populations, *Stud. Avian Biol.* 13, 58–64.

Karr, J.R., Nichols, J.D., Klimkiewicz, M.K. and Brawn, J.D. (1990) Survival rates of birds of tropical and temperate forests: will the dogma survive? *Am. Nat.* 136, 277–291.

Karr, J.R., Robinson, S.K., Blake, J.G. and Bierregaard, R.O. Jr. (1990) Birds of four Neotropical forests, in A.H. Gentry (ed.) *Four Neotropical rainforests*, Yale University Press, New Haven, Conn. pp. 237–269.

Kays, R.W. (1999) Food preferences of kinkajous (*Potos flavus*): a frugivorous carnivore, *J. Mammal.* 80, 589–599.

Kays, R.W. and Gittelman, J.L. (1995) Home range size and social behavior of kinkajous (*Potos flavus*) in the Republic of Panama, *Biotropica* 27, 530–534.

Kazuhiko, H., Wajirou, S. and Satohiko, S. (1997) Impacts of secondary seed dispersal and herbivory on seedling survival in *Aesculus turbinata*, *J. Veg. Sci.* 8, 735–742.

Keast, A. (ed.) (1990) *Biogeography and ecology of forest bird communities*, SPB Academic Publishing bv, The Hague.

Keith, P. (1997) L'inventaire des poissons d'eau douce en guyane: bilan et perspectives, *Cybium*, 21 (1, suppl.), 17–25.

Keith, P., Le Bail, P.Y. and Planquette, P. (2000) *Atlas des poissons d'eau douce de Guyane* (tome 2, fasc. 1), Patrimoines naturels, MNHN, SNP, 43.

Kelly, D.L. (1985) Epiphytes and climbers of a Jamaican rain forest: vertical distribution, life forms and life histories, *J. Biogeogr.* 12, 223–241.

Kendrick, R.E. and Kronenberg, G.H.M. (1994). Photomorphogenesis in plants. 2nd edition. Kluwer Academic Publishers, Dordrecht.

Kessler, P. (1995*a*) Revierverhalten, Nahrungstrategie und Habitatpräferenzen des Rothandtamarins (*Saguinus midas midas*) in Französich-Guayana, Diplomarbeit. Anthropologisches Institut und Museum der Universität Zürich.

Kessler, P. (1995*b*) Preliminary field study of the Red-Handed Tamarin, *Saguinus midas*, in French Guiana, *Neotrop. Primates* 3, 184.

Kessler, P. (1998) Primate densities in the natural reserve of Nouragues, French Guiana, *Neotrop. Primates* 6, 45–46.

Kiltie, R.A. (1981a) Distribution of palm fruits on a rain forest floor: why white-lipped peccaries forage near objects, *Biotropica* 13, 141–145.

Kiltie, R.A. (1981b) Stomach contents of rain forest peccaries (*Tayassu tajacu* and *T. pecari*), *Biotropica* 13, 234–236.

Kiltie, R.A. and Terborgh, J. (1983) Observations on the behavior of rain forest peccaries in Peru: why do white-lipped peccaries form herds? *Z. Tierpsychol.* 62, 241–255.

King, D.A. (1994) Influence of light level on the growth and morphology of saplings in a Panamanian forest *Am. J. Bot.* 81, 948–957.

Kinzey, W.G. and Norconk, M.A. (1993) Physical and chemical properties of fruit and seeds eaten by *Pithecia* and *Chiropotes* in Surinam and Venezuela, *Int. J. Primatol.* 14, 207–227.

Kira, T. and Yoda, K. (1989) Vertical stratification in microclimate, in H. Lieth and M.J.A. Werger (eds.), *Tropical rain forest ecosystems: biogeographical and ecological studies*, Elsevier, Amsterdam, pp. 55–71.

Knight, R.S. and Siegfried, W.R. (1983) Inter-relationships between type, size and colour of fruits and dispersal in Southern African trees, *Oecologia* 56, 405–412.

Koivula, K., Orell, M., Rytkönen, S. and Lahti, K. (1995) Fatness, sex and dominance; seasonal and daily body mass changes in Willow Tits, *J. Avian Biology* 26, 209–216.

Kok, P.J.R. (2000) A survey of the Anuran fauna of Montagne Belvédère, County of Saül,French Guiana: field list with comments on taxonomy and ecology, *British Herpetol. Soc. Bull.* 71, 6–28.

Kollmann, J. and Schill, H.P. (1996) Spatial patterns of dispersal, seed predation and germination during colonisation of abandoned grassland by *Quercus petraea* and *Corylus avellana*, *Vegetatio* 125, 193–205.

Koop, H. and Sterck, F.J. (1994) Light penetration through structurally complex forestcanopies: an example of a lowland tropical rainforest, *Forest Ecol. Manage.* 69, 111–122.

Krebs, J.R. (1973) Social learning and the adaptative significance of mixed-species flocks of chikadees (*Parus* spp.), *Can. J. Zool.* 51, 1275–1288.

Krebs, J.R. and Davies, N.B. (1987) *An introduction to behavioural ecology*, Blackwell, Oxford.

Krebs, J.R., Clayton, N.S., Healy, S.D., Cristol, D.A., Patel, S.N. and Jolliffe, A.R. (1996) The ecology of the avian brain: food-storing memory and the hippocampus, *Ibis* 138, 34–46.

Krebs, J.R., McRoberts, M.H. and Cullen, J.M. (1972) Flocking and feeding in the Great Tit Parus major. An experimental study, *Ibis* 113, 59–72.

Krijger, C.L., Opdam, M., Théry, M. and Bongers, F. (1997) Courtship behaviour of manakins affects seed bank in the tropical rain forest of French Guiana, *J. Trop. Ecol.* 13, 631–636.

Kull, O. and Niinemets, U. (1993) Variation in leaf morphometry and nitrogen concentration in *Betula pendula* Roth, *Corlus avellane* L. and *Lonicera xylosteum* L., *Tree Physiology* 12, 311–318.

Kullander, S.O. and Nijssen, H. (1989) *The Cichlids of Surinam*. E.J. Brill Ed., Leiden.

Lacey, C.J. (1974) Rhizomes in tropical eucalypts and their role in recovery from fire damage, *Aust. J. Sci.* 22, 29–38.

Lacey, C.J. (1983) Development of large plant-like lignotubers in *Eucalyptus botryoides* in relation to environment factors, *Aust. J. Bot.* 31, 105–118.

Lacey, R.E. and Whealan, P.I. (1976) Observations on the ecological significance of vegetative reproduction in the Katherine-Darwin region of the Northern Territory, *Aust. Forest* 36, 131–139.

Lack, D. (1968) *Ecological adaptations for breeding birds*, Metheun Press, London.

Laman, T. (1996) *Ficus* seed shadows in a Bornean rain forest, *Oecologia* 107, 347–355.

Larpin, D. (1993) *Les formations ligneuses sur un inselberg de Guyane française, étude floristique, structurale et dynamique*, Thèse de Doctorat, Université Paris 6.

Larpin, D. (1994) *Les réserves naturelles des Nouragues et de La Trinité*, Rapport scientifique (Ministère de l'Environnement, DIREN Guyane, SEPANGUY).

Larpin, D., Sarthou, C. and Tardy, C. (2000) Dynamique de la végétation sur l'inselberg des Nouragues (Guyane française) à différentes échelles de temps (pluriannuelle à plurimillénaire), in M. Servant and S. Servant-Vildary (eds.), *Dynamique à long terme des écosystèmes forestiers tropicaux*, MAB-UNESCO, IRD, CNRS, pp 189–197.

Laurance, W.F., Ferreira, L.V., Rankin-de-Merona, J.M. and Hutchings, R.W. (1998) Influence of plot shape on estimates of tree diversity and community composition in Central Amazonia, *Biotropica* 30, 662–665.

Lawton, R.O. and Putz, F.E. (1988) Natural disturbance and gap-phase regeneration in a wind-exposed tropical cloud forest, *Ecology* 69, 764–777.

Lazarus, J. (1972) Natural selection and functions of flocking in birds: a reply to Murton, *Ibis* 114, 556–558.

Lazarus, J. (1979) The early warning function of flocking in birds: an experimental study with captive Quelea, *Anim. Behav.* 27, 855–865.

Le Bail, P.Y., Keith P. and Planquette, P. (2000) *Atlas des poissons d'eau douce de Guyane* (tome 2, fasc. II), Patrimoines naturels (MNHN/SPN) 43 (II).

Le Bail, P.Y., Margueridon, A., Cauty, C., Planquette, P., Prévost, E. and Loir, M. (1989) Biologie de la reproduction de *Myleus ternetzi*, *Aquat. Liv. Resour.* 2, 175–184.

Le Bail, P.Y., Planquette, P. and Géry, J. (1984) Clé de détermination des poissons continentaux et côtiers de guyane. I- Clé simplifiée des familles, *Bull. de Liaison INRA-CRAAG* 6, 1-63.

Le Bail, P.Y., Planquette, P. and Géry, J. (1984) Clé de détermination des poissons continentaux et côtiers de guyane. III- Clé simplifiée des espèces. Familles généralement inféodées au milieu marin, *Bull. de Liaison INRA-CRAAG* 8, 1-67.

Le Bail, P.Y., Planquette, P. and Géry, J. (1984) Clé de détermination des poissons continentaux et côtiers de guyane. IV- Espèces dulçaquicoles non siluriformes. *Bull. de Liaison INRA-CRAAG* 9, 1–97.

Le Corff, J. (1996) Establishment of chasmogamous and cleistogamous seedlings of an ant-dispersed understorey herb, *Calathea micans* (Maranthaceae), *Am. J. Bot.* 83, 155–161.

Leaver, L. and Daly, M. (1998) Effects of food preference on scatterhoarding by kangaroo rats (*Dipodomys merriami*), *Behaviour* 135, 823–832.

Lecomte, F., Boujard, T., Meunier, J.F., Renno, J.F. and Rojas-Beltran, R. (1993) The growth of *Myleus rhomboidalis* (Cuvier, 1817) (Characiforme, Serrasalmidae) in two rivers of French Guiana. *Rev. Ecol.- Terre Vie* 48, 431–444.

Lecomte, F., Meunier, F.J. and Rojas-Beltran, R. (1985) Mise en évidence d'un double cycle de croissance annuel chez un Silure de Guyane, *Arius couma* (VaL, 1839) (Teleostei, Siluriforme, Ariidae) à partir de l'étude squelettochronologique des épines des nageoires, *C.R. Acad. Sci., Paris,* Série III 300, 181–184.

Lecomte, F., Meunier, F.J. and Rojas-Beltran, R. (1989) Some data on the growth of *Arius proops* (Ariidae, Siluriforme) in the estuaries of French Guiana, *Aquat. Living Resour.* 2, 63–68.

Ledru, M.P., Blanc, P., Charles-Dominique, P., Fournier, M., Martin, L., Riéra, B. and Tardy, C. (1997) Reconstitution palynologique de la forêt guyanaise au cours des 3000 dernières années, *C.R. Acad. Sci. Paris* Sér. IIa 324, 469–476.

Lee, D.W. (1987) The spectral distribution of radiation in two Neotropical rainforests, *Biotropica* 19, 161–166.

Lee, D.W. and Graham, R. (1986) Leaf optical properties of rainforest sun and extreme shade plants. *Am. J. Bot* 73,1100–1108.

Lee, D.W., Bone, R.A., Tarsis, S.L. and Storch, D. (1990) Correlates of leaf optical properties in tropical forest sun and extreme-shade plants. *Am. J. Bot.* 77, 370–380.

Lieberman, M, Lieberman, D, Peralta R and Hartshorn, G.S. (1995) Canopy closure and the distribution of tropical forest tree species at La Selva, Costa Rica, *J. Trop. Ecol.* 3, 347–358

Leigh, E.G. Jr (1999) *Tropical forest ecology. A view from Barro Colorado Island,* Oxford University Press, New York, Oxford.

Leigh, E.G., Rand, A.S. and Windsor, D.M. (1983) *The ecology of a tropical forest. Seasonal rhythms and long-term changes,* Smithsonian Institution Press, Washington.

Leite, Y.L.R., Costa, L.P. and Stallings, J.R. (1996) Diet and vertical space use of three sympatric opossums in a brazilian Atlantic forest reserve, *J. Trop. Ecol.* 12, 435–440.

Leite, Y.L.R., Stallings, J.R. and Costa, L.P. (1994) Partição de recursos entre espécies simpátricas de marsupiais na reserva biológica de poço das antas, Rio de Janeiro, *Rev. Bras. Biol.* 54, 525–536.

Lescure, J. (1972) Contribution à l'étude des amphibiens de Guyane française. II. *Leptodactylus fuscus* (Schneider), *Ann. Mus. Nat. Hist. Nat. Nice* 1, 91–100.

Lescure, J. (1975) Biogéographie et écologie des amphibiens de Guyane française, *C.R. Soc. Biogéogr.* 440–442, 68–82.

Lescure, J. (1976) Contribution à l'étude des amphibiens de Guyane française. VI. Liste préliminaire des Anoures, *Bull. Mus. Nat. Hist. Nat.* Paris, 3e série, 377, *Zoologie* 265, 475–525.

Lescure, J. (1981a) Contribution à l'étude des amphibiens de Guyane française. VII. Les *Eleutherodactylus* (Anura, Leptodactylidae), *Rev. fr. Aquariol. Herpétol.*, 8, 25–32.

Lescure, J. (1981b) Contribution à l'étude des amphibiens de Guyane française. VIII. Validation d'*Atelopus spumarius* Cope, 1871, *Bull. Mus. Nat. Hist. Nat.*, Paris 4, 893–910.

Lescure, J. (1982) Les amphibiens Anoures de la forêt Guyanaise (région de Trois Sauts, Guyane française), *Mém. Mus. Nat. Hist. Nat.*, Paris Serie A, Zoologie 132, 43–51.

Lescure, J. (1984) Las larvas de Dendrobatidae, I Reunion Iberoamericana de Zoologos de Vertebrados, *Amer. Cons. Zool.* Vert., 37–45.

Lescure, J. (1986) Amphibiens et reptiles de la bande côtière en Guyane française. *Le littoral guyanais. Fragilité de l'environnement,* Congrès Sepanguy-Sepanrit, Cayenne 1985, 111–118.

Lescure, J. and Marty, C. (2000) Atlas des Amphibiens de Guyane, Patrimoines Naturels, 45.

Lescure, J., Marty, V., Marty, C., and Auber-Thomay, M. (1996) Contribution à l'étude des amphibiens de Guyane française. XI. Les *Phrynohyas* (Anura, Hylidae), *Rev. fr. Aquariol. Herpétol.*, 23, 69–76.

Lescure, J., Marty, V., Marty, C., Starace, F., Auber-Thomay, M. and Letellier, F. (1995) Contribution à l'étude des amphibiens de Guyane française. X. Les *Phyllomedusa* (Anura, Hylidae), *Rev. fr. Aquariol. Herpétol.*, 22, 35–50.

Levey, D.J. and Byrne, M.M. (1993) Complex ant-plant interactions : rain forest ants as secondary dispersers and post-dispersal seed predators, *Ecology* 74, 1802–1812.

Levey, D.J. and Cipollini, M.L. (1998) A glycoalkaloid in ripe fruit deters consumption by cedar waxwings, *Auk* 115, 359–367.

Levey, D.J. and Stiles, F.G. (1994) Birds: Ecology, behavior, and taxonomic affinities.in McDade, L.A., Bawa, K.S., Hespenheide, H.A., and Hartshorn, G.S. (eds.), *La Selva. Ecology and natural history of a Neotropical rain forest,* University of Chicago Press, Chicago and London, pp. 217–228.

Levey, D.J., Moermond, T.C., and Denslow, J.S. (1994) Frugivory: An overview, in L.A. McDade, K.S. Bawa, H.A. Hespenheide and G.S. Hartshorn (eds.), *La Selva. Ecology and natural history of a neotropical rain forest,* University of Chicago Press, Chicago and London, pp 282–294.

Lieberman, D., Hall, T.B., Swaine, M.D. and Lieberman, M. (1979) Seed dispersal by baboons in the Shai Hills, Ghana, *Ecology* 60, 65–75.

Lieberman, M. and Lieberman, D. (1986) An experimental study of seed ingestion and germination in a plant-animal assemblage in Ghana, *J. Trop. Ecol.* 2, 113–126.

Lieberman, M., Lieberman, D., Peralta, R. and Hartshorn, G.S. (1995) Canopy closure and the distribution of tropical forest tree species at La Selva, Costa Rica, *J. Trop. Ecol.* 11, 161–178.

Linares, O.F. (1984) Conclusions, in O.F. Linares and A.J. Ranere (eds.), *Adaptative radiations in prehistoric Panama,* Peabody Museum Monographs n°5. Harvard Univ. Cambridge, Massachusetts, pp 233–247.

Loir, M., Cauty, C., Planquette, P. and Le Bail, P.-Y. (1989a) Comparative study of the male reproductive tract in seven families of South-American catfishes, *Aquat. Living Resour.* 2, 45–56.

Lorenzo, J.L. (1994) La première colonisation du Nouveau monde, *Bull. Soc. Préhist. Fr.*, 91, 4/5, 342–346.

Lovejoy, T.E., Bierregaard Jr., R.O., Rylands, A.B., Malcom, J.R., Quintela, C.E., Harper, L.H., Brown Jr., K.S., Powell, A.H., Powell, G.V.N., Schubart, H.O.R. and Hays, M.B. (1985) Edge and other effects of isolation on Amazon forest fragments, in M.E. Soulé (ed.), *Conservation biology. The science of scarcity and diversity*, Sinauer Associates, Inc., Sunderland, Massachussetts, pp. 257–285.

Lowe-McConnell, R.H. (1964) The fishes of the Rupununi savanna district of British Guiana, South America. Part 1. Ecological groupings of fish species and effects of the seasonal cycle on the fish,*J. Linnean Soc. (Zool.)* 45, 103–144.

Lowe-McConnell, R.H. (1998) Freshwater fishes of northern South America: a need for field guides,*Environ. Biol. Fish.* 53, 111–115.

Lugo, A.E. and Scatena, F.N. (1996) Background and catastrophic tree mortality in tropical moist, wet, and rain forests, *Biotropica* 28, 585–599.

Lythgoe, J.N. (1979). *The ecology of vision.* Oxford University Press, Oxford.

MacArthur, R.H. (1972) *Geographical ecology: patterns in the distribution of species,* New York, Harper and Rowe.

MacArthur, R.M., Recher, H. and Cody, M. (1966) On the relation between habitat selection and species diversity, *Am. Nat.* 100, 319–332.

MacDonald, I.M.V. (1997) Field experiments on duration and precision of grey and red squirrel spatial memory, *Anim. Behav.* 54, 879–891.

Machado, S.R. (1993) A new genus of Amazonian vine snake (Xenodontinae: Alsophiini), *Acta Biol. Leopoldensia* 15, 99–108.

Mack, A. (1990) Is frugivory limited by secondary compounds in fruits? *Oikos* 57, 135–138.

Maley, J. (1996) Fluctuations majeures de la forêt dense humide africaine au cours des vingt derniers millénaires, in C.M. Hladik, A. Hladik, H. Pagézy, O. Linares, and A. Froment (eds.), *L'alimentation en forêt tropicale. Interactions bioculturelles et perspectives de développement*, UNESCO, CNRS, ORSTOM, pp. 31–52.

Mares, M.A., Ernest, K.A. and Gettinger, D.D. (1986) Small mammal community structure and composition in the Cerrado province of Central Brazil,*J. Trop. Ecol.* 2, 289–300.

Martinez del Rio, C. and Restrepo, C. (1993) Ecological and behavioral consequences of digestion in frugivorous animals, in T.H. Fleming and A. Estrada (eds.), *Frugivory and seed dispersal: Ecological and evolutionary aspects.* Advances in Vegetation Science 15, Reprinted from *Vegetatio* 107/108. Kluwer Academic Publishers, Dordrecht, Boston, London, pp. 205–216.

Marty, C. and Gaucher, P. (1999) *Guide sonore des amphibiens Anoures de Guyane. (Sound guide to the tailless amphibians of French Guiana).* (compact disc), Sittelle Ceba, Le Verdier, F-38710 Mens.

Massary, J.C. de and Lescure, J. (1998) Présence de *Ctenophryne geayi* (Amphibia: Microhylidae) en Guyane française, *Rev. fr. Aquariol. Herpétol.* 25, 53–58.

Masson, D., Breuil, A., Breuil M., Leboulanger, F., Leuge, F. and Masson, C. (1994)*La place des Chiroptères dans la dissémination, par endophytosporie, des plantes forestières de la Guadeloupe,* Report, SFEPM, SRETIE, Parc National de la Guadeloupe.

Master, T.C., Frankel, M. and Russell, M. (1993) Benefits of foraging in mixed-species wader aggregations in Southern New Jersey salt marsh,*Colonial Waterbirds* 16, 149–157.

Mattheck, C. and Kubler, H. (1995) *Wood: the internal optimisation of trees*, Springer, Berlin.

Mattison, C. (1995). *The encyclopedia of snakes*, Blandford, London.

Maury-Lechon, G. and Poncy, O. (1986) Dynamique forestière sur 6 hectares de forêt dense humidede Guyane française, à partir de quelques espèces de forêt primaire et de cicatrisation, *Mém. Mus. Natn. Hist. Nat. (A)*, 132, 211–242.

McDiarmid, R.W., T'Shaka Touré and Savage, J.M. (1996) The proper name of the neotropical tree Boa often referred to as *Corallus enydris* (Serpentes: Boidae), *J. Herpetol.* 30, 320–326.

McFarland Symington, M. (1988) Food competition and foraging party size in the black spider monkey (*Ateles paniscus chamek*), *Behaviour* 105, 117–134.

McGraw, W.S. (1996) Cercopithecid locomotion, support use, and support availability in the Tai forest, Ivory Coast, *Am. J. Phys. Anthrop.* 100, 507–522.

McKey, D. (1975) The ecology of coevolved seed dispersal systems, in L.E. Gilbert, and P.D Raven (eds.), *Coevolution of animals and plants*, Univ. of Texas Press, Austin, pp. 15–191.

McMahon, T.A. (1973) Size and shape in biology: elastic criteria impose limits on biological proportions, and consequently on metabolic rates, *Science* 179, 1201–1204.

Medem, F. (1983) *Los crocodylia de Sur America*, Bogota, 2.

Mendel, F.C. (1976) Postural and locomotor behavior of *Alouatta palliata* on various substrates, *Folia primatol.* 26, 36–53.

Mendes Pontes, A.R. (1994) Environmental determinants of primate abundance in Maraca Island, Roraima, Brazilian Amazonia. Masters thesis, University of Cambridge.

Mérigoux, S., Ponton, D. and de Mérona, B. (1998) Fish richness and species-habitat relationship in two coastal streams of French Guiana, South America, *Environ. Biol. Fish.* 51, 25–39.

Métrailler, S. and Le Gratiet, G. (1996) *Tortues continentales de Guyane française*, Bramois, P.M.S.

Meunier, F.J., Keith, P. and Le Bail, P.-Y. (1999) Conserver les specificités de l'Ichtyofaune guyanaise, *JATBA, Rev. Ethnobiol.* 40, 239–259.

Meunier, F.J., Rojas-Beltran, R., Boujard,T. and Lecomte, F. (1997) Rythmes saisonniers de la croissance chez quelques Téléostéens de Guyane française, *Rev. Hydrobiol. Trop.* 27, 423–440.

Miles, M.A., de Souza, A.A. and Póvoa, M.M. (1981) Mammal tracking and nest location in Brazilian forest with an improved spool-and-line device, *J. Zool. London* 195, 331–347.

Milési, J.P., Egal, E., Ledru, P., Vernhet, Y., Thiéblemont, D., Cocherie, A., Tegyet, M., Martel-Jantin, B. and Lagny, P. (1995) Les minéralisations du Nord de la Guyane française dans leur cadre géologique, *Chronique Rech. Minière* 518, 5–58.

Milliken, W. (1998) Structure and composition of one hectare of central amazonian *terra firme* forest. *Biotropica* 30, 530–537.

Milton, K. (1980) *The foraging strategy of howler monkeys. A study in primate economics*, Columbia University Press, New York.

Milton, K. (1981) Food choice and digestive strategies of two sympatric primate species, *Am. Nat.* 117, 476–495.

Milton, K. and Mc Bee, R.H. (1983) Rates of fermentative digestion in the howler monkey, *Alouatta palliata* (Primates: Ceboidea), *Comp. Biochem. Phys.* A 74, 29–31.

Mitchell, P.L. and Whitmore, T.C. (1993) *Use of hemispherical photographs in forest ecology.* Oxford Forestry Institute, Occasional Papers, Oxford, UK.

Mittermeier, R.A. (1978) Locomotion and posture in *Ateles geoffroyi* and *Ateles paniscus*, *Folia primatol.* 30, 161–193.

Mittermeier, R.A. and van Roosmalen, M.G.M. (1981) Preliminary observations on habitat utilization and diet in eight Surinam monkeys, *Folia Primatol.* 36, 1–39.

Moermond, T.C. and Denslow, J.S. (1983) Fruit choice in neotropical birds: effects of fruit type and accessibility on selectivity, *J. Anim. Ecol.* 52, 407–420.

Moermond, T.C., Denslow, J.S., Levey, D.J. and Santana, C.E. (1986) The influence of morphology on fruit choice in neotropical birds, in A. Estrada and T.H. Fleming (eds.), *Frugivores and seed dispersal.* Junk, Dordrecht, pp. 137–146.

Morat, P., Jaffre, T., Veillon, J.M. and Mackee, H.S. (1986) Affinités floristiques et considérations sur l'origine des maquis miniers de la Nouvelle Calédonie, Bull. Mus. Nat. Hist. Nat. Paris, 4ème Sér., 8(B), *Adansonia* 2, 133–182.

Moreno, E. and Carrascal, L.M. (1991) Patch residence time and vigilance in birds foraging at feeders: implications of bill shape, *Ethol. Ecol. Evol.* 3, 345–350.

Mori, S.A. (1991) The Guayana lowland floristic Province, *C.R. Somm. Séan. Soc. Biogéogr.* 67, 67–75.

Mori, S.A. and Boom, B.M. (1987) The forest, in S.A. Mori, and Coll., (eds.), The Lecythidaceae of a lowland neotropical forest: La Fumée Mountain, French Guiana, *Mem. New York Botanical Garden* 44, 9–29.

Mori, S.A., Boom, B.M., de Carvalho, A.M. and dos Santos, T.S. (1983) Southern Bahian moist forests, *Bot. Rev.* 49, 155–232.

Morris, D.O. (1962). The behavior of the green acouchi (*Myoprocta pratti*) with special reference to scatterhoarding, *Proceed. Zool. Soc. London* 139, 701–732.

Morrison, D. (1978) Foraging ecology and energetics of the frugivorous bat *Artibeus jamaicensis*, *Ecology* 59, 716–723.

Morrison, D. (1980) Foraging and day-roosting dynamics of canopy fruit bats in Panama, *J. Mammal.* 61, 29–39.

Morse, D.H. (1970) Ecological aspects of some mixed-species foraging flocks of birds,*Ecol. Monogr.* 40, 119–168.

Morse, D.H. (1977) Feeding behavior and predator avoidance in heterospecific groups, *Bioscience* 27, 332–339.

Morse, D.H. (1985) Flocking, in B. Campbell and E. Lack (eds.), *A dictionary of birds.*, T and A.D. Poyser, Calton., pp. 226–228.

Morton, E.S. (1973) On the evolutionary advantages and disadvantages of fruit eating in tropical birds,*Am. Nat.* 107, 8–22.

Moynihan, M. (1962) The organization and probable evolution of some mixed species flocks ofneotropical birds, *Smithsonian Miscellaneous Collections* 143, 1–140.

Muller, S. and Isbrücker, I.J.H. (1993) *Lithoxus boujardi* (Siluriformes, Loricariidae), une espèce nouvelle du bassin de l'Approuague, Guyane Française, *Cybium* 17, 71–76.

Munn, C.A. (1985) Permanent canopy and understory flocks in Amazonia: species composition and population density, in P.A. Buckley, E.S. Morton, R.S. Ridgely and F.G. Buckley (eds.),*Neotropical ornithology*, AOU Ornithological Monographs, 36, pp. 683–712.

Munn, C.A. (1986) Birds that cry wolf, *Nature* 319, 143–145.

Munn, C.A. and Terborgh, J. (1979) Multi-species territoriality in neotropical foraging flocks, *Condor* 81, 338–347.

Munsell, A.H. (1975) Munsell book of color. Munsell Color Co. Baltimore.

Murray, B.G.Jr (1985) Evolution of clutch size in tropical species of birds, in P.A. Buckley, E.S. Morton, R.S. Ridgely and F.G. Buckley (eds.), *Neotropical ornithology*, AOU Ornithological Monographs, 36, pp. 505–519.

Myers, C.W. and Cadle, J.E. (1994) A new genus for South American snakes related to*Rhadinaea obtusa* Cope (Colubridae) and resurrection of *Taeniophallus* Cope for the '*Rhadinaea*' *brevirostris* group, *Novitates, Am. Mus. Nat. Hist.*, New York, 3102, 1–33.

Nelson, J. (1994) *Fishes of the World.* 3rd edition, John Wiley & Son, New York, USA.

Nicotra, A.B., Chazdon, R.L. and Iriarte, S.V.B. (1999) Spatial heterogeneity of light and woody seedling regeneration in tropical wet forest. *Ecology* 80, 1908-1926.

Niinemets, U. (1997) Distribution patterns of carbon and nitrogen as affected by tree dimensions and relative light conditions in the canopy of *Picea abies*, *Trends Ecol. Evol.* 11, 144–154.

Niklas, K.J. (1992) *Plant biomechanics: an engineering approach to plant form and function,* University of Chicago Press, Chicago.

Norberg, U.M. (1990) *Vertebrate flight. Mechanics, physiology, morphology, ecology and evolution. Zoophysiology, vol. 27*, Springer-Verlag, Berlin, Heidelberg, New York, London, Paris, Tokyo, Hong Kong.

Notman, E., Gorchov, D.L. and Cornejo, F. (1996) Effect of distance, aggregation, and habitat on levels of seed predation for two mammal-dispersed neotropical rain forest tree species, *Oecologia* 106, 221–227.

Oates, J.F., Whitesides, G.H., Davies, A.G., Waterman, P.G., Green, S.M., Dasilva, G.L. and Mole, S. (1990) Determinants of variation in tropical forest primate biomass: new evidence from West Africa,*Ecology* 71, 328–343.

Oberbauer, S.F. and Strain, B.R. (1986) Effects of canopy position and irradiance on the leaf physiology and morphology of*Pentaclethra macroloba* (Mimosaceae), *Am. J. Bot.* 73, 409–416.

Oberbauer, S.F., Clark, D.A., Clark, D.B. and Quesada, M. (1989) Comparative analysis of photosynthetic light environemtns within the crowns of juvenile rain forest trees. *Tree Physiology* 5, 13-23

Oldeman, R.A.A. (1974a) Ecotopes des arbres et gradients écologiques verticaux en forêt guyanaise,*Rev. Ecol.-Terre Vie* 28, 487–520.

Oldeman, R.A.A. (1974b) L'architecture de la forêt guyanaise, *Mém. ORSTOM* 73, 1–204.

Oldeman, R.A.A. (1978) Architecture and energy exchange of dicotiledonous trees in the forest, in R.B. Tomlinson and M.H Zimmermann (eds.), *Tropical trees as living systems*, Cambridge University Press, pp. 535–560.

Olson, S.L. (1985) The fossil record of birds, in D.S. Farner, J.R. King, and D.C. Parkes (eds.),*Avian biology, VIII*, Academic Press, New York, pp. 80–238

Orian, G.H. (1969) The number of bird species in some tropical forests, *Ecology* 50, 783–801.

Ouboter, P.E. (1993) Introduction, in P.E. Ouboter (ed.) *The freshwater ecosystems of Suriname*, Kluwer Acad. Publi., Dordrecht, pp 1–6.

Pack, K.S., Henry, O., and Sabatier, D. (1999) The insectivorous-frugivorous diet of the golden-handed tamarin (*Saguinus midas midas*) in French Guiana, *Folia Primatol.* 70, 1–7.

Pascal, M., Hostache, G., Tessier, C., and Vallat, P. (1994) Cycle de reproduction et fécondité de l'Atipa, *Hoplosternum littorale* (Siluriforme) en Guyane française, *Aquat. Liv. Resour.* 7, 25–37.

Patton, J.L., Silva, M.N., Lara, M.C. and Mustrangi, M.A. (1997) Diversity, differentiation, and the historical biogeography of nonvolant small mammals of the Neotropical forests, in W.F. Laurance and R.O. Bierregaard (ed.), *Tropical forest remnant: Ecology, management, and conservation of fragmented communities,* University of Chicago Press, pp. 455–465.

Pearcy, R.W. (1987) Photosynthetic gas exchange responses of Australian tropical forest trees in canopy, gap, and understory micro-environments,*Funct. Ecol.* 1, 169–178.

Pearcy, R.W. (1990) Sunflecks and photosynthesis in plant canopies.*Ann. Rev. Plant Physiol. Plant Mol. Biol.* 41, 421–453.

Pearson, D.L. (1977) A pantropical comparison of bird community structure on six lowland forest sites,*Condor* 79, 232–244.

Pélissier, R. (1998) Tree spatial patterns in three contrasting plots of a southern Indian tropical moist evergreen forest. *J. Trop. Ecol.* 14, 1–16.

Pélissier, R. and Riéra, B. (1993) Dix ans de dynamique d'une forêt dense humide de Guyane,*Rev. Ecol. - Terre Vie* 48, 21–33.

Pena-Claros, M. (2001) *Secondary forest succession. Processes affecting the regeneration of Bolivian tree species.* PhD. Thesis, Utrecht University.

Peres, C.A. (1993) Notes on the ecology of buffy saki monkeys (*Pithecia albicans*, Gray 1860): a canopy seed predator, *Am. J. Primatol.* 31, 129–140.

Peres, C.A. (1997) Primate community structure at twenty western Amazonian flooded and unflooded forests,*J. Trop. Ecol.* 13, 381–405.

Phillips, O.L. (1996) Long-term environmental change in tropical forests: increasing tree turnover,*Biol. Conserv.* 23, 235–248.

Phillips, O.L. and Gentry, A.H. (1994) Increasing turnover through time in tropical forests, *Science* 263, 954–958.

Phillips, O.L., Hall, P., Gentry, A.H., Sawyer, S.A. and Vasquez, R. (1994) Dynamics and species richness of tropical forests, *Proceed. Nat. Acad. Sci USA* 91, 2805–2809.

Pianka, E.R. (1966). Latitudinal gradients in species diversity: a review of concepts, *Am. Nat.* 100, 33–46.

Pianka, E.R. (1973) The structure of lizard communities, *Annu. Rev. Ecol. Syst.* 4, 53–74.

Pickett, S.T.A. and White, P.S. (eds.) (1985) *The ecology of natural disturbance and patch dynamics.* Academic Press, New York.

Planquette, P., Keith, P. and Le Bail, P.Y. (1996) *Atlas des poissons d'eau douce de Guyane* (tome 1). Collection du Patrimoine naturel, vol. 22. IEGB-MNHN, INRA, CSP, Min. Env., Paris.

Platt, W.J. and Strong, D.R. (1989) Special feature: gaps in forest ecology, *Ecology* 70, 535–576.

Poncy, O., Riéra, B., Larpin, D., Belbenoit, P., Jullien, M., Hoff, M. and Charles-Dominique, P. (1998) The permanent field research station 'Les Nouragues' in the tropical rainforest of French Guiana: current projects and preliminary results on tree diversity, structure and dynamics, in F. Dallmeier and J.A Comiskey (eds.), *Forest Biodiversity in North, Central and South America and the Caribbean: Research and Monitoring,* Man and the Biosphere Series 21 UNESCO and The Parthenon Publishing Group, Carnforth, Lancashire, UK., pp. 385–410,

Poncy, O., Sabatier, D., Prévost, M.F. and Hardy, I. (2001) The lowland high forest: structure and tree species diversity, in F. Bongers, P. Charles-Dominique, P.M. Forget, and M. Théry (eds.),*Nouragues: Dynamics and plant-animal interactions in a neotropical rainforest,* Kluwer Academic Publisher, Dordrecht.

Pons, T.L., Schieving, F., Hirose, T. and Werger, M.J.A. (1989) Optimalization of leaf nitrogen allocation for canopy photosynthesis in *Lysimachia vulgaris*, in H. Lambers, M.L. Cambridge, H. Konings and T.L. Pons (eds.), *Causes and consequences of variation in growth rate and productivity of higher plants,* SPB Academic Publishing BV, The Hague, pp. 175–186

Ponton, D. and Copp, G.H. (1997) Early dry-season community structure and habitat use of young fish in tributaries of the river Sinnamary (French guiana, South America) before and after hydrodam operation, *Environ. Biol. Fish.* 50, 235–256.

Ponton, D. and de Mérona, B., (1998) Fish life-history tactics in a neotropical river with a highly stochastic hydrological regime: the Sinnamary River, French Guiana, South America, *Pol Arch. Hydrobiol.* 45, 201–224.

Ponton, D. and Tito De Morais, L. (1997) Les stratégies de reproduction et les premiers stades de vie des poissons du fleuve Sinnamary (Guyane française): une revue bibliographique,*Rev. Hydrobiol. Trop.* (1994) 27, 441–465.

Poorter, L. (1998) *Seedling growth of Bolivian rain forest tree species in relation to light and water availability.* PhD thesis, Utrecht University, Utrecht.

Poorter, L. (1999). Growth responses of 15 rainforest tree species to a light gradient: the relative importance of morphological and physiological traits, *Funct. Ecol.* 13, 396-410.

Poorter, L. (2000) Effects of seasonal drought on gap and understorey seedlings in a Bolivian moist forest*J. Trop. Ecol.* 16, 481–498.

Poorter, L., Oberbauer, S.F. and Clark, D.B. (1995) Leaf optical properties along a vertical gradient in a tropical rain forest canopy in Costa Rica. *Am. J. Bot.* 82, 1257–1263.

Poorter, L., Jans, L., Bongers, F. and Van Rompaey, R.S.A.R. (1994) Spatial distribution of gaps along 3 catenas in the moist forest of Tai National Park, Ivory-Coast, *J. Trop. Ecol.* 10, 385–398.

Popma, J. and Bongers, F. (1988) The effect of canopy gaps on growth and morphology of seedlings of rain forest species, *Oecologia* 75, 625–632.

Popp, J.W. (1988) Scanning behaviour of finches in mixed-species groups, *Condor* 90, 510–512.

Porembski, S. (1996) Notes on the vegetation of inselbergs in Malawi, *Flora* 191, 1–8.

Porembski, S. and Barthlott, W. (1992) Struktur und Diversität der Vegetation westafrikanischer Inselberge, *Geobot. Kolloq* 8, 69–80.

Porembski, S., Barthlott, W., Dörrstock, S. and Biedinger, N. (1994) Vegetation of rock outcrops in Guinea: granite inselbergs, sandstone table mountains and ferricretes - remarks on species numbers and endemism, *Flora* 189, 1–12.

Porembski, S., Brown G. and Barthlott, W. (1995) An inverted latitudinal gradient of plant diversity in shallow depressions on Ivorian inselbergs, *Vegetatio* 117, 151–163.

Porembski, S., Martinelli, G., Ohlemüller, R. and Barthlott, W. (1998) Diversity and ecology of saxicolous vegetation mats on inselbergs in the Brazilian Atlantic rainforest, *Divers. Distrib.* 4, 107–119.

Powell, G.V.N. (1974) Experimental analysis of the social value of flocking by starlings (*Sturnus vulgaris*) in relation to predation and foraging, *Anim. Behav.* 22, 501–505.

Powell, G.V.N. (1977) *Socioecology of mixed species flocks in a neotropical forest,* Unpublished Ph.D. dissertation, University of California, Davis.

Powell, G.V.N. (1979) Structure and dynamics of interspecific flocks in a mid-elevation neotropical forest, *Auk* 96, 375–390.

Powell, G.V.N. (1985) Sociobiology and adaptative significance of interspecific foraging flocks in the Neotropics, in P.A. Buckley, E.S. Morton, R.S. Ridgely and F.G. Buckley (eds.), *Neotropical ornithology*, AOU Ornithological Monographs, 36, pp. 713–732.

Powell, G.V.N. (1989) On the possible contribution of mixed species flocks to species richness in neotropical avifaunas, *Behav. Ecol. Sociobiol.* 24, 387–393.

Powers, M.E. (1983) Grazing responses of tropical freshwater fishes to different scales of variation in their food, *Environ. Biol. Fish.* 9, 103–115.

Prance, G.T. (1973) Phytogeographic support for the theory of Pleistocene forest refuges in the Amazon Basin, based on evidence from distribution patterns in Caryocaraceae, Chrysobalanaceae, Dichapetalaceae and Lecythidaceae, *Acta Amazonica* 3, 5–28.

Prance, G.T. (1982) Forest refuges: evidence from woody angiosperms, in G.T. Prance (ed.), *Biological diversification in the tropics*, Columbia University Press, New York, pp. 137–157.

Prance, G.T. (1983) A review of phytogeographic evidences for Pleistocene climate changes in the neotropics, *Ann. Mo Bot. Gard.* 69, 594–624.

Prance, G.T. (1996) Islands in Amazonia, *Philos. T. Roy. Soc. B* 351, 823–833.

Prévost, M.F. (1981) Mise en évidence d'espèces pionnières dans le sol de forêt primaire en Guyane, *Turrialba* 31, 121–127.

Prévost, M.F. (1983) Les fruits et les graines des espèces végétales pionnières de Guyane française, *Rev. Ecol.-Terre Vie* 38, 121–145.

Prévost, M.-F. and Sabatier, D. (1996) Variations spatiales de la richesse et de la diversité du peuplement arboré en forêt guyanaise, in J.L. Guillaumet, M.Belin and H. Puig (eds.), *Phytogéographie tropicale, réalités et perspectives*, Actes du Coll. Intern. Phytogéographie tropicale, Paris 6-8 juillet 1993, ORSTOM, Colloques et Séminaires, pp. 263–277.

Price, M.V. and Jenkins, H.H. (1986) Rodents as seed consumers and dispersers, in D.R. Murray (ed.), *Seed dispersal*, Academic Press, Sidney, pp. 191–235.

Prum, R.O. (1992) Syringeal morphology, phylogeny, and evolution of the Neotropical manakins (Aves: Pipridae), *Am. Mus. Novit.* 3043.

Puaux, (1993) Sinnamary, Barrage de Petit Saut, in G. Mazière (ed.), *Bilan scientifique de la région Guyane, DRAC*, Service régional de l'Archéologie, pp. 31–43.

Puig, H., Blasco, F. and Bellam, M.P. (1981) Carte de la végétation d'Amérique du Sud. *Service de la carte international du tapis végétal*, Toulouse, UNESCO, Paris, 189 p, 1 carte.

Pujos, M. and Odin, G.S. (1986) La sédimentation au quaternaire terminal sur la plate-forme continentale de la Guyane française, *Oceanol. Acta.* 9, 363–382.

Pulliam, H.R. (1973) On the advantages of flocking, *J. Theor. Biol.* 38, 419–422.

Putz, F.E. and Chan, H.T. (1986) Tree growth, dynamics and productivity in a mature mangrove forest in Malaysia, *Forest Ecol. Manag.* 17, 211–230.

Puyo, J. (1949) *Faune de l'Empire français. XII. Poissons de la Guyane française*. ORSTOM, Paris, 280 pp.

Raaimakers, D., Boot, R.G.A., Dijkstra, P., Pot, S. and Pons, T.L. (1995) Photosynthetic rates in relation to leaf phosphorus content in pioneer versus climax tropical rain forest trees,*Oecologia* 102, 120–125.

Rankin, J.M. (1978) The influence of seed predation and plant competition on tree species abundances in two adjacent tropical rain forest, University of Michigan, Ann Arbor.

Rankin-de-Merona, J.M., Hutchings, R.W. and Lovejoy, T.E. (1990) Tree mortality and recruitment over a five-year period in undisturbed upland rainforest of the Central Amazon, in A.H. Gentry (ed.),*Four neotropical rainforests*, Yale Univ. Press, New Haven, pp. 573–584.

Rankin-de-Merona, J.M., Prance, G.T., Silva, M.F., Rodrigues, W.A. and Uehling, M. (1992) Preliminary results of a large-scale tree inventory of upland rain forest in the central Amazon, *Acta Amazonica* 22, 493–534.

Redford, K.H., MacLean Stearman, A. and Trager, J.C. (1989) The kinkajou (*Potos flavus*) as a myrmecophage, *Mammalia* 53, 132-134.

Reid, F.A. (1997) *A field guide to the mammals of Central America and Southeast Mexico,* Oxford University Press.

Reifsnyder, W.E. Furnival, G.M. and Horowitz, J.L. (1971) Spatial and temporal distribution of solar radiation beneath forest canopies. *Agr. Meteorol.* 9, 21–37.

Renno, J.F., Berrebi, P., Boujard, T. and Guyomard, R. (1990) Intraspecific differentiation of*Leporinus friderici* (Anostomidae, Pisces) in French Guiana and Brazil: a genetic approach to the refuge theory,*J. Fish Biol.* 36, 85–95.

Renno, J.F., Guyomard, R., Boujard, T. and Bastide, G. (1989) Evidence for genetic isolation among four morphological species of *Leporinus* (Anostomidae, Pisces) in French Guiana, *Aquat. Liv. Resour.* 2, 127–134.

Richards, P.W. (1952) *The tropical rain forest*, Cambridge University Press, Cambridge.

Richards, P.W. (1996) *The tropical rainforests, an ecological study*. 2nd ed., Cambridge Univ. Press, Cambridge.

Ricklefs, R.E. and Schluter, D. (eds.), (1993) *Species diversity in ecological communities. Historical and geographical perspectives*, The University of Chicago Press, Chicago and London.

Ridgely, R.S. and Tudor, G. (1994) *The birds of South America, Vol. II, the suboscine Passerines*, Oxford University Press, Oxford.

Riéra, B. (1983) *Chablis et cicatrisation en forêt guyanaise (piste de St Elie).* Thèse 3e cycle, Université Paul Sabatier, Toulouse, 191 p.

Riéra, B. (1995) Rôle des perturbations actuelles et passées dans la dynamique et la mosaïque forestière, *Rev. Ecol. - Terre Vie* 50, 15–20.

Riéra, B. (1995) Role of past and present disturbances in forest dynamics and structure,*Rev. Ecol.- Terre Vie* 50, 209–222.

Riéra, B. and Joly, A. (1996) *Eperua falcata*: un cas de distribution spatiale en relation avec les changements climatiques récents, *Actes Symp. Dynamique à long terme des écosystèmes forestiers intertropicaux*, ORSTOM/CNRS, 247–248.

Riéra, B., Puig, H. and Lescure, J.P. (1990) La dynamique de la forêt naturelle, *Bois et forêts Trop.* 219, 69–78.

Rijkers, T. (2000) *Leaf function in tropical rain forest canopy trees: the effect of light on leaf morphology and physiology in different-sized trees.* PhD thesis, Wageningen University, Wageningen.

Rijkers, T., Pons, T.L. and Bongers, F. (2000a) The effect of tree height and light availability on photosynthetic traits of four neotropiocal species differing in shade tolerance. *Funct. Ecol.* 14, 77–68.

Rijkers, T., Pons, T.L. and Bongers, F. (2000b) Photosynthetic induction in three shade tolerant tree species: comparing understorey with gap grown saplings in a French Guiana rain forest,*Oecologia 125:331-340.*

Robinette, W.L., Loveless, C.M. and Jones, D.A. (1974) Field tests of strip census methods,*J. Wildlife. Manage.* 38, 81–96.

Robinson, J.G and Redford, K.H. (1986) Body size, diet, and population density ofneotropical mammals, *Am. Nat.* 128, 665–680.

Robinson, J.G. (1986) Seasonal variation in use of time and space by the wedge-capped capuchin monkey,*Cebus olivaceus* Smiths, *Contrib. Zool.* 43, 1–60.

Robinson, J.G. and Janson, C.H. (1987) Capuchins, squirrel monkeys and atelines: socioecological convergence with Old World Monkeys, in B.B. Smuts, D.L. Cheney, R.M. Seyfarth, R.W. Wrangham, and T.T. Struhsaker (eds.), *Primate Societies*, The University of Chicago Press, Chicago, London, pp. 69–82.

Robinson, S.K. and Terborgh, J. (1990) Bird communities of the Cocha Cashu Biological Station in Amazonian Peru, in A.H. Gentry (ed.) *Four neotropical rainforests*, Yale University Press, New Haven, Conn, pp. 199-216.

Robinson, S.K., Terborgh, J. and Munn, C.A. (1990) Lowland tropical forest bird communities of a site in Western Amazonia, in A. Keast (ed.), *Biogeography and ecology of forest bird communities*. SPB Academic Publishing bv, The Hague, The Netherlands, pp 229–258.

Rodrigues, M., Olmos, F. and Galetti, M. (1993) Seed dispersal by tapir in Southeastern Brazil, *Mammalia* 57, 460-462.

Rodriguez, M.A. and Lewis, W.M. (1997) Structure of fish assemblages along environmental gradients in floodplain lakes of the Orinoco river, *Ecol. Monog.* 67 109–128.

Rodríquez, L.O. and Duellman, W.E. (1994) *Guide to the frogs of the Iquitos region, Amazonian Peru, Univ. Kansas Publ. Mus. Nat. Hist.* 22.

Rojas-Beltran, R. (1984) Clé de détermination des poissons continentaux et côtiers de guyane, II- Siluriformes, *Bull. Liaison INRA-CRAAG* 7, 1–63.

Rojas-Beltran, R. (1986) Evolution du peuplement ichtyologique d'un petit cours d'eau temporaire de la savane littorale de Guyane, *Cybium* 10, 263–277.

Rollfinke, B.F. and Yahner, R.H. (1991) Flock structure of wintering birds in an irrigated mixed oak-forest, *Wilson Bull.* 103, 282–285.

Room, P.M., Maillette, L. and Hanan, J.S. (1994) Modular and metamer dynamics and virtual plants, *Advances in Ecological Research* 25, 105–157.

Roosvelt, A.C., Lima da Costa, M., Lopes Machado, C., Michab, M., Mercier, N., Villadas, H., Feathers, J., Barnett, W., Imazio da Silveira, M., Henderson, A., Silva, J., Chernoff, B., Reese, D.S., Holman, J.A., Toth, N. and Schick, K. (1986) Paleoindian cave dwellers in the Amazon: the peopling of the Americas, *Science* 72, 373–384.

Rosenberger, A.L. (1992) Evolution of feeding niches in New World monkeys, *Am. J. Phys. Anthrop.* 88, 525–562.

Rosenberger, A.L. and K.B. Strier (1989) Adaptive radiation of the ateline primates, *J. Hum. Evol.* 18, 717–750.

Roubach, R. and Saint-Paul, U. (1994) Use of fruit and seeds from Amazonian inundated forests in feeding trials with *Colossoma macropomum* (Cuvier, 1818) (Pisces, Characidae), *J. Appl. Ichthyol.* 10, 134–140.

Rousteau, A. (1990) *Essai sur le polymorphisme des plantules d'Angiospermes*, Thèse de Doctorat, Université Paris 6.

Runkle, J.R. (1981) Gap regeneration in some old-growth forests of the eastern United States, *Ecology* 62, 1041–1051.

Rylands, A.B., Mittermeier, R.A. and Rodriguez Luna, E. (1995) A species list for the New World primates (Platyrrhini): Distribution by country, endemism, and conservation status according to the Mace-Lande system, *Neotrop. Primates* 3 (suppl.), 113–160.

Sabatier, D. (1983) *Fructification et dissémination en forêt guyanaise. L'exemple de quelques espèces ligneuses*, Thèse 3ème cycle, U.S.T.L., Montpellier.

Sabatier, D. (1985) Saisonnalité et déterminisme du pic de fructification en forêt guyanaise, *Rev. Ecol.-Terre Vie* 40, 289–320.

Sabatier, D. and Prévost, M.F. (1990a) Variations du peuplement forestier à l'échelle stationnelle; le cas de la Station des Nouragues en Guyane française, Atelier MAB/UNESCO, IUFRO, FAO, 12–16 mars 1990, Cayenne, 169–187.

Sabatier, D. and Prévost, M.-F. (1990b) Quelques données sur la composition floristique et la diversité des peuplements forestiers de Guyane française, *Bois. et Forêts Trop.* 219, 31-55.

Sabatier, D., Grimaldi, M., Prévost, M.-F., Guillaume, J., Godron, M., Dosso, M. and Curmi, P. (1997) The influence of soil cover organization on the floristic and structural heterogeneity of a Guianan rain forest, *Plant Ecol.* 131, 81-108.

Saldarriaga, J. G. (1994) *Estudios en la Amazonia Colombiana, V Recuperacion de la selva de 'Tierra Firme' en el alto rio Negro Amazonia colombiana-venezolana*, Tropenbos Foundation, Wageningen.

Sallabanks, R. and Courtney, S.P. (1992) Frugivory, seed predation and insect-vertebrate interactions, *Annu. Rev. Entomol.* 37, 377–400.

Sarthou, C. (1992) *Dynamique de la végétation pionnière sur un inselberg en Guyane française*, Thèse de Doctorat, Université Paris 6.

Sarthou, C. and Grimaldi, C. (1992) Mécanismes de colonisation par la végétation d'un inselberg granitique en Guyane française, *Rev. Ecol. - Terre Vie* 47, 329–349.

Sarthou, C. and Villiers, J.F. (1998) Epilithic plant communities on inselbergs in French Guiana, *J. Veg. Sci.* 9, 847–860.

Sarthou, C., Thérézien, Y., and Couté, A. (1995) Cyanophycées de l'inselberg des Nouragues (Guyane française), *Nova Hedwigia* 61, 85–109.

Sastre, C. (1977) Quelques aspects de la phytogéographie des milieux ouverts guyanais,*Publ. Lab. Zool. Ecole Norm.* 9, 67–74.

Sasvari, L. (1979) Observational learning in Great, Blue, and Marsh Tit, *Anim. Behav.* 27, 767–771.

Sasvari, L. (1992) Great Tits benefit from feeding in mixed-species flocks: a field experiment,*Anim. Behav.* 43, 289–296.

Schnell, R. (1966) Problèmes phytogéographiques, écologiques et économiques de la Caatinga brésilienne, *JATBA*. 13 (1,2,3), 59–90.

Schnell, R. (1987) *La Flore et la Végétation de l'Amérique Tropicale, Tome 1,* Masson, Paris.

Schoetlle, A.W. (1994) Influence of tree size on shoot structure and physiology of*Pinus contorta* and *Pinus aristata, Tree Physiol.* 14, 1055–1068.

Schomburgk, R.H. (1841) *Fishes of Guyana,* Naturalist's library, Edinburgh, 31.

Schupp, E.W. (1992) The Janzen-Connell model for tropical tree diversity: population implications and the importance of spatial scale, *Am. Nat.* 140, 526–530.

Schupp, E.W. and Fuentes, M. (1995) Spatial patterns of seed dispersal and the unification of plant population ecology, *Ecoscience* **2**, 267-275.

Schupp, E.W., Howe, H.F., Augspurger, C.K. and Levey, D.J. (1989) Arrival and survival in tropical treefall gaps, *Ecology* 70, 562–564.

Schwartz, D. (1993) Les retombées de litière en tant que source de carbone et de l'azote du sol. Quantification et périodicité des apports en relation avec les caractéristiques climatiques et édaphiques dans deux parcelles de forêt dense à Dimonika (Mayombe, Congo), in B. Cros, J. Diamouangana, and M. Kabala (eds.),*Echanges forêt-atmosphère en milieu tropical humide. Recueil des travaux effectués dans le Mayombe,* UNESCO, Paris, pp. 141–158.

Schwartz, D., De Foresta, H., Deschamps, R. and Lanfranchi, R. (1990) Découverte d'un premier site de l'âge de fer ancien (2110 BP) dans le Mayombe congolais. Implications paléobotaniqueset pédologiques, *C.R. Acad. Sci. Paris*, série II, 1293–1289.

Shephard, V.E. and Chapman, C.A. (1998) Dung beetles as secondary dispersers: impact on seed predation and germination, *J. Trop. Ecol.* 14, 199–215.

Sibley, C.G. and Ahlquist, J.E. (1990) *Phylogeny and the classification of birds. A study in molecular evolution,* Yale University Press, New Haven & London.

Siegel, S. (1956) *Nonparametric statistics for the behavioral sciences*, McGraw-Hill, New York.

Simmen, B. (1991). *Stratégies alimentaires des primates néotropicaux en fonction de la perception des produits de l'environnement végétal,* Thèse de Doctorat, Université Paris-Nord, Villetaneuse.

Simmen, B. (1992) Competitive utilization of *Bagassa* fruits by sympatric howler and spider monkeys, *Folia Primatol.* 58, 155–160.

Simmen, B. and Sabatier, D. (1996) Diets of some French Guianan primates: food composition and food choices, *Int. J. Primatol.* 17, 661–693.

Simmen, B., Hladik, C.M. and Martin, R.D. (1995) Sweet and bitter taste discrimination and energy requirements in nonhuman primates, *Chem. Senses* 20, 153 (abstract).

Simmen, B., Julliot, C., Bayart, F. and Pagès, E. (1998) Densités de primates en forêt guyanaise. Test d'une méthode d'estimation par transect, *C.R. Acad. Sci. Paris* 321, 699–704.

Simmen, B., Julliot, C., Bayart, F. and Pagès-Feuillade, E. (2001) Diet and population densities of th primate comunity in relation to fruit supplies, in F. Bongers, P. Charles-Dominique, P.M. Forget, and M. Théry (eds.), *Nouragues: Dynamics and plant-animal interactions in a neotropical rainforest,* Kluwer Academic Publishers, Dordrecht.

Simmons, N. and Voss, R.S. (1998) The mammals of Paracou, French Guiana : a neotropical lowland rainforest fauna, Part 1 Bats, *B. Am. Mus. Nat. Hist*, 237, 219 p.

Simms, E.L. (1992) Costs of plant resistance to herbivory, in R.S. Fritz and E.L. Simms (eds.),*Plant resistance to herbivores and pathogens. Ecology, evolution and genetics*, The University of Chicago Press, Chicago, London, pp. 392–425.

Skutch, A.F. (1985) Clutch size, nesting success, and predation on nests of neotropical birds, reviewed, in P.A. Buckley, E.S. Morton, R.S. Ridgely and F.G. Buckley (eds.),*Neotropical ornithology*, AOU Ornithological Monographs, 36, pp. 575–594.

Smallwood, P.D. and Peters, W.D. (1986) Grey squirrel food preferences: the effects of tannin and fat concentration, *Ecology* 67, 168–174.

Smith, C.C. and Reichman, O.J. (1984) The evolution of food caching by birds and mammals,*Annu. Rev. Ecol. Syst.* 15, 329–351.

Smith, D.C. and Buskirk, J. (1988) Winter territoriality and flock cohesion in the Black-capped chikadee, Parus atricapillus, *Anim. Behav.* 36, 466– 476.

Smith, H. (1995). Physiological and ecological function within the phytochrome family,*Annu. Rev. Plant Physiol. Mol. Biol.* 46, 289-315.
Smythe, N. (1970) Relationships between fruiting seasons and seed dispersal methods in aneotropical forest, *Am. Nat.* 104, 25–35.
Smythe, N. (1978) The natural history of the Central American agouti (*Dasyprocta punctata*), *Smithsonian Contributions Zoology* 257, 1–52.
Smythe, N., Glanz, W. E. and Leigh, E. G., Jr. (1982) Population regulation in some terrestrial frugivores, in E.G. Leigh, Jr., A.S. Rand and D.M. Windsor (eds.) *The ecology of a tropical forest*, Smithsonian Institution Press, Washington, D.C., pp. 227-252.
Snow, B. and Snow, D. (1988) *Birds and berries*, Poyser, Calton, UK.
Snow, D. (1981) Tropical frugivorous birds and their food plants: a world survey, *Biotropica*, 13, 1–14.
Snow, D. (1982) *The Cotingas*. British Museum, London, and Oxford University Press, Oxford.
Soné, K. and Kohno, A. (1996) Application of radiotelemetry to the survey of acorn transportation by*Apodemus* mice, *Ecol. Res.* 11, 187–192.
Sork, V.L. and Boucher, D.H. (1977) Dispersal of sweet pignut hickory in a year of low fruit production, and the effect of predation by a curculionid beetle, *Oecologia (Berl.)* 28, 289–299.
Soubies, F. (1980) Existence d'une phase sèche en Amazonie brésilienne, datée par la présence de charbons dans les sols (- 5000-3000 ans B.P.), *Cahiers ORSTOM*, série Géologie, XI, pp. 133–148.
Sourd, C. and Gautier-Hion, A. (1986) Fruit selection by a forest guenon. *J. Anim. Ecol.* 55, 235–244.
Spichiger, R., Bertoni, B.S., and Loiseau, P.A. (1992) The forests of the Paraguayan Alto Parana,*Candollea* 47, 219–250.
Spironello, W.R. (1999) *The Sapotaceae community ecology in a central amazonian forest: effects of seed dispersal and seed predation*, Ph.D. dissertation, Department of Anatomy, Churchill College, Cambridge.
Stapanian, M.A. and Smith, C.C. (1984) Density–dependent survival of scatterhoarded nuts: an experimental approach, *Ecology* 65, 1387–1396.
Stapanian, M.A. and Smith, C.C. (1986) A model for seed scatterhoarding: coevolution of fox squirrels and black walnut, *Ecology* 59, 884–896.
Starace, F. (1998) *Guide des serpents et amphisbènes de Guyane française*. Guyane, IBIS rouge editions.
Steel, M.A., Knowles, T., Briddle, K. and Simms, E.L. (1993) Tannins and partial consumption of acorns: Implications for dispersal of oaks by seed predators, *Am. Midl. Nat.* 130, 229–238.
Steele, M.A., Hadj, C.L.Z., and Hazeltine, J. (1996) Caching and feeding decisions by *Sciurus carolinensis*: responses to weevil-infested acorns, *J. Mammal.* 77, 305–314.
Sterck, F.J. (1997) *Trees and light: Tree development and morphology in relation to light availability in a tropical rain forest in French Guiana*, Ph.D.Thesis, Wageningen University, Wageningen, the Netherlands.
Sterck, F.J. (1999) Crown development in tropical rain forest trees in gaps and understorey,*Plant Ecol.* 143, 89–98.
Sterck, F.J. and Bongers, F. (1998) Ontogenetic changes in size, allometry, and mechanical design of tropical rain forest trees, *Am. J. Bot.* 85, 266–272.
Sterck, F.J. and Bongers, F. (2001) Crown development in tropical rain forest trees: patterns with tree height and light availability, *J. Ecol. 89:1-13*.
Sterck, F.J., Clark, D.B., Clark, D.A. and Bongers, F. (1999) Fluctuating light environments and crown plasticity of saplings in a tropical rain forest. *J. Trop. Ecol.* 15, 83–95.
Steyermark, J.A., Berry, P.E. and Holst, B.K. (1995) Introduction, in P.E. Berry, B.K Holst, and K. Yatskievych, (eds.), *The flora of the Venezuelan Guayana, Vol. 1.*, Missouri Botanical Garden, Timber Press.
Stiles, E.W. and White, D.W. (1986) Seed deposition patterns: influence of season, nutrients, and vegetation structure, in A. Estrada and T.H.Fleming (eds.), *Frugivores and seed dispersal*, Junk, Dordrecht, pp. 45–54.
Stiles, F.G. (1985) On the role of birds in the dynamics of Neotropical forests, in A.W. Diamond and T.E. Lovejoy (eds.), *Conservation of tropical forest birds.* ICBP Technical Paper n° 4, Cambridge, pp. 49–59.
Stöcklin, J. and Bäumler, E. (1996) Seed rain, seedling establishment and clonal growth strategies on a glacier foreland, *J. Veg. Sci.* 7, 45–56.
Stouffer, P.C. and Bierregaard, R.O. (1995) Use of amazonian forest fragments by understorey insectivorous birds, *Ecology* 76, 2429– 2445.
Strier, K.B. (1992) Atelinae adaptations: Behavioral strategies and ecological constraints.*Am. J. Phys. Anthrop.* 88, 515-524.
Struhsaker, T.T. (1975) *The Red Colobus Monkey*, University of Chicago Press, Chicago.
Suhonen, J. (1993a) Risk of predation and foraging sites of individuals in mixed-species tit flocks,*Anim. Behav.* 45, 1193-1198.
Suhonen, J. (1993b) Predation risk influences the use of foraging sites by tits, *Ecology* 74, 1197-1203.

Sussman, R.W. and Phillips-Conroy, J.E. (1995) A survey of the distribution and density of the primates of Guyana, *Int. J. Primatol.* 16, 761–791.

Sutton, S.L., Whitmore, T.C., and Chadwich, A.C. (1983) *Tropical rain forest: ecology and management*, British Ecological Society Special Publ. Ser. n° 2. Blackwell Scientific Publications, Oxford.

Svenning, J.-C. (2000) Small canopy gaps influence plant distributions in the rain forest understorey.*Biotropica* 32, 252–261.

Tanaka, H. and Nakashizuka, T. (1997) Fifteen years of canopy dynamics analysed by aerial photographs in a temperate deciduous forest, Japan, *Ecology* 78, 612–620.

Tardy C., Vernet, J.L., Solari, M.E., Wengler, L., Fournier, M., Leprun, J.C., Pessenda, L.C.R., Servant, M., Sifeddine, A., Soubiès, F. and Turcq, B. (2000) Feux, sols et écosystèmes forestiers tropicaux, in M. Servant et S. Servant-Vildary (eds.), *Dynamique à long terme des écosystèmes forestiers tropicaux* Actes Symp. Internat., Paris 20-22 mars 1996, pp 343–348.

Tardy, C. (1998) *Paléofeux et environnements forestiers de Guyane française au Tardiglaciaire et à l'Holocène, Approches chronologique et anthracologique,* Thèse de Doctorat, Université Montpellier II.

Tejerina-Garro, F.L., Fortin, R. and Rodriguez, M.A. (1998) Fish community structure in relation to environmental variation in floodplain lakes of the Araguaia River, Amazon Basin,*Env. Biol. Fish.* 51, 399-410.

Telewski, F.W. (1995) Wind-induced physiological and developmental responses in trees, in M.P. Coutts and J. Grace (eds.), *Wind and trees,* Cambridge University Press, Cambridge, pp. 237–263.

Ter Steege, H. (1993) *Hemiphot, a programme to analyse light, and light quality and vegetation indices from hemispherical photographs,* Tropenbos Documents 3, Tropenbos Foundation, Wageningen , the Netherlands.

Ter Steege, H. and Cornelissen, J.H.C. (1989) Distribution and ecology of vascular epiphytes in lowland rain forest of Guyana, *Biotropica* 21, 331–339.

Terborgh, J. (1983) *Five new world primates. A study in comparative ecology,* Princeton University Press, Princeton, N.J.

Terborgh, J. (1986) Community aspects of frugivory in tropical forests, in A. Estrada and T.H. Fleming (eds.), *Frugivores and seed dispersal,* Dr. W. Junk Publishers, Dordrecht, pp. 371–384.

Terborgh, J. (1990) Mixed flocks and polyspecific associations: costs and benefits of mixed groups to birds and monkeys, *Am. J. Primatol.* 21, 87-100.

Terborgh, J., Robinson, S.K., Parker III, T.A., Munn, C.A. and Pierpont, N. (1990) Structure and organization of an Amazonian forest bird community, *Ecol. Monogr.* 60, 213–238.

Terwilliger, V.J. (1978) Natural history of Baird's tapir on Barro Colorado Island, Panama Canal Zone, *Biotropica* 10, 211-220

Théry, M. (1987) Influence des caractéristiques lumineuses sur la localisation des sites traditionnels, parade et baignade des Manakins (Passeriformes, Pipridae), *C.R. Acad. Sci.* série III 304, 19-24.

Théry, M. (1990a) *Ecologie et comportement des oiseaux Pipridae en Guyane: leks, frugivorie et dissémination des graines,* Thèse de Doctorat, Université Paris 6.

Théry, M. (1990b) Influence de la lumière sur le choix de l'habitatet le comportement sexuel des Pipridae (Aves: Passeriformes) en Guyane française, *Rev. Ecol. - Terre Vie* 45, 215–236.

Théry, M. (1990c) Display repertoire and social organization of the White-fronted and White-throated Manakins, *Wilson Bull.* 102, 123–130.

Théry, M. (1992) The evolution of leks through female choice: differential clustering and space utilization in six sympatric manakins, *Behav. Ecol. Sociobiol.* 30, 227–237.

Théry, M. (1997) Wing-shape variation in relation to ecology and sexual selection in five sympatric lekking manakins (Passeriformes: Pipridae), *Ecotropica* 3, 9–19.

Théry, M. (2001) Forest light and its influence on habitat selection, *Plant Ecol.* (in press).

Théry, M. and Larpin, D. (1993) Seed dispersal and vegetation dynamics at a cock-of-the-rock's lek in the tropical rain forest of Frech Guiana, *J. Trop. Ecol.* 9, 109–116.

Théry, M. and Vehrencamp, S.L. (1995) Light patterns as cues for mate choice in the lekking White-throated Manakin *Corapipo gutturalis, Auk* 112, 133–145.

Théry, M., Erard, C. and Sabatier, D. (1992) Les fruits dans le régime alimentaire de *Penelope marail* (Aves, Cracidae) en forêt guyanaise: frugivorie stricte et sélective? *Rev. Ecol.- Terre Vie* 47, 383–401.

Theunissen, P.A., and Wildschut, J.T. (1970) Vegetation and flora of the savannas in the Brickheuvel Nature Reserve, Northern Suriname, *Verh. Kon. ned. Akad. Wet.* 59, 1–58.

Thiollay, J.M. (1984) Pressions de prédation comparées sur trois groupements d'oiseaux en forêt tropicale,*Gerfaut* 74, 209-225.

Thiollay, J.M. (1986) Structure comparée du peuplement avien dans trois sites de forêt primaire en Guyane, *Rev. Ecol.- Terre Vie* 41, 59–105.

Thiollay, J.M. (1988) Diversité spécifique et coexistence en forêt tropicale: l'occupation des milieux par une guilde d'oiseaux insectivores en Guyane française, *Rev. Ecol. - Terre Vie* 43, 59–92.

Thiollay, J.M. (1989) Area requirements for the conservation of rain forests raptors and game birds in French Guiana, *Conserv. Biol.* 3, 128–137.

Thiollay, J.M. (1990) Comparative diversity of temperate and tropicalforest bird communities: the influence of habitat heterogeneity, *Acta Oecol.* 11, 887–911.

Thiollay, J.M. (1991a) Tropical rain forest communities: emerging patterns and processes of organization. *Acta XX Congr. Intern. Ornith. (Christchurch)* 1489–1496.

Thiollay, J.M. (1991b) Foraging, home range use and social behaviour, of a group-living rain forest raptor, the Red-throated Caracara, *Daptrius americanus*, *Ibis* 133, 382-393.

Thiollay, J.M. (1992) Influence of selective logging on bird species diversity in aguianan rain forest, *Conserv. Biol.* 6, 47-63.

Thiollay, J.M. (1994) Structure, density and rarity in an Amazonian rainforestbird community, *J. Trop. Ecol.* 10, 449–481.

Thiollay, J.M. and Jullien, M. (1998) Flocking behaviours of foraging birds in a neotropical rain forest and the antipredator defence hypothesis, *Ibis* 140, 382–394.

Thomas, D.W. (1982) *The ecology of an african savanna fruit bat community: resource partitioning and role in seed dispersal.* PhD Thesis, University of Aberdeen.

Thomas, M., Chrétien, D., Florion, A. and Terver, D. (1996a) Real-time detection of potassium cyanide pollution in surface waters using electric organ discharges wawe emitted by the tropical fish,*Apteronotus albifrons*, *Environ. Technol.* 17, 561-574.

Thomas, M., Florion, A., Chrétien, D. and Terver, D. (1996b) Real-time biomonitoring of water contamination by cyanide based on analysis of the continuous electric signal emitted by a tropical fish:*Apteronotus albifrons*, *Wat. Res.* 30, 3083-3091.

Thomas, O. (1888) Catalogue of the Marsupialia and Monotremata in the collection of the British Museum (Natural History). British Museum (Natural History), London, in D.E. Wilson and D.M. Reeder (ed.), *Mammal species of the world*, 2nd ed. (1993) Washington, DC: Smithson. Inst. Press.401 pp.

Thomas, S.C. (1995) Ontogenetic changes in leaf size in Malaysian rain forest trees,*Biotropica* 27, 427–434.

Tito De Morais, L. and Lauzanne, L. (1997) Zonation longitudinale des peuplements ichtyques avant mise en eau de la retenue de Petit-Saut (Guyane française) *Rev. Hydrobiol. tropic.* (1994) 27, 467-483.

Tito De Morais, L., Lointier, M. and Hoff, M. (1995) Extent and role for fish populations of riverine ecotones along the Sinnamary river (French Guiana), *Hydrobiologia* 303, 163-179.

Tomlinson, P.B. (1987) Architecture of tropical plants, *Annu. Rev. Ecol. Syst.* 18, 1–21.

Torquebiau, E.F. (1988) Photosynthetically active radiation environment, patch dynamics and architecture in a tropical rainforest in Sumatra, *Aust. J. Plant Physiol.* 15, 327–342.

Tostain, O., Dujardin, J.L., Erard, C. and Thiollay, J.M. (1992) *Oiseaux de Guyane*, Société d'Etudes Ornithologique, MNHN, Brunoy.

Trail, P.W. (1985) Territoriality and dominance in the lek-breeding Guianan cock-of-the-rock, *National Geographic Research* 1, 112-123.

Trail, P.W. and Adams, E.S. (1989) Active mate choice at cock-of-the-rock leks: tactics of sampling and comparison, *Behav. Ecol. Sociobiol.* 25, 283-292.

Turnbull, M. H., and D.Y. Yates (1993). Seasonal variation in the red/far-red ratio and photon flux density in an Australian sub-tropical rainforest. *Agricultural and Forest Meteorology* 64, 111–127.

Tutin, C.E.G., Williamson, E.A., Rogers, M.E. and Fernandez, M. (1991) A case study of a plant-animal relationship: *Cola lizae* and lowland gorillas in the Lopé Reserve, Gabon, *J. Trop. Ecol.* 69, 751-763.

Twidale, C.R. (1980) The evolution of bornhardts, *Am. Sci.* 70, 268–276.

Twidale, C.R. and Bourne J.A. (1978) Bornhardts, *Z. Geomorphol. N. F.* Suppl. 31, 111–137.

Upton G.J.G. and Fingleton, B. (1985) *Spatial data analysis by example*, Volume I, Wiley, New York.

Valencia, R., Balslev, H. and Paz y Mino, G. (1994) High tree-alpha-diversity in Amazonian Ecuador,*Biodiv. Conserv.* 3, 21-28.

Valverde, T and Silvertown, J. (1997) Canopy closure rate and forest structure. *Ecology* 78, 1555–1562.

Van der Hammen, T. (1974) The Pleistocene changes of vegetation and climate in tropicalSouth America, *J. Biogeogr.* 1, 3–26.

Van der Maarel, E. (1979) Transformation of cover-abundance values in Phytosociology and its effects on community similarity, *Vegetatio* 39, 97–114.

Van der Meer, P.J. (1995) *Canopy dynamics of a tropical rain forest in French Guiana*, PhD thesis, Wageningen Agricultural University, Wageningen.

Van der Meer, P.J. (1997). Vegetation development in canopy gaps in a tropical rain forest in French Guiana. *Selbyana* 18, 38-50.

Van der Meer, P.J. and Bongers, F. (1996a) Patterns of tree-falls and branch-falls in a tropical rain forest in French Guiana, *J. Ecol.* 84, 19–29.

Van der Meer, P.J. and Bongers, F. (1996b) Formation and closure of canopy gaps in the rain forest of Nouragues, French Guiana, *Vegetatio* 126, 167–179.

Van der Meer, P.J., Bongers, F. , Chatrou, L., and Riéra, B. (1994) Defining canopy gaps in a tropical rain forest: effects on gap size ans turnover time. *Acta Oecol.* 15, 701-714.

Van der Meer, P.J., Sterck, F.J. and Bongers, F. (1998) Tree seedling performance in canopy gaps in a tropical rain forest at Nouragues, French Guiana. *J. Trop. Ecol.* 14, 119–137.

Van der Pijl, L. (1957) The dispersal of plants by bats (Chiropterochory), *Acta Bot. Neerl.* 6, 291–315.

Van der Pijl, L. (1969) Evolutionary action of tropical animals on the reproduction of plants,*Biol. J. Linn. Soc.* 1, 85–96.

Van der Pijl, L. (1972) *Principles of dispersal in higher plants*, Springer Verlag, Berlin.

Van der Pijl, L. (1982) *Principles of dispersal in higher plants,* 3rd edn, Springer, Berlin.

Van Roosmalen, M.G.M. (1980) *Habitat preference, diet, feeding strategy and social organization of the black spider monkey (Ateles paniscus paniscus* Linnaeus *1758) in Surinam*, Ph.D. thesis, Agricultural University of Wageningen, Leersum, Neitherland.

Van Roosmalen, M.G.M. (1985) Habitat preferences, diet, feeding strategy and social organization of the black spider monkey (*Ateles paniscus* Linnaeus 1758) in Surinam, *Acta Amazonica (suppl.)* 15, 1–238.

Van Roosmalen, M.G.M. and Klein, L.L. (1988) The spider monkeys, genus *Ateles*, in R.A. Mittermeier, A.B. Rylands, A. Coimbra-Filho, and G.A.B. Fonseca (eds.), *Ecology and behaviour of neotropical primates*, World Wildlife Fund, Washington, D.C., pp. 455–537.

Vander Wall, S.B. (1990) *Food hoarding in animals*, University of Chicago Press, Chicago.

Vander Wall, S.B. (1991) Mechanisms of cache recovery by yellow pine chipmunks, *Anim. Behav.* 41, 851–864.

Vander Wall, S.B. (1993a) A model of caching depth: Implications for scatterhoarders and plant dispersal,*Am. Nat.* 141, 217–232.

Vander Wall, S.B. (1993b) Cache site selection by chipmunks (*Tamias* spp.) and its influence on the effectiveness of seed dispersal in Jeffrey pine (*Pinus jeffreyi*), *Oecologia* 96, 246–252.

Vander Wall, S.B. (1993c) Seed water content and the vulnerability of buried seeds to foraging rodents,*Am. Midl. Nat.* 129, 272–281.

Vander Wall, S.B. (1994) Seed fate pathways of antelope bitterbrush: dispersal by seed-caching yellow pine chipmunks, *Ecology* 75, 1911–1926.

Vander Wall, S.B. (1995a) Dynamics of yellow pine chipmunk (*Tamias amoenus*) seed caches: underground traffic in bitterbrush seeds, *Ecoscience* 2, 261–266.

Vander Wall, S.B. (1995b) The effects of seed value on the caching behavior of yellow pine chipmunks,*Oikos* 74, 533–537.

Vander Wall, S.B. (1995c) The influence of substrate water on the ability of rodents to find buried seeds,*J. Mammal.* 76, 851–856.

Vander Wall, S.B. and Peterson, E. (1996) Associative learning and the use of cache markers by yellow pine chipmunks (*Tamias amoenus*), *Southwest. Nat.* 41, 88–90.

Vander Wall, S.B. and Joyner, J.W. (1998) Recaching of Jeffrey pine (*Pinus jeffreyi*) seeds by Yellow pine chipmunks (*Tamias amoenus*): potential effects on plant reproductive success, *Can. J. Zool.* 76, 154–162.

Vanzolini, P.E. (1973) Paleoclimates, relief and species multiplication in Equatorial forest, in B.J Meggers., E.S Ayensu, and W.P. Duckworth (eds.), *Tropical forest ecosystem in Africa and South America: a comparative review,* Smithsonian Press, Washington D.C., pp. 255–258.

Vázquez-Yanes, C. and Orozco-Segovia, A. (1984) Ecophysiology of seed germination in the tropical humid forests of the world: a review, in E. Medina, H. A. Mooney, and C. Vazquez-Yanes (eds.),*Physiological ecology of plants of the wet tropics*, Junk, The Hague, pp 37–50.

Vázquez-Yanes, C. and Orozco-Segovia, A. (1993) Patterns of seed longevity and germination in the tropical rainforest, *Annu. Rev. Ecol. Syst.* 24, 69-87.

Vázquez-Yanes, C., Orozco-Segovia, A., François G. and Trero, L. (1975) Observations on seed dispersal by bats in a tropical humid region in Veracruz, Mexico. *Biotropica* 7, 73–76.

Vercauteren Drubbel, R. and Gautier, J. (1993) On the occurrence of nocturnal and diurnal loud calls, differing in structure and duration, in red howlers (*Alouatta seniculus*) of French Guiana, *Folia Primatol.* 60, 195–209.

Vidal, N., de Massary, J.C., and Marty, C. (1999) Nouvelles espèces de serpents pour la Guyane française, *Rev. fr. Aquariol. Herpétol.* 25, 131–134.

Vié, J.-C. (1998) *Translocation de deux espèces de primates en Guyane Française: le singe hurleur roux (Alouatta seniculus) et le saki à face pâle (Pithecia pithecia),* PhD thesis, Université Montpellier II.

Vié, J.-C., Volobouev, V., Patton, J.L., and Granjon, L. (1996) A new species of*Isothrix* (Rodentia: Echimyidae) from French Guiana, *Mammalia* 60, 393–406.

Vilhena Vialou, A. and Vialou, D. (1994) Les premiers peuplements préhistoriques du Mato Grosso,*Bull. Soc. Préhist. Fr.*, 91, 257–263.

Vitt, L.J., Caldwell, J.P., Wilbur, H.M., and Smith, D.C. (1990) Amphibians as harbingers of decay, *Bioscience,* 40, 418.

Vooren, F. 1986. Nature and origin of tree and branch-fall in Tai National Park, Ivory Coast, *Netherlands J. Agri. Sci.* 34, 112–115.

Voss, R.S. and Emmons, L. H. (1996) Mammalian diversity in neotropical lowland rainforests: a preliminary assessment, *Bull. Amer. Mus. Nat. Hist.* 230, 1–230.

Voss, R.S., Lunde, D. P. and Simmons, N. P. (in press) Mammals of Paracou, French Guiana: a Neotropical rain forest fauna, part 2. nonvolant species, *B. Am. Mus. Nat. Hist.*

Voysey, B. (1996) *Seed dispersal by gorillas in the Lopé Reserve, Gabon*, Ph. D. Thesis, University of Edinburgh.

Waite, T.A. and Reeve, J.D. (1997) Multistage scatter-hoarding decisions in the grey jay (*Perisoreus canadensis*), *Bird Behav.* 12, 7–14.

Walker, P.L. and Cant, J.G.H. (1977) A population survey of kinkajous (*Potos flavus*) in a seasonally dry tropical forest, *J. Mammal.* 58 100-102.

Waterman, P.G. and Kool, K.M. (1994) Colobine food selection and plant chemistry, in A.G. Davies and J.F. Oates (eds.), *Colobine monkeys: their ecology, behaviour and evolution*, Cambridge University Press, Cambridge, pp. 251–284.

Weitzman, A.L. (1993) Speciation, diversity, and relationships of Theaceae and Bonnetiaceae in the Neotropics, in H. Balslev (ed.), *Neotropical montane forests, biodiversity and conservation*, Abstracts from a Symposium at The New York Botanical Garden, June 21–26, 1993, AAU Reports 31.

Welden, C.W., Hewett, S.W., Hubbell, S.P. and Foster, R.B. (1991) Sapling survival, growth and seedling establishment: relationship to canopy height in a neotropical forest, *Ecology* 72, 35–50.

Westcott, D. (1993) Habitat characteristics of lek sites and their availability for the ochre-bellied flycatcher, *Mionectes oleagineus, Biotropica* 25, 444-451.

Westhoff, V. and van der Maarel, E. (1973) The Braun-Blanquet approach, in R.H. Wittaker (ed.),*Ordination and classification of plant communities*, Junk, The Hague, pp. 617–737.

Wheelwright, N.T. (1985) Fruit size, gape with, and the diets of fruit-eating birds, *Ecology* 66, 808–818.

Wheelwright, N.T. and Janson, C.J. (1985) Colors of fruit displays of birds dispersed plants in two tropicalforest, *Am. Nat.* 126, 777–799.

White, F. (1976) The underground forests of Africa: a preliminary review, *Gard. Bull. Sing.* 29, 57–71.

White, J. (1979) The plant as a metapopulation,*Annu. Rev. Ecol. Syst.* 10, 109–145.

White, J. (1984) Plant metamerism, in R. Dirzo and J. Sarukhan (eds.), *Perspectives on Plant Population Ecology*, Sinauer Associates, Sunderland, Massachusetts, pp. 15–47.

White, P.S. and Pickett, S.T.A. (1985) Natural disturbance and patch dynamics, in Pickett, S.T.A. and White, P.S. (eds.), *The ecology of natural disturbance and patch dynamics.* Academic Press, New York, pp. 3–13

Whitmore, T.C. (1978) Gaps in the forest canopy, in P.G. Tomlinson and M.H. Zimmerman (eds.), *Tropical trees as living systems*, Cambridge University Press, pp. 639–655.

Whitmore, T.C., Brown, N., Swaine, M.D., Kennedy, D., Goodwin-Baily,C.I., and Gong, W.K. (1993) Use of hemispherical photographs in forest ecology: measurement of gap size and radiation totals in a Bornean rain forest. *J. Trop. Ecol.* 9, 131-151

Wiley, R.H. (1980) Multispecies antbird societies in lowland forests of Surinam and Ecuador: stable membership and foraging differences, *J. Zool. London* 191, 127–145.

Williams, S.L. and Genoways, H.H. (1980) Results of the Alcoa Foundation, Surinam expeditions, II Additional records of bats (Mammalia: chiroptera), *Ann. Carnegie Mus.* 49, 213–236.

Willis, E.O. (1981) Diversity in adversity: the behavior of two subordinate ant birds,*Arquiv. Zool. S. Paulo* 30, 159–234.Williams, S.L. and Genoways, H.H. (1980) Results of the Alcoa Foundation, Surinam expeditions. II. Additional records of bats (Mammalia: chiroptera) *Ann. Carnegie Mus.* 49, 213–236.

Willis, E.O. and Oniki Y. (1978) Birds and army ants, *Annu. Rev. Ecol. Syst.* 9, 243–263.

Willson, M.F. (1986) Avian frugivory and seed dispersal in eastern North America, *Curr. Ornithol.* 3, 223–279.

Willson, M.F. (1991) Birds and fruits: How does this mutualism matter?*Acta 20th Congr. Int. Ornithol.* 3, 1630–1635.

Willson, M.F. and Whelan, C.J. (1990) The evolution of fruit color in fleshy-fruited plants.*Am. Nat.* 136, 790–809.

Wilson, D. E. and Reeder, D.M. (1993) *Mammal species of the world*, 2nd ed. Washington, DC, Smithson. Inst. Press.

Wolf, J.H.D. (1994) Factors controlling the distribution of vascular and non-vascular epiphytes in the northern Andes, *Vegetatio* 112, 15–28.

Yanhong, T and Naoki, K (1997) A measuring system for characterizing spatial and temporal variation of photon flux density within plant canopies. *Forest Ecol. Manag.* 97, 85–90.

Yasuda, M., Nakagoshi, N. and Takahashi, F. (1991) Examination of the spool-and-line method as a quantitative technique to investigate seed dispersal by rodents, *Jpn. J. Ecol.* 41, 257–262.

Yoda, K. (1974) Three dimensional distribution of light intensity in a tropical rain forest of West Malaysia,*Jpn. J. Ecol.* 24, 247–254.

Youlatos, D. (1993) Passages within a discontinuous canopy: Bridging in the red howler monkey (*Alouatta seniculus*). *Folia primatol.* 61, 144-147.

Youlatos, D. (1994) *Maîtrise de l'espace et accès aux ressources chez le Singe hurleur roux (Alouatta seniculus) de la Guyane française. Etude morpho-fonctionnelle.* Thèse de Doctorat, Université Paris 7.

Youlatos, D. (1998) Positional behavior of two sympatric guianan capuchin monkeys, the brown capuchin (*Cebus apella*) and the wedge-capped capuchin (*Cebus olivaceus*). *Mammalia* 62, 351-365.

Young, T.P. and Hubbell, S.P. (1991) Crown asymmetry, treefalls, and repeat disturbance of broad-leaved forest gaps, *Ecology* 72, 1464–1471.

Zanata, A.M. (1997) *Jupiaba*, a new genus of Tetragonopterinae with spine-like pelvic bones (Characidae, Characiformes). *Iheringia*, ser. *Zool.* 83, 99–136.

Zhang, S.-Y. (1994) *Utilisation de l'espace, stratégies alimentaires et rôle dans la dissémination des graines du singe capucin Cebus apella (Cebidae, Primate) en Guyane Française*, Thèse de Doctorat, Université Paris 6.

Zhang, S.-Y. (1995) Activity and ranging patterns in relation to fruit utilization by brown capuchins (*Cebus apella*) in French Guiana, *Int. J. Primatol.* 16, 489–507.

Zhang, S.-Y. and Wang, L.-X. (1995) Comparison of three fruit census methods in French Guiana,*J. Trop. Ecol.* 11, 281–294.

Zhang, S.-Y. and Wang, L.-X. (2000) Following of Brown capuchin monkeys by White hawks in French Guiana, *Condor* 102, 198–201.

Zimmerman, B.L. and Rodrigues, M.T. (1990) Frogs, snakes, and lizards of the INPA-WWF Reserves near Manaus, Brazil, in A. Gentry (ed.), *Four neotropical rainforests*, Yale Univ. Press, London, pp. 426–454.

Zylberberg, L. and Meunier, F.J. (1996) Ultrastructural data on the melanophores associated with the cellular elasmoid scales in *Leporinus friderici* (Teleostei: Ostariophysi, Anostomidae): their putative participation in scale matrix formation,*J. Morph.* 228, 155–164.

INDEX

INDEX

Bongers *et al.* (eds.), Nouragues. Dynamics and plant-animal interactions in a neotropical rainforest, 417-421.
2001 Kluwer Academic Publishers. Printed in the Netherlands.

Cover and parts design

Marjolein de Vette, Beeldgroep, Wageningen University.

Front cover

Background. Light enters through a hole in the forest canopy. *Photo Patrick Jansen.*
Upper left. Adult female of *Corallus hortulanus*, the Cook's boa. This species was frequently seen at the campsite. Nouragues, 1998. *Photo Marga Born.*
Upper right. The cock-of-the-rock (*Rupicola rupicola*). *Photo Patrick Jansen.*
Lower left. Adult male of *Hyla geographica.* This beautiful tree frog breeds in the inundated edges of slow running creeks. Nouragues, 1998. *Photo Marga Born.*
Lower right. Colourful fungi cover dead wood on the forest floor. *Photo Pierre Charles-Dominique.*

Back cover

Upper left. *Micrastur. Photo Pierre Charles-Dominique.*
Upper right. The arboreal tree frog *Osteocephalus taurinus* is a common species in the Nouragues forest. *Photo Gerrie Offerman.*

Part 1. Light enters through a hole in the forest canopy. *Photo Patrick Jansen.*
Part 2. The Venus-shoe orchid *Phragmipedium lindleyanum* near the inselberg. *Photo Pierre Charles-Dominique.*
Part 3. View from the inselberg looking over the campsite and a liana forest. *Photo Pierre Charles-Dominique.*
Part 4. *Artibeus obscurus*, a frugivorous bat. *Photo Pierre Charles-Dominique.*
Part 5. The saprophytic orchid species *Wullschlaegelia calcarata* in fruiting stage, visited by flies. Nouragues, 1991. *Photo Marga Born.*
Appendices. Adult male of *Epipedobates femoralis*, carrying tadpoles on its back. These tadpoles were released at the campsite in a plastic basket filled with water. Nouragues, 1991. *Photo Marga Born.*
References. The flower of the epiphyte *Cavendishia callista* (Ericaceae). *Photo Pierre Charles-Dominique.*
Index. *Micrastur. Photo Pierre Charles-Dominique.*

Bongers *et al.* (eds.), Nouragues. Dynamics and plant-animal interactions in a neotropical rainforest.
2001 Kluwer Academic Publishers. Printed in the Netherlands.

MONOGRAPHIAE BIOLOGICAE

1. *Physiologia Comparata et Oecologia*, Vol. I. 1948 out of print
2. *Physiologia Comparata et Oecologia*, Vol. II. 1952 out of print
3. *Physiologia Comparata et Oecologia*, Vol. III. 1954 out of print
4. *Physiologia Comparata et Oecologia*, Vol. IV. 1957 out of print
5. (1) R. A. Kennedy: *The Metzner Theory of Urine Formation.* 1957
   (2) M. D. Grmek: *On Ageing and Old Age.* Basic Problems and Historic Aspects of Gerontology and Geriatrics. 1958
   (3–4) O. Bassir: *Biochemical Aspects of Human Malnutrition in the Tropics.* 1962 out of print
6. F. S. Bodenheimer: *Animal Ecology Today.* 1958 out of print
7. R. R. Gates: *Taxonomy and Genetics of Oenothera.* 1958 ISBN 90-6193-061-8
8. A. Keast, R. L. Crocker and C. S. Christian (eds.): *Biogeography and Ecology in Australia.* 1959 out of print
9. S. Stanković: *The Balkan Lake Ohrid and Its Living World.* 1960 out of print
10. E. Rivnay: *Field Crop Pests in the Near East.* 1962 ISBN 90-6193-063-4
11. M. Kozhov: *Lake Baikal and Its Life.* 1962 ISBN 90-6193-064-2
12. M. S. Mani: *The Ecology of Plant Galls.* 1964 out of print
13. E. van Oye (ed.): *The World Problem of Salmonellosis.* 1964 ISBN 90-6193-066-9
14. D. H. S. Davis (ed.): *Ecological Studies in Southern Africa.* 1964 out of print
15. J. van Mieghem and P. van Oye (eds.): *Biogeography and Ecology in Antarctica.* 1965 ISBN 90-6193-067-7
16. H. Boyko (ed.): *Salinity and Aridity.* New Approaches to Old Problems. 1966 out of print
17. M. J. Coe: *The Ecology of the Alpine Zone of Mount Kenya.* 1967 ISBN 90-6193-068-5
18. E. J. Fittkau, J. Illies, H. Klinge, G. H. Schwabe and H. Sioli (eds.): *Biogeography and Ecology in South America*, Vol. I. 1968 out of print
19. E. J. Fittkau, J. Illies, H. Klinge, G. H. Schwabe and H. Stioli (eds.): *Biogeography and Ecology in South America*, Vol. II. 1969 ISBN 90-6193-071-5
20. R. N. Kaul (ed.): *Afforestation in Arid Zones.* 1970 out of print
21. R. Battistini and G. Richard-Vindard (eds.): *Biogeogaphy and Ecology of Madagascar.* 1972 ISBN 90-6193-073-1
22. J. E. Bishop: *Limnology of a Small Malayan River Sungai Gombak.* 1973 ISBN 90-6193-074-X
23. M. S. Mani (ed.): *Ecology and Biogeography in India.* 1974 ISBN 90-6193-075-8
24. E. K. Balon and A. G. Coche (eds.): *Lake Kariba.* A Man-made Tropical Ecosystem in Central Africa. 1974 ISBN 90-6193-076-6
25. W. D. Williams (ed.): *Biogeography and Ecology in Tasmania.* 1974 ISBN 90-6193-077-4
26. A. de Vos: *Africa, the Devastated Continent?* Man's Impact on the Ecology of Africa. 1975 ISBN 90-6193-078-2
27. G. Kuschel (ed.): *Biogeography and Ecology in New Zealand.* 1975 ISBN 90-6193-079-0
28. I. Prakash and P. K. Ghosh (eds.): *Rodents in Desert Environments.* 1975 ISBN 90-6193-080-4

MONOGRAPHIAE BIOLOGICAE

29. J. Rzóska (ed.): *The Nile.* Biology of an Ancient River. 1976 ISBN 90-6193-081-2
30. G. Kunkel (ed.): *Biogeography and Ecology in the Canary Islands.* 1976 ISBN 90-6193-082-0
31. M. J. A. Werger (ed.): *Biogeography and Ecology of Southern Africa*, 2 vols. 1978 ISBN 90-6193-083-9
32. C. Serruya (ed.): *Lake Kinneret* [Lake of Tiberias / Sea of Galilee]. 1978 ISBN 90-6193-085-5
33. Ph. D. Mordukhaí-Boltovskoi (ed.): *The River Volga and Its Life.* 1979 ISBN 90-6193-084-7
34. Ch. W. Heckman: *Rice Field Ecology in Northeastern Thailand.* The Effect of Wet and Dry Seasons on a Cultivated Aquatic Ecosystem. 1979 ISBN 90-6193-086-3
35. M. Kalk, A. J. McLachlan and C. Howard-Williams (eds.): *Lake Chilwa.* Studies of Change in a Tropical Ecosystem. 1979 ISBN 90-6193-087-1
36. B. R. Allanson (ed.): *Lake Sibaya* [South Africa]. 1979 ISBN 90-6193-088-X
37. H. Löffler (ed.): *Neusiedlersee.* The Limnology of a Shallow Lake in Central Europe. 1979 ISBN 90-6193-089-8
38. J. Rzóska: *Euphrates and Tigris.* Mesopotamian Ecology and Destiny. With Contributions by J. F. Talling and K. E. Banister. 1980 ISBN 90-6193-090-1
39. K. A. Brodsky: *Mountain Torrent of the Tien Shan.* A Faunistic-Ecology Essay. Translated from Russian by V. V. Golosov. 1980 ISBN 90-6193-091-X
40. M. S. Mani and L. E. Giddings (eds.): *Ecology of Highlands.* 1980 ISBN 90-6193-093-6
41. A. Keast (ed.): *Ecological Biogeography of Australia*, 3 vols. 1981 ISBN 90-6193-092-8
42. J. L. Gressit (ed.): *Biogeography and Ecology of New Guinea*, 2 vols. 1982 ISBN 90-6193-094-4
43. H. Heathwole, T. Done and E. Cameron: *Community Ecology of a Coral Cay.* A Study of One-Tree Island, Great Barrier Reef, Australia. 1981 ISBN 90-6193-096-0
44. P. S. Maitland (ed.): *The Ecology of Scotland's Largest Lochs.* Lomond, Awe, Ness, Morar and Shiel. 1981 ISBN 90-6193-097-9
45. K. Müller (ed.): *Coastal Research in the Gulf of Bothnia.* 1982 ISBN 90-6193-098-7
46. G. K. Rutherford (ed.): *The Physical Environment of the Faeroe Islands.* 1982 ISBN 90-6193-099-5
47. J. I. Furtado and S. Mori (eds.): *Tasek Bera* [Malaysia]. The Ecology of a Freshwater Swamp. 1982 ISBN 90-6193-100-2
48. G. R. South (ed.): *Biogeography and Ecology of the Island of Newfoundland.* 1983 ISBN 90-6193-101-0
49. J. A. Thornton (ed.): *Lake McIlwaine.* The Eutrophication and Recovery of a Tropical African Man-made Lake. 1982 ISBN 90-6193-102-9
50. R. W. Edwards and M. P. Brooker: *The Ecology of the Wye* [U.K.]. 1982 ISBN 90-6193-103-7
51. T. Petr (ed.): *The Purari* [Papua New Guinea]. Tropical Environment of a High Rainfall River Basin. 1983 ISBN 90-6193-104-5

MONOGRAPHIAE BIOLOGICAE

52. H. Kuhbier, J. A. Alcover and C. Guerau d'Arellano Tur (eds.): *Biogeography and Ecology of the Pityusic Islands.* 1984 ISBN 90-6193-105-3
53. J.-P. Carmouze, J.-R. Durand and C. Lévêque (eds.): *Lake Chad.* Ecology and Productivity of a Shallow Tropical Ecosystem. 1983 ISBN 90-6193-106-1
54. S. Horie (ed.): *Lake Biwa* [Japan]. 1984 ISBN 90-6193-095-2
55. D. R. Stoddart (ed.): *Biogeography and Ecology of the Seychelles Islands.* 1984 ISBN 90-6193-107-X
56. H. Sioli (ed.): *The Amazon.* Limnology and Landscape Ecology of a Mighty Tropical River and Its Basin. 1984 ISBN 90-6193-108-8
57. C. H. Fernando (ed.): *Ecology and Biogeography in Sri Lanka.* 1984 ISBN 90-6193-109-6
58. S. J. Casper (ed.): *Lake Stechlin* [Germany]. A Temperate Oligotrophic Lake. 1985 ISBN 90-6193-512-1
59. U. Th. Hammer: *Saline Lake Ecosystems of the World.* 1985 ISBN 90-6193-535-0
60. B. R. Davies and K. F. Walker (eds.): *The Ecology of River Systems.* 1986 out of print
61. P. de Deckker and W. D. Williams (eds.): *Limnology in Australia.* 1986 ISBN 90-6193-578-4
62. Y. Yom-Tov and E. Tchernov (eds.): *The Zoogeography of Israel.* The Distribution and Abundance at a Zoogeographical Crossroad. 1988 ISBN 90-6193-650-0
63. F. D. Por: *The Legacy of Tethys.* An Aquatic Biogeography of the Levant. 1989 ISBN 0-7923-0189-7
64. B. R. Allanson, R. C. Hart, J. H. O'Keeffe and R. D. Robarts: *Inland Waters of Southern Africa.* An Ecological Perspective. 1990 ISBN 0-7923-0266-4
65. F. di Castri, A. J. Hansen and M. Debussche (eds.): *Biological Invasions in Europe and the Mediterranean Basin.* 1990 ISBN 0-7923-0411-X
66. R. W. Edwards, A. S. Gee and J. H. Stoner (eds.): *Acid Waters in Wales.* 1990 ISBN 0-7923-0493-4
67. S. K. Jain and L. W. Botsford (eds.): *Applied Populations Biology.* 1992 ISBN 0-7923-1425-5
68. C. Dejoux and A. Iltis (eds.): *Lake Titicaca* [Peru/Bolivia]. A Synthesis of Limnological Knowledge. 1992 ISBN 0-7923-1663-0
69. R. B. Wood and R. V. Smith (eds.): *Lough Neagh.* The Ecology of a Multipurpose Water Resource. 1993 ISBN 0-7923-2112-X
70. P. E. Ouboter (ed.): *The Freshwater Ecosystems of Suriname.* 1993 ISBN 0-7923-2408-0
71. M. A. Brunt and J. E. Davies (eds.): *The Cayman Islands.* Natural History and Biogeography. 1994 ISBN 0-7923-2462-5
72. V. Fet and K. Atamuradov (eds.): *Biogeography and Ecology of Turkmenistan.* 1994 ISBN 0-7923-2738-1
73. F. D. Por: *The Pantanal of Mato Gross (Brazil).* World's Largest Wetlands. 1995 ISBN 0-7923-3481-7
74. D. S. Edwards, W. E. Booth and S. C. Choy (eds.): *Tropical Rainforest Research – Current Issues.* Proceedings of the Conference held in Bandar Seri Begawan. 1996 ISBN 0-7923-4038-8

MONOGRAPHIAE BIOLOGICAE

75. R. W. Schmieder (ed.): *Rocas Alijos*. 1996 ISBN 0-7923-4056-6
76. Z. Mianping: *An Introduction to Saline Lakes on the Qinghai-Tibet Plateau*. 1996 ISBN 0-7923-4098-1
77. C. W. Heckman: *The Pantanal of Poconé*. Biota and Ecology in the Northern Section of the World's Largest Pristine Wetland. 1998 ISBN 0-7923-4863-X
79. J. T. Lehman (ed.): *Environmental Change and Response in East African Lakes*. 1998 ISBN 0-7923-5118-5
80. F. Bongers, P. Charles-Dominique, P.-M. Forget and M. Théry (eds.): *Nouragues*. Dynamics and Plant-Animal Interactions in a Neotropical Rainforest. 2001 ISBN 1-4020-0123-1

KLUWER ACADEMIC PUBLISHERS – DORDRECHT / BOSTON / LONDON